Naturally Occurring Phorbol Esters

Editor

Fred J. Evans, Ph.D., M.P.S., F.L.S.
Senior Lecturer in Pharmacognosy
School of Pharmacy
University of London
London
England

CRC Press, Inc.
Boca Raton, Florida

Library of Congress Cataloging-in-Publication Data
Main entry under title:

Naturally occurring phorbol esters.

Bibliography: p.
Includes index.
1. Phorbol esters. 2. Euphorbiaceae.
3. Thymelaeaceae. 4. Chemistry, Pharmaceutical.
5. Botanical chemistry. 6. Pharmacognosy.
I. Evans, Fred J., 1943- .
RS431.P46N37 1986 581.1'92 85-25485
ISBN 0-8493-5117-0

Direct all inquiries to CRC Press, Inc., 2000 Corporate Blvd., N.W., Boca Raton, Florida, 33431.

International Standard Book Number 0-8493-5117-0

Library of Congress Card Number 85-25485
Printed in the United States

THE EDITOR

Fred J. Evans, Ph.D., is Senior Lecturer in Pharmacognosy at The School of Pharmacy, University of London, England. Dr. Evans was graduated from the University of London with a first class honors B. Pharmacy degree in 1967, and the following year was registered as a pharmaceutical chemist with the Pharmaceutical Society of Great Britain. In 1971 he received the degree of Ph.D. in Pharmacognosy from the University of London.

From 1967 to 1970 Dr. Evans was assistant lecturer in Natural Product Chemistry at Brighton Polytecnic, from where he transferred to the School of Pharmacy as lecturer in Prof. J.W. Fairbairn's department. He has at various times been a visiting professor to the University College Hospital, Ibadan, the Chemistry Department of the University of Qatar in Doha, and the National Research Centre in Cairo.

Dr. Evans was elected a Fellow of the Linnean Society in 1974 and became a member of the Phytochemical Society in 1973 and the National Institute of Medical Herbalists in 1976.

He is best known for his work in toxicology concerned with the chemistry and mechanism of action of phorbol esters and has published over 150 papers. He is known also for a number of books and articles concerned with medicinal plants and herbs and has made several excursions to underdeveloped countries for the collection of botanical samples.

CONTRIBUTORS

Alastair Aitken, Ph.D.
Lecturer
Department of Pharmaceutical Chemistry
University of London
London
England

Mustafa M. El-Missiry, Ph.D.
Associate Professor
Pharmaceutical Sciences Laboratory
National Research Center
Dokki, Cairo
Egypt

Fred J. Evans, Ph.D., M.P.S., F.L.S.
Senior Lecturer in Pharmacognosy
University of London
London
England

Anne R. Kinsella, Ph.D.
Scientific Officer
Department of Biochemical Genetics
Paterson Laboratories
Christie Hospital and Holt Radium Institute
Manchester
England

Alan Radcliffe-Smith, B.Sc.
The Herbarium
Royal Botanic Gardens
Kew Surrey
Richmond, England

Abdel-Fattah Rizk, Ph.D.
Professor
Scientific Applied Research Centre
Qatar University
Doha, Qatar

Richard J. Schmidt, Ph.D., M.P.S., F.L.S.
Lecturer in Pharmacognosy
The Welsh School of Pharmacy
University of Wales Institute of Science and Technology
Honorary Scientific Officer
Department of Dermatology
University of Wales College of Medicine
Cardiff
Wales

TABLE OF CONTENTS

Chapter 1

ENVIRONMENTAL HAZARDS OF DITERPENE ESTERS FROM PLANTS

Fred J. Evans

TABLE OF CONTENTS

I. INTRODUCTION

Plant species of the families Euphorbiaceae, which contains about 300 genera and 7000 species, and of the Thymelaeaceae with 55 genera and about 500 species, are distributed worldwide both in the tropical and temperate areas. The Euphorbiaceae in particular is a large and unwieldy family and has provided problems for botanists and taxonomists alike due to the great variation of form exhibited. Individual species of these families have long been noted for their toxicological effects on animals and man,[1-2] including the induction of inflammation to skin[3] and mucus membranes,[4] conjunctivitis in the eyes,[3-5] production of scouring in animals,[6] and for their purgative actions in man.[7] They have also been used as fish poisons[8] and many species are used as an ingredient of arrow poisons.[9] With such diverse toxicological activities it is not surprising that plants of the Euphorbiaceae and Thymelaeaceae have been utilized as drugs in both traditional and alternative medicine. Their uses in medicine have been equally diverse, including their use as treatments for growths and tumors,[10] migraine, parasite infestations, bacterial infections including venereal disease, skin conditions, and as purgatives and abortifacients.[11-15] In conventional Western medicine many of these plants were considered to be too toxic for human use and were eventually removed from modern pharmacopoeias.[16] The health hazards to animals and man presented by these plants is considerable even when they are considered as poisonous plants. However, these dangers would be no worse than those of other noxious plant families such as the Apocyanaceae,[17] Solanaceae,[18] and Scrophulariaceae[19] were it not for their involvement in the promotion of cancer in mammalian systems.[20]

The isolation and final structure elucidation of tetradecanoyl phorbolacetate from a member of the genus *Croton* of the family Euphorbiaceae in 1968 by Hecker[21] initiated intensive chemical and biochemical research into many species of this large family. This soon included investigations of the closely related family Thymelaeaceae. Tetradecanoylphorbolacetate (TPA) was the first pure tumor-promoting agent isolated and is still the most widely used compound of a series of similar esters for tumor-promoting and related studies.[22]

Considering the worldwide distribution of plants of the Euphorbiaceae and Thymelaeaceae, little work has been carried out to date to substantiate the possible role that these species play in the etiology of human cancer. It is possible to speculate that exposure to these plants, plant parts, or processed materials may be responsible in part for tumor formation in certain communities around the world. One of the main problems associated with such an investigation is the fact that unlike chemical carcinogens,[23] no valid and rapid in vitro tests presently exist for tumor-promoting agents.[24] Furthermore, the phenomenon of tumor promotion as demonstrated on mouse skin[25] may not be applicable to the human situation. Nevertheless, it has recently been realized that for the majority of human cancers, solitary carcinogenesis engineered by a single chemical carcinogen may be the exception rather than the rule.[26] The majority of human cancers may result from exposure to more than one carcinogenic risk factor. This process is known as syncarcinogenesis.[27] It is also likely that a number of human tumors are due to exposure to small quantities of a carcinogen and a tumor promoter. This is known as cocarcinogenesis.[28] Several groups of compounds are known to be tumor-promoting agents, but the diterpene esters of the plant families Euphorbiaceae and Thymelaeaceae remain not only the most potent agents known, but also the most widely studied in animal systems.

Human exposure to the naturally occurring tumor-promoting diterpenes can be considered as arising from several sources: first, from the cultivation of species of these families in gardens and greenhouses, second, as an occupational exposure of workers involved in processing these plants for economic products (including medicines), third, by the consumption of food products including honey, milk, or meat derived ultimately from a plant source, and finally, by the use of these plants in alternative forms of medicine such as

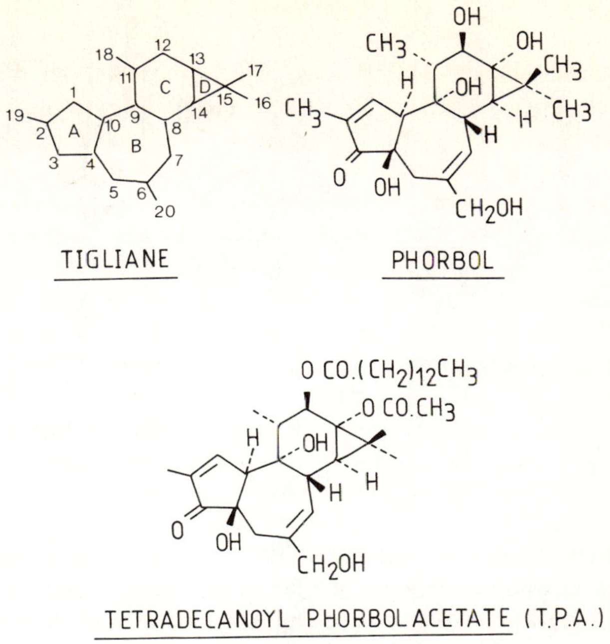

TIGLIANE

PHORBOL

TETRADECANOYL PHORBOLACETATE (T.P.A.)

FIGURE 1.

homeopathy, herbal, or folk medicine.[10-15,29-31] In addition, many species of the herbaceous varieties are common weeds on agricultural, waste, and domestically cultivated areas.[6,32] Pollen and dried powdered plant parts form a normal component of atmospheric dust at certain times of the year. This is certainly the case for plants of the family Compositae which produce sequeterpene lactones responsible for an allergenic reaction known as allergic contact dermatitis.[33] As with many other areas of environmental toxicology, an investigation of the health hazards presented by these plants requires a multidisciplinary approach including aspects of botany and taxonomy, organic chemistry, biochemistry, and pharmacology.

II. NOMENCLATURE OF THE DITERPENES

Diterpene tumor-promoting agents are traditionally known as esters of phorbol. TPA, sometimes called phorbol myristate acetate (PMA), is the most potent tumor-promoting agent known.[34] This diterpene is the active principle of *Croton tiglium* and was used in experimental cancer biochemistry in the impure form as seed oil or resin for many years before its chemical structure was known.[20,24] In recent years this compound has been the object of an increasing number of publications,[22,35-37] and it is a diester of the diterpene phorbol. Phorbol was originally isolated in 1931 by Bohm et al.,[38] but its structure remained unknown until 1968 when Hecker's group at Heidelberg deduced its absolute configuration from X-ray analysis.[39] Publication of the structure of the diesters of *Croton tiglium,* including TPA, quickly followed.[21]

The term "phorbol" or "phorbols" is loosely used today by biologists to describe a family of naturally occurring compounds correctly referred to as the tigliane diterpenes (Figure 1). These substances are widely distributed in plant species of the families Euphorbiaceae and Thymelaeaceae.[40,41] The hydrocarbon tigliane is a perhydrocyclopropabenzazulene which, when systematically named by IUPAC rules, is 1,1aα, 1bβ,2,3,4,4aβ,6,7,7aα,7bα,8,9,9aα-tetradecahydro-1,1,3ϵ,6ϵ,8α-pentamethyl-5H-cyclopropa[3,4]benz-[1,2-e]azulene. In 1967, Hecker[42] recommended a simplified nomenclature based upon tigliane. According to this generally accepted system, phorbol becomes

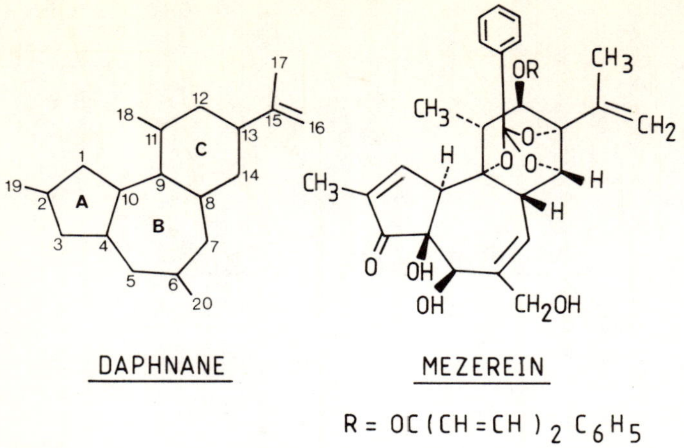

DAPHNANE

MEZEREIN

R = OC(CH=CH)$_2$C$_6$H$_5$

FIGURE 2.

4,9,12β,13,20-penta-hydroxy-1,6-tigliadien-3-one. The structural formulae of tigliane is conventionally drawn with the A ring to the left of the chirality center at C-14 (Figure 1). Tiglianes therefore consist of a 5-membered ring A, normally *trans-* linked to a 7-membered ring B. Ring C is 6-membered and the cyclopropane ring D is linked to it in the *cis-*configuration. Polyhydroxylated tiglianes in the esterified form have been detected in many plant species, although phorbol itself is restricted at the present time to only seven individual species of the genera *Croton, Sapium,* and *Euphorbia.*[43-48] The more common derivatives of tigliane found in plants include 12-deoxyphorbol,[49] 4-deoxyphorbol,[50] 12-deoxy-16-hydroxyphorbol,[51] and 16-hydroxyphorbol.[52] The tigliane diterpenes, as first elucidated by Hecker, are accordingly a large and biologically important new group of tetracyclic diterpenes. (See Chapter 7.)

Although phorbol was the first and most significant compound to be isolated, and gave rise to the well-known tigliane derivatives, these compounds themselves belong to a larger family of diterpenes, many of which may be biosynthetically related,[53] but all of which are constituents of plants from the families Euphorbiaceae and Thymelaeaceae. Certain of these diterpenes share with the tigliane esters the ability to promote tumor-formation in mammalian systems[54] or cytotoxic actions, and some may be phytoallexins.[56] The closest group of compounds, chemically, to the tiglianes are those based upon the hydrocarbon daphnane (Figure 2). The daphnane diterpenes are tricyclic products in which the D ring of the tigliane nucleus has opened out to form an isopropenyl side chain. Daphnane diterpenes were identified in 1970 by Stout et al.[57] as the poisonous principles of mezeron bark. This material is the commercially available mixture of the bark of *Daphne mezereum, D. laureola,* and *D. gnidium* of the family Thymelaeaceae. The first compound from mezeron was daphnetoxin, but later work has shown that the daphnane diterpenes are a widely distributed group of compounds in species of the Euphorbiaceae[58] as well as the Thymelaeaceae. This group includes the well-known agent mezerein isolated by Ronlar and Wickberg[59] in 1970 which is used in tumor-promoting studies[60] together with the phorbol esters. Daphnane diterpenes sometimes occur naturally as the *O*-acyl esters, but more commonly they are isolated as the unique ortho ester derivatives. These substances are divided into classes depending upon the nature of the nucleus, giving rise to the groups known as daphnetoxin,[61], 12-hydroxydaphnetoxin,[62], 1-alkyldaphnane,[63] and resiniferonol.[64] Many of these compounds have pronounced toxicological effects, including the ability to induce acute inflammation of skin[3] and to act as fish poisons,[8] and some have reputed antileukemic activities.[65] (See Chapter 8.)

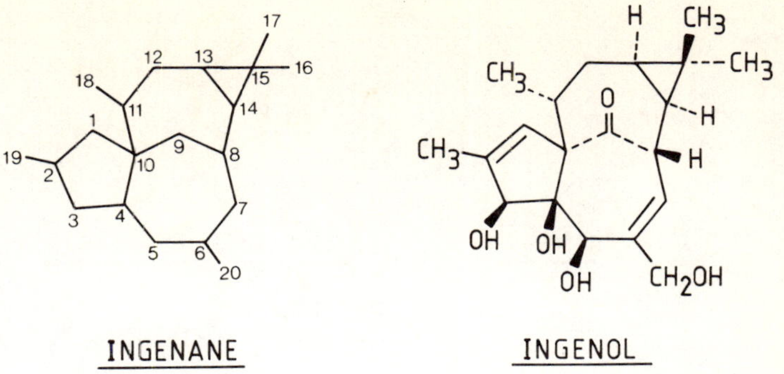

INGENANE

INGENOL

FIGURE 3.

The ingenane diterpenes are a group of compounds which have been shown to be naturally occurring tumor-promoting agents.[66] They are related to the tigliane group, occur in the same plant families and have many biological activities in common with them. They are distinct from the phorbol esters, however, on a structural basis. In comparison with the phorbol esters little work has been done as yet on the ingenane esters, possibly because of their chemical instability,[67] in comparison to compounds such as TPA or mezerein. Ingenane esters are difficult to isolate and purify and are sensitive to hydrolysis and transesterification reactions during separation. There is the possibility that some isolated products could be artifacts of the extraction and separation procedures. Major tumor-promoting studies have been done with semisynthetic ingenane derivatives,[68] although a number of highly active natural products have been available.[66,67,69] The first ingenane diterpene obtained from plants was ingenol (Figure 3), isolated in the esterified form from *Euphorbia lathyris.*[66] Ingenol is a polyhydroxylated ingenane derivative and ester moities are commonly found at the secondary hydroxyl on carbons 3 and 5 as well as the primary hydroxyl at C-20. In 1970, Zechmeister et al.[70] obtained the absolute configuration of ingenol, as its synthetic acetate by triple product methods. In the ingenane hydrocarbon nucleus ring, C is 7-membered, C-8 being linked to C-10 by means of a keto bridge. Esters of polyhydroxylated ingenanes are the most abundant compounds detected to date in poisonous species of the genus *Euphorbia.*[71] Their distribution in the environment, particularly in ornamental plants, drug plants, and herbaceous types which commonly invade cultivated land as weeds, still requires extensive investigation.[72] Besides ingenol itself, several other derivatives of ingenane have more recently been structure elucidated. These include 5-deoxyingenol,[73] 20-deoxy-ingenol,[74] 16-hydroxyingenol,[75] 13-hydroxyingenol,[76] and 13,19-dihydroxyingenol.[77] (See Chapter 9.)

Thus, the tumor-promoting and pro-inflammatory diterpenes sometimes referred to as "phorbols" belong to three closely related families of diterpenes. These are known chemically as the tigliane, daphnane, and ingenane groups. However, these toxic substances in turn are also closely related on a chemical and biosynthetic basis to further groups of diterpenes found within the plant family Euphorbiaceae. Many of these related compounds exist naturally in the oxygenated form and are esterified at one or more positions. They are collectively known as the macrocyclic diterpenes. (See Chapter 6.) The first member of this group of natural products was isolated in 1937 by Dublyanskaya,[78] but because of their apparent lack of toxicity they were only of chemical interest until 1970. Kupchan et al.[79] demonstrated that macrocyclic diterpenes were potentially antileukemic in action when tested against mouse P-388 lymphocytic leukemia in vivo. The macrocyclic diterpenes are divided into groups according to the chemical nature of their hydrocarbon nuclei. To date six distinct chemical groups have been identified as occurring in plants of the family Euphorbiaceae.

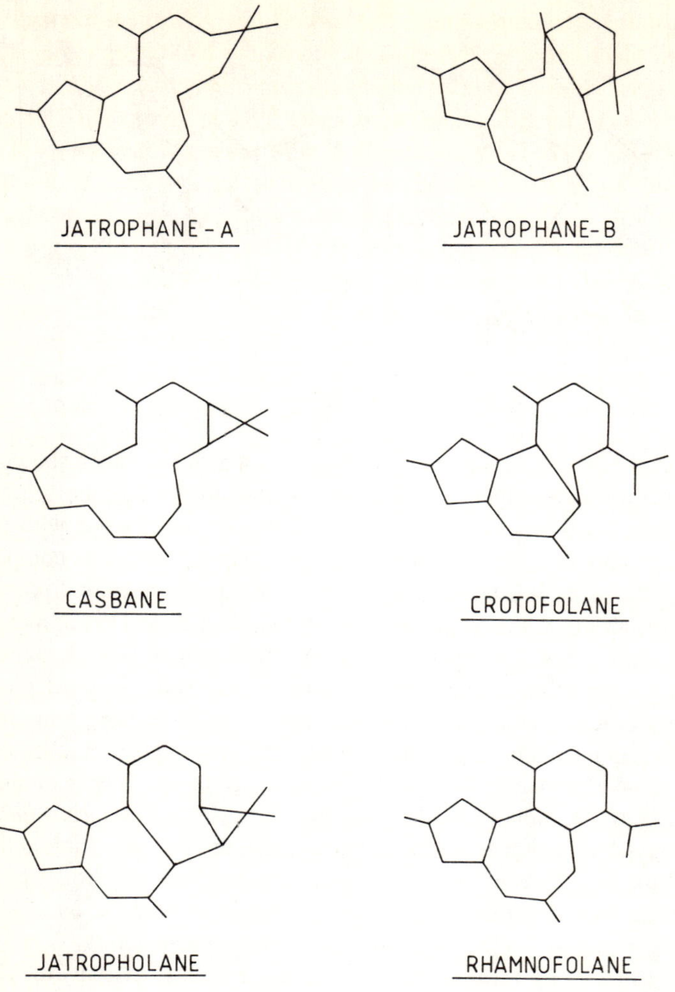

FIGURE 4. The macrocyclic hydrocarbon nuclei from the family Euphorbiaceae.

These are the casbane, jatrophane, lathyrane, jatropholane, crotofolane and rhamnofolane diterpenes (Figure 4). Many species of the Euphorbiaceae in particular have been used in traditional medicine. It is interesting to note that these plants contain not only derivatives of the phorbol type, but also the nontoxic macrocyclic esters.[80]

III. MICROMETHODS OF CHEMICAL DETECTION

The environmental hazards presented by plants which contain tumor-promoting diterpenes was apparent from the classical tumor-promoting experiments carried out by Berenblum.[81] Most of the early research (both chemical and toxicological)[82] centered around an investigation of *Croton tiglium,* until Roe and Peirce[83] attempted in 1961 to screen a number of different species of the genus *Euphorbia* for their ability to act as tumor promoters. With the original elucidation of the structure of phorbol esters in 1968, rapidly followed by the identification of numerous tumor-promoting and pro-inflammatory diterpenes of the tigliane, daphnane, and ingenane classes, it became possible to investigate various species of plants using microchemical methods for the presence of these toxic compounds. Many of the esterified forms of diterpenes are unstable and either hydrolyze or rearrange during extraction. Furthermore, configurational changes and transesterification reactions may occur during attempts to crystallize the compounds. The earliest microchemical methods of detection of diterpenes from plant extracts, therefore, involved the hydrolysis of the natural esters to their parent polyols followed by acetylation with acetic anhydride/pyridine (2/1) to the stable acetates.[84] Subsequent identifications were made by thin layer (TLC) or gas liquid chromatographic (GLC) analysis in combination with mass spectrometry (MS) and circular dichroism. Using these micromethods of detection, the presence of diterpenes could be confirmed in as little as 25 g of fresh plant material, either fruits, leaves, or herbs, or in as little as 5 mℓ of fresh plant latex. It is a feature of several genera of the family Euphorbiaceae[85] and the genus *Euphorbia,*[86] in particular, that they secrete a milky or sometimes translucent latex or juice into special structures known as lactifers.[87] Diterpene esters of the phorbol and related types are concentrated in this latex and it provides an ideal material for chemical-toxicological screening due to the absence of interfering plant pigments such as the chlorophylls. Latex may be readily obtained from suitable plants by cutting the stem or leaf surface and allowing the fluid to drip out into a small volume of methanol or ethanol as a preservative. The enzymes, proteins, and rubber polymers of the latex coagulate in this mixture, leaving an ethanolic supernatant which contains the diterpene esters.

This material may be dried by low temperature vacuum distillation or stored in solution in a deep freeze for as long as 5 years before deterioration is evident. Fresh plant material may be ground up in alcohol and stored in a similar manner or the alcoholic extract may be evaporated to dryness as before. Whether samples are dried before storage or kept as alcoholic extracts, the headspace of the storage vial should be flushed with nitrogen gas before freezing. Dried plant parts may be extracted in the same way although several hours may be required to extract the diterpenes into the alcohol, due to the relative hardness of the dried plant tissues. Seeds provide an ideal source of diterpenes, which are concentrated in the fixed oil.[88] As little as 1 g of seeds may be sufficient for the detection of toxic diterpenes. Postharvest chemical changes in natural compounds are a common problem to be overcome when analyzing dried plant samples. However, if all extracts are hydrolyzed and acetylated before microanalysis, then at least the parent polyols present in the species may be identified. In terms of pinpointing potential toxicological hazards presented by a particular species, this may be all that is required. The sensitivity of microanalysis for the detection of diterpene esters may be illustrated by the work of Marsh et al.[89] concerning the nature of toxins sequestered by plant feeding species of moths. *Hyales euphorbia* is a species of moth whose larvae feed on plants of the genus *Euphorbia.* These larvae are apparently unacceptable to birds which normally prey on insect caterpillars or adults. Using micromethods of TLC analysis, extracts of larvae, pupae, and adults were shown to contain microgram quantities of toxic ingenane derivatives. These compounds were sequesterd or stored during the feeding phase of the insect's life and served not only to protect the larvae from their predators, but were carried over into adult life for the same purpose. Furthermore, extracts of these insects were shown to have potent antitumor and cytotoxic activities. Chemical analysis revealed the presence of macrocyclic diterpenes apparently sequestered in a similar manner.

Microchemical analysis is a rapid and sensitive manner by which one may detect polyols of the tigliane, ingenane, and daphnane types. Such results would suggest that particular plants are potentially toxic hazards and capable of acting as tumor-promoting agents.[90] Nevertheless, not all of these parent diterpenes are tumor-promoting and the magnitude of the environmental hazards presented by any species may only be fully realized by biological testing.

The earliest microchemical method for diterpenes or their stable acetates was the TLC technique of Evans and Kinghorn.[91] This method was later applied to an investigation of abut 60 different species of plants of the genera *Euphorbia* and *Elaeophorbia.*[71] The method consisted of a hydrolysis and acetylation procedure for the free polyols followed by TLC

Table 1
MIGRATION RATES OF DITERPENES BY TLC[91]

	R_f -values						
	1	2	3	4	5	6	7
Silica gel G							
Chloroform/ether (95/5)	0.16	0.15	0.21	0.07	0.26	0.08	0.44
Ether/ethylacetate/hexane (1/1/1)	0.48	0.51	0.52	0.26	0.49	0.34	0.65
Hexane/isopropylalcohol (2/1)	0.64	0.68	0.55	0.44	0.57	0.60	0.72
Silica gel H							
Chloroform/ethylacetate (2/3)	0.49	0.50	0.53	0.39	0.55	0.41	0.59
Alumina E							
Toluene/ethylacetate (9/2)	0.05	0.06	0.12	0.01	0.11	0.02	0.32
Chloroform/acetone/benzene (95/5/50)	0.31	0.33	0.51	0.04	0.54	0.10	0.69
Chloroform	0.53	0.53	0.72	0.10	0.76	0.20	0.80
Hexane/ether/benzene (1/2/1)[a]	0.06	0.10	0.11	0.01	0.07	0.02	0.37
Benzene/hexane/ether/ethyl/acetate (20/40/15/30)[a]	0.30	0.38	0.52	0.03	0.45	0.11	0.70
Chloroform/ether/benzene (1/3/3)[a]	0.22	0.29	0.36	0.02	0.43	0.07	0.74
Ethyl acetate/benzene (1/3)[b]	0.35	0 39	0.59	0.05	0.51	0.11	0.86

Note: Compounds: (1) phorboltriacetate, (2) 12-deoxyphorbol diacetate, (3) 4-deoxy-4α-phorbol triacetate, (4) 4α-phorbol triacetate, (5) 4α-phorbol tetraacetate, (6) crotophorbolone acetate, (7) ingenol triacetate.

[a] Plates eluted three times.

[b] Plates eluted four times.

identifications in combination with direct insertion MS. Identifications were based upon migration rates, measured as the R_f-values and color reactions (Table 1) to a series of spray reagents together with the characteristic mass spectra produced by acetates of this series of compounds. Migration rates of the diterpene acetates was governed by the number and configuration of hydroxyl groups in the nucleus and to a lesser extent by the number of acetate groups (Table 2). One of the major problems associated with the identification of various naturally occurring esters of these diterpenes by TLC was the fact that compounds differing only in the nature of their acyl substituents at a particular carbon atom tend to migrate as inseparable zones when adsorption methods were used. Thus, mono-, di-, and triesters of a common parent polyol were readily separated into bands or zones, but each band was a mixture of a number of closely related compounds.[92] TLC has been used for the detection of microquantities of natural esters in the partition mode.[93] This technique was initially described for mono- and diesters of 12-deoxyphorbol, and involved their identification on three separate adsorbents coated with 20% diethylene glycol or propylene glycol. The stationary phase was applied to the adsorbent by developing the plates their full length in an acetone solution of glycol. After removal from the tanks the plates were air dried to leave a deposit of glycol on the thin-layer particles. Plants from which diterpenes have been isolated normally contain a large number of derivatives. For example, one species of *Euphorbia* was found to contain about 30 different derivatives of the tigliane and daphnane types.[94,95] The migration rates of esters of 12-deoxyphorbol, when separated by partition methods,[93] were dependent upon the chain length of the C-13 esterifying acid and on the presence or absence of a primary hydroxyl group at C-20 of the nucleus (Table 3). This technique has more recently been applied to the detection of several other groups of compounds, including the daphnane,[96,97] ingenane,[67] and macrocyclic[98] esters.

GLC was used to determine the composition of mixtures of fatty acids[99] produced by the hydrolysis of naturally occurring esters. However, this method was not initially applied to

Table 2
COLOR REACTIONS OF DITERPENES TO SPRAY REAGENTS[91]

	Silica gel			Alumina		
	A	B	D	A	B	C
Phorbol triacetate						
Daylight	Orange	Purple-brown	Red-purple	Orange	Blue-gray	Pink
UV light	Orange	Blue	Yellow-brown	Yellow-brown	Blue-brown	Yellow-brown
12-Deoxyphorbol diacetate						
Daylight	Red-brown	Gray-brown	Orange-brown	Red-brown	Gray-green	Red-brown
UV light	Pink-brown	Dull red	Pink	Orange	Blue-green	Dull red
4-Deoxy-4α-phorbol triacetate						
Daylight	Yellow-brown	Purple-blue	Mauve	Yellow	Blue-purple	Yellow-brown
UV light	Yellow	Blue	Yellow-brown	Yellow-brown	Blue-brown	Light yellow
4α-Phorbol triacetate						
Daylight	Blue-gray	Blue	Blue	Yellow-brown	Blue-black	Brown
UV light	Yellow-brown	Dark blue	Pink	Yellow-brown	Dark blue	Orange
4α-Phorbol tetraacetate						
Daylight	Gray	Blue	Gray-blue	Olive-brown	Blue-black	Olive-brown
UV light	Yellow	Dark blue	Faint pink	Yellow-brown	Dark-blue	Yellow-brown
Crotophorbolone acetate						
Daylight	Pink	Purple	Olive brown	Orange	Blue-green	Pink
UV light	Orange	Blue	Orange	Yellow-brown	Blue	Orange
Ingenol triacetate						
Daylight	Olive-brown	Gray-brown	Yellow-brown	Yellow-brown	Gray-brown	Olive-brown
UV light	Yellow	Yellow	Yellow	Yellow	Yellow-brown	Light-yellow

Note: All plates were viewed in daylight and UV light at 366 nm.

Reagents: A = 60% sulfuric acid (plates heated for 15 min at 110°C), B = 5% vanillin in concentrated H_2SO_4 (plates heated 5 min at 110°C), C = 1% anisaldehyde, 2% H_2SO_4 in glacial acetic acid (plates heated 10 min at 110°C), D = methanol/H_2SO_4 (1/1) (plates heated 15 min at 110°C).

Table 3
MIGRATION OF ESTERS BY PARTITION TLC[93]

Solvent system	1	2	3	4	5	6	7	8	9
Cyclohexane/ethylacetate (80/20)	0.89	0.70	0.66	0.83	0.78	0.13	0.13	0.18	0.40
Benzene	0.84	0.82	0.69	0.85	0.79	0.24	0.28	0.32	0.65
Methylethylketone/cyclohexane (15/85)	0.83	0.67	0 64	0.94	0.72	0.16	0.18	0.23	0.40

Compounds of 12-deoxyphorbol: (1) phorbol-myristate, acetate; (2) acetate, tigliate; (3) acetate, isobutyrate; (4) acetate, α-methylbutrate; (5) acetate, dodecanoate; (6) dodecanoate; (7) isobutyrate; (8) α-methylbutyrate; (9) dodecenoate.

Table 4
GLC OF DITERPENE ACETATES[100]

Column	Compound relative retention				
	1	2	3	4	5
10% SE-30 and 0.05% EGS Oven 213°; N_2 flow, 60 mℓ/min Codeine retention time, 17.0 min	2.78	6.16	3.95	4.65	3.23
10% SE-52 Oven 228°; N_2 flow, 60 mℓ/min Codeine retention time, 29.1 min	2.52	5.47	3.59	4.16	2.95
2% QF-1 Oven 186°; N_2 flow, 60 mℓ/min Codeine retention time, 3.65 min	5.58	24.4	12.6	17.1	13.9
2% OV-17 Oven 220°; N_2 flow, 60 mℓ/min Codeine retention time, 12.0 min	2.52	7.33	4.17	5.25	4.24

Compound: (1) ingenol triacetate, (2) phorbol triacetate, (3) 12-deoxyphorbol diacetate, (4) 4-deoxy-4α-phorbol triacetate, (5) crotophorbolone acetate.

the parent diterpenes because of their thermo-instability and high molecular weight. GLC has the advantage that it provides both qualitative and quantitative data and may be adapted for GLC/MS analysis. A qualitative and quantitative GLC method[100] has been described for the semisynthetic acetates of a number of diterpene alcohols. For this study four columns of various polarity were used for the determinations at oven temperatures of between 186 and 228°C, and a flow rate of nitrogen gas up to 60 mℓ/min. Retention times were measured relative to codeine (Table 4). Quantitative estimations from plant material were achieved using a mixed column of 10% SE-30 and 0.05% EGS. Percentage standard errors of the estimated relative weight ratios of the diterpenes to codeine varied from 0.9 to 1.27% (Table 5) for up to 42 estimations. When plant latex was assayed, the total recovery of the added ingenol was 98.3 ± 3%.

Perhaps the technique which in the future will largely replace both TLC and GLC for the detection of naturally occurring diterpenes is that of high pressure liquid chromatography (HPLC). This method is carried out at room temperature and is therefore suitable for thermo-unstable large molecular weight esters of these compounds. Futhermore, because separations are achieved without exposure to air, contrary to TLC, autoxidation reactions should be negligible. The method provides both qualitative and quantitative data for the naturally

Table 5
QUANTITATIVE GLC OF DITERPENE ACETATES[100]

	Weight ratio relative to codeine	n[a]	Standard error (%)
Ingenol triacetate	1.11	42	1.09
Phorbol triacetate	0.72	30	1.22
12-Deoxyphorbol diacetate	0.21	21	0.91
4-Deoxy-4α-phorbol triacetate	0.85	30	1.27
Crotophorbolone acetate	—	—	—

[a] n = number of determinations.

occurring esters rather than for their semisynthetic derivatives. HPLC is extremely sensitive because of the UV means of detection based upon the strong UV absorbance of the ester carbonyl functions in the molecular structure. The first reported HPLC separation of these compounds was by Rao et al.[101] in 1974. In this short communication the parent diterpene phorbol was separated from phorbol myristate acetate on a 0.5 m × 2.6 mm column of Sil-X at room temperature using isopropyl alcohol/diethyl ether/methanol as solvent. At an inlet pressure of 350 psi phorbol the more polar compound was eluted within 5 min. A similar method was later published[102] for phorbol diesters of differing acyl group molecular weights and also for possible metabolites of phorbol from mammalian systems. This technique employed a Spherisorb 5 μm silica column 2.5 m in length and used 2,2,4-trimethyl pentane/isopropanol (50/50) in a linear gradient at a flow rate of 2.2 mℓ/min as solvent. Phorbol, the last compound to be eluted, had a retention time of about 20 min. It was claimed that the use of tritiated TPA could increase the sensitivity of the method to subnanogram levels for the critical evaluation of the biological fate of phorbol-ester tumor promotors. HPLC methods have also been applied to the detection and separation of ingenane derivatives[76] and to a series of daphnane and tigliane esters for structure/activity studies.[103]

IV. ISOLATION OF DITERPENES

A large number of diterpene esters have now been isolated from plant sources for biochemical, toxicological, and structure/activity studies in both in vivo and in vitro systems. Most of the problems associated with the purification of larger (gram or milligram) quantities of these compounds are concerned with their toxic nature, instability during purification, and their poor separating qualities by conventional methods. The acute effect of the diterpenes on contact with mammalian skin is the induction of erythema. The irritating redness develops within 2 to 3 hr of exposure and persists for 24 hr or more. Adequate safety precautions should be used by all workers in contact with phorbol-type compounds, and this is especially necessary in the organic chemistry laboratory where large qauntities of materials may be handled. These compounds are sensitive to heat, light, oxygen, and alkaline and acid conditions, and appropriate precautions should be taken during their isolation.

Extractions of plant material of between 500 g and 20 kg[104] in weight are best carried out at room temperature by means of cold solvent maceration for up to 1 week. Methanol or acetone, freshly redistilled, are the solvents of choice for this process. After filtration to remove plant debris, the solution may be evaporated to a soft extract by reduced pressure distillation below 45°C. The diterpenes are neutral lipid components of plants and complex separation methods are required for their purification. Separation procedures have been based upon a combination of chromatographic and partition methods. The soft extract may be redissolved in a mixture of methanol and water. This polar phase is initially partitioned against hexane or petroleum spirit to remove waxes, triterpenoids, and some macrocyclic diterpenes, and then partitioned against diethyl ether. The ether phase contains the bulk of the diterpene esters from the plant extract. This phase is washed by partition against 1% sodium carbonate solution to remove pigments and acidic materials. It is dried by shaking with anhydrous sodium sulfate and evaporated below 45°C as before to produce a cream- to yellow-colored friable solid. This resin possesses potent vesicant properties on the skin, and is a mixture of closely related diterpene esters. Further separation of this mixture may be achieved by chromatographic methods. A dry column method[105,106] has been advocated for the separation of these compounds. In this technique silica gel as the adsorbent is used in column form inside a tank containing a solvent mixture which percolates up the adsorbent mass. The technique is in some ways analogous to TLC, but the resolutions achieved are inferior. Phorbol, a parent polyol, has been isolated from croton oil using silica gel column chromatography,[107] and it is claimed that this technique is more rapid than crystallization methods for hydrolyzed fractions of the oil. An interesting new form of chromatography[108] known as droplet countercurrent chromatography (DCC) has been applied to the separation of phorbol from 4α-phorbol after hydrolysis of mixtures of natural esters. In this method a mixture of hexane-ether-*n*-propanol-95% ethanol (4/8/3/5) were used as solvents. The mixture was allowed to come to equilibrium and the lower phase was run out and used as the mobile solvent for the separation, while the upper phase was used as the stationary solvent in the tubes of the DCC apparatus. Although separation of the two tigliane isomers was not 100% satisfactory, this is the first instance of the application of DCC to the separation of diterpenes of this class. The method has proven extremely successful with several other classes of organic products including alkaloids,[109] flavonoids,[110] and glycosides,[111] and the literature of Reference 108 should provide a valuable basis for the future development of DCC in the diterpene area.

Mixtures of diterpene esters have traditionally been fractionated by conventional gradient elution column chromatography using Florosil[112] as adsorbent. Attempts using silica gel as adsorbent have led to transesterification occurring on-column[21,34] and the subsequent formation of artifacts. Repeated separations using a number of different columns[44] have also proved useful for the purification of a major component of a complex mixture. HPLC has also been used for the purification of 12-deoxyphorbol, ingenol, and resiniferonol esters (see Section III). More recently, the new technique of centrifugal liquid chromatography (CLC) has been applied to the separation of mixtures of esters of 4-deoxyphorbol, 20-deoxyphorbol, 5-hydroxyphorbol, and phorbol.[113] This method has the advantage that separation which would normally require several days to complete by conventional gradient elution column methods may be achieved in a matter of hours. The technique involves the preparation of a disc of adsorbent which spins at a constant velocity on a turntable. Mixtures of plant esters are applied in solution to the center of the spinning disc and separate outward by centrifugal forces when a solvent of increasing polarity is applied as a gradient. The apparatus is provided with a fraction collector and a UV detector and is a convenient and rapid alternative to column chromatography.

Initial separations by column chromatography or other methods provide fractions which are mixtures of closely related compounds. Further separations of these mixtures require the application of partition techniques. Liquid-liquid partion separations based upon the principles of countercurrent have been utilized on a large scale. For hand-operated machines, O'Keefe type distributions have been described and those of the Craig type[34] are used for automatic machines. Final purifications of individual compounds is best achieved by means of preparative layer chromatography (PLC). Layers of 0.5 to 1.0 mm thickness, buffered at pH 7.0 with phosphate buffer, may be used for the preparation of 1 to 200 mg of pure compounds. The separations may be of the adsorption type on silica gel or of the partition type using ethylene or propylene glycols as a stationary phase.[93]

V. INVOLVEMENT IN TUMOR PROMOTION

Cancers in humans induced by exposure to chemical substances in the environment have long been recognized as an environmental hazard. As early as 1761[114] the involvement of snuff use in the formation of nasal cancers was documented. The existence of chemical carcinogens was first verified in 1914 by Yamagiwa and Ichikawa[115] when they demonstrated that applications of coal tar to skin resulted in the formation of skin cancer. The tar solutions induced inflammation due to the acid and phenolic components, and irritation and tumor production became linked in biochemical terms. The hypothesis of two-stage carcinogenesis[123,124] was later put forward as an explanation for tumor development on the basis that agents which induce hyperplasia frequently induce tumor formation because of the presence of latent tumor cells in the skin. "Initiation" is used to describe exposure to a carcinogen which induces the formation of latent tumor cells, while "promotion", the second stage of carcinogenesis, was thought to be due to exposure to a nonspecific cancer-promoting agent or hyperplasic substance. The work of Berenblum[20,118] and others demonstrated for the first time the phenomenon of tumor promotion. These workers observed that applications of a chemical carcinogen, for example, dimethylbenzanthracene, to the skin of mice in a single dose resulted in the formation of tumors unless the dose was reduced below that of a minimum threshold. This was termed the initiating or subthreshold dose of the carcinogen. Berenblum used croton oil, the oil of the seeds of *Croton tiglium*, as a promoter and applied this substance weekly over a period of 20 or 30 weeks. When croton oil was applied in this manner to the back of mice after a subthreshold dose of the initiator, malignant tumors formed, whereas in the experimental group of animals which received the initiating dose of carcinogen but not croton oil, no tumors developed. The oil was not acting as a carcinogen because if the initiator was omitted no sarcomas developed on the animals, and only a few papillomas were observed. If the order of application was reversed no tumors developed, and if the intervals between applications of croton oil were extended from a weekly to a monthly basis, no tumors developed either. The action of the initiator was irreversible while that of the promoter was reversible. Any delay of even up to 1 year between the application of the initiator and the promoter had no effect upon the development of malignant growths. This observation may be of particular significance in terms of understanding tumor formation in humans in that minute traces of initiating carcinogens are present in our environment from food, water, and the atmosphere, but at the present time the nature and occurrence of promoting agents is largely unknown.

Chemical carcinogens may therefore be termed first-order carcinogenic risk factors in the environment, while tumor-promoting agents are known as second-order risk factors. Although a great many first-order risk factors have been identified,[119] and considerable effort has been expended on their detection methods,[120] only a few human cancers are known to be directly attributed to these agents alone. Examples include blood cancers of personnel in X-ray departments, lung carcinoma of asbestos workers, and liver carcinoma in workers from the polymer industry. As explained by Hecker[121] this is surprising because it is possible that about every third or fourth person will be affected by cancer during his or her lifetime. Second-order carcinogenic risk factors possibly play a major role in the etiology of human cancer despite the fact they are not tested for by industrial manufacturers in the cosmetic, processed food, and pharmaceutical industries. Tumor promoters are therefore noncarcinogenic materials which have the ability to promote malignancies subsequent to exposure to small doses of a carcinogen.[22,122,123] Substances other than the diterpenes of the phorbol ester type which are known to be second-order carcinogenic risk factors include phenols,[124] esters of certain fatty acids,[125] surface active agents[126] (which may well include detergents), anthralin,[127] iodacetic acid,[128] tobacco,[129] and some steroidal hormones.[130] Such promoting agents are active in the dose range 0.1 to 1.0 μmol per application, whereas TPA, the active diterpene of *Croton tiglium,* is active at dose levels of 0.002 μmol.[131]

Table 6
TUMOR-PROMOTING ACTIVITY OF *EUPHORBIA* LATICES[83]

Plant species	No. of papillomas[a]
Euphorbia abyssincia	28
E. canariensis	103
E. candelabrum	68
E. cooperi	4
E. grandidens	163
E. obovalifolia	40
E. tirucalli	358
E. triangularis	1
E. wulfenii	69
E. ingens	54

[a] Three groups of 20 mice received 300 μg DMBA as an initiator followed by weekly applications of 1% acetone solution of plant extract.

Other species of *Croton* have been shown to produce similar compounds. A recent study[132,133] concerning *Croton flavens* concluded that this species was directly involved in the etiology of esophageal cancer in a human population. It was suggested that the phorbol derivatives produced by this plant comprised a synergistic component of the overall carcinogenic load of the environment. This study was carried out on the island of Curaçao northwest of Venezuela, where statistical evidence indicated that between the years 1936 to 1965 21 per 100 deaths were due to esophagal cancer. In this area the roots of *Croton flavens* are chewed in a similar manner to the roots of *Acorus calamus*[134] in the north of England, tobacco worldwide, and *Areca* nut[135] in parts of India and Africa. The aerial parts of *Croton flavens* are also used to produce a bush tea. An extract of this plant was shown to be a tumor-promoting agent when applied to the skin of mice. From this extract six novel tigliane esters were isolated by chemical methods. These compounds were based upon 16-hydroxy and 4-deoxy-16-hydroxy phorbol. Certain of the *Croton* factors were cryptic promoters in that they were acylated at the C-20 position, while others were directly active in the biological test system used. Cryptic tumor-promoting agents[136] are compounds which, because of the acyl group present at C-20 of the phorbol nucleus, are not biologically active in current test systems. However, the C-20 primary acyl moiety is chemically unstable in the presence of acid and alkali and is susceptible to hydrolysis, possibly even by lipase enzymes in vivo, thereby liberating the tumor promoting and pro-inflammatory diesters of phorbol. Two compounds isolated[133] from *Croton flavens, Croton* factors F_1 and F_2, exhibited promoting activity comparable to TPA. Thus, in the case of esophageal cancer in Curaçao the second-order risk factors were established, but the nature of the initiator or initiators was unknown. It was suggested that drinking water on the island contaminated with gasoline, a source of PAH-type initiators, might be responsible, although the possibility of some kind of nitrosamine as a putative initiator was not ruled out.

A study of the closely related genus *Euphorbia* by Roe and Peirce in 1961[83] clearly indicated that this genus of plants contains tumor-promoting agents which possibly constitute an environmental hazard in terms of animal and human malignancies. The botanical relationship between *Croton* and *Euphorbia* led these workers to propose that there was a similarity between the toxic constituents of the two genera. This was later confirmed on a chemical basis.[123] In this work[83] the tumor-promoting capabilities of ten species of *Euphorbia* (Table 6) were investigated. DMBA, 300 μg per animal, was used as the initiator, followed by weekly doses of a 1% acetone solution of an extract of *Euphorbia* latices. Twenty mice were used in each group of experimental animals consisting of ten male and ten female animals. Numerous papillomas and two squamous cell carcinomas were induced on the backs of 101 strain mice during this investigation. There was a good correlation between the degree of epidermal hyperplasia and the tumor-promoting actions. *Euphorbia tirucalli* was the most potent species investigated in this exercise.

The importance of this early paper on the genus *Euphorbia* in terms of environmental problems associated with the genus was later emphasized by the study of Ott and Hecker[137] and by Uphadhyay and others.[138] *Euphorbia coerulescens* Haw. is a species of plant used economically for the production of honey. It was believed that an investigation of the honey produced by bees using this plant as a pollen and nectar source might allow conclusions to be drawn as to the stability of phorbol and related esters in commercial food products. The main *Euphorbia* diterpene isolated from the honey was 12-deoxyphorbol, a parent polyol which was inactive in tumor-promoting experiments. This product was possibly produced as a result of hydrolysis of active esters from the plant. Also detected in the honey were two esters of 12-deoxyphorbol, one of which, when compared in Berenblum-type experiments, was comparable to TPA at about four times the dose. Minor amounts of tumor-promoting agents were present in the honey, which possibly constituted a health hazard over a period of time. Independently it was shown that the honey source plant *Euphorbia sequieriana* Neck. secreted esters of the diterpene ingenol in honey produced from its nectar.[138] The ingenol esters are another group of tumor-promoting diterpenes related to the phorbol esters.

Several other genera of the families Euphorbiaceae and Thymelaeaceae have also been shown to possess tumor-promoting activities. The genus *Pimelea* is noted for its toxic effects in animals. An interesting connection between the ingestion of *Pimelea trichostachya* and a disease in cattle known as St. George's disease was made by Kelly and Bick.[139] Several fractions from this plant produced sustained contractions of bovine pulmonary vein. The LD_{50} values in mice and their acute toxicity in calves also correlated. These results provided evidence for the role of pulmonary venous hypertension in the pathogenesis of *Pimelea* poisoning in the bovine. St. George's disease in cattle is due to several related species of this genus.[140] Chemical investigations have been carried out on *Pimelea simplex*[141] and *P. prostrata*[142-144] and esters of the tigliane and daphnane types have been isolated from those plants. *Pimelea* factors P_1 and P_2, both of which are of the daphnane type of compounds, have been shown to be promoting agents in Berenblum-type experiments.[122] Similar compounds have also been obtained from *Pimelea linifolia*[145] and *P. ligustrina*.[146] Tumor-promoting data are currently available for the constituents of 21 species of the families Euphorbiaceae and Thymelaeaceae (Table 7). These genera include *Croton, Euphorbia, Hippomanne, Hura, Pimelea,* and *Daphne* species, and the compounds tested range from tigliane and both ingenane and daphnane derivatives.[121-123] The parent polyols are thought to be inactive but weak promoting activity has been demonstrated by Armuth and others[147] in internal organs of mice correctly initiated. There is also some doubt concerning the tumor-promoting actions of the daphnane class,[148] including mezerein.

Traditional Berenblum-type experiments normally require at least 20 weeks to obtain a result. It is not surprising, therefore, that full promoting data are not available for most plants of the Euphorbiaceae. The vast majority of investigations of these species have used a rapid form of pro-inflammatory assay[149] in combination with chemical analysis for the identification of the diterpenes present. It is possible on a structure/activity basis to suggest which of the many species presently under investigation may be tumor-promoting agents in the environment, but this is no substitute for definite biological data. Several rapid in vitro techniques have been proposed for the assessment of tumor-promoting activities. Such tests include induction of ornithine decarboxylase,[150] induction of terminal differentiation in hu-

Table 7
PLANTS OF THE EUPHORBIACEAE AND THYMELAEACEAE FOR WHICH BERENBLUM-TYPE DATA ARE AVAILABLE

Plant species	Class of promoting agent	Ref.
Croton tiglium	Tigliane	21
C. flavens	Tigliane	133
C. sparciflorus	Tigliane	43
Daphne mezerium	Daphnane	57
Euphorbia biglandulosa	Tigliane, ingenane	73, 205
E. coerulescens	Tigliane	46, 137
E. cooperi	Tigliane	51, 206
E. esula	Ingenane	207
E. ingens	Ingenane	75
E. lactea	Ingenane	69
E. lathyris	Ingenane	66
E. resinifera	Tigliane, ingenane	64
E. serrata	Ingenane	208
E. tirucalli	Tigliane, ingenane	45
E. triangularis	Tigliane	49, 51
E. unispina	Tigliane, daphnane	64, 94
E. virgata	Ingenane	209
Euphorbium	Tigliane, ingenane	192
Hippomane mancinella	Tigliane, daphnane	210
Hura crepitens	Daphnane	131
Pimelea prostrata	Daphnane, tigliane	122, 142,
P. simplex		144, 145

man promyelocytic leukemia cells,[151] inhibitory effect on terminal differentiation of Friend virus transformed cells,[152] enhancement of clones of SV cells,[153] and enhancement of mutagenesis in bacteria.[154] Specific receptor assays[68,155-158] are also available for phorbol esters, and in the future assays based upon the activation of proteinkinase[159-161] might be applied to this problem. Many of these biochemical methods have only been applied to a limited structural range of esters, mainly TPA and mezerein, but their application to the screening of plant extracts and their products for environmental health purposes has been lacking in the current literature.

A short-term biological assay was devised for the detection of diterpene ester tumor-promoting agents and applied to the extracts of a number of plants of the family Euphorbiaceae.[162,163] This technique involved the utilization of Epstein-Barr virus (EBv) expression in EBv genome-carrying human lymphoblastoid cells. *n*-Butyrate was used as an inducer of the test and EBv-nonproducer Raji cells as the indicator. After cultivation for 48 hr at 37°C, the ratio of EBv early antigen (EA) to expressing cells was determined by immunofluorescence. Substances such as anthralin, phenol, and Tween® did not significantly react to this assay. Four species of the genus *Croton* were investigated, including *Croton tiglium, Jatropha curcus,* and four species of the genus *Euphorbia* including *Euphorbia lathyris* (Table 3).

All of the plant extracts with the exception of *Euphorbia hirta* and *Croton classifolius* produced a strongly positive reaction, and on this basis must be considered to possess tumor-promoting capabilities. These species are all ornamental varieties cultivated in Japan and the Far East. While investigating possible routes through which active diterpene esters might gain access to humans, the soil underneath which plants of the Euphorbiaceae and Thymelaeaceae were growing were investigated by the EBv expression assay.[164] Soil samples were collected, extracted with ether, and the residue produced after evaporation of solvent was assayed as before[163] (Table 8). The results suggested a possible interaction between

Table 8
EBv-EA DATA FOR SPECIES OF THE FAMILY EUPHORBIACEAE

Species	EBv-EA[163,164,167] (%)
Plant extracts[163]	
Croton tiglium (5.0 μg/mℓ)	67.9
C. megalocarpus (10 μg/mℓ)	16.4
C. macrostachyus (10 μg/mℓ)	32.9
C. classifolius (10 μg/mℓ)	<0.1
Euphorbia laythris (5.0 μg/mℓ)	20.2
E. tirucalli (10 μg/mℓ)	24.6
E. pulcherrima (10 μg/mℓ)	1.3
E. hirta (10 μg/mℓ)	<0.1
Jatropha curcas (10 μg/mℓ)	14.0
Aleurites fordii (tung oil[167]) (50 μg/mℓ)	34.2
Soil samples[164]	
Sapium japonica (20 μg/mℓ)	26.5
S. sebriferum (20 μg/mℓ)	23.8
Codiaeum variegatum (20 μg/mℓ)	12.0
Daphne odora (20 μg/mℓ)	15.6
Euphorbia laythris (20 μg/mℓ)	3.6
Edgeworthia papyrifera (20 μg/mℓ)	5.9
Gardening soil (20 μg/mℓ)	0.1

plant-derived diterpene esters and EBv-associated disease, particularly nasopharyngeal carcinoma. A new view of the etiology of nasopharyngeal carcinoma[165] suggested that there was a geographical similarity between the distribution of *Croton tiglium* and other members of the family Euphorbiaceae and nasopharyngeal carcinoma in the southern regions of China. *Fusobacterium* and other common microbial flora of the mouth and nasopharynx in man produced *n*-butyric acid in laboratory culture. The seeds of several Euphorbiaceous plants are used as a herbal drug in the south of China and it was hypothesized that nasopharyngeal cancer was intitiated by persisting EBv together with the combined effects of bacterial fatty acids and promoters of plant origin. An analogous proposal[166] was later put forward concerning Burkitt's lymphoma in parts of tropical Africa and the use of herbal drugs of this family which are effective inducers of EBv-associated antigens. A further point of contact between humans and diterpene esters in the etiology of cancer was later investigated by Ito et al.[167] Tung oil, the oil from the seeds of *Aleurites fordii* (Chinese tung oil tree), when used in combination with *n*-butyrate was shown to be an effective activator of EBv in Raji cells. Tung oil increased the yield of infectious EBv by about five times. Because tung oil is used to manufacture paints, varnishes, and anticorrosive materials, it was suggested that this material might be implicated in tumor formation in persons involved in these industries. Tigliane-type diterpenes which possess EBv-inducing activity were also isolated from *Sapium sebriferum,* a common roadside and garden tree in Japan.[168]

The acute effect of extracts from these plants is the induction of erythema[169,170] on mammalian skin. Combined chemical and toxicological studies have clearly demonstrated[170] that the pure diterpenes have both pro-inflammatory and tumor-promoting activities. However, although all tumor-promoting diterpenes isolated to date are pro-inflammatory agents, several irritant phorbol and daphnane diterpenes are not promoting agents.[171,172] The use of an erythema assay[149] for extracts of plants from these two noxious families does provide a rapid means of pinpointing species which are potentially hazardous in terms of human cancer. About 60 species of the genus *Euphorbia,* chosen so as to be representative of many divisions within this large group of over 1600 species, were investigated by Kinghorn and Evans[173] for their ability to induce erythema of mouse skin. This is the largest screen to date of plants of the *Euphorbia* genus for their toxicological properties. In this investigation 5 $\mu\ell$ of an acetone extract of plant latex or herbaceous material was applied to mouse ears. Six dilutions of extract varying by a logarithmic factor or two were applied to six mice per group. Positive redness or irritancy was assessed on an all-or-nothing basis. The irritant dose 50% (ID_{50}) was calculated together with standard deviations as described by Hecker and others.[174] This assay was later modified to enable statistical assessment by means of a computer program.[149] Assessments of erythema were made at 4 and 24 hr after application because in some species of plants the difference in inflammation induced at these two time intervals was marked (Table 9). Rapidly acting irritants were found in species of the section Tithymalus, subsections Galarrhaei Boiss. and Esculae Boiss., and in the section Euphorbium, subsection Polygonae Borg. of the botanical classification of Pax and Hoffmann.[85] Certain plants failed to demonstrate irritant effects on mammalian skin. These plants were of the section Poinsettia Boiss. and of the section Anisophyllum Haw. Roep., subsection Hypericifolia Boiss. *Euphorbia hirta* belongs to this group and is one of the few *Euphorbia* species to be used in medicine for the treatment of asthma.[175] Of the 60 species investigated in this study, only 7 were not biologically active, and the European spurges commonly found both wild and as garden ornamental plants were equally as toxic as species from tropical regions of the world. A similar screen of *Euphorbia* plant extracts from plants growing in the Azarbaijan province of Iran was later carried out by Upadhyay and others.[176] In this study 15 species were tested for their irritant effects on mouse ears (Table 10) and their ID_{50} varied from 0.24 to 100 μg per ear. In this instance no differences were noted for the magnitude of the irritant effects when measured at 4 and 24 hr. It was not surprising, therefore, that ingenane esters were chemically detected in these plants because ingenol derivatives have long-term irritant effects.

The assessment of ornithine decarboxylase-inducing activity in mouse skin has been applied to a number of phorbol and ingenol esters.[150] This activity is believed to correlate well with their tumor-promoting actions. The daphnane diterpenes have been less investigated in this respect, although Mufson and others[177] have reported ornithine decarboxylase-inducing activity by the daphnane ortho ester mezerein. Esters of the daphnane class are known to be constituents of *Daphne odora* Thunb. of the family Thymelaeaceae. Seven ortho esters were recently obtained from this species and these included odoracin, odoratrin, and gniditrin.[178] These compounds exhibited potent ornithine decarboxylase-inducing activity, which suggested that *Daphne odora* is a plant capable of acting as a tumor-promoting agent. This plant is used as a fish poison in the Far East.

VI. ETHNOBOTANY

Studies concerning the ethnobotanical uses of plants of the Euphorbiaceae and Thymelaeaceae are of considerable interest to toxicologists involved with environmental aspects of tumor promotion. Original observations have been rare in recent years possibly because of the need to acquire first hand information from people living in underdeveloped or geographically inaccessible regions. This type of work is nevertheless of increasing urgency because of the rapid development and industrialization of primitive societies. One such brief study concerning the genus *Euphorbia* native to Nigeria was reported in 1975.[179] The latices, together with botanical specimens of plants for record purposes, were collected and the traditional uses and any toxic effects of these plants were catalogued from direct conversations with native populations in the villages. Botanical authentications of collected materials were made in collaboration with faculties of botany in local universities and government institutions. Samples were later analyzed for the presence of diterpenes of the tigliane, daphnane, and ingenane types (Table 11). The local uses of plants are often related to their toxicity

Table 9
PRO-INFLAMMATORY ACTIVITY OF *EUPHORBIA* LATICES[173]

Species[173]	ID_{50} μg/5 μℓ/ear 4 hr	24 hr	Species[173]	ID_{50} μg/5 μℓ/ear 4 hr	24 hr
E. androsaemifolia	2.2	35.6	*E. myrsinites*	1.3	1.1
E. antiquorum	2.2	14.1	*E. neriifolia*	22.5	50.0
E. balsamifera	—	—	*E. nivulia*	2.0	5.0
E. biglandulosa	2.5	5.0	*E. nubica*	—	—
E. canariensis	0.9	2.5	*E. paganorum*	0.2	2.8
E. candelabrum	1.6	2.5	*E. palustris*	28.3	79.0
E. characias	2.8	2.5	*E. pentagona*	28.3	35.6
E. coerulescens	7.1	22.2	*E. pilosa*	2.8	5.6
E. cooperi	1.4	2.4	*E. poissonii*	0.1	17.8
E. coralloides	0.5	3.5	*E. polyacantha*	0.4	79
E. cyparissias	1.6	1.4	*E. polychroma*	14.1	44.8
E. deightonii	0.8	2.0	*E. pseudograntii*	5.0	1.4
E. desmondi	0.9	2.2	*E. pulcherrima*	—	—
E. erythreae	2.0	4.0	*E. resinifera*	0.2	—
E. fortissima	3.1	17.8	Euphorbium	1.0	—
E. franckiana	17.8	22.1	*E. robbiae*	5.0	6.3
E. geniculata	—	—	*E. royliana*	7.9	25.1
E. helioscopia	1.6	1.32	*E. sikkimensis*	14.1	6.3
E. hiberna	4.0	1.33	*E. stenoclada*	—	—
E. kamerunica	4.4	3.5	*E. subthorpii*	4.5	4.0
E. kotschyana	1.4	20.0	*E. tirucalli*	1.6	2.8
E. lactea	2.0	2.0	*E. triangularis*	7.0	40.0
E. lateriflora	4.4	14.1	*E. unispina*	3.1	9.0
E. lathyris	4.5	6.3	*E. wulfenii*	1.25	5.5
E. ledienii	5.0	17.7	*Elaeophorbia drupifera*	1.5	7.0
E. memoralis	7.9	4.0	*El. grandifolia*	3.1	5.6
E. milii	50.0	32.5	Candle plant	0.2	—

Table 10
PRO-INFLAMMATORY ACTIVITY OF IRANIAN *EUPHORBIA* SPECIES

Species[176]	ID_{50} μg/5 μℓ/ear 24 hr	Diterpene present
E. dulcis	34.8	Ingenol
E. esula	0.20	Ingenol
E. herbecarpa	100	Ingenol
E. megalantha	100	—
E. myrsinites	100	Ingenol
E. orientalis	100	—
E. paulestris	100	Ingenol
E. peplus	100	Ingenol
E. piploid	100	—
E. segitalis	133	Ingenol
E. sequieriana	1.14	Ingenol
E. serrata	1.40	Ingenol
E. sovitzii	224	Ingenol
E. striatella	0.25	Ingenol
E. virgata	6.5	Ingenol

Table 11
GEOGRAPHICAL DISTRIBUTION AND LOCAL USES OF NIGERIAN *EUPHORBIA* SPECIES[179]

Species	Where collected	Diterpenes (% w/w latex)	Uses and effects of the plant
Euphorbia poissonii Pax. — Tinya (Hausa), Zuzuk (Kaje); candle plant	Northern Nigeria Katsina district	Tigliane and unknown diterpenes (0.7%)	Insecticide and arrow poison; can cause blindness and irritates skin and mucous membranes
E. unispina N.E. Br. — Tinya (Hausa); candle plant	Benue plateau	Tigliane and unknown diterpenes (0.2%)	Fish and arrow poison, skin irritant
E. paganorum A. Chev.	North central state Zaria district	Tigliane diterpenes (1.51%)	Skin irritant
E. balsamifera Ait. — Aguwa (Hausa)	Northern Nigeria Sokoto and Katsina areas	Traces of tigliane diterpenes	Treatment for gingivitis, gonorrhea, and insect bites, ordeal poison
E. desmondi Keay and Milne — Redhead; Kirana (Hausa), Panjam (Birom)	Benue plateau	Ingenane diterpenes (1.05%)	Fish poison, hedging plant, skin irritant
E. deightonii Croizat — Kirana (Hausa)	Western Nigeria Ibadan area	Ingenane diterpenes (0.96%)	Skin irritant
E. kamerunica Pax. — Kirana (Hausa); Riyip (Birom)	Area of Vom	Ingenane diterpenes (1.01%)	Treatment for scabies, ringworm, snake bite; fish poison; causes blindness; fencing material
E. lateriflora Schum and Thonn. — Hdatsartsa (Hausa)	Kano and Katsina areas	Unknown diterpenes	Treatment for intestinal parasites; removes spelks; irritant
Elaeophorbia drupifera Stapf; *Elaeophorbia grandifolia* Croizat	Western Nigerian coastal region	Ingenane diterpenes (0.33—0.47%)	Treatment for insect and snake bites, warts, and ringworm; severe skin blistering agent

— for example, their use as fish poisons[8,58,146] or arrow poisons.[9] Normally the less toxic the latex of the plants the more varied are their uses. *Euphorbia kameranica* is mainly used in Nigeria as an arrow poison and as a protective fence around villages to keep domestic animals inside and predators outside,[179] while *Euphorbia balsamifera* is used as an antiseptic, a hemostatic, and purgative[13] in traditional medicine due to its comparative lack of toxicity. Literature dating from the colonial period and more recently updated is of particular use in identifying hazardous species of plants. The ethnobotany of the Euphorbiaceae of West Africa,[13,15,30,180-182] eastern and southern Africa,[31,87,183,184] the Indian subcontinent,[14,20] Samoa,[185] Asia,[29] British and European species,[6] Italian species,[186] and the *Euphorbias* of the New World has been extensively compiled by von Reis Altshul.[1] A review of the toxicological properties of the Euphorbiaceae was published in 1979.[187] In recent reviews, Hecker made a study of the uses of plants from both the Euphorbiaceae and the Thymelaeaceae[121-123] (Table 12). Particularly highlighted was the use of plants of the Euphorbiaceae in commerce. *Aleurites fordii*[52] is used for the production of a drying oil known as tung oil and for the manufacture of cattle cake.[167] Several species of *Euphorbia,* including *E. lathyris, E. ingens,* and *E. cooperi,* produced copious quantities of latex which is rich in polymers and hydrocarbons. The polymers are used for chewing gum and the rubber industries,[121-123] while the hydrocarbons are suitable for industrial cracking in the manufacture of petroleum. It has been proposed[188,189] that certain *Euphorbia* species, because of their ability to grow in arid

Table 12
ECONOMIC USES OF SOME PLANTS OF THE FAMILIES EUPHORBIACEAE AND THYMELAEACEAE[121-123]

	Use of the plant	Diterpene group
Euphorbia species		
E. tirucalli L.	Rubber and petroleum production, natural hedging	Tigliane Ingenane
E. triangularis Desf.	Chewing gum and nectar production	Tigliane
E. resinifera Berg.	External application as an ointment, counterirritant	Tigliane Ingenane Daphnane
E. cooperi N.E.Br.	Rubber and nectar production	Tigliane
E. unispina N.E.Br.	Tobacco snuff, fish and arrow poison	Tigliane, daphnane
E. poisonii Pax.	Arrow, fish, and cattle poison	Tigliane, and daphnane
E. myrsinites L.	Medicine, horticulture	Ingenane
E. kansui Liou	Used in Chinese medicine	Ingenane
E. ingens E. Mey.	Rubber production, arrow poison	Ingenane
E. laythris L.	Purgative drug, soap, and petroleum production, horticulture	Ingenane
E. milii Ch.	Horticulture	Ingenane
E. esula L.	Folk medicine	Ingenane
E. marginata Pursh.	Horticulture, nectar production	Ingenane
E. cyparissias	Homeopathy, alternative medicine	Ingenane
Croton species		
Croton tiglium L.	Purgative, abortifacient, counterirritant	Tigliane
C. sparsiflorus Morong.	Purgative drug	Tigliane
C. oblongifolius Roxb.	Alternative medicine, homeopathy, purgative	Tigliane
C. flavens L.	Roots used for chewing, tea from leaves	Tigliane
Aleurites species		
A. fordii Hemsl.	Tung oil production	Tigliane
Hippomane species		
H. mancinella L.	Used for timber construction	Tigliane Daphnane
Pimelea species		
P. prostrata Willd.	Cattle poison	Daphnane, tigliane
P. simplex F. Muell.	Cattle poison	Daphnane diterpenes
Daphne species		
D. racemosa Griseb.	Purgative drug	Daphnane
D. mezereum L.	Folk medicine, horticulture, protected wild plant	Daphnane
Gnidia species		
G. lamprantha Gilg.	Folk remedy	Daphnane
G. subcordata Meissn.	Purgative drug	Daphnane
Lasiosiphon species		
Lasiosiphon burchelli Meissn.	Cattle poison	Daphnane
Excoecaria species		
Excoecaria agalocha L.	Fish and arrow poison	Daphnane
Hura species		
H. crepitens L.	Horticulture	Daphnane
Synaptrolepsis species		
Synaptolepsis kirkii Oliv.	Antiepileptic, emetic	Daphnane
Daphnopsis species		
Daphnopsis racemosa Griseb.	Purgative drug	Daphnane
Stillingia species		
S. officinalis	Alternative medicine	Daphnane, tigliane

areas, might be used for desert reclamation projects.[190] This may provide an economic crop from inclement land in Africa and parts of the U.S. *Hippomane mancinella*[122] is used for the manufacture of timber for the furniture and construction industries and several Euphorbiaceae are of considerable commercial value in the food industry. *Manihot* is used to produce cassava and tapioca, *Ricinus* is used for the production of castor oil and cattle cake,[191] and several other species produce proteins, fats, and oils for human and animal consumption, as well as being source plants for the production of honey and honey products.[137,138] Direct exposure to tumor-promoting agents could possibly result from the industrial use of such plants.

Further direct exposure to tumor-promoting agents from plants may occur as a result of the use of many of these species in medicine. Traditionally, *Croton tiglium* oil has been used in Western medicine as a purgative and as a counterirritant to the skin.[16] It has also been illicitly recommended as an abortifacient. This drug was removed after the appearance of the British Pharmaceutical Codex of 1949 because it was considered too toxic for human use. Nevertheless, *Croton* oil continued to be used medicinally both in the prescribed form and as a "home remedy" for many years. The isolation of TPA and other phorbol diesters from this plant has effectively removed the plant oil from normal medicinal use in Britain and the U.S. Euphorbium[175] is a dried plant latex most probably produced from the fresh latex of *Euphorbia resinifera,* which has been allowed to sun-dry on the plant after injury. The medicinal properties of this material have long been appreciated, and in fact the whole of the genus *Euphorbia* owes its name to King Juba II of Mauritania (25 B.C.) who honored a succulent species prized for its curative powers by naming it after his physician, Euphorbos, who was equally fleshy.[13] Euphorbium has been used as a purgative, as a counterirritant, and for the treatment of parasitic infections. Both ingenane and daphnane diterpenes have been isolated from this commercial drug material[192] and its use in Western medicine is now contraindicated.

Besides the use of a few members of the Euphorbiaceae in past pharmacopoeias, perhaps the aspect of the use of these plants as drugs which causes most concern at the present time is their use in "alternative" or "herbal" medicine in Britain, Europe, and the U.S. In alternative forms of medicine many people, often with chronic conditions such as rheumatism or nervous diseases, are increasingly becoming disillusioned with the benefit-vs.-side effects relationships of modern chemical drug therapy, and are turning to other philosophies for relief. These include homeopathy, aromatherapy, health foods, vitamins, and plant-based remedies. A very wide range of plant materials are now being used and are available to the individual through health food stores, mail order catalogues, or by prescription from a "herbalist" or herbal practitioner. A number of plants of the Euphorbiaceae and Thymelaeaceae are used in these cases including *Daphne, Gnidia,* and *Euphorbia* species for the treatment of growths and tumors. Some of these plants have been shown to contain tumor-promoting agents. However, not all members of the Euphorbiaceae are necessarily poisonous. For example, *Euphorbia hirta* or *E. pillulifera,* which is a popular remedy in India, is also included in the *British Herbal Pharmacopoeia*[193] for the treatment of bronchitic asthma, laryngeal spasm, upper respiratory catarrh, and intestinal amebiasis. The drug is said to have antiasthmatic, spasmolytic, and expectorant properties. When this plant was investigated for the presence of tumor-promoting diterpenes it was found to be negative.[173] Queens root, *Stillingia officinalis,* of the family Euphorbiaceae is a component of herbal mixtures available in the U.S.[122,123] and is also included in the *British Herbal Pharmacopoeia.*[194] It is specifically indicated for the treatment of urinary calculus and is claimed to have diuretic, diaphoretic, litholytic, and antilithic actions. Tigliane and daphnane diterpenes have been isolated from this species[195] and on this basis in the U.S. efforts have been made to remove the drug from sale. Not all phorbol and related esters are tumor-promoting agents and their structure/activity relationships are now well understood. For example, phorbol triesters, 4α-phorbol esters, 20-deoxyphorbol esters, some daphnane esters,[122] and the macrocyclic diterpenes[53] do not possess this toxic activity. It would be premature for governments in European countries and the U.S. to prevent the use of a particular species of Euphorbiaceae in herbal medicine simply because of the reputation of the family. These remedies do, however, represent a possible health hazard in terms of human cancer promotion and should be fully investigated by both biological and chemical methods before decisions are made as to their allowable use in alternative medicine.

A considerable research effort has been made on both the families Euphorbiaceae and Thymelaeaceae as sources of antitumor drugs. In a review in 1966 Norman Farnsworth[55] noted the antineoplastic effects of 11 species of the family Euphorbiaiceae from folklore tradition, including *Bridelia orata, Emblica officinalis, Macaranga triloba, Mallotus philippensis,* and *Piscaria species.* Hartwell[10] published a survey on the use of plants of the Euphorbiaceae in the treatment of growths, warts, and tumors. This survey included *Acalypha,* 9 species (Mexico, Guatemala, and Antilles); *Actinostemon anisandrus* (Argentina); *Aleurites moluccana* (Japan); *Baliosperum,* species (India, Japan); *Breynia,* 2 species (India); *Chrozophora tinctoria* (Mexico, South America); *Colliguaja,* 2 species (South America); *Croton,* 7 species (Africa, India, South America); *Elaeophorbia drupifera* (Africa); *Euphorbia,* 79 species (worldwide); *Excoecaria,* 2 species (Australia, Indonesia); *Gelonium multiflorum* (Argentina); *Gymnathes lucida* (Cuba); *Hippomane mancinella* (Cuba, Paraguay); *Jatropha,* 9 species (Mexico, South America, East Indies); *Mallotus,* 2 species (Japan, India); *Manihot,* various species (South America, Africa); *Mercurialis,* 2 species (not recorded); *Microdesmis puberula* (Nigeria); *Pedilanthus,* 4 species (Mexico, South America); *Phyllanthus,* 3 species (India, South America, Mexico); *Ricinus,* 2 species (India, U.S., South America, China); *Sapium,* 5 species (U.S., Cuba, South America); *Stillingia sylvetica* (U.S.); *Terminallis bellerica* (India); and *Tragia hepetaefolia* (Mexico). Several of these species are finding their way into mixtures used by therapists of the alternative medicine type for the treatment of malignancies in humans, and because of their extreme toxicity and potential promoting abilities, are hazardous. Antitumor drugs of the macrocyclic classes have been isolated from plants of the Euphorbiaceae, and one of these compounds known as euphornin,[196] is undergoing clinical trials at the present time. Kupchan and others[79] isolated diterpenes of the macrocylic type from *Jatropha* species. These compounds were of the jatrophan hydrocarbon class of diterpenes and were active in vivo against P-388 lymphocytic leukemia in mice. Torrance et al.[197] later isolated related antitumor diterpenes from *Jatropha macrorhiza* of the family Euphorbiaceae, and Abo and Evans[198] demonstrated that macrocyclic ingol diterpenes from *Euphorbia* species were cytotoxic against TLX/5 lymphoma cells in vitro. Two compounds of the 1-alkyl-daphnane type were isolated from *Gnidia subcordada,*[69] known as gnidimacrin and its 20-*O*-palmitate. These substances were antileukemic in action when tested in vivo. The 12-hydroxy-daphnetoxins, gnididin, gnidicin, and gniditrin from *Gnidia lampranthra*[199] and *G. latifolia*[200] were also shown to be antileukemic agents. The co-occurrence of tumor-promoting tigliane or ingenane diterpenes in the same plants as antitumor macrocyclic or 1-alkyl-daphnane diterpenes may be explained on the basis that possibly these compounds are biosynthetically related and produced by plants via the same biochemical pathway. (See Chapter 4.)[53] However, even should antitumor agents be produced from members of the Euphorbiaceae for chemotherapy in the pure form, the use of the plants as teas, extracts, or formulations without previous promoting or pro-inflammatory tests is obviously not warranted.

Indirect exposure to plants of the families Euphorbiaceae and Thymelaeaceae could result from the use of these plants in veterinary medicine[122] and from the ingestion of meat or milk products produced from animals previously in contact with them. In several areas of the world, euphorbiaceous plants are habitually used as ornamental species in gardens, parks, and homes. *Jatropha gossipifolia* is a small ornamental shrub common in India, *Excoecaria agallocha*,[2] known as the "blinding tree" because of its vesicant properties particularly to the eyes, and other *Excoecaria* and *Sapium* species are also common wild and cultivated species of the same area. *Euphorbia* species such as *E. ingens*, *E. tirucalli*, and *E. royleana* are cultivated in gardens in Egypt and parts of tropical Africa. Such succulent varieties are favored in temperate zones by cactus growers for greenhouse cultivation. Herbaceous *Euphorbia* varieties such as *E. laythris*, *E. milii*, *E. myrsinites*, *E. wulfenii*, and many others are used as border plants in gardens in the United Kingdom, Japan, and the U.S., often deliberately planted to give ground cover and deter vermin such as moles.[202] A species of *Poinsettia* known as the Christmas rose *(E. pulcherrima)* is a popular house plant during December. Some botanical gardens, such as the Royal Botanical Gardens at Kew, London, contain large collections of the Euphorbiaceae in greenhouses and borders and these gardens are open to the public. Gardeners and members of the public come into regular contact with cultivated species, and it has been demonstrated that even the soil in which such plants grow contain tumor-promoting diterpenes.[164] However, according to Saffiotti[203] the problems presented by such contact are slight on the basis that promoters have to be regularly applied before tumors develop. Inflammatory conditions of the skin, eyes, and mucous membranes are more likely to result from occasional contact with such plants. This is also true of species which are found as weeds in cultivated areas. These plants invade arable land as secondary growth.[6,30-32]

Euphorbia helioscopia is an annual herb found as a common weed on cultivated land in the British Isles. This plant is known as the sun spurge, or by local names such as "wolf's milk" and "cat's milk". From this plant a series of esters of 12-deoxyphorbol were isolated and their irritant potency assessed.[204] The plant is known to be responsible for poisoning in both animals and children.[6] *Daphne* species such as the spurge olive *(D. mezereum)* and the spurge laurel *(D. laureola)* are also common poisonous British plants, and in these cases the red and black berries produced are attractive to children who mistake them for currants and fatalities have been recorded[6] after ingestion. Members of the Euphorbiaceae and Thymelaeaceae families are commonly involved in animal poisoning in many areas of the world (see Table 12). *Euphorbia coerulescens* Haw., known as the blue euphorbia or sweetnoors, is even cultivated in the Noorsveld, South Africa as an animal feedstuff, whereas its hybrid, known locally as sournoors, is too toxic for this purpose. A chemical investigation[48] of true sweetnoors demonstrated the presence of phorbol and 12-deoxyphorbol esters, but the low pro-inflammatory activity of sweetnoors in comparison to sournoors was due not only to the low yields of tigliane esters, but also to the presence of cryptic irritants which exhibit a C-20 acyl function.

Although the toxic effects of plants from these two families are well documented, the significance of plants of the Euphorbiaceae and Thymelaeaceae in human cancer remains unclear at present. Several points of direct and indirect exposure to these plants by humans and animals have been recorded, but clearly each case must be assessed separately. Any hazards in terms of human tumor production must be less than, for example, exposure to *Nicotiana tabacum* and its products, but may nevertheless be significant as a contribution to the overall carcinogenic load of the environment. Perhaps the greatest contribution that persistent study of these plants and their diterpenes have made is to the elucidation of the mechanisms of action of promotion and in assisting the identification of other unknown promoting agents in the environment.

REFERENCES

1. **von Reis Altschul, S.,** *Drugs and Foods from Little Known plants,* Notes in Harvard University Herbaria, Harvard University Press, Boston, 1973, 143.
2. **Chopra, R. N., Chopra, I. C., Handa, K. L., and Kapur, L. D.,** *Indigenous Drugs of India,* 2nd ed., H. N. Dhur & Sons, Calcutta, 1958.
3. **Evans, F. J., Kinghorn, A. D., and Schmidt, R. J.,** Some naturally occurring skin irritants, *Acta Pharmacol. Toxicol.,* 37, 250, 1975.
4. **Lampe, K. and Fagerström,** *Plant Toxicity and Dermatitis,* Williams & Wilkins, Baltimore, 1968.
5. **Kingsbury, J.,** *Poisonous Plants of the United States and Canada,* Prentice-Hall, Englewood Cliffs, N.J., 1964.
6. **Forsyth, A. A.,** *British Poisonous Plants,* Ministry of Agriculture, Fisheries, and Food, Her Majesty's Stationery Office, London, 1968, 74.
7. **Trease, G. E. and Evans, W. C.,** *Pharmacognosy,* 10th ed., Bailliere Tindall, London, 1972, 473.
8. **Ohigashi, H., Kawazu, K., Koshimizu, K., and Mitsui, T.,** A piscicidal constituent of *Sapium japonicum, Agric. Biol. Chem.,* 36, 2529, 1972.
9. **Castagnou, R., Baudriment, R., and Gauthier, J.,** *Compt. Rend.,* 260, 4109, 1965.
10. **Hartwell, J. L.,** Plants used against cancer. A survey, *Lloydia,* 32, 153, 1969.
11. **Watt, J. M. and Breyer-Brandwijk, M. G.,** *The Medicinal and Poisonous Plants of Southern and Eastern Africa,* 2nd ed., E. & S. Livingstone, Edinburgh, 1962, 395.
12. **Up Hof, Th. J. C.,** *Dictionary of Economic Plants,* H. R. Engelmann, London, 1959, 151.
13. **Seghal, L. and Paliwal, G. S.,** *Bot. J. Linn. Soc.,* 68, 173, 1974.
14. **Dymock, W., Warden, C. J. H., and Hooper, D.,** *Pharmacographica Indica,* Vol. 1 (iii), Hamdard National Foundation, Karachi, Pakistan, 1890, 247 (reprinted 1972).
15. **Ainslie, J. R.,** A List of Plants in Native Medicine in Nigeria, Inst. pap. no. 7, Oxford University, Oxford, England, 1937, 27.
16. *British Pharmacopoeia,* The Pharmaceutical Press, London, 1949.
17. **Doull, J., Klaassen, C. D., and Amdur, M. O., Eds.,** *The Basic Science of Poisons,* 2nd ed., Bailliere Tindall, London, 1980, 1.
18. **Frohne, D. and Pfänder,** *Poisonous Plants* (translated by N. G. Bisset), Wolfe Scientific, London, 1984, 1.
19. **Singh, B. and Rastogi, R. P.,** Cardenolides, glycosides and genins, *Phytochemistry,* 9, 315, 1970.
20. **Berenblum, I. and Shubik, P.,** The role of Croton oil applications, associated with a single painting of a carcinogen in tumour induction of the mouse skin, *Br. J. Cancer,* 1, 379, 1947.
21. **Hecker, E.,** Co-carcinogenic priniciples from the seed oil of *Croton tiglium* and from other Euphorbiaceae, *Cancer Res.,* 28, 2338, 1968.
22. **Blumberg, P. M.,** In vitro studies on the mode of action of the phorbol esters, potent tumour promoters. I and II., *CRC Crit. Rev. Toxicol.,* 8(2), 153, 1980; 8(3), 199, 1981.
23. **Dipaolo, J. A. and Casto, B. C.,** *In vitro* transformation of mammalian cells, sequential treatment with diverse agents, *Carcinogenesis,* Vol. 2, *Mechanisms of Tumour Promotion and Cocarcinogenesis,* Slaga, T. J., Sivak, A., and Boutwell, R. K., Eds., Raven Press, New York, 1978, 517.
24. **Bohrman, J. S.,** Identification and assessment of tumour-promoting and co-carcinogenic agents: state of the art *in vitro* methods, *CRC Crit. Rev. Toxicol.,* 11(2), 121, 1982.
25. **Mottram, J. C.,** A developing factor in experimental blastogenesis, *J. Pathol. Bacteriol.,* 56, 181, 1944.
26. **Weinstein, I. B., Wigler, M., and Pietropoole, C.,** The action of tumour promoting agents in cell culture, in *Origins of Human Cancer, Book B, Mechanisms of Carcinogenesis,* Heath, H. H., Watson, J. D., and Winston, J. F., Eds., Cold Spring Harbor Laboratory, Cold Spring Harbor, N.Y., 1977, 751.
27. **Likhacher, A. Y.,** Combined effects of the carcinogenic substances, *Vopr. Oncol.,* 14, 114, 1968.
28. **Kidd, J. G. and Rous, P.,** The carcinogenic effect of a papilloma virus on the tarred skin of rabbits, *J. Exp. Med.,* 68, 529, 1938.
29. **Perry Lily, M.,** *Medicinal Plants of East and South-East Asia,* MIT Press, Cambridge, Mass., 1980, 140.
30. **Dalziel, J. M.,** *The Useful Plants of West Tropical Africa,* Crown Agents, London, 1937, 141.
31. **Verdcourt, B. and Trump, E. C.,** *Common Poisonous Plants of East Africa,* Collins, London, 1959, 50.
32. **Tackholm, V.,** *Students Flora of Egypt,* 2nd ed., Cooperative Printing Co./Cairo University Press, Beirut, 1974, 314.
33. **Evans, F. J. and Schmidt, R. J.,** Plants and plant products that induce contact dermatitis, *Planta Med.,* 38, 289, 1980.
34. **Hecker, E. and Schmidt, R.,** Phorbol esters, the irritants and co-carcinogens of *Croton tiglium* L., *Fortschr. Chem. Organ. Naturst.,* 31, 377, 1974.
35. **Van Duuren, B. L.,** Tumor-promoting agents in two stage carcinogenesis, *Prog. Exp. Tumor Res.,* 11, 31, 1969.

36. **Slaga, T. J., Sivak, A., and Boutwell, R. K., Eds.,** *Carcinogenesis*, Vol. 2, *Mechanisms of Tumor-Promotion and Cocarcinogenesis*, Raven Press, New York, 1978, 1.
37. **Slaga, T. J., Ed.,** Mechanisms of tumor promotion series, *Tumor-Promotion and Skin Carcinogenesis*, CRC Press, Boca Raton, Fla., 1982, 1.
38. **Bohm, R., Flaschentrager, B., Lendle, L.,** Über die Wirksamkeit von Substanzen aus dem Kroton-O1, *Arch. Exp. Pathol. Pharmacol.*, 177, 212, 1935.
39. **Hoppe, W., Brandl, F., Strell, I., Röhrl, M., Gassmann, I., Hecker, E., Bartsch, H., Kreibich, G., Szczepanski, Ch. V.,** X-ray structure analysis of Neophorbol, *Angew. Chem. Int. Ed.*, 6, 809, 1967.
40. **Evans, F. J. and Soper, C. J.,** The tigliane, daphnane and Ingenane diterpenes, their chemistry, distribution and biological activities, *Lloydia*, 41, 193, 1978.
41. **Evans, F. J. and Taylor, S. E.,** Pro-inflammatory, tumour-promoting and anti-tumour diterpenes of the plant families Euphorbiaceae and Thymelaeaceae, *Fortschr. Chem. Organ. Naturst.*, 44, 1, 1983.
42. **Hecker, E., Bartsch, H., Bresch, H., Gschwendt, M., Härle, E., Kreibich, G., Kubinyi, H., Schairer, H. U., Szczepanski, Ch. V., and Theilmann, H. W.,** Structure and stereochemistry of the tetracyclic diterpene phorbol from *Croton tiglium*, *Tetrahedron Lett.*, 3165, 1967.
43. **Upadhyay, R. R. and Hecker, E.,** A new cryptic irritant and co-carcinogen from the seeds of *Croton sparciflorus*, *Phytochemistry*, 15, 1070, 1976.
44. **Ohigashi, H. and Mitsui, T.,** Studies on the biologically active substances of *Sapium japonicum*, *Bull. Inst. Chem. Res. Kyoto Univ.*, 50, 239, 1972.
45. **Fürstenberger, G. and Hecker, E.,** New highly irritant euphorbia factors from the latex of *Euphorbia tirucalli* L., *Experientia*, 33, 986, 1977.
46. **Evans, F. J.,** A new phorbol triester from the latices of *Euphorbia franckiana* and *E. coerulescens*, *Phytochemistry*, 16, 395, 1977.
47. **Taylor, S. E., Evans, F. J., Gafur, M. A., and Choudhury, A. K.,** Sapintoxin-D, a new phorbol ester from *Sapium indicum*, *J. Nat. Prod.*, 44, 729, 1981.
48. **Evans, F. J.,** The irritant toxins of the blue *Euphorbia*, *Toxicon*, 16, 51, 1978.
49. **Gschwendt, M. and Hecker, E.,** Tumour-promoting compounds from *Euphorbia triangularis*, mono and di-esters of 12-deoxyphorbol, *Tetrahedron Lett.*, 3509, 1969.
50. **Fürstenberger, G. and Hecker, E.,** The new 4-deoxyphorbol and its highly unsaturated irritant diesters, *Tetrahedron Lett.*, 925, 1977.
51. **Gschwendt, M. and Hecker, E.,** Tumorpromovierende diterpenfettsäureester aus *Euphorbia trangularis* und *E. cooperi*, *Fette, Seifen, Anstrichm.*, 73, 221, 1971.
52. **Okuda, T., Yoshida, T., Koike, S., and Toh, N.,** The toxic constituents of the fruits of *Aleurites fordii*, *Chem. Pharm. Bull.*, 22, 971, 1974.
53. **Adolf, W. and Hecker, E.,** Diterpenoid irritants and cocarcinogens in Euphorbiaceae and Thymelaeaceae. Structural relationships in view of their biogenesis, *Isr. J. Chem.*, 16, 75, 1977.
54. **Hecker, E.,** New phorbol esters and related cocarcinogens, *Proc. 10th Int. Cancer Congress*, Year Book Medical Publishers, Chicago, 1971, 213.
55. **Farnsworth, N. R.,** Biological and chemical screening of plants, *J. Pharm. Sci.*, 55, 225, 1966.
56. **Crombie, L., Knaen, G., Pattanden, G., and Whybrow, D.,** Total synthesis of the macrocyclic diterpene *(s)*-casbene, the putative biogenetic precursor of the lathyrane, tigliane, ingenane and related terpenoid structures, *J. Chem. Soc.*, 8, 1711, 1980.
57. **Stout, G. H., Balkenhol, W. G., Poling, M., and Hickernell, G. L.,** The isolation and structure of Daphnetoxin, the poisonous principle of *Daphne* species, *J. Am. Chem. Soc.*, 92, 1070, 1970.
58. **Sakata, K., Kawazu, K., Mitsui, T., and Masaki, N.,** The structure and stereochemistry of huratoxin, a piscicidal constituent of *Hura crepitans*, *Tetrahedron Lett.*, 1141, 1971.
59. **Ronlan, A. and Wickberg, B.,** The structure of mezerein, a major toxic principle of *Daphne mezereum* L., *Tetrahedron Lett.*, 4261, 1970.
60. **Silaga, T. J., Fischer, S. M., Weeks, C. E., and Klein-Szanto, A. J. P.,** Multistage chemical carcinogenesis in mouse skin. Biochemistry of normal and abnormal epidermal differentiation, Bernstein, I. A. and Seije, M., Eds., University of Tokyo Press, Tokyo, 1980, 193.
61. **Sakata, K., Kawazu, K., and Mitsui, T.,** Studies on the piscicidal constituent of *Hura crepitans*. II. Chemical structure of huratoxin, *Agric. Biol. Chem.*, 35, 2113, 1971.
62. **Nyborg, J. and LaCour, T.,** X-ray diffraction study of molecular structure and conformation of mezerein, *Nature (London)*, 257, 824, 1975.
63. **Kupchan, S. M., Shizuri, Y., Murae, T., Sweeny, J. G., Haynes, H. R., Shen, M. S., Barrick, J. C., Bryan, R. F., van der Helm, D., Wu, K. K.,** Gnidimacrin and gnidimacrin-20-palmitate, novel macrocyclic antileukemic diterpenoid esters from *Gnidia subcordata*, *J. Am. Chem. Soc.*, 98, 5719, 1976.
64. **Hergenhahn, M., Adolf, W., and Hecker, E.,** Resiniferatoxin and other novel polyfunctional diterpenes from *Euphorbia resinifera* and *E. unispina*, *Tetrahedron Lett.*, 1595, 1975.
65. **Kupchan, S. M. and Baxter, R. L.,** Mezerein, antileukemic principle isolated from *Daphne mezereum*, *Science*, 187, 652, 1974.

66. **Adolf, W., Opferkuch, H. J., and Hecker, E.,** Über die diterpenoiden Inhaltsstoffe des Samenöls von *Euphorbia lathyris* und ihre tumorpronovierende Wirkung, *Fette Seifen Anstrichm.*, 70, 850, 1968.
67. **Abo, K. A. and Evans, F. J.,** Ingenol esters from the pro-inflammatory fraction of *Euphorbia kamerunica*, *Phytochemistry*, 21, 725, 1981.
68. **Shoyab, M. and Todaro, G. J.,** Specific high affinity cell membrane receptors for biologically active phorbol and ingenol esters, *Nature (London)*, 288, 451, 1980.
69. **Upadhyay, R. R. and Hecker, E.,** Diterpene esters of the irritant and co-carcinogenic latex of *Euphorbia lactea*, *Phytochemistry* 14, 2514, 1975.
70. **Zechmeister, K., Brandl, F., Hoppe, W., Hecker, E., Opferkuch, H. J., and Adolf, W.,** Structure determination of the new tetracyclic diterpene ingenol-triacetate with triple product methods, *Tetrahedron Lett.*, 4075, 1970.
71. **Evans, F. J. and Kinghorn, A. D.,** A comparative phytochemical study of the diterpenes of some species of the genera *Euphorbia* and *Elaeophorbia* (Euphorbiaceae), *Bot. J. Linn. Soc.*, 74, 23, 1977.
72. **Sayed, M. D., Rizk, A. R., Hammouda, F. M., El-Missiry, M. M., Williamson, E. M., and Evans, F. J.,** Constituents of Egyptian Euphorbiaceae. IX. Irritant and cytotoxic ingenane esters from *Euphorbia paralias* L., *Experientia*, 36, 1206, 1980.
73. **Evans, F. J. and Kinghorn, A. D.,** A new ingenol type diterpene from the irritant fraction of *Euphorbia myrsinites* and *E. biglandulosa*, *Phytochemistry*, 13, 2324, 1974.
74. **Uemura, D., Ohwaki, H., Hirata, Y., Chen, Y. P., and Hsu, H. Y.,** Isolation and structures of 20-deoxyingenol new diterpene derivatives and ingenol derivative obtained from Kansui, *Tetrahedron Lett.*, 2527, 1974.
75. **Opferkuch, H. J. and Hecker, E.,** New diterpenoid irritants from *Euphorbia ingens*, *Tetrahedron Lett.*, 261, 1974.
76. **Uemura, D. and Hirata, Y.,** New diterpene 13-oxy-ingenol isolated from *Euphorbia kansui*, *Tetrahedron Lett.*, 2529, 1974.
77. **Ott, H. H. and Hecker, E.,** Highly irritant ingenane type diterpene esters from *Euphorbia cyparissias*, *Experientia*, 37, 88, 1981.
78. **Dublyanskaya, N. F.,** Über en biologischen effekt der die toxizität von *E. lathyris* bedingen den fraktionen, *Pharm. Pharmacol.*, 11, 50, 1937.
79. **Kupchan, S. M., Sigel, C. W., Matz, M. J., Renauld, J. A. S., Haltiwanger, R. C., and Bryan, R. F.,** Jatrophone, a novel macrocyclic diterpenoid tumour inhibitor from *Jatropha gossypiifolia*, *J. Amer. Chem. Soc.*, 92, 4476, 1970.
80. **Adolf, W., Hecker, E., Balmain, A., Lohmme, M. F., Nakatani, G., Ourisson, G., Ponsinet, R., Pryce, R. J., Santhanakrishnan, T. S., Matyukhina, L. G., Saltikova, I. A.,** Euphorbiasteroid (epoxylathrol) a new tricyclic diterpene from *Euphorbia lathyris*, *Tetrahedron Lett.*, 2241, 1970.
81. **Berenblum, I.,** The co-carcinogenic action of *Croton* resin, *Cancer Res.*, 1, 44, 1941.
82. **Mottram, J. C.,** A developing factor in blastogenesis, *J. Path. Bacteriol.*, 56, 181, 1944.
83. **Roe, F. J. C. and Peirce, W. E. H.,** Tumor promotion by *Euphorbia* latices, *Cancer Res.*, 21, 338, 1961.
84. **Evans, F. J. and Kinghorn, A. D.,** A screening method for co-carcinogens, *J. Pharm. Pharmacol.*, 25, 145P, 1973.
85. **Pax, F. and Hoffmann, H.,** Euphorbiaceae, In, *Die Natürlichen Pflanzenfamilien*, Vol. 19C, 2nd ed., Engelmann, Leipzig, 1931, 208.
86. **Boissier, E.,** Euphorbieae, in *Prodromus systematis naturalis regni vegetabilis*, 15, 3, 1862.
87. **White, A., Dyer, R. A., and Shoane, B. L.,** *The Succulent Euphorbieae*, Vols. 1 and 2, Abbey Gardens Press, Pasadena, Calif., 1941.
88. **Farnsworth, N. R., Blomster, R. N., Messmer, W. M., King, J. C., Persinos, G. J., and Wilkes, J. D.,** A phytochemical and biological review of the genus *Croton*, *Lloydia*, 32, 1, 1969.
89. **Marsh, N., Rothschild, M., and Evans, F. J.,** A new look at *Lepidoptra* toxins, in *The Biology of Butterflies*, Academic Press, London, 1984, 135.
90. **Evans, F. J. and Kinghorn, A. D.,** Ingenol from *Euphorbia desmondi*, *Phytochemistry*, 13, 1011, 1974.
91. **Evans, F. J. and Kinghorn, A. D.,** Thin-layer chromatographic behaviour of the acetates of some polyfunctional diterpene alcohols of toxicological interest, *J. Chromatogr.*, 87, 443, 1973.
92. **Abo, K. and Evans, F. J.,** The composition of a mixture of ingol-esters from *Euphorbia kamerunica*, *Planta Med.*, 43, 392, 1981.
93. **Evans, F. J., Schmidt, R. J., and Kinghorn, A. D.,** A microtechnique for the identification of diterpene ester inflammatory toxins, *Biomed. Mass Spectr.*, 2, 126, 1975.
94. **Schmidt, R. J. and Evans, F. J.,** The succulent *Euphorbias* of Nigeria. II. Aliphatic diterpene esters from *Euphorbia poissonii*, Pax. and *E. unispina*, N. E. Br., *Lloydia*, 40, 225, 1977.
95. **Evans, F. J. and Schmidt, R. J.,** The succulent *Euphorbias* of Nigeria. III. Structure and potency of the aromatic diterpenes of *Euphorbia poissonii* Pax., *Acta Pharmacol. Toxicol.*, 45, 181, 1979.
96. **Schmidt, R. J. and Evans, F. J.,** The structure and potency of the tinyatoxins, *J. Pharm. Pharmacol.*, 27, 50P, 1975.

97. **Evans, F. J. and Schmidt, R. J.,** Two new toxins from the latex of *Euphorbia poissonii, Phytochemistry,* 15, 33, 1976.
98. **Abo, K. and Evans, F. J.,** A triester of ingol from the latex of *Euphorbia kamerunica, J. Nat. Prod.,* 45, 365, 1981.
99. **Hecker, E., Bresch, H., Szczepanski, Ch. V.,** Co-carcinogen A, der erstereine hochaktive Winkstoff aus Crotonöls, *Angew. Chem.,* 76, 225, 1964.
100. **Kinghorn, A. D. and Evans, F. J.,** A quantitative GLC method for phorbol and related diterpenes as their acetates, *J. Pharm. Pharmacol.,* 26, 408, 1974.
101. **Rao, G. H. R., Jachimowicz, A. A., and White, J. G.,** Rapid separation of tumor-promoting agents, phorbol and phorbol myristate acetate by high pressure liquid chromatography, *J. Chromatogr.,* 96, 151, 1974.
102. **Berry, D. L.,** Separation of a series of phorbol-ester tumour-promoters by LC, *Chromatogr. Rev.,* 3(2), 5, 1977.
103. **Driedger, P. E. and Blumberg, P. M.,** Structure-activity relationships in chick embryo fibroblasts for phorbol-related diterpene esters showing anomalous activities *in vivo, Cancer Res.,* 40, 339, 1980.
104. **Rizk, A. M., Hammouda, F. M., Ismail, S. E., El-Missiry, M. M., and Evans, F. J.,** Irritant resiniferonol derivatives from Egyptian *Thymelea hirsuta* L., *Experientia,* 40, 808, 1984.
105. **Ocken, P. R.,** Dry column chromatographic isolation of fatty acid esters of phorbol from *Croton* oil, *J. Lipid Res.,* 10, 460, 1969.
106. **Rizk, A. M., Hammouda, F. M., El-Missiry, M. M., Radwan, H. M., and Evans, F. J.,** Macrocyclic diterpene esters from *Euphorbia royleana, Phytochemistry,* 23, 2377, 1984.
107. **Cairnes, D. A., Mirvish, S. S., Wallcave, L., Nagel, D. L., and Smith, J. W.,** A rapid method for isolating phorbol from *Croton* oil, *Cancer Lett.,* 14, 85, 1981.
108. **Marshall, G. T. and Kinghorn, A. D.,** Isolation of phorbol and 4α-phorbol from *Croton* oil by droplet counter-current chromatography, *J. Chromatogr.,* 206, 421, 1981.
109. **Tani, C., Nagakura, N., and Kuriyama, C.,** Studies on the alkaloids of Papaveraceous plants. XXXI. Separation by droplet counter-current chromatography, the alkaloids of *Corydalis ophiocarpa* Hook et Thoms., *Yakugaku Zasshi,* 98, 1243, 1978.
110. **Hostettmann, K., Hostettmann-Kaldes, M., and Nakanishi, K.,** Droplet counter current chromatography for the preparative isolation of various glycosides, *J. Chromatogr.,* 170, 355, 1979.
111. **Hostettmann, K., Hostettmann-Kaldes, M., and Sticher, O.,** Preparative scale separation of xanthones and iridoid glycosides by droplet counter current chromatography, *Helv. Chim. Acta,* 62, 2079, 1979.
112. **Kinghorn, A. D. and Evans, F. J.,** Skin irritants of *Euphorbia fortissima, J. Pharm. Pharmacol.,* 27, 329, 1975.
113. **Edwards, M. C., Taylor, S. E., Williamson, E. M., and Evans, F. J.,** New phorbol and deoxyphorbol esters: isolation and relative potencies in inducing platelet aggregation and erythema of skin, *Acta Pharmacol. Toxicol.,* 53, 177, 1983.
114. **Redmond, D. E.,** Tobacco and cancer: the first clinical report, 1761, *N. Engl. J. Med.,* 282, 18, 1970.
115. **Yamagiwa, K. and Ichikawa, K.,** Experimental study of the pathogenesis of carcinoma, *J. Cancer Res.,* 3, 1, 1918.
116. **Rous, P. and Kidd, J. G.,** Conditional neoplasms and subthreshold neoplastic states. A study of tar tumors in rabbits, *J. Exp. Med.,* 73, 365, 1941.
117. **Friedwald, W. F. and Rous, P.,** The initiating and promoting elements in tumor production, *J. Exp. Med.,* 80, 101, 1944.
118. **Berenblum, I.,** The co-carcinogenic action of *Croton* resin, *Cancer Res.,* 1, 44, 1941.
119. **Schramm, T. and Teichmann, B.,** Some problems involved in testing chemicals, including new drugs, for carcinogenicity, in *Evaluation of Embryotoxicity, Mutagenicity and Carcinogenicity Risks in New Drugs,* Beneova, O., Rychter, Z., and Helinek, R., Eds., University Karlova, Praha, Czechoslovakia, 1979, 293.
120. **Berwald, Y. and Sachs, T.,** *In vitro,* transformation of normal cells to tumour cells by carcinogenic hydrocarbons, *J. Natl. Cancer Inst.,* 35, 641, 1965.
121. **Hecker, E.,** Co-carcinogens and tumor promoters of the diterpene ester type as possible carcinogenic risk factors, *J. Cancer. Res. Clin. Oncol.,* 99, 103, 1981.
122. **Hecker, E.,** New toxic, irritant and co-carcinogenic diterpene esters from Euphorbiaceae and Thymelaeaceae, *Pure Appl. Chem.,* 49, 1423, 1977.
123. **Hecker, E.,** Structure activity relationships in diterpene esters irritant and co-carcinogenic to mouse skin, *Carcinogenesis,* Vol. 2, *Mechanisms of Tumor-Promotion and Co-Carcinogenesis,* Slaga, T. J., Sivak, A., and Boutwell, R. K., Eds., Raven Press, New York, 1978, 11.
124. **Boutwell, R. K. and Bosch, D. K.,** Tumor-promoting action of phenol and related compounds for mouse skin, *Cancer Res.,* 19, 413, 1959.
125. **Holsti, P.,** Tumor promoting effects of some long chain fatty acids in experimental skin carcinogenesis in the mouse, *Acta Pathol. Microbiol. Scand.,* 46, 51, 1959.

126. **Boutwell, R. K. and Bosch, D. K.,** Studies on the role of surface active agents in the formation of skin tumors in mice, *Am. Assoc. Cancer Res.*, 2, 190, 1957.
127. **Van Duuren, B. L. and Goldschmidt B. M.,** Structure activity relationships of tumor-promoters and co-carcinogens and interaction of phorbo myristate acetate and related esters with plasma membranes, in *Carcinogenesis*, Vol. 2, *Mechanisms of Tumor-Promotion and Co-Carcinogenesis*, Slaga, T. J., Sivak, A., and Boutwell, R. K., Eds., Raven Press, New York, 1978, 491.
128. **Gwynn, R. H. and Salaman, N. H.,** Studies on carcinogenesis. SH-reactors and other substances tested for co-carcinogenic action in mouse skin, *Br. J. Cancer*, 7, 482, 1953.
129. **Van Duuren, B. L., Sivak, A., Langseth, L., Goldschmidt, B. M., and Segal, A.,** Initiators and promoters in tobacco carcinogenesis, *Natl. Cancer Inst. Monogr.*, 28, 173, 1964.
130. **Yager, J. D. and Yager, R.,** Oral contraceptive steroids as promoters of hepatocarcinogenesis in female Sprague-Dawley rats, *Cancer Res.*, 40, 3680, 1980.
131. **Hecker, E.,** Co-carcinogens and co-carcinogenesis, in *Handbuch der Allgemeine Pathologie*, Vol. 6, Grundmann, E., Ed., Springer Verlag, Berlin, 1975, 651.
132. **Hecker, E. and Weber, J.,** Co-carcinogens from *Croton flavens* L. and the incidence of esophageal cancer in Curacao, in *Proc. 7th Int. Symp. Biol. Characterization of Human Tumors*, Davis, W., Ed., Excerpta Medica, Budapest, 1977, 52.
133. **Weber, J. and Hecker, E.,** Co-carcinogens of the diterpene ester type from *Croton flavens* and esophageal cancer in Curaçao, *Experientia*, 34, 1595, 1978.
134. *Federal Register*, 09, 33, 6937, May 1968.
135. **Ashby, J., Styles, J. A., and Boyland, E.,** Betal nuts, arecaidine and oral cancer, *Lancet*, 1, 112, 1979.
136. **Hecker, E.,** Isolation and characterisation of the co-carcinogenic principles of *Croton* oil, *Methods of Cancer Research*, Vol. 6, Busch, H., Ed., Academic Press, London, 1971.
137. **Ott, H. H. and Hecker, E.,** Are co-carcinogens of *Euphorbia* species present in their nectar collected by honey bees? Japanese-German workshop on chemical, viral and environmental carcinogenesis *Dtsch. Krebsforsch.*, May 1980.
138. **Upadhyay, R. R., Islampanah, S., and Davoodi, A.,** Presence of a tumor promoting factor in honey. 4th Asian Cancer Conf., Bombay, December 1979, *Gann*, 71 (Abstr. 15), 557, 1980.
139. **Kelly, W. R. and Bick, I. R. C.,** Some *in vivo* and *in vitro* properties of various fractions of *Pimelea trichostachya*, *Res. Vet. Sci.*, 20, 311, 1976.
140. **Kelly, W. R. and Seawright, A. A.,** *Pimelea* spp. poisoning of cattle, *Effects of Poisonous Plants on Livestock*, Academic Press, New York, 1978, 293.
141. **Roberts, H. B., McClure, T. H., Ritchie, E., Taylor, W. C., and Freeman, P. W.,** The isolation and structure of the toxin of *Pimelia simplex* responsible for St. George's disease of cattle, *Aust. Vet. J.*, 51, 325, 1975.
142. **Cashmore, A. R., Seelye, R. N., Cairn, B. F., Mack, H., Schmidt, R., and Hecker, E.,** The structure of prostratin, a toxic tetracyclic diterpene ester from *Pimelea prostrata*, *Tetrahedron Lett.*, 1737, 1976.
143. **McCormick, I. R. N., Nixon, P. E., and Waters, T. N.,** On the structure of prostratin, an X-ray study, *Tetrahedron Lett.*, 1735, 1976.
144. **Zayed, S., Hafez, A., Adolf, W., and Hecker, E.,** New tigliane and daphnane derivatives from *Pimelea prostrata* and *P. simplex*, *Experientia*, 33, 1554, 1977.
145. **Zayed, S., Adolf, W., Hafez, A., and Hecker, E.,** New highly irritant 1-alkyl daphnane derivatives from several species of thymelaeaceae, *Tetrahedron Lett.*, 3481, 1977.
146. **Tyler, M. I. and Howden, M. E. H.,** Piscicidal constituents of *Pimelea*, *Tetrahedron Lett.*, 689, 1981.
147. **Armuth, V., Berenblum, I., Adolf, W., Opferkuch, H. J., Schmidt, R., and Hecker, E.,** Systematic promoting action and leukemogenesis in SWR mice by phorbol and structurally related polyfunctional diterpenes, *J. Cancer Res. Clin. Oncol.*, 95, 19, 1979.
148. **Slaga, T. J., Fischer, S. M., Weeks, C. E., Nelson, K., Mamrack, M., and Klein-Szanto, A. J. P.,** Specificity and mechanisms of promoter inhibitors in multistage promotion, *Carcinogenesis*, Vol. 7, Hecker, E., Fusenig, N. E., Kunz, W., Marks, R., and Thielmann, H. W., Eds., Raven Press, New York, 1982, 19.
149. **Evans, F. J. and Schmidt, R. J.,** An assay procedure for the comparative irritancy testing of esters in the tigliane and daphnane series, *Inflammation* 3, 215, 1978.
150. **O'Brien, T. G., Simsiman, R. C., and Boutwell, R. K.,** Induction of the polyamine biosynthetic enzymes in mouse epidermis by tumor-promoting agents, *Cancer Res.*, 35, 1662, 1975.
151. **Herberman, E. and Callaham, M. F.,** Induction of terminal differentiation in human promyelocytic leukemia cells by tumor-promoting agents, *Proc. Natl. Acad. Sci. U.S.A.*, 76, 1293, 1979.
152. **Yamasaki, H., Finach, E., Nudel, U., Weinstein, I. B., Rifkind, R. A., and Marks, P. A.,** Tumor-promoters inhibit spontaneous and induced differentiation of murine erythroleukemia cells in culture, *Proc. Natl. Acad. Sci. U.S.A.*, 74, 3451, 1977.
153. **Sivak, A. and Van Duuren, B. L.,** A cell culture system for the assessment of tumor-promoting activity, *J. Natl. Cancer Inst.*, 44, 1091, 1970.

154. **Soper, C. J. and Evans, F. J.**, Investigations into the mode of action of the co-carcinogen, 12-*O*-tetradecanoyl phorbol-13-acetate using auxotrophic bacteria, *Cancer Res.*, 37, 2487, 1977.
155. **Horowitz, A. D., Greenebaum, E., and Weinstein, I. B.**, Identification of receptors for phorbol-ester tumor-promoters in intact mammalian cells and of an inhibitor of receptor binding in biologic fluids, *Proc. Natl. Acad. Sci. U.S.A.*, 78, 2315, 1981.
156. **Sando, J. J., Hilfiker, M. L., Salomon, D. S., and Farrar, J. J.**, Specific receptors for phorbol-esters in lymphoid cell populations, role in enhanced production of T-cell growth factor, *Proc. Natl. Acad. Sci. U.S.A.*, 78, 1189, 1981.
157. **Jaken, B., Tashjian, A. H., and Blumberg, P. M.**, Characterization of phorbol-ester receptors and their down-modulation in GH_4C_1 rat pituitary cells, *Cancer Res.*, 41, 2175, 1981.
158. **Driedger, P. E. and Blumberg, P. M.**, Specific binding of phorbol ester tumor promoters, *Proc. Natl. Acad. Sci. U.S.A.*, 77, 567, 1980.
159. **Kikkawa, U., Takai, Y., Tanaka, Y., Miyake, R., and Nishizuka, Y.**, Protein kinase C as a possible receptor protein of tumour-promoting phorbol-esters, *J. Biol. Chem.*, 258, 11442, 1983.
160. **Gilmore, T. and Martin, G. S.**, Phorbol-ester and diacylglycerol induce protein phosphorylation at tyrosine, *Nature (London)*, 306, 487, 1983.
161. **Naka, M., Nishikawa, M., Adelstein, R. S., and Hidaka, H.**, Phorbol ester-induced activation of human platelets is associated with protein kinase C phosphorylation of myosin light chains, *Nature (London)*, 306, 490, 1983.
162. **Ito, Y., Yanase, S., Fujita, J., Harayama, T., Takashima, M., and Imanaka, H.**, A short term *in vitro* assay for promoter substances using human lymphoblasoid cells latently infected with Epstein-Barr virus, *Cancer Lett.*, 13, 29, 1981.
163. **Ito, Y., Kawanishi, M., Harayama, T., and Takabayashi, S.**, Combined effect of the extracts from *Croton tiglium, Euphorbia lathyris*, or *E. tirucalli* and *n*-butyrate on Epstein-Barr virus expression in human lymphoblastoid P3HR-1 and RAJI cells, *Cancer Lett.*, 12, 175, 1981.
164. **Ito, Y., Ohigashi, H., Kashimizu, K., and Yi, Z.**, Epstein-Barr virus activating principle in the ether extracts of soils collected from under plants which contain active diterpene esters, *Cancer Lett.*, 19, 113, 1983.
165. **Hirayama, T. and Ito, Y.**, A new view of the etiology of nasopharyngeal carcinoma, *Prev. Med.*, 10, 614, 1981.
166. **Ito, Y., Kishishita, M., Morigaki, T., Yanase, S., and Hirayama, T.**, Induction and intervention of Epstein-Barr virus expression in human lymphoblastoid cell lines, a simulation model for study of cause and prevention of nasopharyngeal carcinoma and Burkitt's lymphoma, *Cancer Campaign*, 5, 255, 1981.
167. **Ito, Y., Yanase, S., Tokuda, H., Kishishita, M., Ohigashi, H., Hirota, M., and Koshimizu, K.**, Epstein-Barr virus activation by tung oil, extracts of *Aleurites fordii* and its diterpene-ester, 12-*O*-hexadecanoyl-16-hydroxyphorbol-13-acetate, *Cancer Lett.*, 18, 87, 1983.
168. **Ohigashi, H., Ohtsuka, T., Hirota, M., Koshimizu, K., Tokuda, H., and Ito, Y.**, Tigliane type diterpene-esters with Epstein-Barr virus inducing activity from *Sapium sebriferum, private communication*, 1983.
169. **Schmidt, R. J. and Evans, F. J.**, Investigations into the skin-irritant properties of resiniferonol orthoesters, *Inflammation*, 3, 273, 1979.
170. **Schmidt, R. J. and Evans, F. J.**, Skin irritant effects of esters of phorbol and related polyols, *Arch. Toxicol.*, 44, 279, 1980.
171. **Thielman, H. W. and Hecker, E.**, Beziehungen zwischen der Struktur von phorbolderivaten und ihren entzündlichen und tumorpromovierenden Eigenschaften, *Fortschritte der Krebsforschung*, Vol. 7, Schmidt, C. G. and Wetter, O., Eds., Schattaurer, Stuttgart, 1969, 171.
172. **Hecker, E.**, New phorbol esters and related cocarcinogens, Proc. 10th Int. Cancer Congr., Houston, Tex., *Oncology 1970*, Vol. 5, Year Book Medical Publishers, Chicago, 1971, 213.
173. **Kinghorn, A. D. and Evans, F. J.**, A biological screen of selected species of the genus *Euphorbia* for skin irritant effects, *Planta Med.*, 28, 325, 1975.
174. **Hecker, E., Immrich, H., Bresch, H., and Schairer, H. U.**, Über die Wirkstoffe des Crotonols. VI. Entzundungsteste am Mauseohr, *Z. Krebsforsch.*, 68, 366, 1966.
175. *British Pharmaceutical Codex*, Pharmaceutical Society of Great Britain, London, 1934, 429.
176. **Upadhyay, R. R., Bakhtavar, F., Mohseni, H., Satar, A. M., Saleh, N., Tafazuli, A., Dizaji, F. M., and Mohaddes, G.**, Screening of *Euphorbia* from Azarbaijan for skin irritant activity and for diterpenes, *Planta Med.*, 38, 151, 1980.
177. **Mufson, R. A., Fischer, S. M., Verma, A. K., Gleason, G. L., Slaga, T. J., and Boutwell, R. K.**, *Cancer Res.*, 39, 4791, 1979.
178. **Ohigashi, H., Hirota, M., Ohtsuka, T., Koshimizu, K., Fujiki, H., Suganuma, M., Yamaizumi, Z., and Sugimura, T.**, Resiniferonol-related diterpene esters from *Daphne odora* Thunb. and their ornithine decarboxylase inducing activity in mouse skin, *Agric. Biol. Chem.*, 46, 2605, 1982.

179. **Evans, F. J. and Kinghorn, A. D.,** The succulent euphorbias of Nigeria. I., *Lloydia,* 38, 363, 1975.
180. **Irvine, F.,** *Woody Plants of Ghana,* Oxford University Press, London, 1961, 1.
181. **Kerharo, J. and Adam, J. G.,** *La Pharmacopée Senegalaise Traditionel,* Vergot Freres, Paris, 1974, 402.
182. **Oliver, B.,** *Medicinal Plants of Nigeria,* Nigerian College of Arts, Science, and Technology, Ibadan, 1960, 26.
183. **White, A., Dyer, R. A., and Sloane, B. L.,** *The Succulent Euphorbiaceae,* Vols. 1 and 2, Abbey Garden Press, Pasadena, Calif., 1941, 1.
184. **Kokaro, J. O.,** *Medicinal Plants of E. Africa,* East Africain Literature Bureau, Kampala, Nairobi, Dar-es-Salam, 1976, 85.
185. **Uhe, G.,** Medicinal plants of Samoa, a preliminary survey of the plants for medicinal purposes in the Samoan Islands, *Econ. Bot.,* 28, 1, 1974.
186. **Gastaldo, P.,** Compendium of the official Italian flora. XII. *Fitoterapia,* 44, 61, 1973.
187. **Kinghorn, A. D.,** Co-carcinogenic irritant Euphorbiaceae, *Toxic Plants,* Kinghorn, A. D., Ed., Columbia University Press, New York, 1979, 137.
188. **Calvin, M.,** Hydrocarbons via photosynthesis, U.S. Energy Research and Development Administration, Contract No. W-7405-ENG-48, ACS Rubber Division of Marketing, San Francisco, Calif., October 1976.
189. **Nielson, P. E., Nishimura, H., Otvos, J. W., and Calvin, M.,** Plant crops as a source of fuel and hydrocarbon-like materials, *Science,* 198, 942, 1977.
190. **Sachs, R. M.,** private communication, 1978.
191. **Lawrence, G.,** *Taxonomy of Vascular Plants,* Macmillan, New York, 1951, 1.
192. **Hergenhahn, M., Kusumoto, S., and Hecker, E.,** Diterpene esters from Euphorbium and their irritant and co-carcinogenic activity, *Experientia,* 30, 1438, 1974.
193. **Evans, F. J.,** Euphorbia, in *The British Herbal Pharmacopoeia,* British Herbal Medical Association, London, 1976, 81.
194. **Evans, F. J.,** *Stillingia, The British Herbal Pharmacopoeia,* Part 2, British Herbal Medical Association, London, 1979, 201.
195. **Adolf, W. and Hecker, E.,** New irritant diterpene esters from the roots of *Stillingia sylvatica* L. (Euphorbiaceae), *Tetrahedron Lett.,* 2887, 1980.
196. **Sanai, R., Rastogi, R. P., Jakupovic, J., and Bohlmann, F.,** A diterpene from *Euphorbia maddeni, Phytochemistry,* 20, 1665, 1981.
197. **Torrance, S. J., Wiedhope, R. M., Cole, J. R., Arora, S. K., Bates, R. B., Beaver, W. A., and Cutler, R. S.,** Anti-tumor agents from *Jatropha macrorhiza* (Euphorbiaceae). II. Isolation and characterisation of jatrophatrione, *J. Org. Chem.,* 41, 1855, 1976.
198. **Abo, K. and Evans, F. J.,** Macrocyclic diterpene esters of the cytotoxic fraction from *Euphorbia kamerunica, Phytochemistry,* 20, 2535, 1981.
199. **Kupchan, S. M., Sweeny, J. G., Baxter, R. L., Murae, T., Zimmerly, V. A., and Sickles, B. R.,** Gnididin, Gniditrin, and Gnidicin. Novel potent antileukemic diterpenoid esters from *Gnidia lamprantha, J. Am. Chem. Soc.,* 97, 672, 1975.
200. **Kupchan, S. M., Shizuri, Y., Sumner, W. C., Haynes, H. R., Leighton, A. P., and Sickles, B. R.,** Isolation and structural elucidation of a new potent anti-leukemic diterpenoid ester from *Gnidia* species, *J. Org. Chem.,* 41, 3850, 1976.
201. *Jatropha, Kew Bulletin,* Natl. Botanic Gdns, Lucknow, India, 1959.
202. **Evans, F. J. and Kinghorn, A. D.,** Letters to the editor, *Sunday Telegraph,* London, March 11, 1973.
203. **Saffiotti, U.,** Are some houseplants and garden plants carcinogenic, *JAMA,* 235, 2344, 1976.
204. **Schmidt, R. J. and Evans, F. J.,** Skin irritants of the sun spurge *(Euphorbia helioscopia* L.), *Contact Dermatitis,* 6, 204, 1980.
205. **Falsone, G. and Crea, A. E. G.,** Drei neue 4-deoxyphorbol-triesters aus *Euphorbia biglandulosa, Liebigs Ann. Chem.,* 1116, 1979.
206. **Gschwendt, M. and Hecker, E.,** Tumour-promoting compounds from *Euphorbia cooperi.* Di- and triesters of 16-hydroxy-12-deoxy-phorbol, *Tetrahedron Lett.,* 567, 1970.
207. **Upadhyay, R. R., Bakhtavar, F., Ghaisarzedeh, M., and Tilabi, J.,** Co-carcinogenic and irritant factors of *Euphorbia esula* L. latex, *Tumori,* 64, 99, 1978.
208. **Upadhyay, R. R., Ansarin, M., Zarintan, and Shakui, P.,** Tumor-promoting constituents of *Euphorbia serrata, Experientia,* 32, 1196, 1976.
209. **Upadhyay, R. R., Samiyeh, R., and Tafazuli, A.,** Tumor-promoting and skin irritant diterpene esters of *Euphorbia virgata* latex, *Neoplasma,* 28, 555, 1981.
210. **Adolf, W. and Hecker, E.,** On the irritant and co-carcinogenic principles of *Hippomane mancinella, Tetrahedron Lett.,* 1587, 1975.

Chapter 2

MULTISTAGE CARCINOGENESIS AND THE BIOLOGICAL EFFECTS OF TUMOR PROMOTERS

Anne R. Kinsella

TABLE OF CONTENTS

I. A BRIEF HISTORY OF CHEMICAL CARCINOGENESIS, COCARCINOGENESIS, AND TUMOR PROMOTION

Fossil records provide clear evidence of osteomas and hemangiomas in the dinosaur population of 80 million years ago.[1] Ramazzini[2] however, was probably the first (1700) to recognize cancer of the breast in nuns as being a tumor related to lifestyle and in 1761, Hill[3] observed the frequent occurrence of cancer of the nasal passages in men who took tobacco snuff. It was, however, Pott[4] (1775) who was the first to recognize an example of chemical carcinogenesis, when soot was held to be responsible for the higher incidence of epithelioma of the scrotum in chimney sweeps. The first preventive measure against chemically induced cancer was in 1778, when the Danish Chimney Sweep's Guild urged its members to bathe daily.[5] In the 1870s occupational skin cancer among oil and tar workers was reported by Volkmann (1875) in Germany and by Bell (1876) in Scotland.[6] Later (1887), the first incidence of skin cancer among operatives in the Lancashire cotton spinning industry, ''mule spinners cancer'', was reported, which was attributed to mineral oil used in lubrication,[6,7] and in 1895, Rehn[8] reported that exposure to aromatic amines caused bladder cancer.

Albeit that many forms of cancer exist, it is skin cancer that has received the most attention from the experimentalists since it was first reported in 1775.[4] In 1915, after attempts by a number of workers, Yamagiwa and Ichikawa[9] of Japan succeeded in producing skin neoplasia experimentally by making repeated topical applications of coal tar to the ears of rabbits. In 1918, Tsutui,[10] a pupil of Yamagiwa, reported the advantages of mouse skin painting as a method of testing carcinogenic tars, and in 1922, Passey[11] confirmed the carcinogenicity of soot by producing mouse skin cancer with ether extracts of tar. Bloch and Dreifuss[12] (1921), investigating the active component of coal tar, showed that it was a neutral compound capable of forming a stable complex with picric acid and probably belonged to the class of cyclic hydrocarbons. Kennaway[13,14] (1924, 1925) produced carcinogenic tars by the pyrolysis, in a hydrogen atmosphere, of several organic materials including acetylene, and again the conclusion was that the carcinogenic agent was an aromatic hydrocarbon. Heiger[15] (1930), however, revealed the similarities in the fluorescent spectra between the products of these carcinogenic tars and synthetic benz(*a*)anthracene derivatives. This, in turn, led to the demonstration by Kennaway and Heiger[16] (1930) of 1,2,5,6-dibenzanthracene as the first synthetic carcinogen, and soon after this, a carcinogenic hydrocarbon isolated from coal tar was identified as 3,4-benzpyrene.[17,18] The evolution of the chemistry of these and related carcinogenic compounds is very adequately reviewed by Haddow and Kon (1947).[6]

Parallel to the studies to elucidate the chemical structures of these compounds, the advent of pure chemicals capable of inducing skin tumors led to studies on the role of stimulation of cell division in the process of tumor formation. Deelman[19,20] (1924, 1927), in experiments involving the treatment of mouse skin with coal tars, showed wounds made in the prepared skin to induce tumors. Thus, the first evidence of the use of a two-stage technique in tumor formation is essentially 60 years old. Twort and Twort[21] demonstrated in 1930 that oleic acid administered 5 times a week for 30 weeks induced tumors in mouse skin previously exposed to benzpyrene 5 times a week for 6 weeks. The choice of oleic acid was based on a previous observation by Twort and Ing[22] that oleic acid induced epidermal hyperplasia.

Further refinements in technique were developed during the early 1940s. Rous and Kidd[23] (1941) and Friedewald and Rous[24] (1944), applying the two-stage technique to the formation of skin tumors in rabbits, showed again that agents and techniques which induced hyperplasia (e.g., turpentine, chloroform, and wounding) could cause tumors to appear. The assumption was that latent tumor cells were present in the epidermis as a result of the previous treatment with methylcholanthrene or tar and that the induction process and subsequent growth stimulatory process (termed, respectively, initiation and promotion) depended on two different classes of stimuli.[24] Much impetus to this analysis of the tumor-formation process was provided by Berenblum in 1941,[25,26] who over a number of the preceding years had studied the role of irritation in carcinogenesis.[27,28] Berenblum[25,26] found that croton oil, the seed oil of the euphorbiaceous *Croton tiglium* discovered by Flaschenträger[29,30] around 1930, was capable of enhancing tumor formation when applied to mouse skin, either together with, or separately from, a subeffective dose of the carcinogenic hydrocarbon 3,4-benzpyrene. Again, this was taken as evidence that the carcinogenic process comprised two separate mechanisms referred to at the time as ''precarcinogenic'' and ''epicarcinogenic'', with a third stage, ''metacarcinogenic'', being responsible for the transformation of benign papillomas into malignant tumors. An important step forward was made in 1944 by Mottram,[31] who demonstrated that a single application of 3,4-benzpyrene, followed by repeated applications of croton oil, was sufficient to lead to tumor formation. Initiation must, therefore, be accomplished almost instantaneously. This observation was confirmed in 1947 by Berenblum and Shubik,[32] who suggested, in addition, that increased tumor induction was only observed when croton oil treatment followed, not preceded, a single application of carcinogen. Again, it was Berenblum and Shubik[33] in 1949 who demonstrated that the administration of croton oil could be delayed for up to 43 weeks after treatment with the carcinogen and still be effective. This provided evidence that the initiation phase was irreversible.

By this time, the terms ''initiation'' and ''promotion'' proposed by Friedewald and Rous[24] were generally accepted as terms to describe the carcinogenic process, the alternatives proposed by Berenblum and Shubik[32] and Mottram[31] having been dispensed with. At this point, it is probably also important to emphasize that the processes of cocarcinogenesis and promotion are not synonymous. Cocarcinogenesis describes a situation in which the response to a carcinogen is increased by a second factor introduced concurrently with the carcinogen, as reported by Berenblum[25,26] in his early studies on mouse skin neoplasia. Promotion, on the other hand, denotes a specific step in the sequence of events leading to skin tumor formation as reported by Mottram[31] and Berenblum and Shubik.[32,33] Promoting agents obviously can be cocarcinogens and vice versa, but the distinction between the two processes must be emphasized and understood in terms of the overall process of carcinogenesis. Also, acceptance of the two-stage theory of carcinogenesis meant that a complete carcinogen had to have both initiating and promoting properties.

To bring about only the first neoplastic change in the skin necessitated the application of a small single dose of carcinogen, and until 1953, all initiating agents were considered to be carcinogenic in their own right. In 1953, Salaman and Roe[34] showed that urethane, which did not produce tumors when applied either as a single high dose or repeatedly, was a powerful initiator when a single treatment was followed by repeated applications of croton oil. Alternate applications of urethane and croton oil had a similar effect. As a consequence of this, urethane was classed as an ''incomplete carcinogen''. In the sequel to these experiments, 29 substances, most of them related either pharmacologically or chemically to urethane, were screened for similar activity.[35] Of these only five, the antileukemic agent triethylene melamine (TEM), urethane derivatives ethyl *N*-methylcarbamate and ethyl *N*-phenylcarbamate, 1,2-benzanthracene, and β-propriolactone, were found to be initiators, but were not carcinogenic in their own right. In turn, only two of these agents, 1,2-benzanthracene and β-propriolactone, induced epidermal hyperplasia. In 1958, Van Esch et al.[36] reported that isopropyl *N*-phenylcarbamate and isopropyl *N*-(3-chlorophenyl) carbamate were also so-called pure initiators, which was not surprising, bearing in mind their relationship to urethane. Meanwhile, new promoters for mouse skin were also being discovered in the form of citrus oils,[37,38] and latices from members of the Euphorbiaceae, the family to which *Croton tiglium* (the source of croton oil) belongs,[39,40] as well as anthralin,[41] used for many years as a treatment for psoriasis. Della Porta et al.[42] (1960) also found that treatment with Tween® 60 alone could induce tumors, based on earlier reports[43] that surface-active agents were tumor promoting. In addition, the antibiotic griseofulvin was shown to promote DMBA (9,10-dimethyl-1,2-benzanthracene)-initiated mouse skin tumors.[44]

Although two-stage carcinogenesis had by now been demonstrated in systems other than mouse skin (such as subcutaneous tissue, muscle,[45-47] liver,[48-50] bladder,[51-52] lung,[53-57] and forestomach[58-59]) it was during the 1960s that two major advances were made which formed the anchor for all subsequent studies. First of all, Boutwell,[60] in the early 1960s, was responsible for breeding strains of mice which were sensitive and resistant to DMBA-croton oil-induced tumor formation by the two stage process, and second, Hecker[61] (Germany) and Van Duuren et al.[62,63] (U.S.) independently isolated and identified the active principles of croton oil.

The need for the development of a more uniform test animal was becoming increasingly obvious. The amount of carcinogen and promoter required for tumor development varied widely from mouse strain to mouse strain and the inherent resistance of many of the mouse strains meant that large doses of initiator and long induction periods were quite standard. The program of selective breeding based on tumor response involved the treatment of Rockland-derived mice with a single application of 75 μg DMBA in 25 μℓ benzene to the back of each mouse followed 1 week later by twice weekly applications of 125 μg of croton oil in 25 μℓ benzene for 12 or 18 weeks. The mice with the earliest and most numerous tumors were bred to obtain the next generation of susceptible mice and the rest discarded. After two generations the mice with no tumors were also bred as a selection for resistance. This process of challenge and selection was continued until a total of 12 selections had been made for susceptibility to tumor formation and 9 selections had been made for resistance. These mice were therefore selected for their sensitivity to both stimuli and were classed as skin tumor-susceptible mice. Eventually, to provide hybrid vigor the susceptible line of mice was crossed with Charles River CD-1 mice, and in later years became referred to as SENCAR (sensitive to carcinogenesis).[64] Such susceptible mice immediately permitted the further delineation of the component steps in tumor formation as demonstrated by Boutwell[60] in experiments to illustrate the qualitative differences between the initiation and promotion processes. Period 1 in the procedure involved a single application of DMBA (50 μg), period 2, repeated applications of croton oil, and period 3 repeated applications of an irritant such as turpentine or acridine, or the process of wound healing. Such experiments indicated the involvement of two components in promotion. It was known that continuous exposure of properly prepared skin to a promoting agent elicited growing papillomas and in sufficiently sensitive mice, carcinomas. However, Boutwell[60] demonstrated that a limited period of promotion would only produce a few tumors, but that the progression toward a high tumor yield could be accompanied by either wounding or repeated applications of a chemical irritant such as turpentine; i.e., "dormant tumor cells" existed that required cellular proliferation for their manifestation. Promotion itself, therefore, came to be thought of as a two-step process consisting of "conversion" and "propagation" events.[60]

It was about this time that Hecker[61] and Van Durren et al.[62,63] isolated and identified the cocarcinogenic principles of croton oil, the promoting agent so widely used in the mouse-skin studies of carcinogenesis. It comprised two factors, both 12,13-diesters of the polyfunctional tetracyclic diterpene alcohol phorbol.[65] Croton oil factor A1, 12-*O*-tetradecanoylphorbol-13-acetate or TPA (Figure 1), was found to be the most abundant and most active of these. The rest of the phorbol-12,13-diesters from croton oil differed from TPA in their carboxylic acid moieties and in the relative positions of their 12,13-diester functions. A high degree of structural specificity was required for the activity of the esters as irritants and promoters[66] and removal of the long chain moiety (decanoate or myristate) from the phorbol ring resulted in the complete loss of biological activity.[67] However, the irritants and promoters of factor groups A and B represented only 50% of the total diterpene esters contained in croton oil. Another 40% were accounted for by the phorbol-12,13,20-triesters.[68]

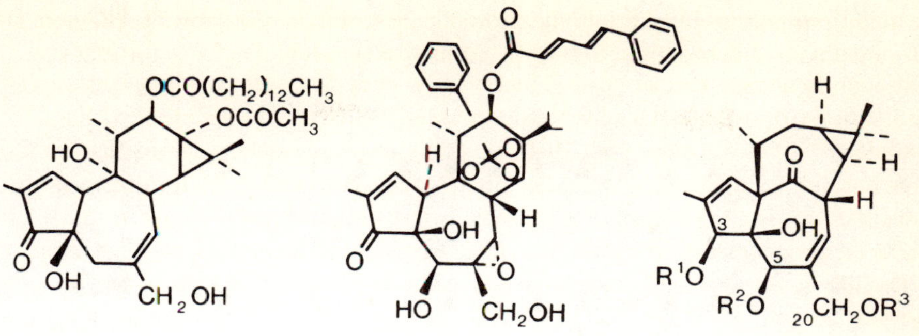

FIGURE 1. A comparison of the structures of TPA, mezerein and ingenol.

Phorbol, the parent alcohol of TPA,[65] is a polyfunctional derivative of the tetracyclic hydrocarbon tigliane and in certain species of both the Euphorbiaceae and Thymelaeaceae a further class of irritant diterpene esters, the tricyclic daphnane derivatives, has been identified, of which the parent alcohol is resiniferonol[66,69,70] (see also mezerein, Figure 1). A third class of irritant diterpene esters of tetracyclic ingenane type was also discovered.[66] Identification of the active principles of croton oil and the related irritant diterpene esters, with varying degrees of promoting activity[68] together with identification of the most appropriate negative control,[66] again paved the way for the detailed delineation of multistage carcinogenesis in vivo in mouse skin, in vitro in a variety of cell culture systems, and in in vivo/in vitro experiments which will be outlined in the next section.

Thus, by the beginning of the 1970s, the two-stage system of carcinogenesis could be summarized as follows: (1) a single appropriately small dose of carcinogen applied to mouse skin causes no tumors (initiating dose), (2) repeated applications of tumor promoter alone cause no tumors, (3) repeated applications of tumor promoter to initiated skin results in the production of many tumors, (4) no tumors arise if the order of the initiating and promoting treatments are reversed,[37] (5) the effect of the initiator is irreversible and promoter treatment can be delayed for almost 1 year after initiation and still be effective,[32,71] (6) interruption of the promoter treatment leads to either the absence of or to less effective induction of tumors (see also Figure 2).

The qualitatively different natures of the processes of initiation and promotion are emphasized by comparison of their biological properties (Table 1), but it is also important to point out the small quantity of chemicals required to produce a high tumor yield when the two processes are combined in an initiation-promotion protocol: at least tenfold less than by repetitive application of DMBA alone.[72] A complete carcinogen, of course, is assumed to have both initiating and promoting activity, and the efficiency of the initiation-promotion protocol in this instance is probably attributable to the fact that DMBA loses its promoting activity at the low levels where it still retains its initiating potential, and that croton oil or TPA is a much more effective promoter than an equivalent amount of DMBA. Therefore, application of the initiation-promotion model to carcinogenesis should relate the biochemical effects of initiators and promoters to a biological end point.

To this end, Colburn and Boutwell[73] (1968) found that the extent to which 7-(2-carboxyethyl)guanine was formed in the DNA of mouse skin correlated well with the degree of initiation. Bowden and Boutwell[74] found that alkylation of DNA correlated well with initiation by *N*-methyl-*N'*-nitro-*N*-nitrosoguanidine (MNNG), which was consistent with the known properties of alkylating agents.[75] Slaga et al.[76,77] found that completely carcinogenic

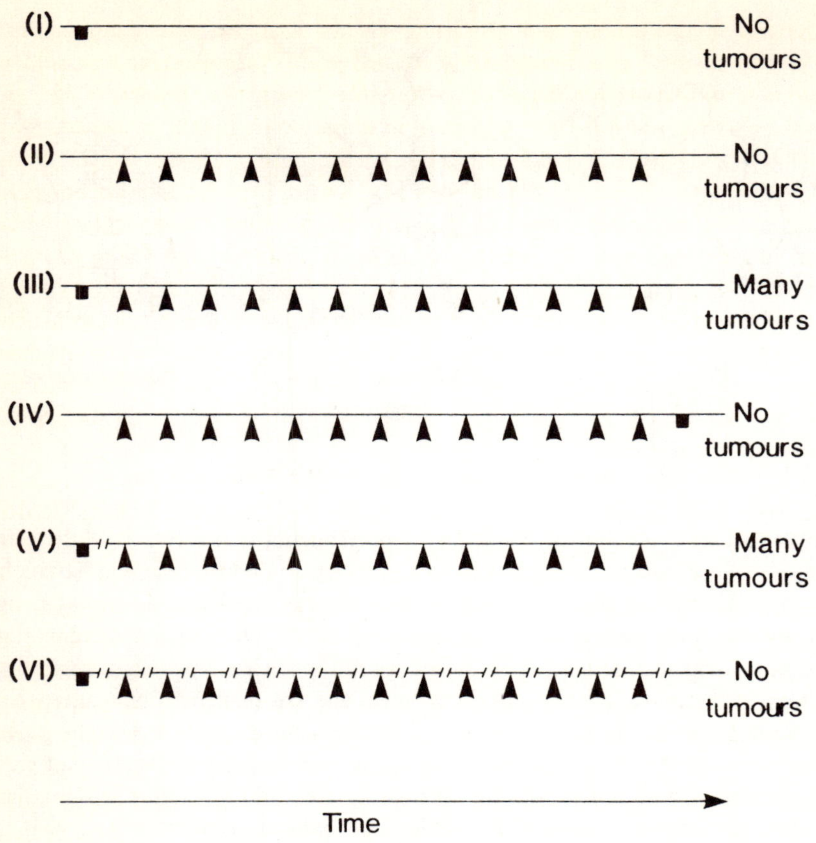

FIGURE 2. Summary of treatment protocols for two-stage carcinogenesis.[60]

Table 1
COMPARISON OF THE BIOLOGICAL PROPERTIES OF INITIATING AND PROMOTING AGENTS

Initiating agents	Promoting agents
Carcinogenic by themselves	Not carcinogenic alone
Must be given before a promoting agent	Must be given after an initiating agent
A single exposure is sufficient	Requires repeated and prolonged exposure
Action is irreversible	Action is reversible up to a certain stage
No apparent threshold	Probable threshold
Yield electrophiles that bind covalently to cellular macromolecules	No evidence of covalent binding
Do not alter gene expression at low initiating doses; no altered morphology detectable	Alter gene expression; induce hyperplasia, enzyme induction, induction of viral genes, gene amplification
Mutagenic in own right	Not mutagenic
DNA binding and resultant repair are detectable responses	Plethora of biological and biochemical responses

doses of several hydrocarbons and alkylating agents acted as gene derepressors, while Hennings and Boutwell,[78] closely followed by Baird et al.,[79] provided the first evidence that a tumor promoter possesses properties of a reversible derepressor. Raineri et al.[80] provided evidence of a possible role for gene activation in tumor promotion as evidenced by histone phosphorylation, and Kreibech et al.[81] and Rohrschneider et al.[82] reported that TPA stimulated phospholipid synthesis. However, the unifying concept that explained the chemical diversity of initiating agents (normally complete carcinogens) was that most yield, generally via metabolism, highly reactive electrophiles that bind covalently to cellular macromolecules. (This property also explained the mutagenic and generally irreversible nature of the action of initiating carcinogens.) No such unifying concept for the biochemical mechanism(s) of promoting agents exists to this day. However, a better understanding of the mechanism of action of tumor-promoting agents and other cofactors in carcinogenesis is of major importance since there is reason to believe that these agents play a decisive role in the causation of certain types of human cancer.

II. A ROLE FOR TUMOR PROMOTION AND COCARCINOGENESIS IN CANCER DEVELOPMENT

To recap slightly, promotion of skin tumors in mice can be induced by treatment with croton oil,[32,60] certain phorbol esters found in croton oil,[61-63,69] some synthetic phorbol esters,[83] a number of phenolic compounds,[84] anthralin,[41] sodium lauryl sulfate,[85] Tween® 60,[42,86] iodoacetic acid,[87] certain *Euphorbia* latices,[39,40] citrus oils,[37,38] the antibiotic griseofulvin,[44] extracts of unburned tobacco,[88] tobacco smoke condensate,[89,90] and fatty acid methyl esters.[91]

The role of inflammation and cellular proliferation in both carcinogenesis (i.e., following treatment with a complete carcinogen) and tumor promotion[60,78,86,92-96] has been studied extensively. Most promoters are irritants that induce hyperplasia,[92,97-100] but not all irritants or inflammatory hyperplasiogenic agents are promoters. However, the isolation of the active components of croton oil and their chemical characterization opened new perspectives. The tumor-promoting abilities of a series of phorbol esters could be correlated with their ability to induce a sustained stimulation of RNA, protein, and DNA synthesis in mouse epidermis,[79,100] and in addition to these changes associated with hyperplasia, they were seen to induce protease activity in skin,[92] increase phospholipid synthesis,[82,101] alter cell morphology,[102] induce elevated levels of ornithine decarboxylase (ODC) activity,[103,104] enhance cellular transformation in in vitro cell culture systems,[105,106] induce plasminogen activator (PA) activity,[107] increase histone synthesis and phosphorylation,[80,108] and increase prostaglandin synthesis.[109] They were also seen to depress isoproterenol-stimulated[110,111] cyclic AMP (cAMP) levels and histidase activity.[112] However, it was recognized by the late 1970s that induction of epidermal cell proliferation by phorbol esters (other agents that induced epidermal proliferation did not promote carcinogenesis), ornithine decarboxylase activity, and dark basal keratinocytes had the best correlation with promoting activity.[100,102,103,113,114]

A. Tumor Promotion and Altered Differentiation

Raick[100] (1973) showed a single application of TPA to mouse skin to induce an ordered sequence of ultrastructural changes in the cells of all layers of the epidermis, which correlated well with the changes in rate of precursor incorporation into protein, RNA, and DNA in the whole skin, and the changes in thickness, number of nucleated cell layers, and mitotic index of the interfollicular epidermis. To emphasize the detail with which this study was done, the precise sequence of events is presented. The incorporation of orotic acid-5-^{3}H into RNA was increased within 1 hr and reached a peak by 3 hr. At 2 hr, an increase in ^{14}C-leucine incorporation into protein was observed, which reached a high plateau by 12 hr.

The incorporation of ^{3}H-thymidine into DNA was markedly depressed for up to 9 hr, rising sharply to reach a first peak by 18 hr, and the mitotic index was depressed at 8 hr, reverting to control levels at 12 to 15 hr. By 16 hr there was a marked increase in the thickness of the interfollicular epidermis and in the number of its nucleated cell layers, and the parallel changes in mitotic index and DNA synthesis suggested that this was due to hypertrophy, not proliferation. Also in the same study,[100] 2 hr after administration of the TPA, the nucleoli of the basal layer were markedly enlarged, with an increase in the fibrillar and granular components. By 4 hr the granular component was further increased, polysomes had appeared, and the intercellular spaces between the basal cells were dilated. By 6 hr the cytoplasm of the basal cells was enlarged and filled with polysomes. By 12 hr the number of polysomes was further increased and the profiles of the granular reticulum and golgi complexes were more numerous so that the stimulated basal cells resembled those of embryonic epidermis. By 24 hr markedly hypertrophic cells and granular mitochondrial inclusions were frequently seen. By 48 hr, flattening, involution of the organelles, and accumulation of numerous keratohyalin granules occurred in the cells of the upper layers. In parallel, granular electron-dense material appeared in the dilated intercellular spaces at 6 hr and increased until 36 hr. The superficial layers, normally flattened, were swollen by 2 hr and by 12 hr resembled the stimulated basal cells. Most important, however, with regard to experiments done later by other workers,[115] dark cells with ultrastructural features distinct from other epidermal cells were observed after administration of TPA.

The ordered sequences of these morphological and metabolic changes were thought by Raick[100] to resemble events seen in other models when a shift from resting phase to active growth occurred. However, it was not taken for granted that all the changes induced in the epidermis in response to TPA were related to cell proliferation. Epidermal cells are, after all, normally renewed rapidly and so one would predict that these changes were only an accentuation of the features normally present in the basal cells. However, in these experiments TPA was seen to induce several kinds of changes that were not evident in normal epidermis, namely, the enlargement and reactivation of the superficial cells and the induction of dark cells. These observations showed that TPA was able to modify differentiation of the epidermal cells, returning them to a less differentiated state, the basal cells resembling those of embryonic skin. Support for this notion came from Balmain,[116] who reported the synthesis of differentiation-associated peptides in response to topical applications of TPA in mouse skin. One of these was characteristic of newborn mouse skin. Raick and Burdzy,[117] in further ultrastructural studies on the epidermis, found that ethyl phenylpropriolate (EPP), an essentially nonpromoting epidermal hyperplasiogenic agent, did not induce the appearance of dark cells, whereas wounding did induce a few dark cells that could be correlated with its weak promoting ability. In addition, a large number of dark cells are to be found in papillomas and carcinomas[100,113] and Klein-Szanto and co-workers[118] reported that TPA induced three to five times the number of dark cells induced by the daphnane derivative mezerein, illustrating the first major difference between these two compounds. However, it was because mezerein was on the whole capable of inducing most of the morphological and biochemical[119] changes in skin and in cell culture induced by TPA, that it was chosen by Slaga and co-workers[115] as a good candidate for studies on the second stage of two-stage promotion.

The various inhibitors of promotion (antiproteases,[92,120] steroidal anti-inflammatory agents,[94,95,121] and vitamin A analogues, retinoids[122,123]) have also provided powerful tools for understanding the mechanism(s) of tumor promotion. Slaga and co-workers[115] showed TPA to be about 50 times more effective as a promoter than mezerein. When 2 μg TPA was given twice weekly for only 2 weeks after DMBA initiation, no tumors were induced, compared with twice weekly treatments for 18 weeks. However, when mezerein was given at a dose of either 1, 2, or 4 μg twice weekly for 18 weeks after the limited TPA treatment, a significant, dose-dependent tumor response was observed. The use of inhibitors of pro-

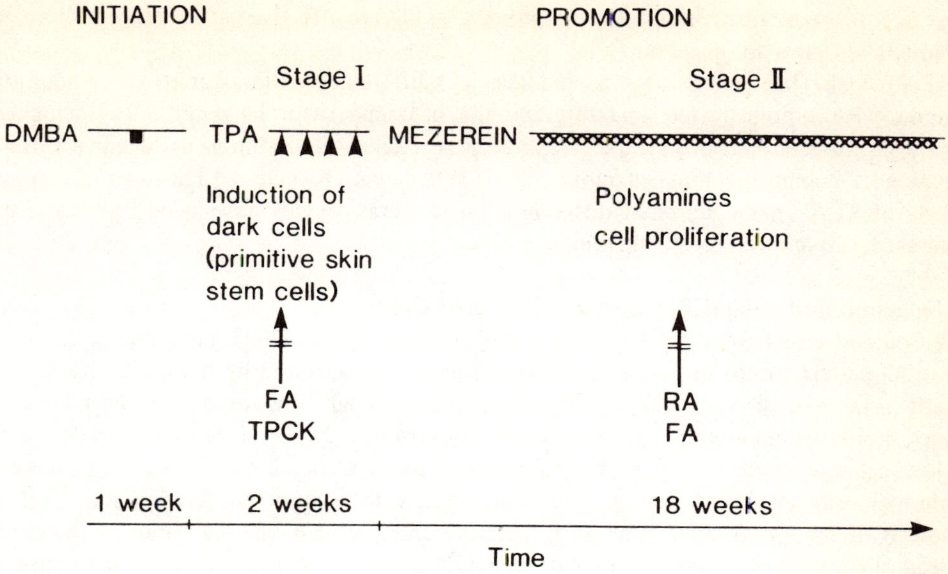

FIGURE 3. A diagram of the various stages of DMBA-initiated skin carcinogenesis using specific inhibitors to define stages I and II of promotion.[115]

motion (see above) showed the synthetic steroidal antiflammatory agent fluocinolone acetonide (FA), probably the most potent inhibitor of mouse skin tumor promotion,[123] to be a potent inhibitor of both stage I and II promotion (Figure 3). The vitamin A analogue retinoic acid (RA) was a potent inhibitor of stage II promotion, but was ineffective as a stage I inhibitor. The synthetic protease inhibitor tosyl phenylalanine chloromethyl ketone (TPCK),[92] on the other hand, specifically inhibited stage I, but not stage II tumor promotion. As one would have predicted, FA and TPCK were both found to inhibit specifically the appearance of dark cells, whereas RA, the potent inhibitor of stage II promotion, did not. TPCK had very little effect on TPA- and mezerein-induced ODC activity. FA also had very little effect on TPA-induced ODC activity; in addition to counteracting the appearance of TPA-induced dark cells, it also suppressed TPA-induced hyperplasia. Interestingly, however, RA seemed to have no effect on TPA or mezerein-induced hyperplasia, although certain retinoids are known to be potent inhibitors of ODC activity.[119,122] These data strongly suggested that the induction of epidermal ODC, followed by increased polyamine synthesis, may be important in stage II promotion.

Thus, the conclusions were that the important event in the first stage of promotion is the induction of dark basal cells which may be primitive stem cells, since they normally occur in large numbers in embryonic skin and are only present in small numbers in adult skin. The important event for second stage promotion (alluded to above) would be the induction of ODC and increased cellular proliferation. This suggests strongly that tumor promoters influence differentiation, at least in the early stage of the process. This was suggested further when dark cells were studied following single topical applications of TPA, mezerein, EPP, anthralin, 12-deoxyphorbol 13-2, 4,6-decatrienate (tri-DPD), 12-deoxyphorbol-13-decanoate (DPD), the nonpromoting derivative of TPA, 4-*O*-methyl TPA, or the calcium ionophore, A21387, to SENCAR epidermis. A good correlation was observed between the dark cell-inducing capacities of these compounds and their respective efficiencies as tumor promoters.[124] Of special note was the case of the two very closely related phorbol compounds, DPD and tri-DPD, which exhibited potent and weak promoting activity, respectively; this correlated well with their dark cell-inducing abilities. Also, at high doses the calcium ionophores, A23187 and 4-*0*-MeTPA, were extremely good inducers of dark cells. Both these agents are ineffective as complete promoters, but are effective as first stage promoters when applied prior to mezerein.

TPA has also been shown to modulate terminal differentiation in a variety of cell culture systems. TPA inhibits the terminal differentiation of Friend erythroleukemic cells,[125] chicken myoblasts,[126] neuroblastoma cells,[127] and adipose cells,[125] and stimulates terminal differentiation of human myeloid leukemia cells[128] and melanoma cells.[129] However, disparate effects of TPA on the terminal differentiation of keratinocytes have been reported (see Reference 130 and references therein).

B. Selection and Clonal Expansion of Initiated Cells

Yuspa and co-workers,[131,132] in studies with primary mouse epidermal cultures, showed the usual pattern of epidermal cell growth in culture characterized by limited proliferation, keratin synthesis, desmosomes, stratification, and terminal differentiation to be markedly altered by lowering the ionic calcium concentration from 1.2 to 2.0 m*M* to 0.02 to 0.1 m*M*. In low calcium medium, desmosome formation and stratification are prevented and the epidermal cells grow indefinitely as a monolayer with a high proliferation rate.[133] The synthesis of keratin proteins continues, but differentiating cells are shed into the medium instead of forming multilayers as they do in medium containing high calcium. Thus, growth in low calcium appears to select for epidermal cells with the characteristics of basal cells. Also, the induction of ODC by TPA, which is thought to be important in the tumor promotion process, could be increased several-fold by growth in low calcium. Morphological differentiation induced by calcium addition was enhanced by TPA and cells with thickened plasma membranes were apparent only 10 hr after TPA treatment, and by 20 hr squamous cells with pycnotic nuclei could be observed detaching from the plate.

Although this population of committed cells is removed more rapidly from the plate, another population of cells is almost unaffected by TPA treatment. This observation is supported by the focal distribution of the tumors produced in mouse skin which indicates that only a fraction of the epidermal cells are involved in promotion; the observation is also supported by results from other laboratories.[130,134] According to Parkinson and co-workers,[134] normal human keratinocyte cultures show a heterogeneous response to TPA in that 90 to 93% of the colony-forming cells lose their colony-forming ability and form cornified envelopes, a late step in keratinocyte terminal differentiation, when treated for 24 hr with doses of 100 n*M* TPA or less; while the remaining 7 to 10% are resistant to doses up to 1000 n*M*. Cultures of squamous carcinoma cell lines and SV40-transformed keratinocytes were found to contain 3 to 17 times more TPA-resistant keratinocytes than normal keratinocyte strains and the size of the TPA-resistant fraction was inversely related to the competence of that cell line to lose colony-forming efficiency when suspended in methocel. This was the first detectable change in an ordered sequence of events resembling terminal differentiation of a keratinocyte, suggesting that transformed human keratinocytes might be less sensitive to TPA-induced terminal differentiation than their normal counterparts. Parkinson's explanation of this[134] is that the resistant cells are not sufficiently committed to terminal differentiation to respond like the rest. Also, since ''initiation'' with reference to the mouse epidermis is essentially irreversible, ''initiated'' keratinocytes might be considered to be stem cells or committed cells that have been altered so that they behave like stem cells. This suggested that TPA, or promoters with a similar mode of action, could select for stem cells that were defective in their commitment to differentiation. Such cells would, therefore, have a greater chance of surviving TPA treatment and expanding at the expense of the normal keratinocyte population during the period of regenerative hyperplasia of two-stage promotion defined by Furstenberger et al.[135] using the nonpromoting, irritant, mitogen phorbol 12-retinoate 13-acetate (RPA). In this agent the antipromoting principle of vitamin A acid is combined with the structure of a phorbol ester tumor promoter to give the most powerful second-stage promoter currently known. This compound is also thought to provide the most appropriate negative control for experiments on the biological effects of tumor promoters, because as an irritant mitogen, it probably has the same mechanism of action as TPA without being either a complete or first-stage promoter.[136] In contrast, 4-*O*-MeTPA which has been, and probably still is, the most commonly used negative control, does not induce skin inflammation and induces epidermal hyperproliferation via a mechanism different from that of TPA.[137] Furstenberger and co-workers[135] confirm the observation that the effects that are probably obligatory and critical for tumor promotion can be brought about by a single application of TPA. They also confirm that subsequent long-term proliferation may only be necessary to make the tumors which are mainly reversibly growing benign papillomas visible to the naked eye. Reddy and Fialkow,[138] however, showed that papillomas induced by repeated carcinogen applications arose from considerably more cells than those induced by an initiation-promotion protocol by exploiting the fact that due to X-chromosome inactivation only one of the two phosphoglycerate kinase (PGK) genes is active in a single somatic cell. Therefore, in mice heterozygous for the usual PGK gene and the variant, there are two cell populations, one producing PGK-B and the other the variant type PGK-A. In such animals, neoplasms with an unicellular origin should exhibit only a single enzyme phenotype (A or B), whereas those with a multicellular origin might exhibit double enzyme phenotypes. Treatment of heterozygous Balb/C mice with a single 200 μg dose of DMBA followed by repeated TPA promotion 3 times a week induced papillomas in 85% of the mice. As anticipated, normal tissues obtained from the mice at autopsy exhibited equal proportions of PGK-B and -A isozyme activities, while 86% of 83 skin papillomas exhibited single enzyme PGK phenotypes, suggesting that most of the neoplasms had developed from relatively few cells. In contrast, papillomas that arose after repeated sequential applications of DMBA in the absence of a promoter had more double enzyme PGK phenotypes, suggesting that these papillomas arose from more cells than the papillomas induced by a single administration of DMBA followed by TPA. These data support the postulate that papillomas induced by an initiation-promotion schedule arise by a qualitatively different mechanism from those induced by complete carcinogens.[139] This in turn is supported by the fact that the papillomas induced by the initiation-promotion schedule regressed and very few developed into carcinomas, while those induced by DMBA treatments alone did not and gave rise to carcinomas.[140] Also, retinoids inhibit the formation of papillomas induced by TPA promotion, but not those induced by DMBA alone.[141]

This difference in the papillomas induced by the two protocols only serves to underline one of the major anomalies of the two-stage mouse skin system on which so many of the studies have been conducted. Promotion of the skin of "initiated" mice by repeated applications of TPA or certain other noncarcinogenic hyperplasiogenic agents only results in the *rapid* production of *benign papillomas*. The initiation event achieved by a single low-dose topical application of a complete carcinogen is assumed to be a mutagenic event;[142] promotion as reviewed above is thought to involve selection and clonal expansion of initiated cells. However, only very long promoter treatments result in the conversion of benign papillomas to carcinomas. Hennings et al.,[143] however, demonstrated that TPA was essentially ineffective in the "conversion" of papillomas to carcinomas, while the initiators urethane, MNNG, and 4-NQO(4-nitroquinoline-*N*-oxide) were effective, suggesting that malignant conversion might result from another genetic change. Indeed, in much earlier experiments Roe et al.[144] reported that treatment of papillomas with urethane increased their conversion to carcinomas. Thus, multistage carcinogenesis in mouse skin can be considered to comprise initiation, promotion, and conversion events, and the relative ineffectiveness of TPA in the conversion of papillomas to carcinomas is probably the result of its inactivity as a mutagen.[145]

A parallel can be drawn between these observations and those made by Kopelovich et al.[146] and Gainer et al.[147] on the effects of TPA on mutant human fibroblast cell strains derived from individuals with hereditary familial polyposis coli (FPC), also known as adenomatosis of the colon and rectum. Familial polyposis coli is transmitted in an autosomal dominant fashion and predisposed individuals develop numerous polyps of the colorectal region which predispose in turn to the development of adenocarcinomas at an earlier age and higher frequency than that observed in the normal population.[148] It was assumed as a working hypothesis in both studies that skin fibroblasts from FPC patients had undergone the first stage of classical two-stage carcinogenesis; i.e., the cells might be considered to carry a mutation analogous to the ''initiation'' event of two-stage carcinogenesis. Evidence that these cells behaved in a more transformed manner than their normal counterparts had been provided by a number of workers with regard to their ability to grow in low serum, susceptibility to transformation by viral and chemical agents, and abnormal migration patterns in collagen gels.[147,149-154] It might be predicted, therefore, that skin fibroblasts from such individuals might be more susceptible to in vitro transformation by tumor promoter alone. Kopelovich et al.[146] were able to transform FPC skin fibroblasts by TPA alone in that the cells were able to grow and form tumors upon injection into the immune privileged anterior chamber of the nude mouse eye, irrespective of whether they grew and formed foci on agar prior to injection. However, the data suggest that prior to injection the cells could not have been fully transformed since they did not grow reproducibly in soft agar nor did they achieve immortality. No tumors were observed on subcutaneous injection. In support of this in our hands,[147] low level transformation was achieved with TPA alone in two out of three FPC strains tested, as assayed by their ability to grow in methyl cellulose medium, but they did not form tumors when injected subcutaneously into nude mice. However, the TPA-treated cells from the FPC patients remained unchanged with regard to their pattern of migration in collagen gels. Consistent with this, none of the isolated anchorage-independent clones showed a stably abnormal migration pattern in collagen gels or developed tumors in nude mice. The same clones, however, did exhibit stable anchorage independence, when after four passages in liquid medium they were plated back into methyl cellulose medium, suggesting that the two assay systems measure different phenomena. Thus, it seems that treatment of certain susceptible strains of FPC skin fibroblasts with the tumor promoter alone brings about only partial transformation, which is entirely consistent with recent data and theories[143,155,156] that suggest that at least three stages might be necessary for complete transformation to occur. In this context, the FPC skin fibroblasts, or perhaps a subpopulation of them, can be considered analogous to the initiated mouse epidermal cells described by Hennings et al.,[143] where treatment with TPA readily brings about the appearance of benign papillomas. The FPC clones which grow either on agar[146] or methylcellulose,[147] but which do not display aberrant migratory behavior in the collagen gel system or produce tumors in nude mice, are probably the in vitro equivalent of benign papillomas. Hennings et al.[143] proposed that a further ''mutational'' change was required in order to convert papillomas to carcinomas and this is probably an accurate assumption for the anchorage-independent FPC skin fibroblasts, TPA only being involved in selection and clonal expansion of cells. However, if the term ''mutation'' as it is applied in Hennings and associates' model for the conversion of papillomas to carcinomas is used in its broadest sense, it could be that chromosomal rearrangements are involved. During the course of the transformation experiments of Gainer et al.,[147] TPA increased the frequency of tetraploidy and aberration frequency in FPC cells and several quadriradial figures were observed. The quadriradial chromosomes all occurred in tetraploid cells and were the result of exchanges between apparently homologous chromosomes. They are probably due in part, therefore, to the increase in chromosome pairing time brought about by TPAs effects on ploidy. One theory of carcinogenesis is that initiation produces a recessive, transforming mutation and that promotion causes this to become homozygous or hemizygous.[157] There are several pathways to this end, such as endoreduplication or polyploidization, followed by chromosome loss or mitotic recombination. It has been demonstrated that TPA, to a limited extent, is capable of causing both these events[147,158-160] as evidenced by an increase in ploidy and quadriradial figures.[161] Thus, this low-level ability of TPA to generate a mutation-like event, could give rise to the low frequency malignant conversion by TPA observed in mouse skin systems. Gainer and co-workers'[147] low recovery of anchorage-independent colonies probably precluded such an observation. Thus, these observations made in vitro in a system other than mouse skin, are not inconsistent with current theories of carcinogenesis and gain some measure of support from recent studies on cells from retinoblastoma[162] and Wilm's tumor patients.[163]

C. Promoter-Induced Chromosomal Alterations

Cytogenetic events have always been considered important in an overall view of carcinogenesis and would seem to have relevance to the later stages, i.e., promotion.[157,164-169] Tumor promoters have been reported to induce sister chromatid exchanges (SCES) variety of established rodent fibroblast cell lines,[157,168-171] and more recently, the induction by TPA of gross chromosome aberrations in human lymphocytes[165] and SCEs in rodent cell lines[170] have been correlated with TPA-induced superoxide anion ($O_2^{\cdot -}$) production. In both cases the chromosomal effect could be inhibited by the external addition of the specific superoxide anion scavenger, superoxide dismutase (SOD). Witz and co-workers[172] previously proposed that TPA-induced tumor promotion is mediated at least in part by the reactive oxygen species produced by the migrating phagocytic cells which infiltrate mouse skin as a result of the promoter-induced inflammatory response. This hypothesis was based on previous studies[173] from the same group that showed that protease inhibitors capable of inhibiting tumor promotion[174,175] also inhibit the stimulation of phagocytic cells, including macrophages and polymorphonuclear leukocytes. Moreover, retinoids, potent inhibitors of stage II promotion (Figure 3), have also been shown to inhibit $O_2^{\cdot -}$ production in response to tumor promoters.[172] Consistent with these observations, Slaga and co-workers[176,177] have shown a number of free radical generating compounds to be tumor promoters in SENCAR mouse skin. SOD levels have also been shown to be reduced on application of TPA to mouse skin[178] and to human lymphocytes and fibroblasts in culture.[158] Kinsella et al.,[158] however, failed to correlate both the $O_2^{\cdot -}$ production and lowering of SOD levels in human leukocytes with the induction of gross chromosomal aberrations. Similarly, the use of an $O_2^{\cdot -}$-generating system failed to induce chromosome aberrations. In the same study, no TPA-induced chromosome aberrations were seen in human fibroblasts following a single treatment with TPA. However, a small but significant percentage increase in the chromosome aberration frequency in human fibroblasts was observed after repeated and prolonged exposure to TPA, coupled with an increased tendency toward polyploidization. Again, TPA had been shown to reduce the SOD levels in these cells. It was postulated that this lowering of the cell's defenses to $O_2^{\cdot -}$ might, in agreement with Emerit and Cerutti,[165] account for the slight, but significant, increase in aberration induction observed in these cells on repeated exposure to TPA. Although it is certain that TPA-induced tumor promotion involves, in part, reactive oxygen species, the data suggest[158] that the chromosomes are not the primary site of action for either $O_2^{\cdot -}$ radicals produced in response to TPA or a pathway in which $O_2^{\cdot -}$ is involved. It seems more likely that the initial effects of the increase in free radicals and the lowering of the levels of the protective enzyme are on proliferation[179] and differentiation[180] via effects on cyclic nucleotide pools[181,182] and proteases. Effects on proliferation and differentiation would be much more consistent with the evidence provided in the literature and suggest a relationship between the type of epigenetic control observed in development and the loss of control observed in malignancy. Only in extreme circumstances where the cells are already transformed or show high levels of spontaneous chromosome aberrations[166,167] or after prolonged and repeated exposure to TPA[147,148] do we see any evidence of promoter-induced aberrations. However, TPA, a putative nonmutagenic substance, was shown by Dzarlieva et al.[166,167] to cause distinct numerical and structural chromosome aberrations in both primary epidermal mouse cultures and a permanent epidermal cell line after short-term treatment with doses as low as 10^{-8} *M*. The cells used were characterized by high levels of spontaneous chromosome aberrations, although TPA consistently induced aberrations well above background level. Therefore, it is tempting to use this to lend support to this author's theory for TPA-induced chromosome aberrations being involved in the late conversion stage of carcinogenesis and predict once more that only in circumstances where the cells are "transformed" or show high levels of spontaneous aberrations do they play a significant role. However, Fusenig[297] is reported to have made similar observations with "normal" cells and in the same system reports that the potent second-stage promoter mezerein does not induce chromosome aberrations. This suggests a role for chromosome aberrations at an early stage in promotion, but since most of the experiments with TPA and mezerein only got to the papilloma or benign tumor stage, this might not be important. Work by Wigley[183] on submandibular salivary gland epithelium in vitro showed TPA to affect early and late stages of chemically induced transformation. TPA shortened the lag-time for the appearance of preneoplastic foci in primary cultures and enhanced their growth rate. In fact, there were similarities between the effects of TPA and the hormones insulin and hydrocortisone in the same cultures. There was also evidence that TPA accelerated late-stage progression of the preneoplastic phenotype in one out of two cell lines investigated. This was in agreement with the data from Colburn et al.,[184] where only a proportion of preneoplastic epidermal cells respond to late-stage promotion. The important conclusion from this is that TPA can promote late-stage progression to a fully transformed tumorigenic phenotype, at least in some systems.

D. Cell Surface Effects and Gene Modulation

In all the discussions so far about tumor promoters and TPA in particular, we have failed to acknowledge that the earliest changes in response to TPA occur at the cell membrane within minutes, in the absence of RNA and protein synthesis. These include increased 2-deoxyglucose uptake, altered cell adhesion and morphology, increased release of arachidonic acid from membrane phospholipids, and inhibition of epidermal growth factor (EGF) binding to its cellular receptor.[185] In fact, it has been speculated that TPA and the phorbol esters might exert their effects on the cell by usurping or disturbing cellular receptor response pathways, thereby eliciting a variety of effects leading to transformation.[186,187] TPA and the phorbol esters share many of the characteristics of hormonal agents in that they show some structural similarities;[188] many of their effects are mediated at nanomolar concentrations via highly specific phorbol ester receptors[189,190] and culminate in the induction of highly pleiotropic and reversible effects.[185] Lee and Weinstein[186] further postulated that the putative receptors were on the cell surface and that the growth regulator might be a polypeptide hormone. In addition to this ability to modulate growth factor binding[191,192] TPA is also known to modulate or uncouple β-adrenergic hormone receptor interactions[193] and modulate or perturb gene expression. Evidence of this has already been provided in the form of the induction of plasminogen activator activity,[107] ornithine decarboxylase activity,[103] and depression of histidase activity,[112] but TPA has also been shown to induce viral genes[194,195] and possibly influence gene amplification.[196,197] Kinsella and co-workers,[198,199] in studies on changes in gene expression induced by tumor promoters, showed TPA and the structurally unrelated but equally potent tumor promoter teleocidin B[200,201] (Figure 4) to induce the increased synthesis of the same two specific polypeptides in three different human fibroblast cell lines, neither of which were induced by the nonpromoting derivative of TPA, 4-*O*-MeTPA, or EGF, which shares a number of its biological effects with TPA and teleocidin B.[187,202-204] The specificity of the response to TPA and teleocidin B is consistent with the view that the pleiotropic effects of these agents are the result of binding either to a specific receptor or to different receptors with a common effector mechanism. Also, although EGF

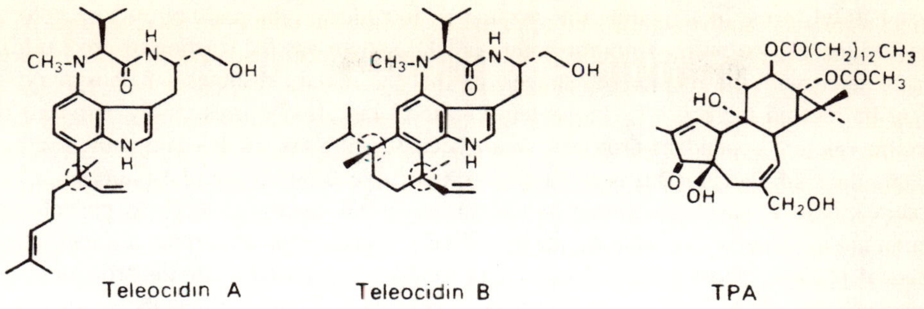

Teleocidin A Teleocidin B TPA

FIGURE 4. The structures of TPA, telecocidin A, and teleocidin B.

shares a number of biological effects with both TPA and teleocidin B (see above) and both TPA and teleocidin B inhibit EGF binding to its cellular receptor,[186,205] the failure of EGF to demonstrate the same highly specific effects on protein synthesis suggests that the changes are probably promoter-specific. Again, it is not certain whether this enhanced peptide synthesis is a necessary early step in promotion or whether it represents some form of promoter-induced differentiation effect. Laskin et al.[206] have reported TPA to induce a marked increase in the synthesis of keratin proteins in mouse epidermis. Such changes have been put forward as representing changes in differentiation state.[207]

Differentiation effect or not, one thing that is becoming increasingly obvious is that promoters and transforming growth factors exert their effects at the cell surface, presumably by modulating normal cellular growth factor control.[208,209] We know that certain sera and physiological fluids are highly inhibitory to 20-^{3}H-phorbol 12,13,-dibutyrate (TPA analogue) receptor binding[210] and that a phorbol ester-binding inhibitory factor has been isolated from human placenta, which might have a possible role in regulation of differentiation.[211] TPA is, therefore, likely to mediate its effect by inhibiting the binding of an endogenous ligand(s) to its own cellular receptor. The modulation of normal cellular control is also confirmed by the effects of TPA on isoproterenol-stimulated cAMP accumulation in human skin fibroblasts and parallel studies in isolated rat hepatocytes.[198,212] These studies by Heyworth et al.[212] showed TPA to inhibit both glucagon and cholera toxin-stimulated cAMP accumulation. Cholera toxin is known to activate adenylate cyclase independently of the glucagon receptor by interacting directly with the guanyl nucleotide regulatory protein involved in coupling the receptor and the catalytic unit of adenylate cyclase. This, together with the observation that ^{125}I-glucagon binding to its receptor is unaffected by TPA, suggested that TPA exerts its effect on this system either at the level of the guanyl nucleotide regulatory protein or the catalytic unit of adenylate cyclase. Recently it has been demonstrated that TPA mediates its effects via a calcium- and phospholipid-dependent protein kinase. Heyworth et al.[212] demonstrated that TPA inhibition of adenylate cyclase depended on both calcium and phosphatidyl serine being present in the incubation mixture. The effect could be mimicked to a degree by diolein, which also activates protein kinase C, but not by 4-*O*-MeTPA. The effect would also appear to involve phosphorylation since no inhibition occurred if γ-ATP replaced ATP in the reaction mixture. This analogue of ATP can be used as a substrate by adenylate cyclase, but not by protein kinases. The contention was and is, therefore, that TPA exerts its effect on adenylate cyclase through activation of protein kinase C and that the site at which protein kinase C exerts its effect is likely to be the guanyl nucleotide regulatory protein.

Many of the biological effects of tumor promoters appear to be mediated via a protein kinase. Nishizuka and co-workers[213] have suggested that protein kinase C is activated in cells by phosphatidylinositol turnover leading to diacylglycerol formation. TPA can substitute for unsaturated diacylglycerol to increase the affinity of protein kinase C for calcium and phosphatidyl serine in activating the enzyme.[214] In addition, the phorbol ester receptor and protein kinase C co-purify from rat brain, and TPA increases the fraction of protein kinase C associated with the membrane, suggesting that the initial substrates for phosphorylation might be located there.[215,216] The protein kinase C can also be irreversibly activated by a specific calcium-dependent protease. Also, because TPA and EGF share some biological effects, the EGF receptor has been studied as a possible substrate for phosphorylation.[217,218] Protein kinase C has been shown in human epidermal carcinoma cells to phosphorylate membrane-associated (as well as purified EGF) receptors at threonine residues at sites identical to those phosphorylated when intact cells are exposed to tumor promoters.[217,218] However, the whole field has recently been reviewed,[219] and I think the main point to emphasize is that the potential importance of protein kinase C in promotion and the process of carcinogenesis comes from the fact that it is a receptor protein and that proteins encoded by cellular oncogenes include a secreted growth factor,[220] a homologue of a growth factor receptor,[221] and several tyrosine-specific protein kinases.[222] It is attractive to suggest, therefore, that the uncontrollable production of an active form of protein kinase C, whether or not it is the product of a cellular oncogene, may "promote" carcinogenesis. The pleiotropic effects of tumor promoters may also be due to the location of protein kinase C at the crossover in pathways of hormone action and cell proliferation, involving calcium, inositol phospholipids, arachidonic acid, and prostaglandins.[219,223] However, the evidence that the final step, in which phosphatidylinositol 4,5-triphosphate breakdown acts as a signal, involves a guanine nucleotide-binding regulatory protein[224] is interesting in light of the guanine nucleotide regulatory protein involved in adenylate cyclase activation (see above).[212] Michell[224] suggests this could be p21, the GTP binding *ras* oncogene product. Certainly, there is evidence that high molecular weight DNA from squamous cell carcinomas induced by sequential treatment of mouse skin with initiators and promoters of carcinogenesis causes morphological transformation of NIH3T3 cells with high frequency due to an activated cellular homologue of the *c-Ha-ras* gene.[225]

E. Relevance to Human Cancer

Even such studies in a defined system[225] tend to mimic the early transfection studies in which single-step infection by tumor viruses and high molecular weight DNA from human tumors and human tumor cell lines were shown to impart morphological transformation to NIH3T3 cells.[226,227] These studies suggest that a single transforming gene is responsible for the carcinogenic process. However, this departure from multistage carcinogenesis is more apparent than real, in that recent studies have shown that each virus carries at least two genes encoding distinct functions, both of which are required to express the tumor phenotype.[228-230] Induction of bursal lymphomas by avian leukosis virus also requires the activation of two separate oncogenes *myc* and *B. lym*,[231-233] as do a promyelocytic leukemia and an American Burkitt's lymphoma, both of which require an altered *myc* and an activated *N-ras* oncogene.[234] Recently, the tumorigenic conversion of primary cell cultures has been shown to require at least two cooperating oncogenes,[156,235] which have been shown to be activated in a variety of systems by either up-regulation[236,237] or mutation.[234,238-240] Therefore, if we can study in defined multistage systems the effects of TPA on these genes or see if TPA can substitute for the immortilization of establishment functions, we can determine whether TPA exerts its effects by unsurping the action of a physiological agent to switch on one of possibly several cellular oncogenes. This in turn should provide insight into how TPA, the phorbol esters, and other promoting agents might mediate their effects on multistage carcinogenesis and aid in the evaluation of their carcinogenic risk to humans.

Table 2
SUMMARY OF EXPERIMENTAL AND ENVIRONMENTAL TUMOR PROMOTERS

Promoting agent	Target tissue	Ref.
Croton oil and diterpenes from plant sources	Skin	25, 63
Anthralin	Skin	42
Fatty acids and fatty acid methyl esters	Skin	93
Certain long-chain alkanes	Skin	272
Phenolic compounds	Skin	86
Surface active agents, e.g., Tween® 60	Skin	43, 44
Citrus oils	Skin, forestomach	38, 39, 58, 59
Tobacco smoke condensate	Skin, lung	90—92
Iodoacetic acid	Skin	89
Benzoyl peroxide	Skin	177
Dihydroteleocidin B, lyngbyatoxin A	Skin	199, 200
Exogenous hormones, e.g., diethylstilbestrol; endogenous hormones	Breast, liver, vagina	263, 266, 267
Endogenous hormones		
Dietary fat	Colon, breast	255—262
Phenobarbitol	Liver	254
Saccharin	Bladder	242—244
Cyclamates	Bladder	242—244
Roussin's red	Esophagus	251, 252
Griseofulvin	Skin	45
Alcohol	Esophagus	268
Asbestos	Lung	269, 270
Tryptophan	Bladder, mammary	246, 248, 263
Cyclophosphamide	Bladder	244
Bilharziasis	Bladder	245
Bile acids	Colon	264, 265
Butylated hydroxytoluene	Liver, lung	249, 253
DDT	Liver	250
EGF	Skin	201
Thioacetamide	Liver	249
Vitamin A deficiency	Bladder	241

III. TEST SYSTEMS USED FOR DETECTING COCARCINOGENS AND TUMOR PROMOTERS

So far in this chapter we have mainly discussed "TPAology" with only limited reference to the equally potent but structurally unrelated tumor promoters teleocidin B and dihydroteleocidin B.[200,201] However, as stated earlier,[45-59] two-stage carcinogenesis has been demonstrated in systems other than mouse skin and has been shown to involve promoting agents other than the phorbol esters (Table 2). Promoting agents found to be effective in tissues other than mouse skin include saccharin and cyclamates[242-244] in the bladder, phenobarbitol and steroids in the liver,[254,266,267] bile acids in the colon,[264,265] prolactin and estrogens in the breast,[263] butylated hydroxytoluene (BHT) in the lung,[253] and Roussin's red in the esophagus.[251,252] Although most of these observations have been made in animal systems, these agents singled out for special attention have their parallel in the human population and have been demonstrated at various times to share properties with TPA.[271-275] Based on the knowledge derived from epidemiological studies, it is thought that the majority of cancers in humans are caused by environmental factors, i.e., exposure to direct or indirect acting carcinogens and promoters related to occupation and lifestyle.[276-279] In fact, promotion is thought to be the rate-limiting step since we are all likely to be initiated due to the large number of initiating agents in our environment. Over the years epidemiological studies have resulted in a growing public awareness, which in turn has led to the rigorous screening of agents used in drugs, foodstuffs, and cosmetics. Public opinion has also resulted in a move away from tests in animals to alternative short-term test systems. In 1975, Ames and co-workers[280,281] developed a mutation test using *Salmonella typhimurium* to detect carcinogens in the environment, which is currently very widely used in research laboratories, industries, and regulatory agencies. The success of this, and related bacterial and mammalian cell mutation assays is based on the concept that carcinogens characteristically react with, modify, induce the repair of, interrupt the replication of, and otherwise alter the structure function and fate of DNA. This in turn is supported by the demonstration that over 90% of the animal carcinogens studied to date are mutagenic in vitro.[282,283] In addition, there are now many other assays, including several based upon mammalian cell transformation in vitro and mutation, which appear to be equally predictive of animal carcinogenicity.[284-286] Cytogenetic damage, e.g., SCEs,[287] chromosome rearrangements,[288,289] polyploidization, and nondisjunction, has also been taken as an end point in short-term assay systems for detecting environmental carcinogens. However, most cocarcinogens and tumor promoters have been discovered using the mouse skin system, mainly because one would predict that the short-term assays listed above for carcinogenic agents would be inappropriate, since these agents are reputed not to interact directly with DNA. Over the years, however, various assay systems for promoters have been proposed: induction of inflammation and hyperplasia,[93,97,99] induction of ODC activity,[103] induction of PA activity,[107] effects on differentiation,[290] enhancement of cellular transformation,[105,106] elimination of metabolic cooperation,[291-294] and induction of SCEs.[157,169-171] Of these, the ability of tumor promoters to inhibit cellular communication was probably pushed the hardest. Trosko and co-workers[295] produced lists of promoter-related compounds and agents that posed a possible threat in the environment, which eliminated metabolic cooperation. However, the validity of this system has been placed in some considerable doubt,[275] and overall, as for carcinogenicity testing, a variety of tests must be employed which take into consideration some of the more specific properties of promoting agents.[296]

ACKNOWLEDGMENTS

The author would like to thank Dr. Margaret Fox for her critical reading of the manuscript.

REFERENCES

1. **Moodie, R. L.,** Pathologic lesions among extinct animals: a study of the evidences of disease millions of years ago, *Surg. Clin. Chicago,* 2, 319, 1918.
2. **Ramazzini, B.,** Diseases of workers, (translation of the Latin text by Wilmer Cave Wright), University of Chicago Press, Chicago, 1940, 191.
3. **Hill, J.,** *Cautions Against the Immoderate Use of Snuff* Baldwin & Jackson, London, 1761, 30.
4. **Pott, P.,** Chirurgical observations relative to cancer of the scrotum, 1775, reprinted *Natl. Cancer Inst. Monogr.,* 10, 7, 1963.
5. **Clemmensen, J.,** On the etiology of some human cancers, *J. Natl. Cancer Inst.,* 12, 1, 1951.
6. **Haddow, A. and Kon, G. A. R.,** Chemistry of carcinogenic compounds, *Br. Med. Bull.,* 4, 314, 1947.
7. **Coombs, M. M.,** Chemical carcinogenesis. A view at the end of the first half century, *J. Pathol.,* 130, 117, 1980.
8. **Rehn, L.,** Blasengeschwulste bei Fuchsin-Arbeitern, *Arch. Klin. Chir.,* 50, 588, 1895.
9. **Yamagiwa, K. and Ichikawa, K.,** Experimentelle studie uber die pathogenese der Epithelialgeschwülste, *Mitt. Med. Fak. Tokio,* 15, 295, 1915; Experimental study of the pathogenesis of carcinoma, *J. Cancer Res.,* 3, 1, 1918.

10. **Tsutui, H.,** Über das kunstlich erzeuyte Cancroid bei der Maus, *Gann*, 12, 17, 1918.
11. **Passey, R. D.,** Experimental soot cancer, *Br. Med. J.*, 2, 1112, 1922.
12. **Bloch, B. and Dreifuss, W.,** Über die experimentelles Erzeugung von Carcinomen mit Lymphdrussen- und Lungenmetastasen durch Teerbestandteile *Schweiz. Med. Wochenschr.*, 51, 1033, 1921.
13. **Kennaway, E. L.,** The formation of a cancer-producing substance from isoprene (2-methylbutadiene), *J. Pathol. Bact.*, 27, 233, 1924.
14. **Kennaway, E. L.,** Experiments on cancer-producing substances, *Br. Med. J.*, 2, 1, 1925.
15. **Heiger, I.,** The spectra of cancer-producing tars and oils and of related substances, *Biochem. J.*, 24, 505, 1930.
16. **Kennaway, E. L. and Heiger, I.,** Carcinogenic substances and their fluorescence spectra, *Br. Med. J.*, 1, 1044, 1930.
17. **Heiger, I.,** Isolation of a cancer-producing hydrocarbon from coal tar, *J. Chem. Soc.*, 395, 1933.
18. **Cook, J. W. and Hewett, C. L.,** Isolation of a cancer-producing hydrocarbon from coal tar, *J. Chem. Soc.*, 398, 1933.
19. **Deelman, H. T.,** Die Entstehung des experimentellen Teerkrebses und die Bedeutung der Zellregeneration, *Z. Krebsforsch.*, 21, 220, 1924.
20. **Deelman, H. T.,** The part played by injury and repair in the development of cancer, *Br. Med. J.*, 1, 872, 1927.
21. **Twort, J. M. and Twort, C. C.,** Comparative activity of some carcinogenic hydrocarbons, *Am. J. Cancer*, 35, 80, 1930.
22. **Twort, C. C. and Ing, H. R.,** Untersuchungen über krebserzeugende Agenzien, *Z. Krebsforsch.*, 27, 309, 1928.
23. **Rous, P. and Kidd, J. G.,** Conditional neoplasms and subthreshold neoplastic states. A study of tar tumours in rabbits, *J. Exp. Med.*, 73, 365, 1941.
24. **Friedewald, W. F. and Rous, P.,** The initiating and promoting elements in tumour production, *J. Exp. Med.*, 80, 101, 1944.
25. **Berenblum, I.,** The co-carcinogenic action of croton resin, *Cancer Res.*, 1, 44, 1941.
26. **Berenblum, I.,** The mechanism of carcinogenesis. A study of the significance of co-carcinogenic action and related phenomena, *Cancer Res.*, 1, 807, 1941.
27. **Berenblum, I.,** The role of irritation in carcinogenesis, *Br. Empire Cancer Campaign*, 15th annu. rep., 225, 1938.
28. **Berenblum, I.,** Irritation and carcinogenesis, *Br. Empire Cancer Campaign*, 16th annu. rep., 212, 1939.
29. **Flaschenträger, B.,** Über den Giftstoff im Crotonöl, Versammlungsberichte, *Z. Angew. Chem.*, 43, 1011, 1930.
30. **Flaschenträger, B. and Falkenhausen, F. V.,** Über den Giftstoff des Crotonöles. 11. Krotophorbolone, *Ann. Chem.*, 514, 252, 1934.
31. **Mottram, J. C.,** A developing factor in experimental blastogenesis, *J. Pathol.*, 56, 181, 1944.
32. **Berenblum, I. and Shubik, P.,** The role of croton oil applications, associated with a single painting of a carcinogen, in tumour induction of the mouse's skin, *Br. J. Cancer*, 1, 379, 1947.
33. **Berenblum, I. and Shubik, P.,** The persistence of latent tumour cells induced in the mouse's skin by a single application of 9:10-dimethyl-1:2-benzanthracene, *Br. J. Cancer*, 3, 384, 1949.
34. **Salaman, M. H. and Roe, F. J. C.,** Incomplete carcinogens: ethyl carbamate (urethane) as an initiator of skin tumour formation in the mouse, *Br. J. Cancer*, 7, 472, 1953.
35. **Roe, F. J. C. and Salaman, M. H.,** Further studies on incomplete carcinogenesis: triethylene melamine (TEM), 1,2-benzanthracene and β-propiolactone, as initiators of skin tumour formation in the mouse, *Br. J. Cancer*, 9, 177, 1955.
36. **Van Esch, G. J., Van Genderen, H., and Vink, H. H.,** The production of skin tumours in mice by oral treatment with urethane, isopropyl-*N*-phenyl carbamate or isopropyl-*N*-chlorophenyl carbamate in combination with skin painting with croton oil and Tween 60, *Br. J. Cancer*, 12, 355, 1958.
37. **Roe, F. J. C.,** The effect of applying croton oil before a single application of 9,10-dimethyl-1,2-benzanthracene (DMBA), *Br. J. Cancer*, 13, 87, 1959.
38. **Roe, F. J. C. and Peirce, W. E. H.,** Tumor promotion by citrus oils: tumors of the skin and urethral orifice in mice, *J. Natl. Cancer Inst.*, 24, 1389, 1960.
39. **Roe, F. J. C. and Peirce, W. E. H.,** Tumour promotion by *Euphorbia* latices, *Cancer Res.*, 21, 338, 1961.
40. **Peirce, W. E. H. and Roe, F. J. C.,** Dose-response in tumour-promotion by *Euphorbia Tirucalli latix*, *Oncologia*, 15, 189, 1962.
41. **Bock, F. G. and Burns, R.,** Tumour promoting properties of anthralin (1,8,9-anthratriol), *J. Natl. Cancer Inst.*, 30, 393, 1963.
42. **Della Porta, G., Shubik, P., Dammert, K., and Terracini, B.,** Role of polyoxyethylene sorbitan monostearate in skin carcinogenesis in mice, *J. Natl. Cancer Inst.*, 25, 607, 1960.

43. **Setala, K., Setala, H., and Holsti, P.,** A new physicochemically well-defined group of tumour-promoting agents for mouse skin, *Science,* 120, 1075, 1954.
44. **Barich, L. L., Schwarz, J., and Barich, D.,** Oral griseofulvin: a co-carinogenic agent to methylcholanthrene-induced cutaneous tumors, *Cancer Res.,* 22, 53, 1962.
45. **Tomatis, L. and Shubik, P.,** Influence of urethane, on subcutaneous carcinogenesis by "Teflon" implants, *Nature (London),* 198, 600, 1963.
46. **Haran-Ghera, B., Trainin, N., Fiore-Donati, L., and Berenblum, I.,** A possible two-stage mechanism in rhabdomyo-sarcoma induction in rats, *Br. J. Cancer,* 16, 653, 1962.
47. **Gottfried, B., Molomut, N., and Patti, J.,** Effect of repeated surgical trauma on chemical carcinogenesis, *Cancer Res.,* 21, 658, 1961.
48. **Odashima, S.,** Development of liver cancers in the rat by 20-methylcholanthrene painting following initial 4-dimethylaminoazobenzene feeding, *Gann,* 50, 321, 1959.
49. **Takayama, S.,** Effect of 4-nitroquinoline-N-oxide painting on azo dye hepatocarcinogenesis in rats, with note on induction of skin fibrosarcoma, *Gann,* 52, 165, 1961.
50. **Nakahara, W., Fukoaka, F., and Sugimura, T.,** Carcinogenic action of 4-nitroquinoline- *N*- oxide, *Gann,* 48, 129, 1957.
51. **Sen Gupta, K. P.,** Hyperplasia of urinary tract epithelium induced by continuous administration of sulphonamide derivatives, *Br. J. Cancer,* 16, 110, 1962a.
52. **Sen Gupta, K. P.,** Tumour promoting action of 4-ethylsulphonylnapthalene-1-sulphonamide, *Nature (London),* 194, 1185, 1962b.
53. **DiPaolo, J. A.,** Effects of oxygen concentrations on carcinogenesis induced by transplacental exposure to urethane, *Cancer Res.,* 22, 299, 1962.
54. **Roe, F. J. C., Salaman, M. H., Cohen, J., and Burgan, J. G.,** Incomplete carcinogens in cigarette smoke condensate: tumour promotion by a phenolic fraction, *Br. J. Cancer,* 13, 623, 1959.
55. **Roe, F. J. C.,** Role of 3,4-benzopyrene in carcinogenesis by tobacco smoke condensate, *Nature (London),* 194, 1089, 1962.
56. **Gellhorn, A.,** The co-carcinogenic activity of cigarette tobacco tar, *Cancer Res.,* 18, 510, 1958.
57. **Wynder, E. L. and Hoffman, D.,** *Acta Pathol. Microbiol. Scand.,* 52, 119, 1961.
58. **Peirce, W. E. H.,** Tumour promotion by lime oil in the mouse forestomach, *Nature (London),* 189, 497, 1961.
59. **Peirce, W. E. H.,** The Role of Citrus Oils in Carcinogenesis, Ph.D. thesis University of London, London, 1963.
60. **Boutwell, R. K.,** Some biological aspects of skin carcinogenesis, *Progr. Exp. Tumor Res.,* 4, 207, 1964.
61. **Hecker, E.,** Co-carcinogenic principles from the seed oil of *Croton Tiglium* and from other Euphorbiaceae, *Cancer Res.,* 28, 2338, 1968.
62. **Van Duuren, B. L. and Orris, L.,** The tumour enhancing principles of *Croton tiglium, Cancer Res.,* 25, 1871, 1965.
63. **Van Duuren, B. L. and Sivak, A.,** Tumour promoting agents from *Croton tiglium* and their mode of action, *Cancer Res.,* 28, 2349, 1968.
64. **DiGiovanni, J., Slaga, T. J., and Boutwell, R. K.,** Comparison of the tumour initiating activity of 7,12-dimethylbenz*[a]*anthracene and benzo*[a]*pyrene in female SENCAR and CD1 mice, *Carcinogenesis,* 1, 381, 1980.
65. **Pettersen, R. C., Ferguson, G., Crombie, L., Games, M. L., and Pointer, D. J.,** The structure and stereochemistry of phorbol, diterpene parent of co-carcinogens of croton oil, *Chem. Commun.,* 716, 1967.
66. **Hecker, E.,** Structure-activity relationships in diterpene esters irritant and co-carcinogenic to mouse skin, in *Carcinogenesis,* Vol. 2, *Mechanisms of Tumour Promotion and Co-Carcinogenesis,* Slaga, T. J., Sivak, A., and Boutwell, R. K., Eds., Raven Press, New York, 1978, 11.
67. **Van Duuren, B L., Tseng, S. S., Segal, A., Smith, A. C., Melchionne, S., and Seidman, I.,** Effects of structural changes on the tumour-promoting activity of phorbol myristate acetate on mouse skin, *Cancer Res.,* 39, 2644, 1979.
68. **Hecker, E.,** Isolation and characterisation of the co-carcinogenic principles from croton oil, in *Methods in Cancer Research,* Vol. 6, Busch, H., Ed., Academic Press, New York, London, 1971, 439.
69. **Adolf, W., Zayed, S., and Hecker, E.,** New, most active 1α-alkyldaphnane-type irritants and tumour promoters from species of the Thymelaceae, in *Carcinogenesis,* Vol. 7, *Co-Carcinogenesis and Biological Effects of Tumour Promoters,* Hecker, E., Fusenig, H. E., Kunz, W., Marks, F., and Thielmann, H. W., Eds., Raven Press, New York, 1982, 49.
70. **Kupchan, S. M. and Baxter, R. L.,** Mezerein:antileukemic principle isolated from *Daphne mezereum, Science,* 187, 652, 1975.
71. **Van Duuren, B. L., Sivak, A., Katz, C., Seidman, I., and Melchionne, S.,** The effect of ageing and interval between primary and secondary treatment in two-stage carcinogenesis in mouse skin, *Cancer Res.,* 35, 502, 1975.

72. **Boutwell, R. K.,** The role of the induction of ornithine decarboxylase in tumour promotion, in *Origins of Human Cancer,* Book B, Hiatt, H. H., Watson, J. D., and Winsten, J. A., Eds., Cold Spring Harbor Laboratories, Cold Spring Harbor, N.Y., 1977, 773.
73. **Colburn, N. H. and Boutwell, R. K.,** The binding of β-propriolactone and some related alkylating agents to DNA, RNA and protein of mouse skin; relation between tumour-initiating power of alkylating agents and their binding to DNA, *Cancer Res.,* 28, 653, 1968.
74. **Bowden, G. T. and Boutwell, R. K.,** Studies on the role of stimulated epidermal DNA synthesis in the initiation of skin tumours in mice by *N*-methyl-*N'*-nitro-*N*-nitrosoguanidine, *Cancer Res.,* 34, 1552, 1974.
75. **Brookes, P. and Lawley, P. D.,** Alkylating agents, *Br. Med. Bull.,* 20, 91, 1964.
76. **Slaga, T. J., Bowden, G. T., Shapas, B. G., and Boutwell, R. K.,** Macromolecular synthesis following a single application of alkylating agents used as initiators of mouse skin tumorigenesis, *Cancer Res.,* 33, 769, 1973.
77. **Slaga, T. J., Bowden, G. T., Shapas, B. G., and Boutwell, R. K.,** Macromolecular synthesis following a single application of polycyclic hydrocarbons used as initiators of mouse skin tumorigenesis, *Cancer Res.,* 34, 771, 1974.
78. **Hennings, H. and Boutwell, R. K.,** Studies on the mechanism of skin tumour promotion, *Cancer Res.,* 30, 312, 1970.
79. **Baird, W. M., Sedgwick, J. A., and Boutwell, R. K.,** Effects of phorbol and four diesters of phorbol on the incorporation of tritiated precursors into DNA, RNA and protein in mouse epidermis, *Cancer Res.,* 31, 1434, 1971.
80. **Raineri, R., Simsiman, R. C., and Boutwell, R. K.,** Stimulation of the phosphorylation of mouse epidermal histones by tumour promoting agents, *Cancer Res.,* 33, 134, 1973.
81. **Kreibech, G., Hecker, E., Suss, R., and Kinzel, V.,** Phorbol ester stimulated choline incorporation, *Naturwissenschaften,* 58, 323, 1971.
82. **Rohrschneider, L. R., O'Brien, D. H., and Boutwell, R. K.,** Stimulation of phospholipid metabolism in mouse skin following phorbol ester treatment, *Biochim. Biophys. Acta,* 280, 57, 1972.
83. **Baird, W. M. and Boutwell, R. K.,** Tumour promoting activity of phorbol and four diesters of phorbol in mouse skin, *Cancer Res.,* 31, 1047, 1971.
84. **Boutwell, R. K. and Bosch, D. K.,** The tumour promoting action of phenol and related compounds for mouse skin, *Cancer Res.,* 19, 413, 1959.
85. **Boutwell, R. K. and Bosch, D. K.,** Studies on the role of surface active agents in the formation of skin tumours in mice, *American Association for Cancer Research,* 2, 190, 1957.
86. **Setälä, K.,** Progress in carcinogenesis, tumour enhancing factors. A bioassay of skin tumour formation, *Progr. Exp. Tumor Res.,* 1, 225, 1960.
87. **Gwynn, R. and Salaman, M. H.,** Studies on co-carcinogenesis, SH-reactors and other substances tested for co-carcinogenic action in mouse skin, *Br. J. Cancer,* 7, 482, 1953.
88. **Bock, F. G., Moors, G. E., and Crouch, S. K.,** Tumour promoting activity of extracts of unburned tobacco, *Science,* 145, 231, 1964.
89. **Van Duuren, B. L., Sivak, A., Langseth, L., Goldschmidt, B. M., and Segal, A.,** Initiators and promoters in tobacco carcinogenesis, *Natl. Cancer Inst. Monogr.,* 28, 173, 1968.
90. **Wynder, E. L. and Hoffman, D.,** A study of tobacco carcinogenesis. X. Tumour promoting activity, *Cancer,* 24, 289, 1969.
91. **Arffman, E. and Galvind, J.,** Tumour promoting activity of fatty acid methyl esters in mice, *Experientia,* 27, 1465, 1971.
92. **Troll, W., Klassen, A., and Janoff, A.,** Tumourigenesis in mouse skin. Inhibition by synthetic inhibitors of proteases, *Science,* 169, 1211, 1970.
93. **Scribner, J. D. and Boutwell, R. K.,** Inflammation and tumour promotion: selective protein induction in mouse skin by tumour promoters, *Eur. J. Cancer,* 8, 617, 1972.
94. **Belman, S. and Troll, W.,** The inhibition of croton-oil promoted skin tumourigenesis by steroid hormones, *Cancer Res.,* 32, 450, 1972.
95. **Slaga, T. J. and Scribner, J. D.,** Inhibition of tumour initiation and promotion by anti-inflammatory agents, *J. Natl. Cancer Inst.,* 51, 1723, 1973.
96. **Scribner, J. D. and Slaga, T. J.,** Multiple effects of dexamethasone on protein synthesis and hyperplasia caused by a tumour promoter, *Cancer Res.,* 33, 542, 1973.
97. **Berenblum, I.,** Irritation and carcinogenesis, *Arch. Pathol.,* 38, 233, 1944.
98. **Saffiotti, U. and Shubik, P.,** Studies on the promoting action of skin carcinogenesis, *Natl. Cancer Inst. Monogr.,* 10, 489, 1963.
99. **Frei, J. V. and Stephens, P.,** The correlation of promotion of tumour growth and induction of hyperplasia in epidermal two-stage carcinogenesis, *Br. J. Cancer,* 22, 83, 1968.
100. **Raick, A. N.,** Ultrastructural, histological and biochemical alterations produced by 12-*O*-tetradecanoyl-phorbol-13-acetate on mouse epidermis and their relevance to skin tumour promotion, *Cancer Res.,* 33, 269, 1973.

101. **Süss, R., Kinzel, V., and Kreibich, G.,** Co-carcinogenic croton oil factor A stimulates lipid synthesis in cell cultures, *Experientia,* 27, 46, 1971.
102. **Raick, A. N.,** Cell differentiation and tumour promoting action in skin carcinogenesis, *Cancer Res.,* 34, 2915, 1974.
103. **O'Brien, T. G., Simsiman, R. C., and Boutwell, R. K.,** Induction of the polyamine biosynthetic enzymes in mouse epidermis by tumour promoting agents, *Cancer Res.,* 35, 1662, 1975.
104. **Yuspa, S. H., Lichti, U., Ben, T., Patterson, E., Hennings, H., Slaga, T. J., Colburn, N. H., and Kelsey, W.,** Phorbol ester tumour promoters stimulate DNA synthesis and ornithine decarboxylase activity in mouse epidermal cell cultures, *Nature (London),* 262, 402, 1976.
105. **Mondal, S., Peterson, A. R., and Brankow, D. W.,** Initiation and promotion in oncogenesis in cell cultures, *American Association for Cancer Research,* 16, 74, 1975.
106. **Mondal, S., Brankow, D. W., and Heidelberger, C.,** Two-stage chemical oncogenesis in cultures of $C_3H10T^1/_2$ cells, *Cancer Res.,* 36, 2254, 1976.
107. **Wigler, M. and Weinstein, I. B.,** Tumour promoter induces plasminogen activator, *Nature (London),* 259, 232, 1976.
108. **Raineiri, R., Simsiman, R. C., and Boutwell, R. K.,** Stimulation of synthesis of mouse epidermal histones by tumour promoting agents, *Cancer Res.,* 37(4), 584, 1977.
109. **Bresnik, E., Meunier, R., and Lamden, M.,** Epidermal prostaglandins after topical application of a tumour promoter, *Cancer Lett.,* 7, 121, 1979.
110. **Belman, S. and Troll, W.,** Phorbol-12-myristate-13-acetate effect on cyclic adenosine 3'-5'-monophosphate levels in mouse skin and inhibition of phorbol myristate acetate-promoted tumourigenesis, *Cancer Res.,* 34, 3446, 1974.
111. **Mufson, R. A., Simsiman, R. C., and Boutwell, R. K.,** The effect of the phorbol ester tumour promoters on the basal and catecholamine stimulated levels of cyclic adenosine 3':5'-monophosphate in mouse skin and epidermis *in vivo, Cancer Res.,* 37, 665, 1977.
112. **Colburn, N. H., Lau, S., and Head, R.,** Decrease of epidermal histidase activity by tumour promoting phorbol esters, *Cancer Res.,* 35, 3154, 1975.
113. **Raick, A. N.,** Cell proliferation and promoting action in skin carcinogenesis, *Cancer Res.,* 34, 920, 1974.
114. **Slaga, T. J., Scribner, J. D., Thompson, S., and Viaje, A.,** Epidermal cell proliferation and promoting ability of phorbol esters, *J. Natl. Cancer Inst.,* 52, 1611, 1974.
115. **Slaga, T. J., Fischer, S. M., Weeks, C. E., Nelson, K., Mamrack, M., and Klein-Szanto, A. J. P.,** Specificity and mechanisms of promoter inhibitors in multi-stage promotion, in *Carcinogenesis,* Vol. 7, Hecker, E., Fusenig, N. E., Kunz, W., Marks, F., and Thielmann, H. W., Eds., Raven Press, New York, 1982, 19.
116. **Balmain, A.,** Synthesis of specific proteins in mouse epidermis after treatment with the tumour promoter, TPA, in *Carcinogenesis,* Vol. 2, *Mechanisms of Tumor Promotion and Co-Carcinogenesis,* Slaga, T. J., Sivak, A., and Boutwell, R. K., Eds., Raven Press, New York, 1978, 153.
117. **Raick, A. N. and Burdzy, K.,** Ultrastructural and biochemical changes induced in mouse epidermis by a hyperplastic agent ethyl phenylpropriolate, *Cancer Res.,* 33, 2221, 1973.
118. **Klein-Szanto, A. J. P., Major, S. M., and Slaga, T. J.,** Induction of dark keratinocytes by 12-*O*-tetradecaonylphorbol-13-acetate and mezerein as an indicator of tumour promoting efficiency, *Carcinogenesis,* 1, 399, 1980.
119. **Mufson, R. A., Fischer, S. M., Verma, A. K., Gleason, G. L., Slaga, T. J., and Boutwell, R. K.,** Effects of 12-*O*-tetradecaonylphorbol-13-acetate and mezerein on epidermal ornthine decarboxylase activity, isoproterenol-stimulated levels of cyclic adenosine 3'-5'-monophosphate and induction of mouse skin tumours, *Cancer Res.,* 39(4), 791, 1979.
120. **Hozumi, M., Ogawa, M., Sugimura, T., Takeuchi, T., and Umezawa, H.,** Inhibition of tumourigenesis in mouse skin by leupeptin a protease inhibitor from actinomycetes, *Cancer Res.,* 32, 1725, 1972.
121. **Schwarz, J. A., Viaje, A., Slaga, T. J., Yuspa, S. H., Hennings, H., and Lichti, U.,** Fluocinolone acetonide: a potent inhibitor of mouse skin tumour promotion and epidermal DNA synthesis, *Chem. Biol. Interact.,* 17, 331, 1977.
122. **Verma, A. K., Rice, H. M., Shapos, B. G., and Boutwell, R. K.,** Inhibition of 12-*O*-tetradecanoyl-phorbol-13-acetate-induced ornithine decarboxylase activity in mouse epidermis by vitamin A analogs (retinoids), *Cancer Res.,* 38, 793, 1978.
123. **Weeks, C. E., Slaga, T. J., Hennings, H., Gleason, G. L., and Bracken, W. M.,** Inhibition of phorbol ester induced tumour promotion by vitamin A analog and antiinflammatory steroid, *J. Natl. Cancer Inst.,* 63, 401, 1979.
124. **Klein-Szanto, A. J. P., Major, S. K., and Slaga, T. J.,** Quantitative evaluation of dark keratinocytes induced by several promoting and hyperplasiogenic agents: their use as an early morphological indicator of tumour promoting action, in *Carcinogenesis,* Vol. 7, Hecker, E., Fusenig, N. E., Kunz, W., Marks, F., and Thielmann, H. W., Eds., Raven Press, New York, 1982, 305.

125. **Diamond, L., O'Brien, T., and Rovera, G.,** Tumour promoters inhibit terminal cell differentiation in culture, in *Carcinogenesis,* Vol. 2, *Mechanisms of Tumor Promotion and Co-Carcinogenesis,* Slaga, T. J., Sivak, A., and Boutwell, R. K., Eds., Raven Press, New York, 1978, 335.

126. **Cohen, R., Pacifici, M., Rubinstein, N., Beihl, J., and Holtzer, H.,** Effect of a tumour promoter on myogenesis, *Nature (London),* 266, 538, 1977.

127. **Ishii, D. B., Fibach, E., Yamasaki, H., and Lucinstein, J. B.,** Tumour promoters inhibit morphological differentiation in cultured mouse neuroblastoma cells, *Science,* 200, 556, 1978.

128. **Huberman, E. and Callahan, M. F.,** Induction of terminal differentiation in human promyelocytic leukaemia cells by tumour promoting agents, *Proc. Natl. Acad. Sci. U.S.A.,* 76, 1293, 1979.

129. **Huberman, E., Heckman, C., and Langenbach, R.,** Stimulation of differentiated functions in human melanoma cells by tumour promoting agents and dimethyl sulphoxide, *Cancer Res.,* 39, 2618, 1979.

130. **Reiners, J. J. and Slaga, T. J.,** Effects of tumour promoters on the rate and commitment to terminal differentiation of subpopulations of murine keratinocytes, *Cell,* 32, 247, 1983.

131. **Kulesz-Martin, M. F., Koehler, B., Hennings, H., and Yuspa, S. H.,** Quantitative assay for carcinogen altered differentiation in mouse epidermal cells, *Carcinogenesis,* 1, 995, 1980.

132. **Yuspa, S. H., Hennings, H., Kulesz-Martin, M., and Lichti, U.,** Study of tumour promotion in a cell culture model for mouse skin — a tissue that exhibits multi-stage carcinogenesis *in vivo,* in *Carcinogenesis,* Vol. 7, Hecker, E., Fusenig, N. E., Kunz, W., Marks, F., and Thielmann, H. W., Eds., Raven Press, New York, 1982, 217.

133. **Hennings, H., Lichti, U., Holbrook, K., and Yuspa, S. H.,** Role of differentiation in determining responses of epidermal cells to phorbol esters, in *Carcinogenesis,* Vol. 7, Hecker, E., Fusenig, N. E., Kunz, W., Marks, F., Thielmann, H. W., Eds., Raven Press, New York, 1982, 319.

134. **Parkinson, E. K., Grabham, P., and Emmerson, A.,** A subpopulation of cultured human keratinocytes which is resistant to the induction of terminal differentiation-related changes by phorbol, 12-myristate 13-acetate: evidence for an increase in the resistant population following transformation, *Carcinogenesis,* 4, 857, 1983.

135. **Fürstenberger, G., Berry, D. L., Sorg, B., and Marks, F.,** Skin tumour promotion by phorbol esters is a two-stage process, *Proc. Natl. Acad. Sci. U.S.A.,* 78, 7722, 1981.

136. **Fürstenberger, G., Richter, H., Argyris, T. S., and Marks, F.,** Effects of the phorbol ester 4-*O*-methyl-12-*O*-tetradecaonyl phorbol-13-acetate on mouse skin *in vivo:* evidence for its uselessness as a negative control compound in studies on the biological effects of tumour promoters, *Cancer Res.,* 42, 342, 1981.

137. **Fürstenberger, G., de Bravo, M., Bertsch, S., and Marks, F.,** The effect of indomethacin on cell proliferation induced by chemical and mechanical means in mouse epidermis *in vivo, Res. Commun. Chem. Pathol. Pharmacol.,* 24, 533, 1979.

138. **Reddy, A. L. and Fialkow, P. J.,** Papillomas induced by initiation-promotion differ from those induced by carcinogen alone, *Nature (London),* 69, 1983.

139. **Scribner, J. D. and Scribner, N.,** Is the initiation-promotion regimen in mouse-skin relevant to complete carcinogenesis, in *Carcinogenesis,* Vol. 7, Hecker, E., Fusenig, N. E., Kunz, W., Marks, F., and Thielmann, H. W., Eds., Raven Press, New York, 1982, 13.

140. **Della Porta, G., Terracini, B., Dammert, K., and Shubik, P.,** Histopathology of tumors induced in mice treated with polyoxyethylene sorbitan monostearate, *J. Natl. Cancer Inst.,* 25, 573, 1960.

141. **Verma, A. K. and Boutwell, R. K.,** Vitamin A acid (retinoic acid), a potent inhibitor of 12-*O*-tetradecaonylphorbol-13-acetate induced ornithine decarboxylase activity in mouse epidermis, *Cancer Res.,* 37, 2196, 1977.

142. **Miller, E. C. and Miller, J. A.,** Searches for ultimate chemical carcinogens and their reactions with cellular macromolecules, *Cancer,* 47, 2327, 1981.

143. **Hennings, H., Shores, R., Wenk, M. L., Spangler, E. F., Tarone, R., and Yuspa, S. H.,** Malignant conversion of mouse skin tumours is increased by tumour initiators and unaffected by tumour promoters, *Nature (London),* 304, 67, 1983.

144. **Roe, F. J. C., Carter, C. L., Mitchley, B. C., Peto, R., and Hecker, E.,** On the persistence of tumour initiation and the acceleration of tumour progression in mouse skin tumourigenesis, *Int. J. Cancer,* 9, 264, 1972.

145. **Kinsella, A. R.,** Investigation of the effects of the phorbol ester TPA on carcinogen-induced forward mutagenesis to 6-thioguanine resistance in V79 Chinese hamster cells, *Carcinogenesis,* 2, 43, 1981.

146. **Kopelovich, L., Bias, N. E., and Helson, L.,** Tumour promoter alone induces malignant transformation of fibroblasts from humans genetically predisposed to cancer, *Nature (London),* 282, 619, 1979.

147. **Gainer, H. St. C., Schor, S. L., and Kinsella, A. R.,** Susceptibility of skin fibroblasts from individuals genetically predisposed to cancer to transformation by the tumour promoter 12-*O*-tetradecanoyl phorbol-13-acetate, submitted, *Int. J. Cancer,* 1984.

148. **Muto, T., Bussey, H. J. R., and Morson, B. C.,** The evolution of cancer of the colon and rectum, *Cancer Res.,* 36, 2251, 1975.

149. **Kopelovich, L., Lipkin, M., Blattner, W. A., Fraumeni, J. F., Lynch, H. T., and Pollack, R. E.,** Organisation of actin containing cables in cultured skin fibroblasts from individuals at high risk of colon cancer, *Int. J. Cancer,* 26, 301, 1980.
150. **Kopelovich, L.,** Hereditary adenomatosis of the colon and rectum: recent studies on the nature of cancer promotion and cancer prognosis in vitro, in *Colorectal Cancer: Prevention and Epidemiology,* Winawer, S., Schottenfeld, D., and Sherlock, P., Eds., Raven Press, New York, 1980, 97.
151. **Kopelovich, L.,** Hereditary adenomatosis of the colon and rectum: relevance to cancer promotion and cancer control in human, *Cancer Genet. Cytogenet.,* 5, 333, 1982.
152. **Miyaki, M., Avamatsu, N., Hirono, U., Ono, T., Tonomura, A., and Utsononuja, J.,** Transformation of fibroblasts from a patient with adenomatosis coli by treatment with chemical compounds, *Gann,* 71, 741, 1980.
153. **Wolman, S. R.,** Determinants in malignant transformation. Studies in hereditary bowel cancer, in *Prevention of Hereditary Large Bowel Cancer,* Alan R. Liss, New York, 1983, 213.
154. **Rasheed, S. and Gardner, M. B.,** Growth properties and susceptibility to viral transformation of skin fibroblasts from individuals at high risk for colo-rectal cancer, *J. Natl. Cancer Inst.,* 66, 43, 1981.
155. **Potter, V. R.,** A new protocol and its rationale for the study of initiation and promotion in rat liver, *Carcinogenesis,* 2, 1375, 1981.
156. **Land, H., Parada, L. F., and Weinberg, R. F.,** Tumorigenic conversion of primary embryo fibroblasts requires at least two co-operating oncogenes, *Nature (London),* 304, 596, 1983.
157. **Kinsella, A. R. and Radman, M.,** Tumour promoter induces sister chromatid exchanges: relevance to mechanisms of carcinogenesis, *Proc. Natl. Acad. Sci. U.S.A.,* 75, 6149, 1978.
158. **Kinsella, A. R., Gainer, H. St. C., and Butler, J.,** Investigation of a possible role for superoxide anion production in tumour promotion, *Carcinogenesis,* 4, 717, 1983.
159. **Gainer, H. St. C. and Kinsella, A. R.,** Analysis of spontaneous, carcinogen-induced and promoter-induced chromosomal instability in patients with hereditary retinoblastoma, *Int. J. Cancer,* 32, 449, 1983.
160. **Kopelovich, L.,** Genetic forms of neoplasia in man: a model for the study of tumour promotion *in vitro,* in *Carcinogenesis,* Vol. 7, Hecker, E., Fusenig, N. E., Kunz, W., Marks, F., and Thielmann, H. W., Eds., Raven Press, New York, 1982, 259.
161. **Chaganti, R. S. K., Schonberg, S., and German, J.,** A many-fold increase in sister chromatid exchanges in Bloom's syndrome lymphocytes, *Proc. Natl. Acad. Sci. U.S.A.,* 71, 4508, 1974.
162. **Cavanee, B. L., Dryja, T. P., Phillips, R. A., Benedict, W. F., Godbout, R., Gallie, B. L., Murphree, A. L., Strong, L. C., and White, E. L.,** Expression of recessive alleles by chromosomal mechanisms in retinoblastoma, *Nature (London),* 305, 799, 1983.
163. **Koufos, A., Hansen, M. F., Lampkin, B. C., Workman, M. L., Copeland, M. G., Jenkins, N. A., and Cavenee, W. K.,** Loss of alleles at loci on human chromosome 11 during genesis of Wilm's tumour, *Nature (London),* 309, 170, 1984.
164. **Levan, A., Levan, G., and Mitelman, F.,** Chromosomes and cancer, *Hereditas,* 86, 15, 1977.
165. **Emerit, I. and Cerutti, P. A.,** Tumour promoter phorbol-12-myristate acetate induces chromosomal damage via indirect action, *Nature (London),* 293, 744, 1981.
166. **Dzarlieva, R. T., Schultz, S., and Fusenig, N. E.,** Spontaneous and co-carcinogen induced chromosome abberrations in primary and permanent mouse epidermal cells, *J. Cancer Res. Clin. Oncol.,* 99, 61, 1981.
167. **Fusenig, N. E. and Dzarlieva, R. T.,** Phenotypic and chromosomal alterations in cell culture as indicators of tumour promoting activity, in *Carcinogenesis,* Vol. 7, Hecker, E., Fusenig, N. E., Kunz, W., Marks, F., Thielmann, H. W., Eds., Raven Press, New York, 1982, 201.
168. **Nagasawa, H. and Little, J. B.,** Effect of tumour promoters, protease inhibitors and repair processes on X-ray induced sister chromatid exchanges, *Proc. Natl. Acad. Sci. U.S.A.,* 76, 1943, 1979.
169. **Gentil, A., Renault, G., and Margot, A.,** The effect of the tumour promoter 12-*O*-tetradecanoylphorbol-13-acetate (TPA) on UV- and MNNG-induced sister chromatid exchanges in mammalian cells, *Int. J. Cancer,* 26, 517, 1980.
170. **Ray-Chaudhuri, R., Currens, M., and Iype, P. T.,** Enhancement of sister chromatid exchanges by tumour promoting chemicals, *Br. J. Cancer,* 45, 769, 1982
171. **Nagasawa, H. and Little, J. B.,** Factors influencing the induction of sister chromatid exchanges in mammalian cells by 12-*O*-tetradecanoyl phorbol-13-acetate, *Carcinogenesis,* 2, 601, 1981.
172. **Witz, G., Goldstein, B. D., Amoruso, M., Stone, D. S., and Troll, W.,** Retinoid inhibition of superoxide anion radical production by human polymorphonuclear leukocytes stimulated with tumour promoters, *Biochem. Biophys. Res. Commun.,* 97, 883, 1980.
173. **Goldstein, B. D., Witz, G., Amoruso, M., and Troll, W.,** Protease inhibitors antagonise the activation of polymorphonuclear leukocyte oxygen consumption, *Biochem. Biophys. Res. Commun.,* 88, 858, 1979.
174. **Golstein, B. D., Witz, G., Amoruso, M., Stone, D. S., and Troll, W.,** Stimulation of polymorphonuclear leukocyte superoxide anion radical production by tumour promoters, *Cancer Lett.,* 11, 257, 1981.
175. **Yavelow, J., Gidlund, M., and Troll, W.,** Protease inhibitors from processed legumes effectively inhibit superoxide anion generation in response to TPA, *Carcinogenesis,* 3, 135, 1982.

176. **Slaga, T. J., Triplett, L. L., Yotti, L. P., and Trosko, J. E.,** Skin tumour promoting activity of benzoyl peroxide a widely used free radical generating compound, *Science*, 213, 1023, 1981.
177. **Klein-Szanto, A. J. P. and Slaga, T. J.,** Effects of peroxide on rodent skin: epidermal hyperplasia and tumour promotion, *J. Invest. Dermatol.*, 79, 30, 1982.
178. **Solanki, V., Rana, R. S., and Slaga, T. J.,** Diminution of mouse epidermal superoxide dismutase and catalase activities by tumour promoters, *Carcinogenesis*, 2, 1141, 1981.
179. **Oberley, L. W., Oberley, T. D., and Buettner, G. R.,** Cell division in normal and transformed cells: the possible role of superoxide and hydrogen peroxide, *Med. Hypotheses*, 7, 21, 1981.
180. **Oberley, L. W., Oberley, T. D., and Buettner, G. R.,** Cell differentiation, ageing and cancer: the possible roles of superoxide and superoxide dismutases, *Med. Hypotheses*, 6, 249, 1980.
181. **Mittal, C. K. and Murad, F.,** Activation of guanyl cyclase by superoxide dismutase and hydroxyl radical: a physiological regulator of guanosine 3′-5′-monophosphate formation, *Proc. Natl. Acad. Sci. U.S.A.*, 74, 4360, 1972.
182. **Novogrodsky, A., Patya, M., Rubin, A. L., and Stenzer, K. H.,** Inhibition of β-adrenergic stimulation of lymphocyte adenylase cyclase by phorbol myristate acetate is mediated via activated macrophages, *Biochem. Biophys. Res. Commun.*, 104, 338, 1982.
183. **Wigley, C. B.,** TPA affects early and late stages of chemically induced transformation in mouse submandibular salivary epithelial cells *in vitro*, *Carcinogenesis*, 4, 101, 1983.
184. **Colburn, N. H., Former, B. F., Nelson, K. A., and Yuspa, S. H.,** Tumour promoter induces anchorage independence irreversibly, *Nature (London)*, 282, 589, 1979.
185. **Weinstein, B., Lee, L. S., Fisher, P. B., Mufson, A., and Yamasaki, H.,** Action of phorbol esters in cell culture: mimickry of transformation, altered differentiation and effects on cell membranes, *J. Supramol. Struct.*, 12, 195, 1979.
186. **Lee, L. S. and Weinstein, I. B.,** Tumour-promoting phorbol esters inhibit binding of epidermal growth factor to its cellular receptors, *Science*, 202, 313, 1978.
187. **Lee, L. S. and Weinstein, I. B.,** Epidermal growth factor, like tumour promoting phorbol esters induces plasminogen activator activity, *Nature (London)*, 274, 696, 1978.
188. **Wilson, S. R. and Huffman, J. C.,** The structural relationship between phorbol and cortisol. A possible mechanism for the tumour promoting activity of phorbol, *Experentia*, 32, 1489, 1976.
189. **Dunphy, W. G., Delclos, K. B., and Blumberg, P. M.,** Characterisation of specific binding of [^{3}H] phorbol 12,13-dibutyrate and [^{3}H] phorbol 12-myristate 13-acetate to mouse brain, *Cancer Res.*, 40, 3635, 1980.
190. **Driedger, P. E. and Blumberg, P. M.,** Specific binding of phorbol ester tumour promoters, *Proc. Natl. Acad. Sci. U.S.A.*, 77, 567, 1980.
191. **Brown, K. D., Dicker, P., and Rozengurt, E.,** Inhibition of epidermal growth factor binding to surface receptors by tumour promoters, *Biochem. Biophys. Res. Commun.*, 86, 1037, 1979.
192. **Shoyab, M., DeLarco, J. E., and Todaro, G.,** Biologically active phorbol esters specifically alter the affinity of epidermal growth factor receptors, *Nature (London)*, 279, 387, 1979.
193. **Garte, S. J. and Belman, S.,** Tumour promoter uncouples β-adrenergic receptor from adenyl cyclase in mouse epidermis, *Nature (London)*, 284, 171, 1980.
194. **Zur Hausen, H. Z., O'Neill, F. J., and Freese, U. K.,** Persisting oncogenic herpes virus induced by a tumour promoter TPA, *Nature (London)*, 272, 373, 1978.
195. **Amtmann, E. and Sauer, G.,** Activation of non-expressed bovine papilloma virus genomes by tumour promoters, *Nature (London)*, 296, 675, 1982.
196. **Varshavsky, A.,** Phorbol ester dramatically increases the incidence of methotrexate-resistant mouse cells: possible mechanism and relevance to tumour promotion, *Cell*, 25, 561, 1981.
197. **Bojan, F., Kinsella, A. R., and Fox, M.,** Effect of tumour promoter 12-*O*-tetradecanoylphorbol-13-acetate on recovery of methotrexate-, *N*-(phosphonacetyl)-L-asparate- and cadmium resistant colony forming mouse and hamster cells, *Cancer Res.*, 43, 5217, 1983.
198. **Kinsella, A. R., Whetton, A. D., Heyworth, C. M., Houslay, M. D., deWynter, E., and Bazill, G. W.,** Membrane mediated responses to 12-*O*-tetradecanoylphorbol-13-acetate in human skin fibroblasts, in *Models, Mechanisms and Etiology of Tumour Promotion*, Börzsönyi, N., Day, N. E., Lapis, K., and Yamasaki, H., Eds., IARC Publ. No. 56, International Agency for Research on Cancer, Lyon, 1984, 177.
199. **Bazill, G. W., deWynter, E., Fujiki, H., and Kinsella, A. R.,** A comparison of the effects of tumour promoters TPA and teleocidin B on gene expression in human skin cell fibroblasts, *Gann*, , 75, 672, 1984.
200. **Fujiki, H., Mori, M., Nakaysu, M., Terada, M., and Sugimura, T.,** A possible naturally occurring tumour promoter, teleocidin B from Streptomyces, *Biochem. Biophys. Res. Commun.*, 90, 976, 1979.
201. **Fujiki, H., Mori, M., Nakaysu, M., Terada, M., Sugimura, T., and Moore, R. E.,** Indole alkaloids, dihydro-teleocidin B, teleocidin and ljngbyatoxin B, a new class of tumour promoters, *Proc. Natl. Acad. Sci. U.S.A.*, 78, 3872, 1981.
202. **Rose, S. P., Stahn, R., Passovey, D. V., and Herschman, R.,** Epidermal growth factor enhancement of skin tumour induction, *Experientia*, 32, 913, 1976.

203. **DiPasquale, A., White, D., and McGuire, P.,** Epidermal growth factor stimulates putresceine transport and ornithine decarboxylase activity in cultivated human fibroblasts, *Exp. Cell Res.*, 116, 317, 1978.
204. **Fisher, P. B., Mufson, R. A., Weinstein, I. B., and Little, J. B.,** Epidermal growth factor-like tumour promoters enhance viral and radiation-induced cell transformation, *Carcinogenesis*, 2, 183, 1981.
205. **Umezawa, K., Weinstein, I. B., Horowitz, A., Fujiki, H., Matsushima, T., and Sugimura, T.,** Teleocidin B and phorbol ester tumour promoters produce similar effects on membranes and membrane receptors, *Nature (London)*, 290, 41, 1981.
206. **Laskin, J. D., Mufson, R. A., Piccini, L., Engelhardt, D. L., and Weinstein, I. B.,** Effects of tumour promoter 12-*O*-tetradecaonylphorbol-13-acetate on newly synthesised proteins in mouse epidermis, *Cell*, 25, 441, 1981.
207. **Roop, D. R., Hawley-Nelson, P., Cheng, C. K., and Yuspa, S. H.,** Keratin gene expression in mouse epidermis and cultured epidermal cells, *Proc. Natl. Acad. Sci. U.S.A.*, 80, 716, 1983.
208. **Reynolds, F. M., Todaro, G. J., Fryling, C., and Stephenson, J. R.,** Human transforming growth factors induce tyrosine phosphorylation of EGF receptors, *Nature (London)*, 292, 259, 1981.
209. **Roberts, A. B., Anzano, M. A., Lamb, L. C., Smith, J. M., Frolik, C. A., Marguardt, H., Todaro, G. J., and Sporn, M. B.,** Isolation from murine sarcoma cells of novel transforming growth factors potentiated by EGF, *Nature (London)*, 295, 417, 1982.
210. **Horowitz, A. D., Greenbaum, E., and Weinstein, I. B.,** Identification of receptor for phorbol ester tumour promoters in intact mammalian cells and of an inhibitor of receptor binding in biological fluids, *Proc. Natl. Acad. Sci. U.S.A.*, 78, 2315, 1981.
211. **Hamel, E., Martel, N., Tayot, J. L., and Yamasaki, H.,** Characterisation of a human placental factor which inhibits specific binding of phorbol esters to cultured cells, in *Models, Mechanisms, and Etiology of Tumour Promotion*, Börzsönyi, N., Day, N. E., Lapis, K., and Yamasaki, H., Eds., IARC Publ. No. 56, International Agency for Research on Cancer, Lyon, 1984, 157.
212. **Heyworth, C. M., Whetton, A. D., Kinsella, A. R., and Houslay, M. D.,** Does the phorbol ester TPA inhibit adenylate cyclase through the action of protein kinase C?, *FEBS Lett.*, 170, 38, 1984.
213. **Kishimoto, A., Takai, Y., Mori, T., Kikkawa, U., and Nishizuka, Y.,** Activation of calcium and phospholipid-dependent protein kinase by diacylglycerol. its possible relation to phospholipid inositol turnover, *J. Biol. Chem.*, 255, 2273, 1980.
214. **Castagna, M., Takai, Y., Kaibuchi, K., Sano, K., Kikkawa, U., and Nishizuka, Y.,** Direct activation calcium-activated, phospholipid dependent protein kinase by tumour promoting phorbol esters, *J. Biol. Chem.*, 257, 7847, 1982.
215. **Niedel, J. E., Kuhn, L. J., and Vandenbark, G. R.,** Phorbol diester receptor co-purifies with protein kinase C, *Proc. Natl. Acad. Sci. U.S.A.*, 80, 36, 1983.
216. **Kraft, A. S. and Anderson, W. B.,** Phorbol esters increase the amount of Ca^{2+}, phospholipid dependent protein kinase associated with plasma membrane, *Nature (London)*, 30, 621, 1983.
217. **Cochet, C., Gill, G. N., Meisenhelder, J., Cooper, J. A., and Hunter, T.,** C-kinase phosphorylates the epidermal growth factor receptor and reduces its epidermal growth factor stimulated tyrosine protein kinase activity, *J. Biol. Chem.*, 259, 2553, 1984.
218. **Iwashito, S. and Fox, C. F.,** Epidermal growth factor and potent phorbol tumour promoters induce epidermal growth factor phosphorylation in a similar but distinctly different manner in human epidermoid carcinoma A431 cells, *J. Biol. Chem.*, 259, 2559, 1984.
219. **Nishizuka, Y.,** The role of protein kinase C in cell surface signal transduction and tumour promotion, *Nature (London)*, 308, 693, 1984.
220. **Waterfield, M. D., Scrace, G. T., Whittle, N., Stroobant, P., Johnsson, A., Wastesone, A., Westermark, B., Heldin, C.-H., Huang, J. S., and Devel, T. F.,** Platelet-derived growth factor is structurally related to the putative transforming protein $p28^{sis}$ of simian sarcoma virus, *Nature (London)*, 304, 35, 1983.
221. **Downward, J., Yarden, Y., Mayes, E., Scrace, G., Totty, N., Stockwell, P., Ullrich, A., Schlessinger, J., and Waterfield, M. D.,** Close similarity of epidermal growth factor receptor and v-erb-B oncogene protein sequences, *Nature (London)*, 307, 521, 1984.
222. **Bishop, J. M.,** Cellular oncogenes and retroviruses, *Annu. Rev. Biochem.*, 52, 301, 1983.
223. **Weinstein, I. B., Horowitz, A. D., Mufson, A., Fisher, P. B., Ivanovic, V., and Greenbaum, E.,** Results and speculations related to recent studies on mechanisms of tumour promotion, in *Carcinogenesis*, Vol. 7, *Co-Carcinogenesis and Biological Effects of Tumour Promoters*, Hecker, E., Fusenig, N. E., Kunz, W., Marks, F., and Thielmann, H. W., Eds., Raven Press, New York, 1982, 599.
224. **Michell, R.,** Oncogenes and inositol lipids, *Nature (London)*, 308, 770, 1984.
225. **Balmain, A. and Pragnell, I. B.,** Mouse skin carcinomas induced *in vivo* by chemical carcinogens have a transforming Harvey-ras oncogene, *Nature (London)*, 303, 72, 1983.
226. **Shih, C., Shilo, B. Z., Goldfarb, M. P., Dannenberg, A., and Weinberg, R. A.,** Passage of phenotypes of chemically transformed cells via transfection of DNA and chromatin, *Proc. Natl. Acad. Sci. U.S.A.*, 76, 5714, 1979.

227. **Cooper, G. M., Okenquist, S., and Silverman, L.,** Transforming activity of DNA of chemically transformed and normal cells, *Nature (London)*, 284, 418, 1980.

228. **Rassoulzadegar, M., Cowie, A., Carr, A., Glaichenhaus, N., Kamen, R., and Cuzin, F.,** The roles of individual polyoma virus early proteins in oncogenic transformation, *Nature (London)*, 300, 713, 1982.

229. **Houweling, A., van den Elsen, P. J., and Van der Eb, A.,** Partial transformation of primary rat cells by the leftmost 4.5% fragment of adenovirus 5 DNA, *Virology*, 105, 537, 1980.

230. **Van den Elsen, P. J., de Pater, S., Houweling, A., van der Veer, J., and van der Eb, A.,** The relationship between region Ela and Elb of human adenoviruses in cell transformation, *Gene*, 18, 175, 1982.

231. **Hayward, N. S., Neel, B. G., and Astrin, S. M.,** Activation of a cellular onc gene by promoter insertion in ALV-induced lymphoid leukosis, *Nature (London)*, 290, 475, 1981.

232. **Payne, G. S., Bishop, J. M., and Varmus, H. E.,** Multiple arrangements of viral DNA and an activated host oncogene in basal lymphomas, *Nature (London)*, 295, 209, 1982.

233. **Cooper, G. M. and Neiman, P. E.,** Two distinct candidate transforming genes of lymphoid leukosis virus induced neoplasms, *Nature (London)*, 292, 857, 1981.

234. **Murray, M. J., Cunningham, J. M., Parada, L. F., Dautry, F., Lebowitz, P., and Weinberg, R. A.,** The HL-60 transforming sequence: a ras oncogene co-existing with altered myc genes in hematopoietic tumours, *Cell*, 33, 749, 1983.

235. **Ruley, E. H.,** Adenovirus early region 1A enables viral and cellular transforming genes to transform primary cells in culture, *Nature (London)*, 304, 602, 1983.

236. **Chang, E., Furth, M. E., Scolnick, E. M., and Lowy, D. R.,** Tumorigenic transformation of mammalian cells induced by a normal human gene homologous to the oncogene of Harvey murine sarcoma virus, *Nature (London)*, 297, 479, 1982.

237. **Alitalo, K., Schwab, M., Lin, C. C., Varmus, H. E., and Bishop, J. M.,** Homogeneously staining chromosomal regions contain amplified copies of an abundantly expressed cellular oncogene c-myc in malignant neuroendocrine cells from a human colon carcinoma, *Proc. Natl. Acad. Sci. U.S.A.*, 80, 1707, 1983.

238. **Tabin, C. J., Bradley, S. M., Bargmann, C. I., Weinberg, R. A., Papageorge, A. G., and Scolnick, E. M., Dhar, R., Lowy, D. R., and Chang, E. H.,** Mechanism of activation of a human oncogene, *Nature (London)*, 300, 143, 1982.

239. **Reddy, E. P., Reynolds, R. K., Santos, E., and Barbacid, M.,** A point mutation is responsible for the acquisition of transforming properties by the T24 human bladder carcinoma oncogene, *Nature (London)*, 300, 149, 1982.

240. **Taub, R., Moulding, C., Battey, J., Murphy, W., Vasicek, T., Lenoir, G. M., and Leder, P.,** Activation and somatic mutation of the translocated c-myc gene in Burkitt lymphoma cells, *Cell*, 36, 399, 1984.

241. **Capurro, P., Angrist, A., Black, J., and Moumgis, B.,** Studies in squamous metaplasia in rat bladder. Effects of hypovitaminosis A, foreign bodies and methylcholanthrene, *Cancer Res.*, 20, 563, 1960.

242. **Hicks, R. M. and Chowaniec, J.,** The importance of synergy between weak carcinogens in the induction of bladder cancer in experimental animals, *Cancer Res.*, 37, 2943, 1977.

243. **Hicks, R. M., Chowaniec, J., and Wakefield, J. St. J.,** Experimental induction of bladder tumors by a two-stage system, in *Carcinogenesis*, Vol. 2, *Mechanisms of Tumor Promotion and Co-Carcinogenesis*, Slaga, T. J., Sivak, A., and Boutwell, R. K., Eds., Raven Press, New York, 1978, 475.

244. **Hicks, R. M., Wakefield, J. St. J., and Chowaniec, J.,** Evaluation of a new model to detect bladder carcinogens or co-carcinogens. Results obtained from saccharin, cyclamate and cyclophosphamide, *Chem. Biol. Interact.*, 11, 225, 1975.

245. **Hicks, R. M., James, C., and Webbe, G.,** The effect of schistosoma haematobium and *N*-butyl-*N*-(4-hydroxybutyl) nitrosamine on the development of urothelial neoplasia in the baboon, *Br. J. Cancer*, 42, 730, 1980.

246. **Dunning, W. F., Curtis, M. R., and Maun, M. E.,** Effect of added diethyl tryptophan on occurrence of 2-acetylaminofluorene-induced liver and bladder cancer in rats, *Cancer Res.*, 10, 454, 1950.

247. **Matsushima, M.,** The role of the promoter L-tryptophan on tumorigenesis in the urinary bladder. II. Urinary bladder carcinogenicity of FANFT (initiating factor) and L-tryptophan (promoting factor) in mice, *Jpn. J. Urol.*, 68, 731, 1977.

248. **Radomski, J. L., Radomski, T., and MacDonald, W. E.,** Carcinogenic interaction between DL tryptophan and 4-aminobiphenyl or 2-napthylamine in dogs, *J. Natl. Cancer Inst.*, 58, 1831, 1977.

249. **Cameron, R., Lee, R., and Farber, E.,** Chemical mitogens as effective alternatives to partial hepatectomy in the new model for the sequential analysis of hepatocarcinogenesis, *Proc. Am. Assoc. Cancer Res.*, 19, 56, 1978.

250. **Peraino, C., Fry, R. J. M., Staffeldt, E., and Christopher, J. P.,** Comparative enhancing effect of phenobarbitol, amobarbitol diphenylhydantoin and dichlorodiphenyl-trichloroethane on 2-acetylaminofluorene induced hepatic tumourigenesis in the rat, *Cancer Res.*, 35, 2884, 1975.

251. **Cheng, S. J., Sala, M., Li, M. H., Wang, M. Y., Pot-Deprun, J., and Chourovlinkov, I.,** Mutagenic, transforming and promoting effect of pickled vegetables from Linxian county, China, *Carcinogenesis,* 1, 685, 1980.
252. **Cheng, S. J., Sala, M., Li, M. H., Courtois, I., and Chouroulinkov, I.,** Promoting effect of Roussin's red identified in pickled vegetables from Linxian China, *Carcinogenesis,* 2, 313, 1981.
253. **Witschi, H. and Lock, S.,** Butylated hydroxytoluene. A possible promoter of adenoma formation in mouse lung, in *Carcinogenesis,* Vol. 2, *Mechanisms of Tumor Promotion and Co-Carcinogenesis,* Slaga, T. J., Sivak, A., and Boutwell, R. K., Eds., Raven Press, New York, 1978, 465.
254. **Peraino, C., Fry, R. J. M., Staffeldt, E. F., and Kisielski, W. E.,** Effects of varying the exposure to phenobaribotol on its enhancement of 2-acetylaminofluorene-induced hepatic tumorogenesis in the rat, *Cancer Res.,* 33, 2701, 1973.
255. **Haenszel, W., Berg, J. W., Segi, M., Kurihara, M., and Locke, F. B.,** Large bowel cancer in Hawaiian Japanese, *J. Natl. Cancer Inst.,* 51, 1765, 1973.
256. **Wynder, E. L.,** The epidemiology of large bowel cancer, *Cancer Res.,* 35, 3388, 1975.
257. **Tannebaum, A.,** The genesis and growth of tumours. III. The effects of a high fat diet, *Cancer Res.,* 2, 460, 1942.
258. **Reddy, B. S., Narisawa, T., Vakusich, D., Weisburger, J. H., and Wynder, E. L.,** Effect of quality and quantity of dietary fat and dimethylhydraozine in colon carcinogenesis in rats, *Proc. Soc. Exp. Biol. Med.,* 151, 237, 1976.
259. **Bansal, B., Rhoads, J. E., and Bansal, S. C.,** Effects of diet on colon carcinogenesis and the immune system in rats treated with 1,2-dimethyl hydrazine, *Cancer Res.,* 38, 3293, 1978.
260. **Benson, J., Lev, M., and Girand, C. G.,** Enhancement of mammary fibroadenomas in the female rat by a high fat diet, *Cancer Res.,* 16, 135, 1956.
261. **Carroll, K. K. and Khor, H. T.,** Effects of dietary fat and dose levels of 7,12-dimethylbenz*(a)*anthracene on mammary tumor incidence in rats, *Cancer Res.,* 30, 2260, 1970.
262. **Gammal, E. B., Carroll, K. K., and Plunkett, E. R.,** Effects of dietary fat on mammary carcinogenesis by 7,12-dimethylbenz*(a)*anthracene in rats, *Cancer Res.,* 27, 1737, 1967.
263. **Dunning, W. F., Curtis, M. R., and Maun, M. E.,** The effect of added tryptophan on the occurrences of diethylstilboeastrol induced mammary cancers in rats, *Cancer Res.,* 10, 319, 1950.
264. **Narisawa, T., Magadia, N. E., Weisburger, J. H., and Wynder, E. L.,** Promoting effect of bile acids on colon carcinogenesis after intra-rectal instillation of *N*-methyl-*N'*-nitro-*N*-nitrosoguanidine in rats, *J. Natl. Cancer Inst.,* 55, 1093, 1974.
265. **Reddy, B. S. and Wynder, E. L.,** Metabolic epidemiology of colon cancer: fecal bile acids and neutral sterols in colon cancer patients and patients with adenomatous polyps, *Cancer,* 39, 2533, 1977.
266. **Schulte-Hermann, R., Ohde, G., Schuppler, J., and Timmermann-Trosiener, I.,** Enhanced proliferation of putative preneoplastic cells in rat liver following treatment with the tumor promoters, phenobarbitol, hexachlorocyclohexane, steroid compounds and nafenopin, *Cancer Res.,* 41, 2556, 1981.
267. **Yager, J. D. and Yager, R.,** Oral contraceptive steroids as promoters of hepatocarcinogenesis in female Sprague-Dawley rats, *Cancer Res.,* 40, 3680, 1980.
268. **Loquet, C., Toussaint, G., and LeTalaer, J. Y.,** Studies on the mutagenic constituents of apple brandy and various alcoholic beverages collected in western France, a high incidence area for esophageal cancer, *Mutat. Res.,* 88, 155, 1981.
269. **Selikoff, I. J.,** Cancer risk of asbestos exposure, in *Origins of Human Cancer,* Vol. C, Hiatt, H. H., Watson, J. D., and Winsten, J. A., Eds., Cold Spring Harbor Laboratories, Cold Spring Harbor, N.Y., 1977, 1765.
270. **Saracci, R.,** Asbestos and lung cancer: an analysis of epidemiological evidence on the asbestos-smoking interaction, *Int. J. Cancer,* 20, 323, 1977.
271. **Van Durren, B. L. and Goldschmidt, B. M.,** Structure activity relationships of tumour promoters and co-carcinogens and interactions of phorbol myristate acetate and related esters with plasma membranes, in *Carcinogenesis,* Vol. 2, *Mechanisms of Tumor Promotion and Co-Carcinogenesis,* Slaga, T. J., Sivak, A., and Boutwell, R. K., Eds., Raven Press, New York, 1978, 491.
272. **Trosko, J. E., Dawson, B., Yotti, L. P., and Chang, C. C.,** Saccharin may act as a promoter by inhibiting metabolic co-operation between cells, *Nature (London),* 285, 109, 1980.
273. **Parry, J. M., Parry, E. M., and Barrett, J. C.,** Tumour promoters induce mitotic aneuploidy in yeast, *Nature (London),* 294, 263, 1981.
274. **Kinsella, A. R.,** Elimination of metabolic cooperation and the induction of sister chromatid exchanges are not properties common to all promoting agents, *Carcinogenesis,* 3, 499, 1982.
275. **Tsutui, T., Maizumi, H., McLachlan, J. A., and Barrett, J. C.,** Aneuploidy induction and cell transformation by diethylstilboestrol: a possible chromosomal mechanism in carcinogenesis, *Cancer Res.,* 43, 3814, 1983.
276. **Crump, K. S., Hoel, D. G., Langley, C. H., and Peto, R.,** Fundamental carcinogenic processes and their impolications to low dose risk assessment, *Cancer Res.,* 36, 2973, 1976.

277. **Doll, R.,** Strategy for detection of cancer hazard to man, *Nature (London)*, 265, 589, 1977.
278. **Siemiatycki, J., Day, N. E., Fabry, J., and Cooper, J. A.,** Discovering carcinogens in the occupational environment: a novel epidemiologic approach, *J. Natl. Cancer Inst.*, 66, 217, 1981.
279. **Sugimura, T.,** Mutagens, carcinogens and tumor promoters in our daily food, *Cancer*, 49, 1970, 1982.
280. **Ames, B. N., McCann, J., and Yamasaki, E.,** Methods for detecting carcinogens and mutagens with *Salmonella*/mammalian-microsome mutagenicity test, *Mutat. Res.*, 31, 347, 1975.
281. **McCann, J. and Ames, B. N.,** Detection of carcinogens as mutagens in the *Salmonella*/microsome test. Assays of 300 chemicals: discussion, *Proc. Natl. Acad. Sci. U.S.A.*, 73, 950, 1976.
282. **McCann, J., Choi, E., Yamasaki, E., and Ames, B. N.,** Detection of carcinogens as mutagens in the *Salmonella*/microsome test: assay of 300 chemicals, *Proc. Natl. Acad. Sci. U.S.A.*, 72, 5135, 1975.
283. **Purchase, I. F. H., Longstaff, E., Ashby, J., Styles, J. A., Anderson, D., Lefevre, P. A., and Westwood, F. R.,** An evaluation of 6 short-term tests for detecting organic chemical carcinogens, *Br. J. Cancer*, 37, 873, 1978.
284. **Heidelberger, C.,** Chemical carcinogenesis, *Annu. Rev. Biochem.*, 44, 79, 1975.
285. **Styles, J. A.,** A method for detecting carcinogenic organic chemicals using mammalian cells in culture, *Br. J. Cancer*, 36, 558, 1977.
286. **Clive, D., Johnson, S. O., Spector, J. F. S., Batson, A. G., and Brown, M. M.,** Validation and characterization of the L5187Y/TK$^{+/-}$ mouse lymphoma mutagen assay system, *Mutat. Res.*, 59, 61, 1979.
287. **Abe, S. and Sasaki, M.,** Chromosome aberrations and sister chromatid exchanges in Chinese hamster cells exposed to various chemicals, *J. Natl. Cancer Inst.*, 58, 1653, 1977.
288. **Evans, H. J. and O'Riordan, M. L.,** Human peripheral blood lymphocytes for the analysis of chromosome aberrations in mutagen tests, in *Handbook on Mutagenicity Test Procedures*, Kilbey, B. L., Legator, M., Nichols, W., and Ramel, C., Eds., Amsterdam Elsevier/North Holland, Amsterdam, 1977, 261.
289. **Heddle, J. A., Benz, R. D., and Countryman, P. I.,** Measurement of chromosomal breakage in cultured cells by the micronucleus technique, in *Mutagen Induced Chromosome Damage in Man*, Evans, H. J. and Lloyd, D. C., Eds., Edinburgh University Press, Edinburgh, 1978, 191.
290. **Yamasaki, H.,** Reversible inhibition of cell differentiation by phorbol esters as a possible mechanism for the promotion phase of carcinogenesis, IARC Publ. No. 27, International Agency for Research on Cancer, Lyons, 1980, 91.
291. **Murray, A. W. and Fitzgerald, D. J.,** Tumour promoters inhibit metabolic co-operation in co-cultures of epidermal and 3T3 cells, *Biochem. Biophys. Res. Commun.*, 91, 395, 1979.
292. **Yotti, L. P., Chang, C. C., and Trosko, J. E.,** Elimination of metabolic co-operation in Chinese hamster cells by a tumour promoter, *Science*, 206, 1089, 1979.
293. **Umeda, M., Noda, K., and Ono, T.,** Inhibition of metabolic cooperation in Chinese hamster cells by various chemicals including tumour promoters, *Gann*, 71, 614, 1980.
294. **Enomoto, T., Sasaki, Y., Shiba, Y., Kanno, Y., and Yamasaki, H.,** Tumour promoters cause a rapid and reversible inhibition of the formation and maintenance of electrical cell coupling in culture, *Proc. Natl. Acad. Sci. U.S.A.*, 78, 5628, 1981.
295. **Trosko, J. E., Yotti, L. P., Warren, S. T., Tsuchimoto, G., and Change, C. C.,** Inhibition of cell-cell communication by tumour promoters, in *Carcinogenesis, Vol. 7, Co-Carcinogenesis and Biological Effects of Tumour Promoters*, Hecker, E., Fusenig, N. E., Kunz, W., Marks, F., and Thielmann, H. W., Eds., Raven Press, New York, 1982, 565.
296. **Peto, J.,** Early and late stage carcinogenesis in mouse skin and in man, in *Models, Mechanisms, and Etiology of Tumour Promotion*, Börzsönyi, N., Day, N. E., Lapis, K., and Yamasaki, H., IARC Publ. No. 56, International Agency for Research on Cancer, Lyons, 1984, 359.
297. **Fusenig, N. E.,** personal communication.

Chapter 3

A REVIEW OF THE FAMILY EUPHORBIACEAE

A. Radcliffe-Smith

TABLE OF CONTENTS

I. INTRODUCTION

The Family Euphorbiaceae is the sixth largest family of flowering plants in the world. Unlike the five larger families Compositae, Orchidaceae, Gramineae, Leguminosae, and Rubiaceae, it is not such a "natural" group, and has by some authorities been regarded as something of a taxonomic dustbin. It has awkward entities not readily accommodated elsewhere placed therein, resulting in a group not readily capable of definition. Nevertheless, there are a number of unifying features which, although found elsewhere as ancillary characteristics in other families somewhat haphazardly distributed, are here usually found together in various combinations and to various degrees.

II. HABIT

Branching is for the most part monopodial in the Euphorbiaceae, although there are exceptions, as, for example, in the genus *Suregada,* where it is sympodial, resulting in leaf-opposed inflorescences. The subterranean parts of the branching system of *Euphorbia dulcis* are also sympodial.

Among perennial forms the caespitose habit is the most common, but some *Euphorbias* (for example, *E.* × *pseudovirgata* and *E. amygdaloides* var. *robbiae*) show a tendency to spread and form mats.

Annual species occur in predominantly perennial genera such as *Euphorbia, Phyllanthus, Croton, Mercurialis,* and *Acalypha,* and while these are generally erect, in *Euphorbia* subgenus *Chamaesyce* they are often prostate.

Among the subshrubs, an ericoid habit with small coriaceous leaves to cut down transpiration is found in genera which occur in areas where this type of habit is characteristic of the vegetation in general. It is seen in species of *Clutia* in South Africa, of *Poranthera* in Australia, and of *Sebastiania* in the drier parts of northeastern Brazil.

The twining habit is characteristic of the tribe Plukenetieae, which includes such genera as *Plukenetia, Tetracarpidium, Pterococcus, Romanoa, Tragia, Cnesmone,* etc., and it is also found in some species of the genus *Dalechampia,* the sole representative of the tribe Dalechampieae.

Lianes are rather rare in the family, but one or two genera include forms that fall into this category, as for example, the genera *Omphalea* and *Manniophyton.*

In the genus *Phyllanthus* a wide variety of habit-form occurs, ranging from delicate annuals with lateral leafy shoots of limited growth subtended by scale leaves known as cataphylls and recalling the sensitive plant *(Mimosa pudica),* to robust shrubs with large photosynthetic lateral cladodes bearing flowers around their edges and with small fugacious leaves. Other *Phyllanthus* species exhibit a scopiform or *Ephedra*-like habit, as do the genera *Calycopeplus* and *Amperea,* in which the terete or triquetrous stems soon become quite leafless and take on the role of photosynthetic organs.

It is, however, in the type genus *Euphorbia* that the widest range of habit-form is demonstrated, from prostrate annuals to large cactiform succulents of a great diversity of structure.

Large underground tubers occur in some *Jatropha* and *Manihot* species.

Shoot-thorns are found in, for example, *Flueggea* species, while in *Bridelia* and *Macaranga* epidermal thorns occur.

Some species of *Macaranga* and *Endospermum* are provided with hollow stems in which ants are able to form colonies.

III. INDUMENTUM

A great variety of hair covering is to be found among members of the family Euphor-

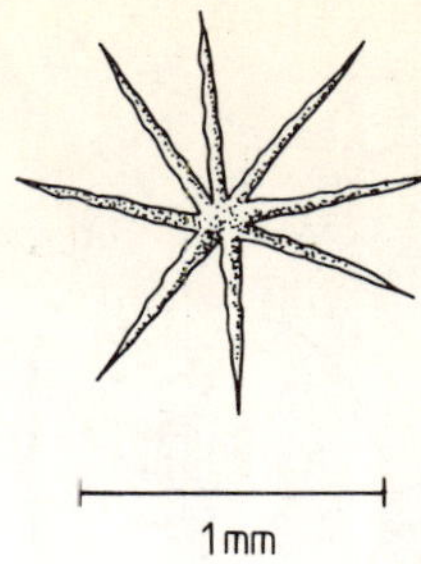

FIGURE 1. Stellate hair from leaf undersurface of *Croton caldensis* Muell. Arg. Rio, Brazil, *Glaziou* 12154.

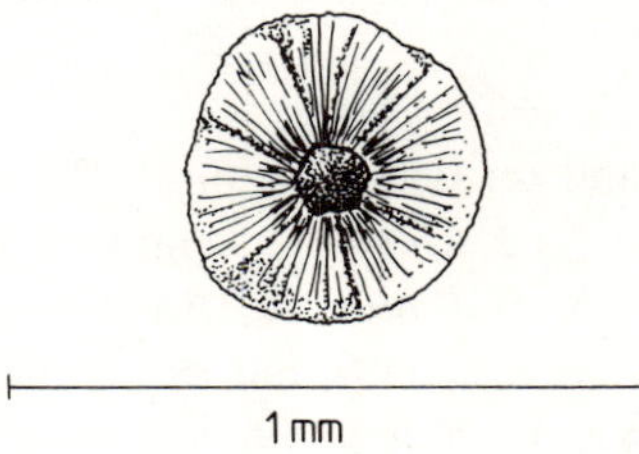

FIGURE 2. Peltate scale from leaf undersurface of *Croton argyrophylloides* Muell. Arg. Bahia, Brazil, *Harley* 15263.

biaceae. Many genera are provided with simple, single-celled trichomes, but in some *Euphorbia* and *Phyllanthus* species, for example, simple multicellular hairs occur. Malpighiaceous hairs (which are bifurcated toward the base) are found in the genera *Ditaxis, Argithamnia, Chiropetalum, Pausandra,* and *Tetrorchidium*. Stalked or sessile stellate hairs (Figure 1) occur in *Croton, Julocroton, Eremocarpus, Chrozophora,* and others. Flattened stellate hairs, with the rays radiating in only one plane and fused together for a part of or almost the whole of their length, constitute peltate scales (Figure 2), and these give rise to an often silvery lepidote indumentum such as may be found in the genera *Hieronyma, Uapaca, Croton, Pseudocroton, Cyrtogonone, Crotonogyne, Leucocroton, Cordemoya, Homonoia, Pera,* and *Ostodes*. Stalked glandular hairs are common in species of *Croton, Jatropha,* etc., but they are completely absent from the large tribe Hippomaneae, which includes such genera as *Hippomane, Sapium, Stillingia, Excoecaria, Sebastiania, Maprounea, Homalanthus, Colliguaya,* etc. Multicellular disciform pellucid glandular hairs occur on the undersurfaces of the leaves especially, in *Macaranga, Hymenocardia, Trewia, Mallotus, Coccoceras,* and some *Acalypha* species. They may be white, yellow, gold, red, or brown in color, and are reminiscent of the leaf-epidermal glands found in the Myricaceae. Stinging hairs like those in the Urticaceae occur in three groups: the tribes Plukenetieae and Dalechampieae and the genus *Cnidoscolus*.

IV. LEAVES

The foliage leaves are alternate in most members of the family, but they may be opposite as in *Trewia* and some *Mallotus* species, or whorled as, for example, in *Mischodon* and *Cocconerion*. In some subgenera of the genus *Euphorbia*, however, alternate, whorled, and

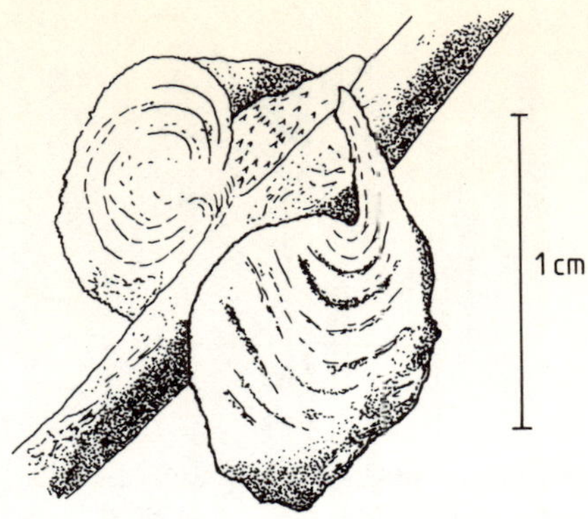

FIGURE 3. Sac-like stipules of *Macaranga saccifera* Pax, housing ant colonies. Yangambi, Zaire, *Léonard* 1676.

opposite leaves may all be found on one individual. Peltate leaves occur in the genera *Jatropha, Macaranga, Manihot,* and *Ricinus.* Compound leaves occur in, for example, *Aristogeitonia, Piranhea, Oldfieldia, Joannesia, Annesijoa, Hevea, Ricinodendron,* and *Leeuwenbergia.* In the Euphorbiaceae, compound leaves may be either ternate or palmate, but never pinnate. Sometimes, as in *Aristogeitonia,* they can be unifoliolate. In some *Jatropha* and *Manihot* species lobed and simple leaves may be found occurring together on the same individual. The cataphylls in *Phyllanthis* have been referred to above.

V. STIPULES

These are very much a characteristic feature of the family, but they are by no means universal; one of the subgenera of the genus *Euphorbia* itself, for example, (i.e., the subgenus *Esula)* is characterized by their absence. Usually the stipules are unremarkable, being rather small subulate or lanceolate structures which are readily deciduous, but in some species of *Antidesma, Drypetes, Givotia, Ricinodendron,* and *Macaranga* they may be large, persistent, and diverse in form. In some *Macaranga* species they may serve to house ant colonies (Figure 3). In many *Jatropha* species the stipules are hair-like, branched, and provided with globose glands at the apices of the segments, or they may consist solely of a mass of sessile glands. In other *Jatropha* species and in some species of *Euphorbia, Phyllanthus,* and *Erythrococca* the stipules are transformed into paired spines, those in one group of succulent *Euphorbias* being situated on specialized horny structures known as spine-shields. Stipels are also found in some genera, for example, *Alchornea* and *Paranecepsia.*

VI. GLANDS

Various different kinds of glands located at a number of points on the plant body are a noteworthy feature of the family Euphorbiaceae. Some of these have already been referred to (the glandular hairs of *Macaranga* and others; the glandular stipules of *Jatropha),* and in addition there are also a number of other types. Thus, in *Ricinus* and some *Jatropha* species there are glands situated on the petioles, while glands of various shapes and sizes occurring at the junction of the petiole and leaf blade are characteristic of a large number of genera, but are particularly noticeable in *Croton,* where they are often yellowish in color and shaped like a saucer, cup, or golf tee, and in *Vernicia,* where they may be large and turbinate. In some *Croton* species, disc-shaped glands may also be found around the leaf margins and situated in the sinuses of the marginal teeth, whereas in a number of other

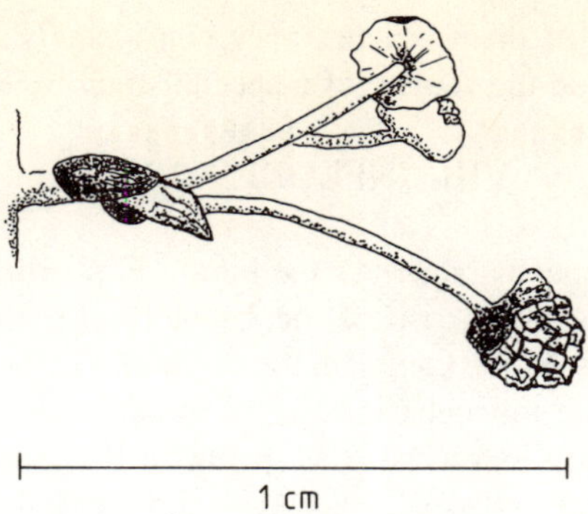

FIGURE 4. Portion of male part of inflorescence of *Mabea taquari* Aubl., showing peduncular gland. Lucie R., Surinam, *Irwin, Prance, Soderstrom & Holmgren* 55365.

genera, particularly in the tribe Acalypheae, the marginal teeth themselves may be found to terminate in small conical glands. Stipitate marginal glands, not only on the leaves but also on the bracts of the inflorescence and often on the calyx lobes as well, are a characteristic feature of the genus *Jatropha.* This may thus be seen to be, of all euphorbiaceous genera, perhaps the most richly endowed with glandular structures of one kind or another.

However, it is perhaps the tribes Hippomaneae and Euphorbieae which possess the best developed inflorescence glands. They are generally found as paired structures at the bases of the inflorescence bracts in the former tribe, as, for example, in the genera *Sapium, Spirostachys,* and *Homalanthus,* although in the genus *Mabea* prominent, often elongate, glandular structures lie along the peduncles of the lateral male branches (Figure 4). These are particularly striking in appearance in herbarium specimens, since they dry a purplish-black, in strong contrast to the tawny or rusty indumentum which is characteristic of many species of this genus. In the tribe Euphorbieae, the inflorescence glands are highly specialized, and are perhaps best dealt with in the section devoted to the inflorescence.

Glands are also found within the floral envelope in a large number of genera, and in the large genera *Phyllanthus* and *Clutia* their number, size, shape, and arrangement vary considerably, giving rise to many possible combinations and permutations which are often diagnostic of species or at least of certain groups of species. As a general rule, the male glands are numerous and free, whereas the females are fused together to form an annular or cupular structure, although there are exceptions to this.

VII. LATEX

This is present in one form or another in many members of the family, but is quite absent from all representatives of the subfamilies Phyllanthoideae and Porantheroideae (in the system of Pax[1]) (which are equivalent to the subfamilies Phyllanthoideae and Oldfieldioideae in the system of Webster[2]).

Milky latex is especially characteristic of the tribes Euphorbieae and Hippomaneae, but its most familiar manifestation for the general reader will, of course, be in the rubber-yielding plants, i.e., species of the genus *Hevea* and some *Manihot* species. The latex of many *Jatropha* species is greenish-yellow and translucent when it flows from a fresh incision, but as it dries and hardens it may become dark reddish-brown and opaque and then resembles dried blood. This fact has earned for many species in Somalia, where the genus is particularly well-represented, the vernacular name ''jilbadig'', meaning ''knee-blood''.

The latex is transported in the plant body in special canals known as laticifers. The starch grains found in the latex of many species vary considerably in shape and size, and this phenomenon has been made the subject of a special study by Mahlberg.[3]

VIII. INFLORESCENCES

No one inflorescence type characterizes the family Euphorbiaceae as it may often do for other families: the umbel in the case of the Umbelliferae, the ebracteate raceme in the Cruciferae, the verticillaster in the Labiatae, the capitulum in the Compositae, the dichasium in the Caryophyllaceae, the monochasium in the Boraginaceae, the spike in the Plantaginaceae, the spikelet in the Gramineae, the spadix in the Araceae, and the ament in the Betulaceae, Corylaceae, Juglandaceae, Fagaceae, and the Salicaceae.

In many Euphorbiaceae, the inflorescence may either be paniculate or spicate, or it may, as is often the case in the subfamily Phyllanthoideae, take the form of a small axillary glomerulus or fascicle; the raceme occurs, for example, in *Croton* and *Manihot,* while the inflorescence of *Jatropha* is basically dichasial. It must be pointed out, however, that the panicles, spikes, and racemes spoken of here are not true ones, for instead of having the flowers singly disposed along the axes, they are for the most part aggregated into tight little cymose clusters or fascicles which are in reality contracted partial inflorescences. These are often irregularly and interruptedly disposed along the axes. Thus, the terms ''pseudopanicles'', ''pseudospikes'', etc. are best used to describe such compound structures. Sometimes the clusters may be bisexual, but more usually they are composed only of male flowers, while the female flowers are solitary to their bracts at the base of a predominantly male inflorescence, as is commonly the case, for example, in the Hippomaneae. Alternatively, they may occur on separate inflorescences on the same plant, as in many *Acalypha* species (the monoecious condition) or on different plants, as in, for example, *Mercurialis* (the dioecious condition). In *Jatropha* and *Cnidoscolus,* however, a solitary female flower terminates each major axis, the male flowers being borne on lateral dichasia, while in *Ricinus* the whole of the upper half of the inflorescence usually consists of female flowers, the males here being relegated to inferior status!

In some groups, the whole inflorescence may mimic a single flower, owing to drastic contraction of the inflorescence axes almost to the vanishing point. Thus, in the genus *Dalechampia* (the sole representative of the tribe Dalechampieae in all the major systems of classification of the family that have been proposed) the closely aggregated male, female, and sterile flowers are subtended by a series of bracts of which two are generally much larger than the rest and may be brightly colored — white, cream, or pink (Figure 5). The genus *Tragia* in the tribe Plukenetieae is generally held to be one of the closest relatives of *Dalechampia* on a number of counts, but its inflorescences are either racemose or pseudoracemose, often with quite elongate axes.

The flowers of *Dalechampia* are provided with perianths, but in the genus *Pera,* the sole representative of the tribe Pereae, (elevated by Klotzsch[4] to the status of a family in its own right), the flowers of some species altogether lack perianths. They are enclosed in small clusters within a pair of adaxially concave bracts which resemble calyx lobes, so that the whole inflorescence resembles nothing so much as a single flower bud.

However, it is in the tribe Euphorbieae, to which the type genus of the family belongs — the genus *Euphorbia* — that the most striking instances of pseudanthia, as inflorescences which resemble single flowers are called, are found. This is because of the extreme simplicity of the flowers, taken in conjunction with the development of nonfloral structural elements to resemble closely the floral whorls in the flowers of other groups. The overall effect is often similar to that produced by the pseudanthial capitulum of the Compositae, except that there floral elements also play their part to produce the pseudanthium.

In *Euphorbia,* a whorl of five bracts has fused together to form an infundibuliform,

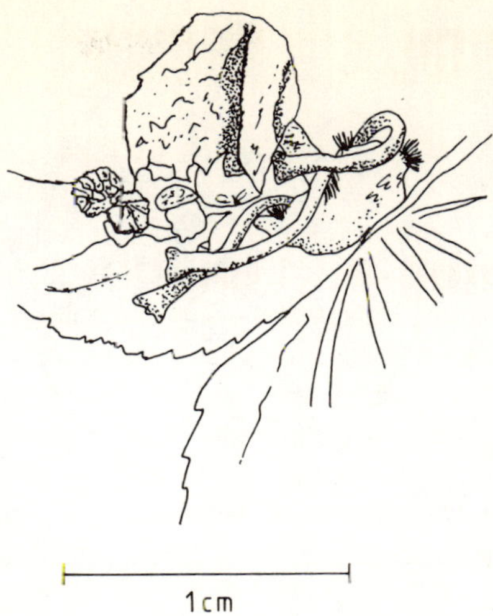

FIGURE 5. Portion of inflorescence of *Dalechampia capensis* Spreng. f., showing only the bases of the two large subtending bracts. Kalambo Falls, Tanzania, *Richards* 19666.

turbinate, campanulate, or urceolate involucre or ''pseudocalyx'', the free portions of which resemble the lobes of a gamosepalous calyx. Alternating with these are the components of a whorl of structures homologous with fused stipules variously termed glands or nectaries, usually four or five in number, although in the subgenus Poinsettia they are reduced to one or absent altogether. They are sometimes either brightly colored or are provided with white or colored appendages, and thus they often have the appearance of petals. The appendages may also have the shape and texture as well as the color of petals, thus heightening the effect. Within the involucre, there are five series of small, usually bracteate male flowers pleiochasially disposed around a central solitary female flower, an arrangement resembling a highly contracted *Jatropha*-like inflorescence. The whole structure is known as a cyathium, and it constitutes an inflorescence type unique in the plant kingdom, and confined to the tribe Euphorbieae (Figure 6).

IX. FLOWERS

The preeminently unifying feature of the family Euphorbiaceae is the unisexual flower, as may have been gathered from a reading of the previous section. The unisexual flower is common to all genera almost without exception. Nevertheless, in some genera nonfunctional stamens known as staminodes may be present in the female flowers, with nonfunctional pistils known as pistillodes in the male flowers. Very occasionally the staminodes may be almost indistinguishable morphologically from true stamens, albeit not functionally, but the pistillodes are always clearly distinguishable from true pistils. For the most part, however, there are not even vestiges of the other sex present at all. Unisexual flowers are not, however, confined to the Euphorbiaceae: they are also found, for example, in the Cucurbitaceae, the Urticaceae, the Moraceae, the Fagaceae, and the Salicaceae.

Reduction of floral parts is a hallmark of the Euphorbiaceae, for the unisexual flowers generally exhibit further stages and degrees of reduction beyond the suppression or elimination of the other sex. A considerable number of genera have apetalous flowers, for example, although in other genera, e.g., *Jatropha* and *Vernicia,* petals are well developed. However, there are some genera in which not only are there no petals, but in which the calyx may

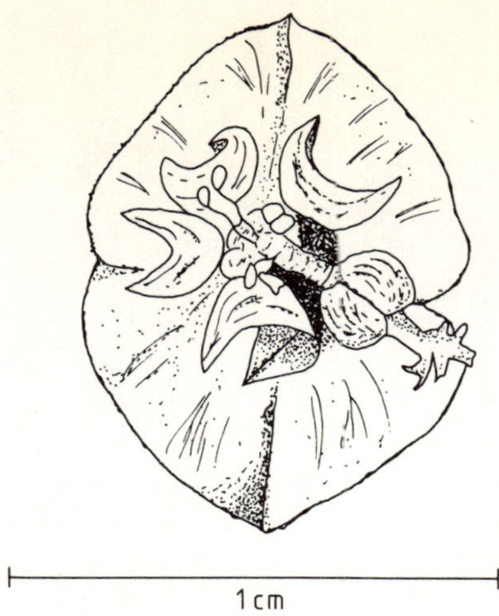

FIGURE 6. Cyathium of *Euphorbia amygdaloides* L. cv. "*purpurea*". Kew.

also be greatly reduced or even absent. Thus, in *Euphorbia, Monadenium,* and *Pedilanthus* the calyx in the female flower is represented merely by a slight expansion of the top of the pedicel at the base of the ovary, and there is no vestige of a calyx at all in the male flowers in these genera. Indeed, the male flowers of *Euphorbia* and its close allies are the simplest of the whole family and among the simplest in the plant kingdom as a whole, for each consists solely of a single, naked stamen borne directly upon the pedicel, with an articulation marking where the pedicel ends and the filament begins. The difference in texture between pedicel and filament becomes more obvious after anthesis, for whereas the stamen shrivels and falls, the pedicel persists until the cyathium breaks up long after the fruiting stage. In *Anthostema,* a more distant relative of *Euphorbia,* a small, toothed, cupuliform calyx surrounds the base of each stamen.

In many ways, the genus *Sapium* and related genera of the tribe Hippomaneae constitute a halfway house between the tribe Euphorbieae with its greatly reduced flowers, and the rest of the family; for while the flowers are reduced with respect to the main mass of the family, they are by no means as far along the road to reduction as are the members of the Euphorbieae. Thus, the male flowers in the Hippomaneae usually consist only of an ill-defined calyx with but a few stamens and no disc or pistillode. Another link between the two tribes lies in the nature of their latex, as was noted above.

The genus *Hura* is rather anomalous among the Hippomaneae, and some authorities have indeed regarded it as constituting a distinct tribe, along with a few satellite genera. In it the inflorescence bracts are completely fused over the male flowers with the axis, and become ruptured by the flowers as they develop (Figure 7). *Hura* is also noteworthy because of its large, umbrella-shaped style and up to 20-locular ovary and fruit.

The androecium in the Euphorbiaceae takes various forms. As referred to above, it may be represented by only a single stamen per flower, but on the other hand there may be anything up to 1000, as in *Ricinus,* the castor oil plant. *Ricinus* is, however, unusual in that the filaments are haphazardly united, thus giving the impression of branching stamens. Three other genera share this feature with *Ricinus,* namely, *Homonoia, Lasiococca* and *Spathiostemon,* but they do not otherwise appear to be particularly closely related to it, for *Ricinus* occupies a rather isolated position in the classification of the family. In a number of genera, e.g., *Phyllanthus, Clutia* and *Ricinocarpus,* the filaments may be fused together to form a column, and in some species of *Phyllanthus* and in the genus *Omphalea* the anthers are

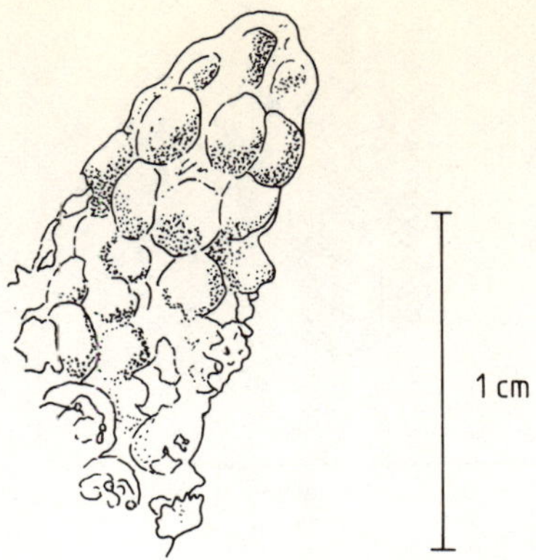

FIGURE 7. Apical portion of male part of inflorescence of *Hura crepitans* L., showing male flowers bursting through the completely fused-over bracts. Para, Brazil, *Prance, Pennington & Silva* 1658.

laterally expanded and contiguous, forming a complete or almost complete ring at the top of the column. In *Fragariopsis,* the anthers are sessile on a swollen, rounded receptacle, so that the androecium presents a strawberry-like appearance; hence the generic name.

The form of the anthers themselves is also quite variable. Thus, in *Claoxylon, Erythrococca, Micrococca,* and their allies, the thecae are separate and erect and look rather like rabbit ears. In *Acalypha* the thecae are also separate, but spreading and wormlike, arising from the apices of the broadly triangular filaments. On this and other counts, *Acalypha* occupies a rather isolated position within the family: although the plants bear some resemblance to certain members of the family Urticaceae, e.g., *Urtica* itself, *Droguetia,* and *Australina* (and the Greek word ακαληφη means "nettle"), there is no real evidence of relationship because the fruit of *Acalypha,* being three chambered and septicidally dehiscent, is typically euphorbiaceous. In *Croton* and a few satellite genera, the stamens are deflexed in the bud at the apex of the filament. In *Monotaxis,* the anther thecae are widely separated from each other on the long arms of the connective at the apex of the filament. In *Poranthera,* the anthers open by pores (hence the generic name) as in the Ericaceae, but longitudinal dehiscence is the rule in the Euphorbiaceae. The anthers in most members of the family are bilocular, but in the genera *Bernardia, Cleidion, Macaranga, Endospermum,* and *Tetrorchidium,* for example, they are quadrilocular.

The stamens are often associated with the floral glands (mentioned above). The staminal column may be surrounded by them, as in *Phyllanthus* and *Clutia* (Figure 8); the free filaments may be intermingled by them, as in *Erythrococca;* the free filaments may arise through the perforations in a completely reticulate glandular receptacle, as in *Argomuellera* and *Pycnocoma;* or they may surround a central discoid gland mass, as in *Drypetes* (this gland mass thus occupies the same position as, although it is not homologous with, a pistillode).

The pollen grains of the Euphorbiaceae fall into a number of basic types. Punt[5] recognizes 76 such types, which may be coterminous with generic limits or not. However, in *Phyllanthus* and *Tragia,* for example, a number of different types occur within the one genus, but it is more often the case that different genera may possess the same pollen type. Punt's[5] "*Croton*-type" is found not only in all species of the genus *Croton* but also in *Julocroton, Crotonopsis,* and *Eremocarpus* of the tribe Crotoneae, and even in some other genera of the subfamily Crotonoideae sensu stricto (so Webster,[2] rather than lato, so Pax[1]) also. Grains of this type are more or less spheroid in shape, have no apertures, and have the tectum usually adorned with clavate or echinate structures.

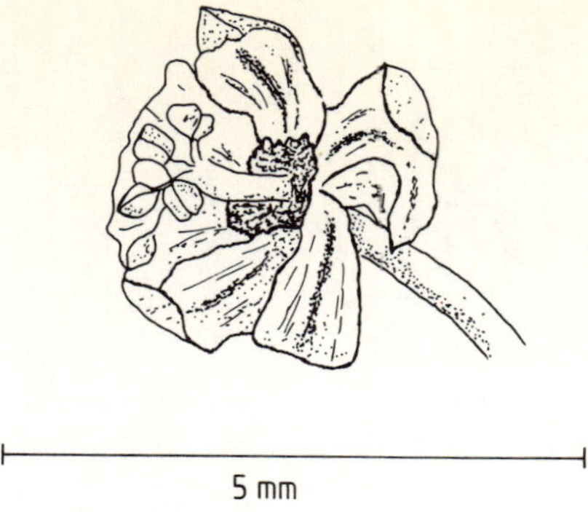

FIGURE 8. Male flower of *Clutia richardiana* Muell. Arg. North Yemen, *Radcliffe-Smith & Henchie* 4689.

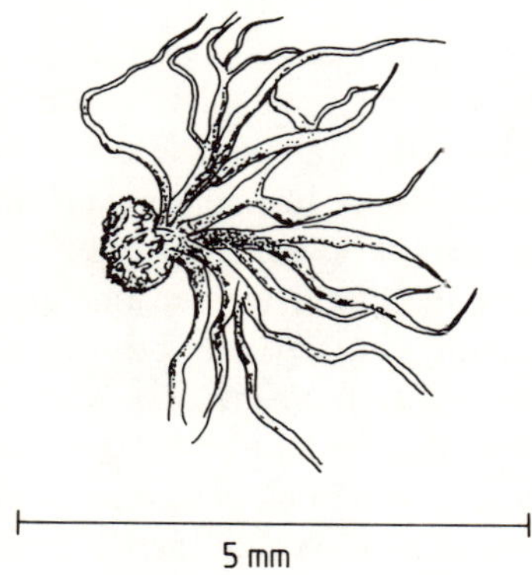

FIGURE 9. Ovary and styles of *Acalypha hispida* Burm. f. Cultivated in Mozambique, *Balsinhas* 1906.

Also, Punt's "*Manihot*-type" of grain (similar to the *Croton*-type but differing in having large pores in the wall of the grain) is not confined to the genus *Manihot,* but is also found in *Cnidoscolus* and *Suregada,* and so one could go on multiplying examples. In general, pollen studies have tended to reinforce traditional concepts of groupings of genera, but there are exceptions, notably the group of Australian genera long referred to as the "Stenolobeae" *(vide infra).*

In the gynoecium, the style presents a wide variety of forms within the family. In a large number of genera, however (e.g., *Euphorbia, Phyllanthus),* there are three short styles, connate or not at the base, which are usually either bilobate, bifid, or bipartite; this may be regarded as the most typical style form in the Euphorbiaceae. However, in the genus *Drypetes,* there is often no style to speak of, for the stigmas may be sessile on the top of the ovary. In *Dalechampia,* on the other hand, the long styles are completely fused into a column which gives no indication of its tripartite nature. The styles of *Acalypha* are noteworthy in that they are deeply laciniate or fimbriate, with slender, filiform segments; they are commonly bright red in color (Figure 9). *Erythrococca* can have fimbriate styles as well, but much smaller than those of *Acalypha* and usually whitish. In a few genera such as *Ricinus* and *Trewia,* the styles are elongate and plumose and may be dark red in color as in these genera, while in *Manihot* they may be botryoidal, i.e., they resemble a small, tight bunch of grapes. In some groups the styles may be undivided, as in the genus *Alchornea* and in most members

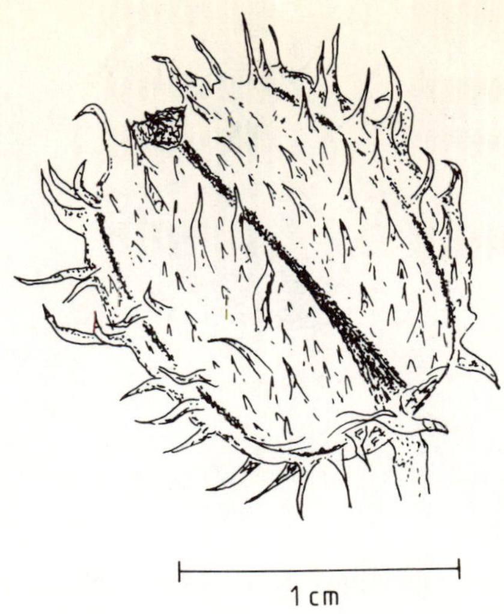

FIGURE 10. Regma of *Ricinus communis* L. Upolu, Samoa, *Cox* 208.

of the tribe Hippomaneae; in some *Sapium* species these undivided styles may be rolled abaxially into tight spirals. The curious umbrella-like style of *Hura* has already been noted. In the tribe Plukenetieae (tribe Acalypheae, subtribe Plukenetiinae, according to Pax[1]) some unusual style forms occur, as in *Astrococcus* and *Sphaerostylis* where they resemble the ovaries of many genera, being subglobose and trilobate or hexalobate and having small apical excrescences which themselves may mimic normal styles; these structures are, however, hollow and open at the top. In *Croton,* the styles may not only be bifid, but in some species they are quadrifid or even multifid.

X. FRUITS

One of the most characteristic features of the family Euphorbiaceae, although not by any means of universal occurrence in the family, is the type of dehiscent fruit known as the regma. It is indeed exemplified by a euphorbiaceous genus, for it was anciently applied as a term to the fruit of *Ricinus* (Figure 10) by the Greeks. The regma is generally trilocular, and is a schizocarpic fruit in which three kinds of dehiscence virtually take place simultaneously: (1) septifragal, or separation of the three mericarps or cocci from the main axis, usually leaving a persistent three-winged or -ridged columella; (2) septicidal, or separation of the three mericarps from each other; and (3) loculicidal, or dehiscence of the mericarps themselves along a median suture line, thus releasing the seeds. The latter phase of the dehiscence may be complete, as in *Phyllanthus,* resulting in two separate valves per mericarp, or it may only be partial, as in most other genera, leaving the pair of valves still joined together at one end.

Each mericarp contains only one or two seeds, for another feature of the family (and one to which there are no exceptions) is that there are never more than two ovules per ovary loculus. The phyllanthoid, porantheroid, and oldfieldioid (bowing for the moment to more than one system) genera (e.g., *Phyllanthus, Andrachne, Poranthera, Oldfieldia)* have two ovules per loculus, whereas the acalyphoid, crotonoid, euphorbioid, and ricinocarpoid genera (e.g., *Acalypha, Ricinus, Jatropha, Croton, Ricinocarpus, Sapium, Euphorbia)* have only one.

However, as was indicated above, not all Euphorbiaceae possess a regma. Some *Croton*

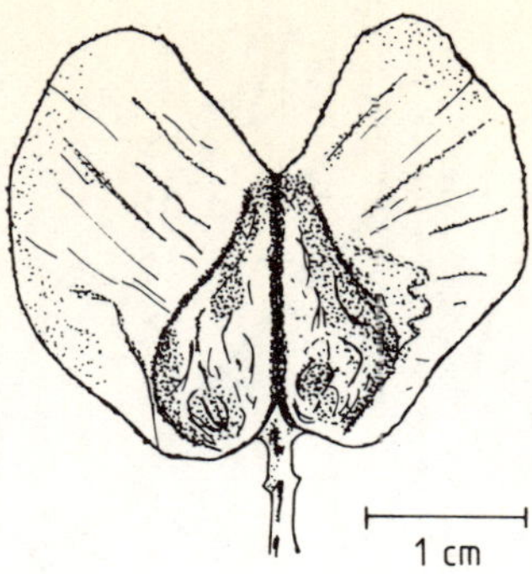

FIGURE 11. Winged fruit of *Hymenocardia acida* Tul. Bururi, Burundi, *Reekmans* 10261.

and *Jatropha* species, for example, have fruits which dehisce only loculicidally and not septicidally, and there are a number of genera in which the fruit is either tardily dehiscent, as in *Pseudolachnostylis;* irregularly dehiscent, as in *Margaritaria;* or completely indehiscent and drupaceous, as in *Drypetes, Uapaca, Antidesma,* and *Ricinodendron.*

A very unusual fruit type is found in the genus *Hymenocardia;* the fruit is bilocular, flattened and two-winged, and it dehisces septifragally into two indehiscent winged mericarps (Figure 11). Before dehiscence they bear some resemblance to the fruits of certain Ulmaceae, and the genus also has other features reminiscent of that family, such as the anther and pollen types. However, the fruits of the Ulmaceae are indehiscent, and furthermore are asymmetrical, whereas those of *Hymenocardia* are bilaterally symmetrical.

The plurilocular fruit of *Hura* has been referred to above. Other genera include species in which the fruits may be plurilocular while the majority of the species are typically trilocular, such as *Phyllanthus* and *Glochidion.*

The fruits of the majority of the Euphorbiaceae are smooth, but a number of genera exhibit various forms of ornamentation. Some species of *Euphorbia* may have the fruits covered or partly covered with hemispherical warts or conical, cylindric, or filiform processes, or the cocci of the fruit may be winged or ridged. The type of ornamentation is generally species diagnostic; however, the majority of *Euphorbia* species have smooth fruits. Warty or echinate fruits also occur in *Mallotus* and are characteristic of *Chaetocarpus.* In *Ricinus,* the narrowly cylindric processes on the fruit are each tipped with a stiff bristle. Six-winged fruits occur in the genus *Manihot.* Horn-like processes, six or more per fruit, are found in the genera *Pycnocoma, Sapium,* and *Sebastiania.*

The genus *Stillingia* is unusual in that while the upper part of the fruit is dehiscent, the lower part is not, and after the dehiscence of the upper part, remains attached to the plant as a triradiate star-like structure.

XI. SEEDS

The seed of *Ricinus* may be taken as typical of a large number of genera in which there is only one ovule per loculus and in which the fruit is dehiscent, although this seed type is by no means universal in that group. It is dorsiventrally flattened, has a ventral raphe, an excrescence at the micropylar end alternatively termed the caruncle or obturator (which can function both as an aid to fertilization and to seed dispersal), and the testa usually has a marbled or mottled pattern. The large genera *Euphorbia, Croton,* and *Jatropha* (Figure 12) mostly have seeds which conform to this type, except that in a number of *Euphorbia* seeds the flattening may not be much in evidence, the caruncle may be absent, and the testa may be concolorous. The seeds of certain *Euphorbia* species, particularly the annuals, may differ

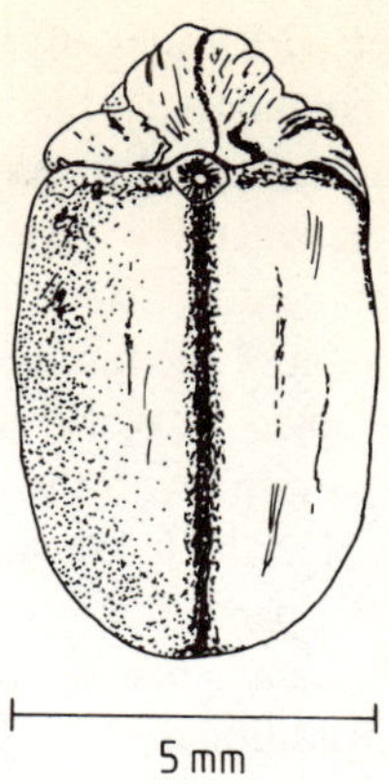

FIGURE 12. Seed of *Jatropha batawe* Pax, showing caruncle. Tana R., Kenya, *Polhill & Paulo* 586.

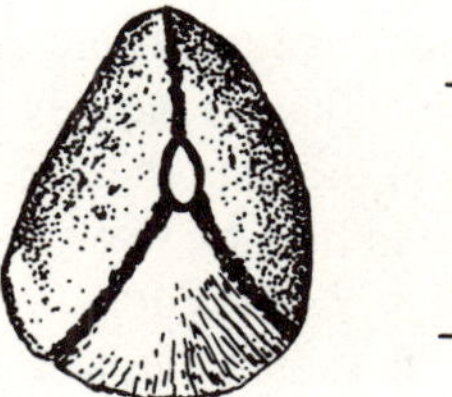

FIGURE 13. Ecarunculate segmentiform seed of *Phyllanthus maderaspatensis* L. Lamu, Kenya, *Rawlins* 363.

from each other considerably with regard to the nature of the testa, however; in some this may be verrucose, rugulose, reticulate, or otherwise ornamented; in others it may be pitted or longitudinally or transversely grooved or otherwise sculptured; and in yet others it may be perfectly smooth. Furthermore, the chalazal end of the *Euphorbia* seed is often truncate, and the micropylar end is often obliquely truncate. The structure of the seed coat as revealed by scanning electron microscopy (SEM) has been made the subject of a special study by Ehler.[6]

The seeds of *Tragia* and *Dalechampia* are almost perfectly spherical, and are often seen to be beautifully patterned when examined with a hand lens. In *Acalypha,* a whitish wing extends partially along the seed in the position occupied by the raphe in other genera. In many *Sapium* species a fleshy or waxy outer seed coat may surround a hard inner one. A thin fleshy outer layer which becomes papery on drying is found, for example, in the seeds of *Chrozophora* and *Caperonia* species. The seeds of *Clutia* species are jet-black and very shiny.

In many genera of the tribe Phyllantheae the seeds, usually six in number per fruit, are characteristically shaped like the segments of an orange, with flat radial surfaces and a convex dorsal surface (Figure 13). The sculpturing and ornamentation of these surfaces may be diagnostic either at or above the species level. All the seeds of this type are without a caruncle.

XII. SYSTEMS OF CLASSIFICATION

These are many and varied, which is hardly surprising in view of the complex distribution of attributes of which the antecedent paragraphs give but a bird's-eye view. In fact, one might go so far as to say that every botanist who has made a particular study of the family has come up with his own particular system. In attempting to resolve the many and fundamental differences which existed between the systems propounded by the two foremost workers on the family in the mid-19th century, Baillon[7] and Mueller of Aargau,[8] Bentham[9] decided to adopt neither for his treatment of the family in his own and the younger Hooker's *Genera Plantarum*. Instead, he attempted to work out a system of his own. Since then, others who have attempted to come to grips with the complexities of character distribution and relationships within the family, Pax,[1] Hurusawa,[10] Airy Shaw,[11] Hutchinson,[12] and Webster,[2] have all developed their own particular systems, parts of which can be correlated with what has gone before and with each other up to a point, but in each of which there are significant new departures from traditional treatments.

The system of Bentham[9] is the one by which the rich herbarium collections at Kew and the British Museum (Natural History) are arranged, and so is the one perhaps most familiar to those who have worked at those institutions. It begins with the tribe Euphorbieae and ends with the Hippomaneae, and herein lies its chief weakness: although these tribes differ considerably from each other in inflorescence structure, in many other features they have much in common, and so in more modern treatments are generally juxtaposed.

It is, however, the system of Pax[1] which has perhaps exercised the most influence in the present century, coming as it does within the important Englerian system of plant classification. In it, Pax[1] recognizes four subfamilies of very unequal size, the Phyllanthoideae with 65 genera, the Crotonoideae with 209, the Porantheroideae with 4, and the Ricinocarpoideae with 5. The two latter small subfamilies are exclusively Australasian, being confined to Australia for the most part, with a few also in Tasmania, New Zealand, and New Caledonia. These differ from all the rest of the family in having narrow cotyledons, no broader than the width of the radicle, whereas the cotyledons of all the other genera are broader than the radicle, often much more so. Mueller of Aargau[8] had this distinction as a primary division, applying the names Stenolobeae and Platylobeae to the two groups, to which he accorded the obsolete rank "tribe-series" (more or less equivalent to subfamily). Bentham[9] accorded tribal status to the Stenolobeae, however, equivalent to his five other tribes Euphorbieae, Buxeae (now Buxaceae), Phyllantheae, Galearieae (now Pandaceae in part), and Crotoneae. Pax[1] still used Mueller's names, but not formally, since there is no hierarchical category between the family and the subfamily in which he could employ them.

Pax's subfamily Phyllanthoideae included Mueller's tribes Phyllantheae and Bridelieae, in both of which there are two ovules per ovary loculus; they differ from each other in that whereas in the Phyllantheae the male calyx lobes are imbricate, in the Bridelieae they are valvate. Bentham did not accord to the Bridelieae any formal recognition. Pax's subfamily Porantheroideae is likewise characterized by having two ovules per loculus, and in fact he regarded it as a sort of stenolobe equivalent of the Phyllanthoideae. In Hutchinson's[12] system, in which the family is split into 40 tribes all having equivalent status (i.e., no category above the tribal rank is recognized), there are three stenolobe tribes and these are dispersed among the platylobe tribes; thus, the tribe Caletieae is his 2nd tribe, the Poranthereae his 9th, and the Ricinocarpeae his 15th. It is thus curious that Webster[2] describes the dismemberment of the Stenolobeae as a radical innovation of his system, when in fact Pax had already hinted at it and Hutchinson had effected it. The chief innovation of Webster's system is his recognition of five subfamilies, but of these only one is due to him (with Köhler[13]), namely, the subfamily Oldfieldioideae, carved out of the Phyllanthoideae chiefly on pollen characters, and with the addition of the tribe Caletieae of the subfamily Porantheroideae. He has four stenolobe tribes in all, as did Pax, instead of Hutchinson's three; Hutchinson sank Mueller's tribe Ampereae, kept up by Pax and resurrected by Webster, into the tribe Ricinocarpeae. Admittedly the distinctions are rather vague: in the Ampereae, the male sepals are "mostly valvate" and the stamens "free or almost free", while in the Ricinocarpeae the male sepals are imbricate and the stamens "mostly connate".

Pax's subfamily Crotonoideae included all the rest of the platylobe tribes, five in all in Mueller's system (Crotoneae, Acalypheae, Hippomaneae, Dalechampieae, and Euphorbieae), in which there is only one ovule per ovary loculus. Pax, however, recognizes 12 crotonoid tribes: the Chrozophoreae, Joannesieae, Pachystromateae, Pereae, Clutieae, Manihoteae, and Gelonieae in addition to Mueller's. In Bentham's system, the tribe Crotoneae had been regarded as being composed of eight subtribes, not all of these being equivalent to Pax's tribes. Thus, for example, Bentham had the Jatropheae, Adrianeae, and Plukenetieae equivalent in rank to the "Eucrotoneae", Chrozophoreae, Acalypheae, Gelonieae, and Hippomaneae (he did not use the ending "-inae" now obligatory as a subtribal ending, but the ending "-eae" for both tribes and subtribes), whereas Pax had the Jatrophinae as a subtribe of his tribe Clutieae, the Plukenetiinae as a subtribe of the tribe Acalypheae, and he dismembered Bentham's Adrianeae by making *Manihot* the sole representative of his tribe Manihoteae, *Pachystroma* the sole representative of his tribe Pachystromateae, and by incorporating *Adriana, Cephalocroton* and *Adenochlaena* into his subtribe Mercurialinae of the tribe Acalypheae. Hutchinson, however, had 28 uniovulate tribes, all but one of them platylobe. Some of these are equivalent to Pax's tribes, or almost so, e.g., the Gelonieae, Manihoteae, Pachystromateae, Hippomaneae (except for *Hura* and *Algernonia,* in Pax's subtribe Hurinae, to which he accords tribal status), Dalechampieae, Pereae, Joannesieae, and Euphorbieae. All the rest, however, are either equivalent to Pax's subtribes or else more or less equivalent to the groupings of genera in the Mercurialinae which Pax termed "series", a category name inadmissible at this hierarchical level under the provisions of the modern nomenclatural code, as its use is now restricted to an infrageneric category. Webster has 35 uniovulate tribes, of which all but two are platylobe. The additional ones have, for the most part, resulted from a further splitting of the tribes recognized by Hutchinson, but the picture is not quite as simple as that, since a considerable amount of reshuffling has been undertaken by Webster. Thus, for example, Hutchinson's tribe Neoboutoneae, which had resulted from a merger of Pax's generum-series Adeliiformes and Neoboutoniiformes of his subtribe Mercurialinae, becomes in Webster's system part of the tribe Adelieae plus part of the subtribe Neoboutoninae of the tribe Aleuritideae. While following Hutchinson in part, he reverts to Pax in part, but only in part, for Pax had had *Aleurites* in quite a different tribe from Neoboutonia.

As an additional complication to the picture as outlined above, not all the authors have stuck wholly to one unvarying system. Pax, for example, made a number of modifications to the system to which he had adhered for the parts of Engler's[14] *Pflanzenreich* by the time the treatment of the family for the second edition of the *Pflanzenfamilien*, which is the version which has been considered here, was ready to go to press.

It can thus be assumed with a fair degree of certainty that the last word upon the classification of the Euphorbiaceae has not yet been said or heard, nor will be for some time to come. It is known to the author that Airy Shaw wished to respond to Webster's proposals with a system of his own fully worked out, but he was not spared to put this into effect. His outline groupings of genera were not formally published, and were largely concerned with the family as represented in southeast Asia. He did, however, make changes at the family level, and it is to these that we must now animadvert as we come to consider relationships outside the family, or rather, relationships of the family Euphorbiaceae to other dicotyledonous families.

In this discussion, the system of Hurusawa[10] has not been considered, since it is presented with a strong regional bias, namely, centered around the family as represented in Japan, Formosa (Taiwan), and the immediately adjacent regions of the Asiatic mainland.

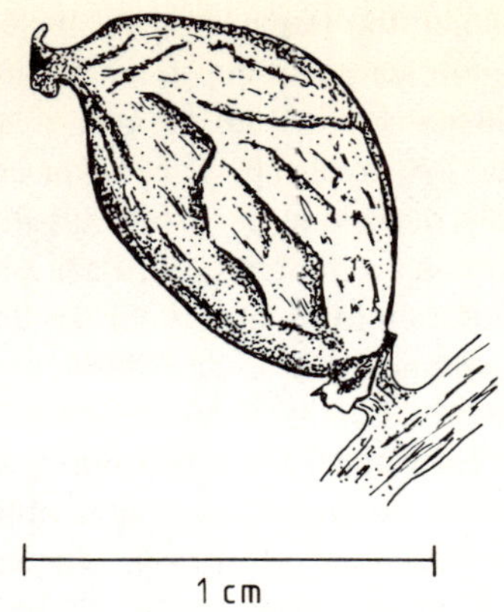

FIGURE 14. Indehiscent fruit of *Antidesma montis-silam* Airy Shaw. Lahad Datu, Sabah, *Muin Chai* in SAN 25061.

XIII. RELATIONSHIPS OF THE FAMILY

Some genera in the Euphorbiaceae show characteristics which are shared by certain members of other plant families, and their inclusion in the Euphorbiaceae has been disputed at various times.

One of these is the paleotropical genus *Antidesma,* which as early as 1825 was separated by Agardh[15] as constituting a monogeneric family, the Stilaginaceae (after *Stilago,* an alternative generic name for *Antidesma),* on account of the unusual fruit, a generally unilocular drupe with an indurated foveolate endocarp (Figure 14). Ignored by Mueller, Bentham, Pax, Hutchinson, Webster, and others, Agardh's treatment received recognition by Airy Shaw[16] in the 7th and 8th editions of Willis' *Dictionary of Flowering Plants and Ferns,* where it is upheld as a distinct family; Airy Shaw considered Antidesma to be intermediate between the Euphorbiaceae and the Icacinaceae (order Celastrales), some genera of which have similar fruits.

Airy Shaw also upheld the inclusion by Pierre[17] of *Microdesmis* and *Galearia* in the family Pandaceae with *Panda,* as does Forman;[18] besides sharing a drupaceous fruit with a massive rugose, muricate, or otherwise ornamented endocarp, these three genera also have in common lateral shoots which bear buds in their ''axils'', thus resembling pinnate leaves. Unfortunately, the genus *Centroplacus* poses a problem here, because while it has many features in common with *Microdesmis*, its fruit is dehiscent and typically euphorbiaceous. Hutchinson and Webster also uphold the inclusion of *Microdesmis* and *Galearia* in the Pandaceae, but they avoid the problem of *Centroplacus* simply by not mentioning it.

Airy Shaw also upheld the separation of the genus *Pera* by Klotzch[4] into a distinct family, the Peraceae. This Neotropical genus, with its flowers enclosed in a pouch of variously connate concave bracts, has dehiscent fruits, but the valves do not have a typically euphorbiaceous appearance. Pax, Hutchinson, and Webster all accord it tribal status only, however.

A number of other genera, hitherto included in the Euphorbiaceae by all authors, were regarded by Airy Shaw as being properly beyond the pale of the family. One of these is the Afromalagasy genus *Androstachys*, with opposite leaves, large connate intrapetiolar stipules, and the stamens borne on an elongate axis, for which he proposed the Androstachydaceae. Another is the Indopacific genus *Bischofia,* with trifoliolate or quinquefoliolate pinnate leaves, flowers in large thyrses, and fruits small globose drupes, which Mueller had as a subtribe of his tribe Phyllantheae; Airy Shaw proposed for this the family Bischofiaceae, and considered it to be closer to the Staphyleaceae (order Sapindales) than to the Euphorbiaceae. A third is the Afroasian genus *Hymenocardia* with its flattened, winged, dicoccous schizocarps, which seems to link the Euphorbiaceae with the Ulmaceae (order Urticales); for this genus he proposed the family Hymenocardiaceae. Fourth, there is the Afromalagasy genus *Uapaca,* with the small, capitate male flowers and the larger solitary female flower, each surrounded by an involucre of 5 to 10 large, showy petaloid bracts, and with pachycaul stems, which Mueller also had as a subtribe of his tribe Phyllantheae. Airy Shaw proposed for this the family Uapacaceae, which he considered to have some connection with the Anacardiaceae, Picrodendraceae, and Pistaciaceae (order Sapindales). Also linking the Euphorbiaceae with the Picrodendraceae in Airy Shaw's opinion, are six genera which share, among other features, compound and/or unifoliolate leaves, namely, the disjunct African genus *Aristogeitonia,* the Mexican *Celaenodendron,* the Brazilian *genus novum affinis Celaenodendro* (unable as yet to be described owing to the incompleteness of the material), the Indo-Sri Lankan *Mischodon,* the African *Oldfieldia* (including *Paivaeusa* and *Cecchia*), and the neotropical *Piranhea.* Webster,[2] however, includes the Picrodendraceae in the Euphorbiaceae in his and Köhler's[13] subfamily Oldfieldioideae, with these six genera and a number of others.

Hyaenanche (Toxicodendron) from South Africa, which Webster also includes in the subfamily Oldfieldioideae, with its leathery leaves, short filaments, and stout styles, provides a link with the Buxaceae. Bentham[9] had this as a tribe of the Euphorbiaceae, which itself links the Euphorbiaceae with the order Celastrales (e.g., the Celastraceae) — by another route, so to speak, from that via *Antidesma* — and also with the Geraniales and the Hamamelidales.

Bridelia, with its valvate sepals, small petals, and large disc provides a link with the order Rhamnales (e.g., the Rhamnaceae). *Drypetes* links the Euphorbiaceae with the Flacourtiaceae and through them to other members of the order Parietales (or Violales). There are also many connections with the Sterculiaceae and through them to other members of the order Malvales (e.g., Malvaceae, Bombacaceae, Tiliaceae, and Elaeocarpaceae). *Jatropha, Cnidoscolus,* and *Manihot* may also be thought of as providing a link with the order Parietales (or Violales), since they have many points in common with the Caricaceae.

Two genera which were included in the Euphorbiaceae up to the time of Bentham (although Mueller[8] previously excluded them) but which are now regarded by common consent as constituting families in their own right, are *Aextoxicon* (Aextoxicaceae) from Chile, a densely lepidote tree with simple racemes and a ruminate endosperm; and *Daphniphyllum* (Daphniphyllaceae) from eastern Asia and Malesia, glabrous exstipulate trees with imperfectly bilocular ovaries and a minute apical embryo.

XIV. GEOGRAPHICAL DISTRIBUTION OF THE FAMILY

The family Euphorbiaceae is a predominantly tropical family. There are very few exclusively extratropical genera. Examples are *Crotonopsis, Eremocarpus* (North America), *Mercurialis* (temperate and warm temperate Eurasia), *Seidelia* and *Leidesia* (South Africa), and *Dysopsis* (temperate and Andean South America).

Only one genus is cosmopolitan, and that is the type genus, *Euphorbia.* It must, however, be pointed out that not all botanists agree that it constitutes a single genus; some would abstract from it one or more of its subgenera (e.g., the subgenera *Chamaesyce,* with all leaves opposite, interpetiolar stipules, C4 metabolism, glands usually appendiculate; and *Poinsettia,* with tightly aggregated terminal cyathial clusters, upper leaves at least partly colored, cyathia usually uniglandular) and make them into genera of more restricted distribution.

Of the tropical genera, relatively few are pantropical. These include *Acalypha, Alchornea, Croton, Dalechampia, Drypetes, Jatropha, Phyllanthus, Sapium,* and *Tragia,* and after *Euphorbia,* include the largest genera in the family.

A considerable number of genera are restricted to only one of the three tropical continents. In this regard, the Neotropical realm is the richest: Pax[1] lists 75 genera as being confined to tropical America, of which the largest and best known is *Manihot,* recently revised by Rogers and Appan[19] for the *Flora Neotropica.* Others in this category include *Hevea, Hippomane, Hura, Mabea, Pedilanthus, Pera,* and *Plukenetia.*

Tropical Asia is the next richest in terms of the number of genera restricted to it. Pax lists 60 exclusively Asiatic genera, including *Aleurites, Aporosa, Baccaurea, Blumeodendron, Homalanthus, Homonoia,* and *Sauropus.*

Although tropical Africa comes third in this respect, nevertheless Pax lists 50 genera as being confined to this region, which also includes the southwest corner of Arabia, once part of the Ethiopian upland. Examples are *Clutia, Erythrococca, Heywoodia, Monadenium, Oldfieldia, Ricinodendron,* and *Zimmermannia.*

However, it is the discontinuous genera which most interest the phytogeographer, because they provide the grist for the mill of much stimulating philosophical discussion which seeks to come up with an explanation of the developments which led to such discontinuities. Land bridges, long-distance dispersal by ocean currents or birds, age and area, continental drift, and plate tectonics have all in turn been employed in these discussions, either one to the exclusion of all the rest or else two or more in combination. However, as will be evident from what follows, the pattern of these discontinuities is too complex for any simple explanation, but possibly the last two scenarios afford the most satisfactory framework upon which to hang the greater part of the known facts.

A few genera occur in both tropical America and Africa. Examples of this discontinuity are *Amanoa, Maprounea,* and *Tetrorchidium.*

The island of Madagascar (Malagasy Republic) is proverbial for the high proportion of endemic taxa in both the plant and animal kingdoms that exist there. In the Euphorbiaceae, there are a small number of endemic genera, e.g., *Cometia, Leptonema,* and *Deuteromallotus.* However, the following are found both in Madagascar and on the African mainland: *Androstachys, Argomuellera, Neopalissya, Pycnocoma, Thecacoris,* and *Uapaca.* Common to the Americas, Africa, and Madagascar are *Caperonia* and *Savia.* Going in the other direction, *Cephalocroton, Cleistanthus, Excoecaria, Givotia, Macaranga, Micrococca,* and *Suregada* are common to Africa, Madagascar, and Asia. *Claoxylon, Sphaerostylis,* and *Synostemon,* however, leave out Africa; they occur in Madagascar, Asia, and Australia. *Glochidion* also leaves out Africa. It otherwise girdles the globe and brings us back to square one, for it occurs in Madagascar, Asia, Australia, Polynesia, and America. *Omphalea* is extremely patchy, being found in tropical America, East Africa, Madagascar, Indochina, West Malesia, Celebes, and Queensland; it was undoubtedly previously more continuous.

Common to Madagascar and the Mascarene Islands is the genus *Lautembergia,* but one genus is endemic to the Mascarenes, namely *Cordemoya. Stillingia* is common to the Mascarenes, Asia, and America. *Wielandia* and *Riseleya* (if the latter is kept distinct from *Drypetes*) are endemic to the Seychelles.

Returning to Africa for a moment, there is a small group of genera which are found both in Africa and Asia, but which do not occur on the oceanic islands of the Indian Ocean. These are *Bridelia, Chrozophora, Flueggea, Hymenocardia, Mallotus, Microdesmis,* and *Pterococcus.* The genus *Antidesma* is found in Africa, Asia, Australia, and Hawaii.

There are fourteen endemic genera in Australia. Nine of these belong to the Stenolobeae, the taxonomic status of which was discussed above. These are *Poranthera, Micrantheum, Pseudanthus, Stachystemon, Ricinocarpus, Bertya, Beyeria, Monotaxis,* and *Amperea.* One further genus of the stenolobes was recently described by Hutchinson[12] (the first from outside Australia), namely *Oreoporanthera,* endemic to New Zealand. The five platylobe Australian endemic genera are *Adriana, Caelebogyne, Calycopeplus, Dissiliaria,* and *Neoroepera.*

Actephila is common to Asia and Australia, and *Petalostigma* to Australia and New Guinea.

There are only two endemic genera in New Guinea, namely, *Annesijoa* and *Neomphalea*. *Alphandia* occurs in New Guinea, the New Hebrides, and New Caledonia, and *Breynia* and *Longetia* are common to Asia, Australia, and New Caledonia. New Caledonia also has six endemic genera: *Bocquillonia, Cocconerion, Dendrophyllanthus, Lasiochlamys, Neoguillauminia,* and *Ramelia*.

Thus, it may be seen that a very involved mosaic of distribution patterns is exhibited by euphorbiaceous genera, and this is also reflected in the distribution of the genera of other large families, and of some smaller ones as well. For much of this information on discontinuous and other distribution patterns the author is indebted to Good's *Geography of Flowering Plants*,[20] which should be consulted for further information on the subject.

XV. ECONOMIC CONSIDERATIONS

Probably the best known economic plant in the Euphorbiaceae is the rubber tree, or more properly the para rubber, *Hevea brasiliensis,* which is the source of most of the world's natural rubber. It is indigenous to the Amazon basin, and is far superior to the other members of the genus *Hevea* both as to yield and quality. In 1876, some 1900 seedlings raised at Kew from some 70,000 seeds from Brazil were dispatched to Ceylon (now Sri Lanka), and that was the start of the rubber plantation industry. The bulk of the world's supply of natural rubber today is obtained from Malaysia, Indonesia, Sri Lanka, Thailand, and Indochina, in that order.

Rubber is also obtained from some species of the genus *Manihot,* such as *M. glaziovii,* the ceara rubber from northeast Brazil. However, the genus is best known on account of the manioc, cassava, or tapioca plant, *M. esculenta,* source of a staple foodstuff for poorer people in many tropical countries. It is not a true species, but rather a cultigen which originated in South America (Rogers and Appan[19]) and has from there been introduced into every part of the Old World tropics, where today the chief countries for large-scale cultivation of the crop are Zaire, Nigeria, India, and Indonesia. Its nearest wild relative is *M. aesculifolia* from Central America.

A serious drawback of cassava cultivation is that it exhausts the soil in which it grows, and crop rotation, green manuring, or the use of fertilizers is essential if soil fertility is to be kept up.

The fresh tubers do not keep for long, and have to be sliced and dried for preservation. A disadvantage of cassava as a foodstuff is the presence of hydrocyanic acid in varying quantities, but since in the so-called ''sweet'' varieties this is mostly concentrated in the inner rind, peeling and washing eliminates most of it. Sun drying and cooking are also effective in this respect. Chips, flour, starch, and sago are produced from the tubers. The ''sweet'' and ''bitter'' varieties were once treated as separate species, but they are now generally considered to be conspecific.[21]

The third major economic euphorbiaceous plant is unquestionably the castor oil plant, *Ricinus communis*. Castor oil is obtained from the seeds, which are very toxic; it is soluble in alcohol, and with a density of approximately 0.97 it is the densest of all plant oils.[22] It has a multitude of uses. It is employed in medicine as a cathartic and in industry in the manufacture of greases and other lubricants, where it is of particular value for machines where the temperature changes are rapid, and for oiling delicate mechanisms such as watches. In India it is employed as an illuminant for lamps; in the tanning industry it is used to preserve both the flexibility and the impermeability of leather; and it is also used in the manufacture of soaps, glycerine, paints, enamels, varnishes, dyes, plastics, rubber, linoleum, polishes, waxes, carbon-paper, and crayons. The oil cake, which is a by-product of the production of oil is poisonous to stock, but it makes a good fertilizer. Brazil is the chief producer of castor oil, followed by India, the U.S.S.R., Thailand, the U.S., and Rumania.

Aleurites moluccana is the source of candlenut oil, also used in the manufacture of soaps, paints, and varnishes; candles shaped from the paste of the kernels were formerly used for illumination; hence the common name. The chief producers are China and the Philippines.

Vernicia is a genus closely related to *Aleurites* and has by some authors been included in it, but it has simple or bifurcated as opposed to stellate hairs, larger flowers, and fewer stamens. It is composed of three species, from each of which is obtained an oil of commercial value used in varnishes, paints, enamels, and lacquers. Tung oil is obtained from *V. fordii*, Chinese wood oil from *V. montana*, and Japanese wood oil from *V. cordata*.

Sapium is a large genus with one species of economic importance, *S. sebiferum*, the Chinese tallow tree. The seeds of this species have a greasy tallow surrounding them which is used in making cosmetics, soap, and candles. A fatty oil called ''stillingia oil'', obtained from the kernels, is superior to linseed oil as a drying oil, and it is also used in the preparation of paints and varnishes, and as an illuminant. The leaves of this tree yield a black dye.

A red or orange dye is obtained from granules produced on the surface of the fruits of *Mallotus philippensis*, the kamala tree, which was formerly extensively used in dyeing silk and wool. It is mostly collected from the wild, the species not being cultivated on a plantation scale. The powder is also used as an antiseptic and as an anthelminthic. The oil from the seeds is employed as a substitute for tung oil.

Blue and purple dyes are obtained from the tournesol, *Chrozophora tinctoria*.

The genus *Jatropha* (the generic name is coined from two Greek words which signify ''doctor'' and ''food'') includes the physic nut, *J. curcas*, a native of tropical America from the seeds of which a powerful purgative is obtained. It was introduced into the Old World by the Portuguese for its seed oil, and it has become naturalized in many areas. A purgative is also obtained from another species, *J. podagrica*, the purging nut, also known as ''Guatemala rhubarb'' or ''white rhubarb''. A green dye is obtained from another species, *J. gossypiifolia*.

The most drastic of all purgatives known to man is also euphorbiaceous in origin and comes from the seeds of a *Croton* species, *C. tiglium*. It is now generally considered unsafe to use, being highly toxic, and it has been dropped from the U.K., U.S., Dutch, and certain other pharmacopoeias. It causes violent evacuation in only the minutest doses, and it may also cause sloughing of the intestinal lining.

The manchineel, *Hippomane mancinella*, from tropical America, the West Indies, and South Florida has a highly poisonous and irritant milky latex which causes temporary blindness if it gets into the eyes. It is a powerful cathartic and vermifuge in very small doses, but large doses can cause fatal enteritis. The wood is valued for cabinet work.

The blinding tree, *Excoecaria agallocha*, from tropical Asia, has a pale yellow acrid latex which blisters the skin, and which can also cause blindness if it enters the eye. The leaves are poisonous to stock.

The timber known in the trade as ''African oak'' comes from the West African tree, *Ricinodendron heudelotii*. In southern Africa another species of *Ricinodendron* is *R. rautanenii*, the manketti nut tree, the seeds of which constitute an important food resource in marginal environments.[23] ''African teak'' comes from *Oldfieldia africana*.

In India, *Trewia nudiflora* yields the timber known to the trade as ''false white teak''.

The arara nut tree, *Joannesia princeps*, from tropical America is grown for a variety of reasons: as an ornamental, as a timber tree, and for the seeds which yield an oil used as a purgative four times more active than castor oil but with a more pleasant smell. It is now cultivated in many tropical countries.

Other euphorbiaceous species, the nuts of which are used locally as a foodstuff, are *Omphalea diandra*, the Jamaican cobnut, from the West Indies and tropical America; *Manniophyton africanum*, the gasso nut, from West Africa and Zaire; and *Tetracarpidium conophorum*, the owusa nut, also from West Africa. The African bomah nut, *Pycnocoma macrophylla,* is used in tanning and making poison, while the Central American pascualito nut, *Garcia nutans,* yields a quick drying oil.

The star gooseberry, *Sauropus androgynus,* is used as a leaf vegetable, both cooked and raw, in southeast Asia. The Otaheite gooseberry or star apple, *Phyllanthus acidus,* the country of origin of which is unknown, is widely cultivated in the tropics for its edible fruit. The Asiatic *Baccaurea sapida* has globose edible yellow fruits, but it must be taken with caution. The fruit of *Emblica officinalis,* the emblic myrobalan, Indian gooseberry or amla, is probably the richest known natural source of vitamin C. It is highly esteemed for making pickles, preserves, and jellies.

The unripe fruits of the sandbox tree, *Hura crepitans,* from tropical America, were formerly used as containers for sand for blotting ink, or as paperweights when filled with lead. Nowadays, the tree is often grown as a shade tree on plantations.

The milk bush, *Euphorbia tirucalli,* previously considered to have been Indian in origin, but now known to have originated in southern Africa,[24] is widely cultivated in the tropics as a hedge plant. Other *Euphorbia* species are more locally employed for this purpose, e.g., *E. buxoides* in the highlands of Papua New Guinea.

The seeds of *Sebastiania pringlei* from Mexico are the well known "jumping beans" of the curiosity trade. They contain the larvae of the moth *Carpocapsa saltitans,* and show the characteristic jerky and hopping movements when warmed up. Various species of the genus *Colliguaya,* from South America, are also jumping bean plants.

XVI. EUPHORBIACEAE IN HORTICULTURE

A large number of temperate species of the genus *Euphorbia,* the spurges, have gained considerably in popularity in recent years for the herbaceous border, rock garden, and for flower arranging. Of these, probably the best are *E. polychroma* from central Europe, with vivid yellow-green raylet leaves in dense heads, and many stems forming thick clumps; *E. characias* ssp. *wulfenii* from the Balkans, forming large clumps with bluish-green stem leaves and cylindrical heads of smaller yellowish cup leaves; *E. amygdaloides* var. *robbiae* from Turkey, a creeping ground-cover-forming perennial with biennial aerial stems which produce a rosette of dark green glossy leaves in the first year, and which elongate and flower in the second, again producing a cylinder of yellowish-green cup leaves; *E. griffithii* from the eastern Himalayas, with red ray leaves; *E. cyparissias,* the "cypress spurge", widespread in Europe but uncommon as a native species in the British Isles, with dense shoots of linear leaves that convey a somewhat conifer-like appearance on the plants; *E. lathyris,* the "caper spurge", also widespread in Europe, and unusual on account of its biennial habit, decussate stem leaves, and spongy fruits; and *E. marginata,* "snow-on-the-mountain" from North America, with white or white-edged upper leaves, which prefers rather warmer conditions than the other species mentioned for optimum growth.[25]

Tropical *Euphorbia* species, which are popular as houseplants or greenhouse plants in more northern latitudes, but which are popular as garden plants in the tropics, include the poinsettia, *E. pulcherrima,* a native of Mexico but now found throughout the tropics, is a shrub to about 10 ft (3 m) with large elliptic crimson, scarlet, yellow, or white upper leaves surrounding the yellow clusters of cyathia, while the lower purple or green leaves are dentate. A close relative of this is the false poinsettia, *E. cyathophora;* it is similar to the poinsettia, but the upper leaves are either mostly red with green tips or else are only red at the base, never wholly red. It is only grown in the tropics. Another Mexican spurge in tropical horticulture, but not related to the poinsettias is *E. fulgens,* the "scarlet plume". It has drooping branches with dark green lanceolate leaves, and the axillary cyathia are furnished with vivid scarlet or orange petaloid glands. *E. cotinoides* and *E. leucocephala* are also commonly cultivated Mexican species.

From the island of Madagascar comes a whole group of spiny spurges with semisucculent stems not found elsewhere, but which are now widely cultivated in the tropics generally on account of their scarlet or creamy-yellow cyathial leaves or "cyathophylls". *E. milii,* the "crown of thorns", is perhaps the best known of these, but others are *E. splendens, E. hislopii, E. decaryi, E. francoisi,* and *E. capsaintemariensis,* the latter being from the very southernmost point of the island. Other cultivated Malagasy spurges related to these, but with fringed stem angles instead of spines, include *E. fournieri.*[26]

However, the part of the world where the greatest profusion of succulent spurges is found which are of interest to the succulent enthusiast, not so much on account of color but rather on account of diversity and weirdness of form, is unquestionably South Africa. Here are found many cactus mimics, such as the almost perfectly spherical *E. obesa,* the strange *E. globosa* with stems like strings of little footballs and hand-shaped cyathial glands covered with chalky-white markings, *E. caput-medusae,* the gorgon's-head spurge, *E. multiformis,* and many other forms, all of which are treated exhaustively by White et al.[27]

The genus *Monadenium,* predominantly from east and south tropical Africa, also includes a multitude of succulent forms, but is not yet as widely known as the above.[28] Another genus of the tribe Euphorbieae fairly widespread in tropical horticulture is the genus *Pedilanthus,* of which perhaps the best known is *P. tithymaloides* ssp. *smallii,* the "Jacob's ladder" (no connection with *Polemonium caeruleum).* This is a native of Florida and Cuba, and it is grown for its strikingly zigzag dark green stems and its pink or coral-red bird's-head-shaped cyathia.[29]

Other genera in other subfamilies and tribes which play their part in tropical horticulture are *Acalypha, Breynia, Codiaeum, Dalechampia,* and *Jatropha.*

Two *Acalyphas* are of note in this connection, namely *A. hispida,* the "red-hot cat's tail", with its long female inflorescences, the color being due to the dense masses of feathery stigmas; and *A. wilkesiana* (including *A. godseffiana* and *A. marginata)* from the Pacific region, the many cultivars of which present a great diversity of form and color.

In the genus *Breynia, B. disticha* var. *disticha* f. *nivosa* is the popular "snow bush", so called because the leaves are often quite white, although they may also be cream-colored or pale pink as a ground color with varying degrees of patterning in the form of spots, flecks, and streaks of purple or other darker hues. This form originated in the New Hebrides.

For a truly protean plant as regards the form, patterning, and color of the leaves, however, there is nothing in the family to touch *Codiaeum variegatum,* the so-called "croton" of the horticulturalist (which is not really closely related to the botanical genus *Croton,* a member of a different tribe). The leaves may be lanceolate, strap-shaped, trilobate, or appendiculate, and may range from plain green to up to four-colored (cream, pink, purple, and green), with the colors in an almost infinite variety of combinations, proportions, and arrangements, analogous in some ways with the genus *Coleus* in the Labiatae. It is native to Malesia, Indonesia, the southwestern Pacific and northern Australia.

In the genus *Dalechampia* with its unusual inflorescences, *D. roezliana* from Mexico is a popular horticultural subject because of its pink inflorescence bracts.

Four New World species of the genus *Jatropha* are commonly cultivated in the tropics. They are *J. gossypiifolia,* which has bronzy young foliage which later turns dark green, and deep crimson or purplish flowers; *J. integerrima,* the fiddle-leaf from Cuba, which has dark green violin-shaped leaves and handsome crimson flowers; *J. multifida,* the coral plant or coral tree, with bright green palmatipartite leaves, the lobes of which are pinnatifid, and clear coral-red flowers and inflorescence branches; and *J. podagrica,* with gouty stems, peltate, lobed leaves, and red flowers.

This completes this review of the family Euphorbiaceae. It is by no means exhaustive. Nevertheless it is hoped this chapter will convey something of the great diversity in form, utility, and interest to be found within the scope and on the fringes of this rather unusual family of flowering plants.

REFERENCES

1. **Pax, F., in Engler, A. and Harms, H.,** *Die Natürlichen Pflanzenfamilien,* 2nd ed., 19c (Tricoccae). 11-233, 1931.
2. **Webster, G. L.,** Conspectus of a new classification of the Euphorbiaceae, *Taxon,* 24(5/6), 593, 1975.
3. **Mahlberg, P.,** Evolution of the laticifer in *Euphorbia* as interpreted from starch grain morphology, *Am. J. Bot.,* 62(6), 577, 1975.
4. **Klotzsch, F.,** Linné's natürliche Pflanzenklasse Tricoccae, *Monatsber. Königl. Akad. Wiss. Berl.,* March, 236, 1859.
5. **Punt, W.,** Pollen morphology of the Euphorbiaceae with special reference to taxonomy, *Wentia,* 7, 1, 1962.
6. **Ehler, N.,** Mikromorphologie der Samenoberflächen der Gattung *Euphorbia, Plant Syst. Evol.,* 126, 189, 1976.
7. **Baillon, H.,** *Étude générale du groupe des Euphorbiacées,* Paris, 1858.
8. **Mueller of Aargau, J., in De Candolle, A.,** *Prodromus Systematis Universalis Regni Vegetabilis,* 15(2), 189, 1866.
9. **Bentham, G., in Bentham, G. and Hooker, J. D.,** *Genera Planterum* 3(1), 239, 1880.
10. **Hurusawa, I.,** Eine nochmalige Durchsicht des herkömmlichen Systems der Euphorbiaceen im weiteren Sinne, *J. Fac. Sci. Univ. Tokyo, Sect. 3 Bot.,* 6(6), 209, 1954.
11. **Airy Shaw, H. K.,** The Euphorbiaceae of Siam, *Kew Bull.,* 26(2), 193, 1972; The Euphorbiaceae of Borneo, *Kew Bull. Add. Ser.,* 4, 4, 1975; The Euphorbiaceae of New Guinea, *Kew Bull. Add. Ser.,* 8, 5, 1980.
12. **Hutchinson, J.,** Tribalism in the Family Euphorbiaceae, *Am. J. Bot.,* 56(7), 738, 1969.
13. **Köhler, E.,** Die Pollenmorphologie der biovulaten Euphorbiaceae und ihre Bedeutung für die Taxonomie, *Grana Palynol.,* 6(1), 26, 1965; Köhler, E. and Webster, G. L., in Webster, G. L., The Genera of Euphorbiaceae in the SE US, *J. Arn. Arb.,* 48, 308, 1967.
14. **Pax, F., in Engler, A.,** *Das Pflanzenreich,* IV (147, i-xvii), 1910—1924.
15. **Agardh, C. A.,** *Aphorismi Botanici,* 14, 199, 1825.
16. **Willis, J. C.,** *A Dictionary of the Flowering Plants and Ferns,* 7th ed., rev. Airy Shaw, H. K., Cambridge University Press, London, 1966; 8th ed., rev. Airy Shaw, H. K., Cambridge University Press, 1973.
17. **Pierre, L.,** Fam. Pandaceae, *Bull. Soc. Linn. Paris,* 2, 1255, 1896; *Bull. Soc. Linn. Paris,* 3, 1327, 1897.
18. **Forman, L. L.,** The reinstatement of *Galearia* Zoll. & Mor. and *Microdesmis* Hook. f. in the Pandaceae, *Kew Bull.,* 20(2), 309, 1966.
19. **Rogers, D. J. and Appan, S. G.,** *Manihot, Manihotoides* (Euphorbiaceae), *Flora Neotrop. Monogr.,* 13, 1973.
20. **Good, R.,** *The Geography of the Flowering Plants,* 2nd ed., Longman, Green & Co., London, 1953.
21. **C.S.I.R.,** *The Wealth of India: Raw Materials* I, A—B, Council of Scientific and Industrial Research, New Delhi, 1948; II, C, 1950; III, D—E, 1952; V, H—K, 1959; VI, L—M, 1962; IX, Rh—So, 1972; X, Sp—W, 1976.
22. **Shishkin, B. K.,** in *Flora URSS,* Komarov, V. L., Ed., 14, 300, 1949.
23. **Howes, F. N.,** Nut, *Encyclopaedia Britannica,* 16, 798, 1970.
24. **Leach, L. C.,** *Euphorbia tirucalli* L.: its typification, synonymy and relationships, *Kirkia,* 9(1), 69, 1973.
25. **Turner, R.,** A review of spurges for the garden, *Plantsman,* 5(3), 129, 1983.
26. **Ursch, E. and Léandri, J.,** Les Euphorbes Malgaches épineuses et charnues du Jardin Botanique de Tsimbazaza, *Mem. Inst. Sci. Mad. Ser. B.,* 5, 109, 1955.
27. **White, A., Dyer, R. A., and Sloane, B. L.,** *The Succulent Euphorbieae (Southern Africa),* Vol. 1 and 2, Pasadena, Calif., 1941, 990 pp.
28. **Bally, P. R. O.,** *The genus Monadenium,* 112 pp., 34 black and white illustrations, 32 color plates (loose), Berne, 1961.
29. **Dressler, R. L.,** The genus *Pedilanthus, Contr. Gray Herb.,* 182, 1957.
30. **Orchard, A. E. and Davies, J. B.,** *Oreoporanthera,* a New Zealand "endemic" plant genus discovered in Tasmania, *Pop. Proc. Roy. Soc. Tasmania,* 119, 61, 1985.

Chapter 4

BIOSYNTHETIC AND CHEMOSYSTEMATIC ASPECTS OF THE EUPHORBIACEAE AND THYMELAEACEAE

Richard J. Schmidt

TABLE OF CONTENTS

I. BIOSYNTHESIS

No detailed studies of the biosynthesis of compounds based on the tigliane, ingenane, daphnane, and related hydrocarbon skeletons appear to have been published. However, there can be little doubt that these compounds are derived via the classical acetate-mevalonate pathway typical of other terpenoid compounds.

The mechanism by which mevalonate is elaborated to the di-, tri-, and tetraprenyl pyrophosphate esters is now well recognized if not completely understood.[1,2] Banthorpe and Charlwood,[1] referring to the biosynthesis of diterpenes, note that "geranyl geranyl pyrophosphate is invariably considered to be the parent of this class...." It is interesting to note, however, that while the four possible isomers ([2Z,6Z]-, [2Z,6E]-, [2E,6Z]-, and [2E,6E]-) of farnesol have been utilized in both hypothetical discussions and in experimental investigations of the biosynthesis of sesquiterpene carbon skeletons,[3-5] discussions of the possible biosynthetic routes to the diterpenes have apparently ignored the possibility that any isomer other than the all-*trans* geranyl-geranyl pyrophosphate may be involved.

Cyclization of a tetraprenyl pyrophosphate ester may occur by one of at least two mechanisms. The long recognized "concertina-like" cyclization initiated by electrophilic attack upon the terminal double bond (and resembling in some ways the cyclization of squalene to the sterols), would appear to be the mechanism by which the kaurane, pimarane, beyerane, trachylobane, and many other tetra- and penta-cyclic diterpenes are produced.[6] Examples of such diterpenes are known from the Euphorbiaceae, but are beyond the scope of this discussion. The second mechanism was recognized more recently. In 1962, several groups reported the isolation of a 14-membered ring diterpene that was evidently not formed by a concertina-like cyclization of a tetraprenyl precursor, but by a head-to-tail condensation. The first to be characterized was cembrene,[7] followed shortly by thunbergene,[8] and then the duvatrienols.[9] Dauben et al.[7] realized the biosynthetic significance of a 14-membered ring diterpene, and this was expanded upon by Erdtman et al.,[10] who recognized that the cembrane (= thunbergane, = duvane) skeleton was very similar to the taxane skeleton of the yew (*Taxus* species, family Taxaceae) alkaloids, and the verticillane skeleton of verticillol, a diterpene alcohol that they had themselves isolated from *Sciadopitys verticillata* Siebold & Zucc. (family Taxodiaceae). Subsequently, a further 14-membered ring diterpene was isolated from a soluble enzyme preparation derived from 60-hr old *Ricinus communis* L. (family Euphorbiaceae) seedlings by Robinson and West.[11] It had been formed from mevalonate together with several other diterpenes and was given the name casbene. Robinson and West[12] also demonstrated that geranyl-geranyl pyrophosphate serves as a precursor for its formation. It thus became evident that an alternative route to the diterpenes must exist, involving a head-to-tail condensation, and resulting in a macrocyclic structure (see Figure 1).

Adolf and Hecker[13] provide a recent and detailed hypothetical discussion of likely biosynthetic relationships between the various diterpenoid compounds of the Euphorbiaceae and Thymelaeaceae. These authors base their hypothesis on several isolated pieces of information. The most significant is the discovery of casbene by Robinson and West.[11,12] While the gross structure of casbene was elucidated, paucity of material precluded a detailed examination of its stereoisomerism. Nevertheless, it was postulated on the basis of its probable biosynthesis via head-to-tail cyclization of geranyl-geranyl pyrophosphate, that the three double bonds in the molecule were all *trans*. Crombie et al.[14] later reported the total synthesis of the all-*trans* 1S,3R(−)-casbene from 1R,3S(+)-*cis*- chrysanthemic acid and claimed that the compound could not be distinguished either by spectroscopic or chromatographic methods from the casbene derived from *Ricinus communis*. Further, the resemblance of casbene to cembrene was evident to Robinson and West.[11,12] Cembrene had been isolated previously from *Pinus albicaulis* Engelm. and other *Pinus* species (family Pinaceae); its

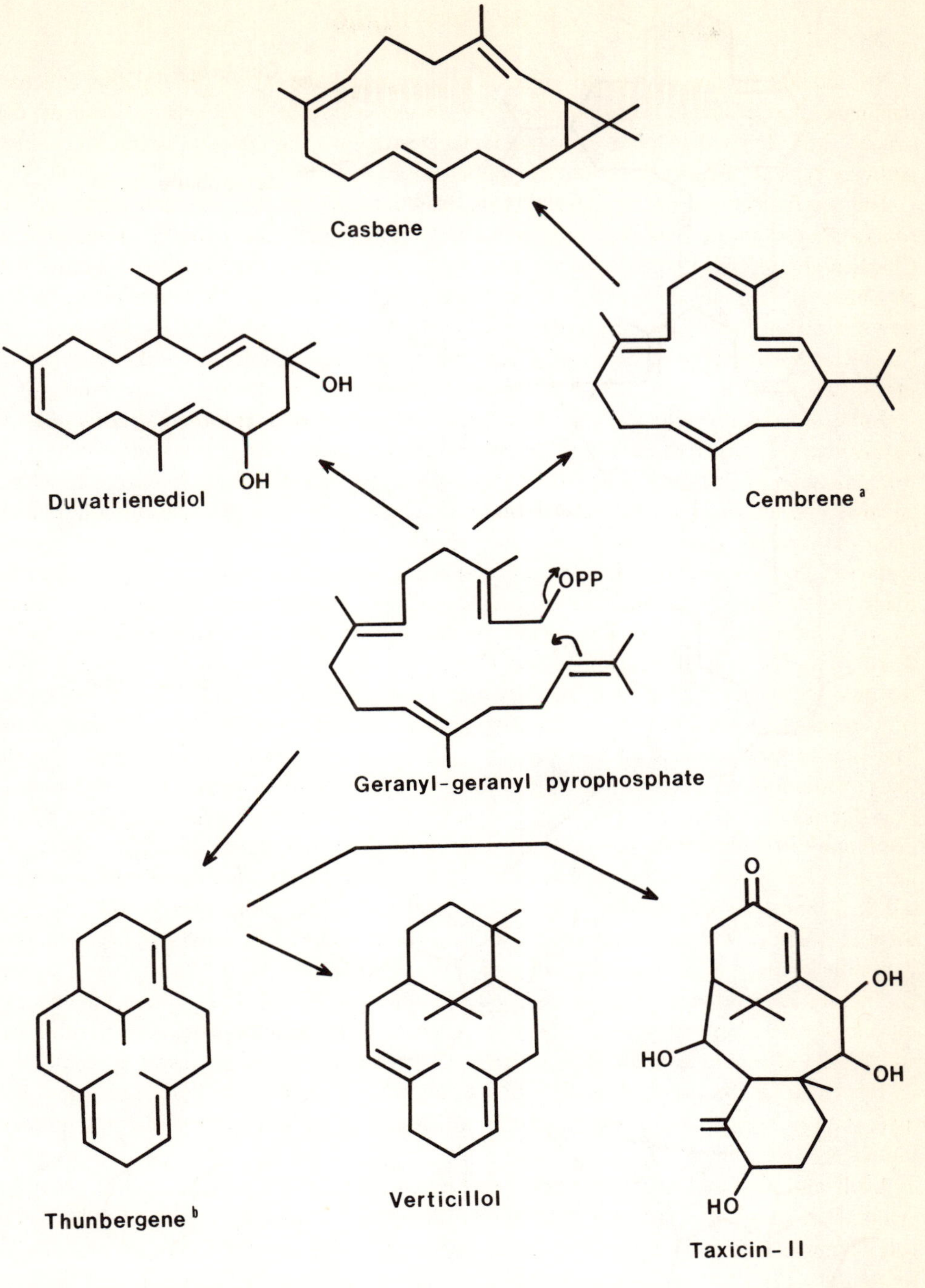

Casbene

Duvatrienediol

OH

OH

Cembrene [a]

OPP

Geranyl-geranyl pyrophosphate

O

OH

HO

OH

HO

Thunbergene [b]

Verticillol

Taxicin-II

FIGURE 1. Putative biogenetic relationships in head-to-tail cyclized tetraprenyl pyrophosphate-derived diterpenes.

structure as determined by X-ray crystallography was shown to be [2E,4Z,7E,11E]-1-isopropyl-4,8,12-trimethyl-2,4,7,11-cyclotetradecatetrene.[15] Subsequently, a very similar compound named cembrene A was found in guggulu resin derived from *Commiphora mukul* Engl. (family Burseraceae). Its similarity to cembrene was recognized but its stereoisomerism was not investigated.[16] Kodama et al.[17] then synthesized [3E,7E,11E]-1-isoprop-15-enyl-4,8,12-trimethyl-3,7,11-cyclotetradecatriene (named in the same manner as cembrene above), and claimed it to be identical in all respects except optical rotation with the compound

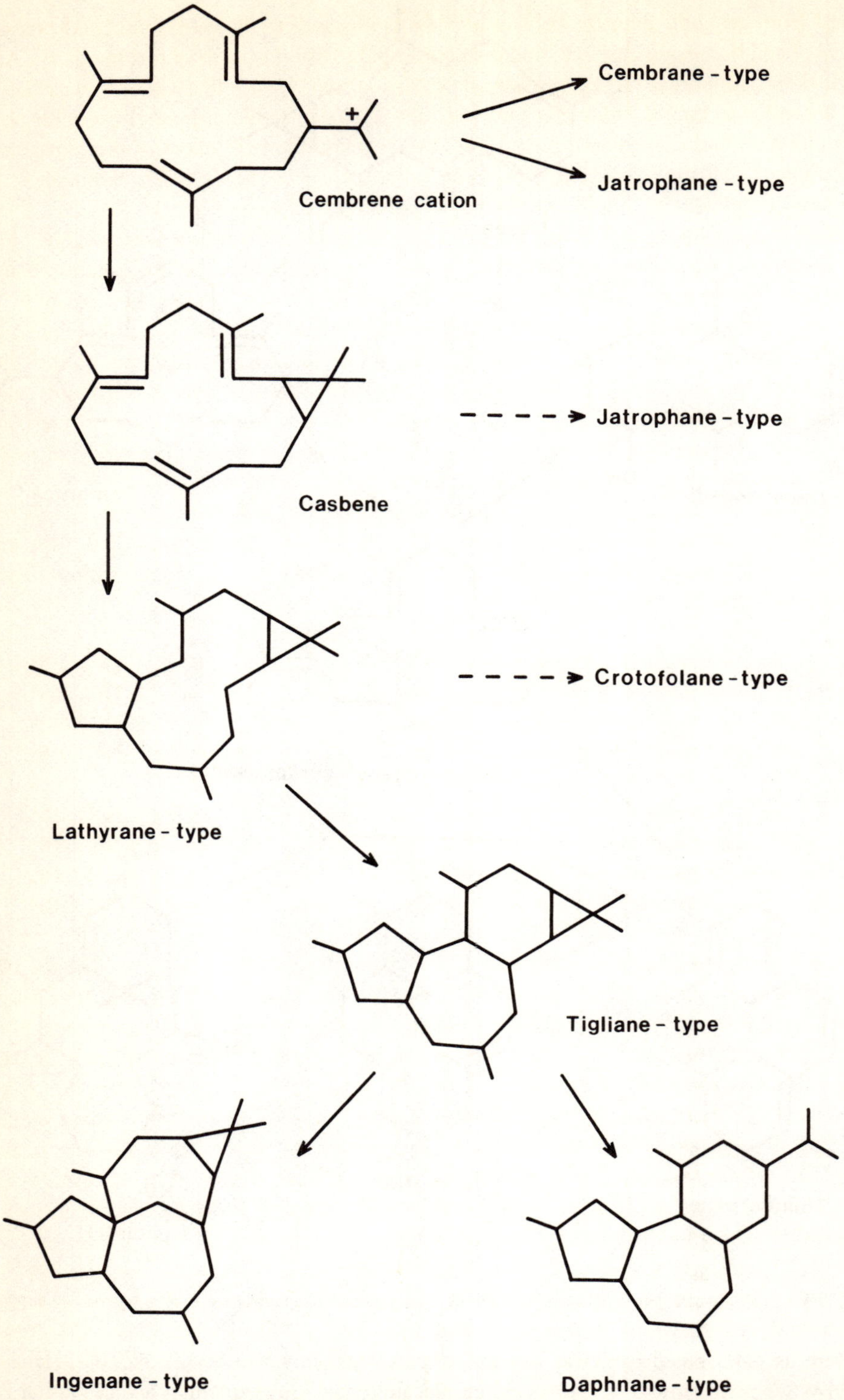

Cembrene cation

Cembrane - type

Jatrophane - type

Casbene

Jatrophane - type

Lathyrane - type

Crotofolane - type

Tigliane - type

Ingenane - type

Daphnane - type

FIGURE 2. Hypothetical biosynthetic pathway to the tiglianes, ingenanes, and daphnanes (after Adolf and Hecker[13]).

isolated from guggulu gum by Patil et al.[16] Notwithstanding the quite unrelated botanical sources of cembrene and casbene, and disregarding the differences in stereochemistry, Adolf and Hecker[13] supported the suggestion of Robinson and West[12] that casbene was derived from a cembrene cation, which in turn was derived from geranyl-geranyl pyrophosphate. Both cembrene and casbene were considered to be plausible precursors of the jatrophanes; cyclization of casbane to the lathyranes and subsequently to the piglianes, daphnanes, and ingenanes completes the Adolf and Hecker hypothesis (see Figure 2).

Certain aspects of the Adolf and Hecker hypothesis deserve further consideration. First, the assumption that geranyl-geranyl pyrophosphate is (invariably) the immediate precursor of the first cyclic carbon skeleton ignores the possibility of *cis/trans* isomerism in this compound. Second, there is no need to postulate that the cembrene cation is a likely precursor to the jatrophane series, since an equally plausible alternative cyclization of the linear precursor will produce the jatrophane skeleton in fewer steps (see Figure 3). Third, and as a consequence of the failure to take into account the possibility of *cis/trans* isomerism of geranyl-geranyl pyrophosphate, the lathyranes should not necessarily be regarded as precursors of the tiglianes, daphnanes, and ingenanes.

Since there are three positions in "geranyl-geranyl pyrophosphate" at which *cis/trans* isomerism may occur, it follows that there are eight possible isomers. To avoid the cumbersome IUPAC nomenclature at this stage, the following trivial nomenclature may be proposed for these isomers:

[2E,6E,10E]-geranyl-geranyl pyrophosphate	or EEE-GG-OPP
[2E,6Z,10E]-geranyl-geranyl pyrophosphate	or EZE-GG-OPP
[2E,6E,10Z]-neryl-geranyl pyrophosphate	or EEZ-NG-OPP
[2E,6Z,10Z]-neryl-geranyl pyrophosphate	or EZZ-NG-OPP
[2Z,6Z,10Z]-neryl-neryl pyrophosphate	or ZZZ-NN-OPP
[2Z,6E,10Z]-neryl-neryl pyrophosphate	or ZEZ-NN-OPP
[2Z,6Z,10E]-geranyl-neryl pyrophosphate	or ZZE-GN-OPP
[2Z,6E,10E]-geranyl-neryl pyrophosphate	or ZEE-GN-OPP

Figure 4 shows the probable route to ZZZ-NN-OPP; equally plausible routes to the other possible isomers may be formulated. However, although mechanistically plausible from a chemical point of view, these routes to the tetraprenyl pyrophosphates take no account of the role and function of the enzyme(s) concerned, nor the mechanism of further cyclization and functionalization of the final diterpenoid compound. Cane[2] notes that there are at least three conflicting theories concerning *cis/trans* isomerism in allylic pyrophosphate metabolism. One involves the direct utilization of the appropriate isomer, avoiding the need to isomerize at a later stage — this is the method proposed here. A second theory involves the participation of a redox system, with isomerizability arising from the generation of an αβ-unsaturated aldehydic intermediate, while the third theory involves *cis/trans* isomerization via an intermediate tertiary allylic pyrophosphate. Cane[2] also notes that there is some experimental evidence to support each of the theories but argues, on the basis of detailed experimental investigations into monoterpene biosynthesis, that the first theory involving the utilization of the appropriate isomer may not be as compelling as previously held. However, when diterpene biosynthesis is considered, and especially the supposed head-to-tail method of cyclization, the alternative theories involving the aldehydic or tertiary allylic intermediates become very much less compelling. These mechanisms for converting geranyl units to neryl units allow only for isomerization at one double bond, i.e., the one adjacent to the pyrophosphate group. The theories cannot be readily applied to the conversion of the all-*trans* geranyl-geranyl pyrophosphate to the all-*cis* neryl-neryl pyrophosphate as would appear to be the requirement in the case of the biosynthesis of tiglianes, ingenanes, and

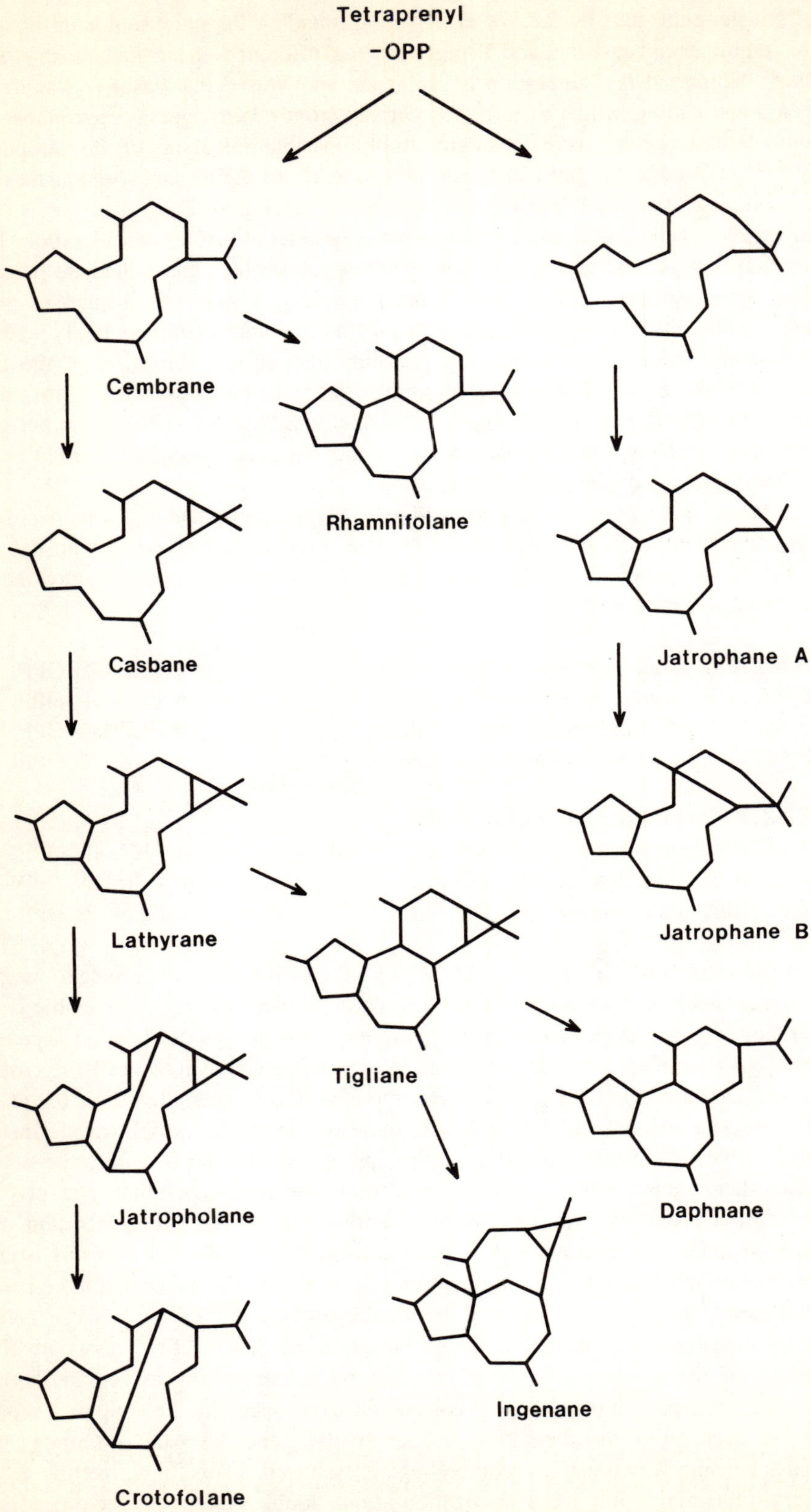

Tetraprenyl
-OPP

Cembrane

Rhamnifolane

Casbane

Jatrophane A

Lathyrane

Jatrophane B

Tigliane

Jatropholane

Daphnane

Ingenane

Crotofolane

FIGURE 3. Proposed alternative modes of head-to-tail cyclization of tetraprenyl pyrophosphate to two separate families of compounds.

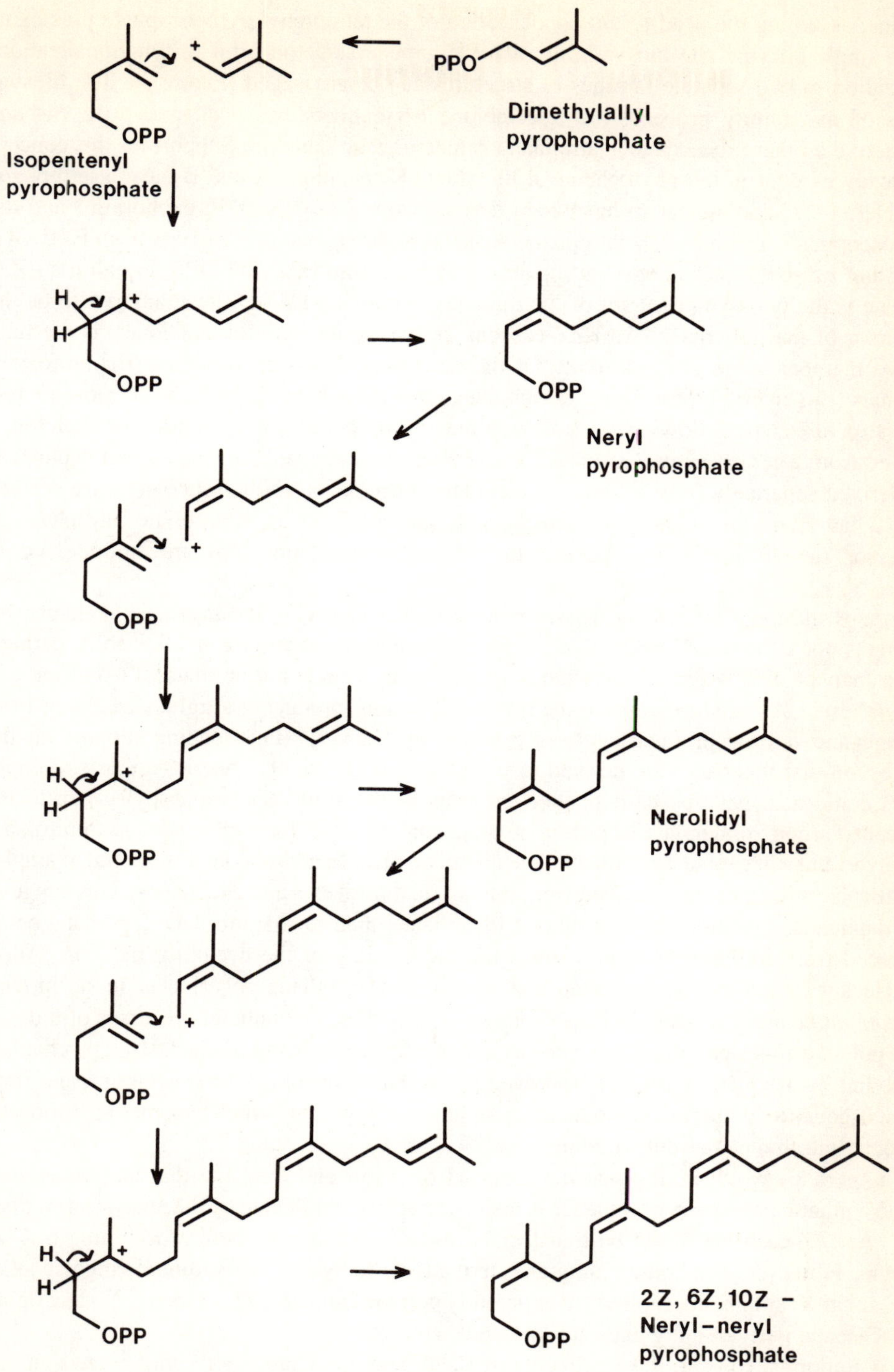

Isopentenyl pyrophosphate

Dimethylallyl pyrophosphate

Neryl pyrophosphate

Nerolidyl pyrophosphate

2Z, 6Z, 10Z - Neryl-neryl pyrophosphate

FIGURE 4. Suggested biosynthetic route to [2Z,6Z,10Z]-neryl-neryl pyrophosphate.

daphnanes. While alternative enzyme-catalyzed isomerization reactions could perhaps be postulated to explain how geranyl-geranyl pyrophosphate may be isomerized, it is difficult to reconcile such an approach with the spatial requirements associated with initial head-to-tail condensation of the tetraprenyl pyrophosphate precursor.

Thus, assuming the head-to-tail condensation of the tetraprenyl pyrophosphate precursor, and a single enzyme site biosynthesis of the diterpene skeleton prior to functionalization, cyclization to two separate families of structures can be envisaged (Figure 3). The biosynthesis of one family proceeds via a cembrane (=thunbergane, =duvane) to a casbane skeleton; the other proceeds to a jatrophane A macrocyclic skeleton. Support for this concept is already evident in the phytochemical literature. Kansuinines A and B have a jatrophane A skeleton,[18,19] and appear to have been derived from ZZZ-NN-OPP; euphornin[20] and the euphoscopins,[21] also having a jatrophane A skeleton, are apparently derived from EZE-GG-OPP; and similarly, jatrophone[22] is apparently derived from EZZ-NG-OPP. Equally, a ZZZ-casbene is the putative precursor of the riglianes, while and EZE-casbene appears to be the precursor of the lathyranes. An EZZ-casbene derivative named crotonitenone[23] is the only known member of its group to date. Of the remaining diterpene skeletons (rhamnifolane, daphnane, ingenane, crotofolane, jatropholane, and jatrophane B), which are more or less rigid structures, the following relationships may be proposed: the rhamnifolane skeleton is derived from a cembrane precursor that has not become a casbane; ingenanes and daphnanes are derived separately from a tigliane precursor; crotofolanes and jatropholanes are derived from a lathyrane precursor; and jatrophane B derives from an appropriate jatrophane A precursor (as discussed by Torrance et al.[24]). These relationships are summarized in Figure 5.

Perhaps the most interesting transformation is that involving the ingenane skeleton. Ingenane is not a "regular" isoprenoid; its formation appears to require an alkyl shift, perhaps in the manner of a Wagner-Meerwein rearrangement. This is not an unusual occurrence in biosynthesis.[1] A transformation in the reverse direction, producing a tigliane skeleton from an ingenane starting product has been reported by Hecker,[25] thus lending support for the suggestion that ingenanes are derived from riglianes in the plant. The biosynthesis of daphnanes could also perhaps be explained in terms of the shift of an isopropyl group to the adjacent carbon, either at the cembrane stage or after the formation of a rhamnifolane. However, this appears to be unlikely. The phytochemical literature contains several examples of tiglianes and daphnanes co-occurring, but not rhamnifolanes and daphnanes. This suggests that daphnanes are more closely related to tiglianes than to rhamnifolanes, perhaps being produced from tigliane precursors following the opening of the cyclopropane ring. Adolf and Hecker[26] have indeed suggested that mancinellin, a tigliane polyol ester found in *Hippomane mancinella* L. (family Euphorbiaceae), may be a biosynthetic precursor of a daphnane polyol orthoester with which it co-occurs, the orthoester being produced by a mechanism involving hydroxylation at C-14, followed by oxidative opening of the cyclopropane ring, and subsequently reduction to produce the daphnane polyol ester which becomes an orthoester by facile intramolecular dehydration.

It is perhaps worth noting that the proposal of Adolf and Hecker[13] that lathyranes may be biosynthetic precursors of tiglianes is not necessarily contradicted by the analysis described here. A ZZZ-casbene would be transformed to a ZZZ-lathyrene before becoming a ZZZ-tigliene. However, the known lathyrane derivatives (lathyrols, bertyadionol, ingols, jolkinols, and the euphohelioscopins) are apparently derived from an EZE-casbene. No analogous ZZZ-casbene derived lathyranes are known at this time.

It is important to echo the sentiments of Banthorpe and Charlwood[1] who noted that "in all biosynthesis studies, one carefully planned phytochemical experiment is worth endless chemical speculation." Crombie et al.[14] reported the total synthesis of [4E,8E,10E]-casbene. Once the other isomers become available, further elucidation of the biosynthetic pathways discussed here might be facilitated.

Notwithstanding the dearth of biosynthetic evidence, there is perhaps sufficient phytochemical information available to enable a genealogical argument to be developed regarding the taxonomic positions of the families Euphorbiaceae and Thymelaeaceae.

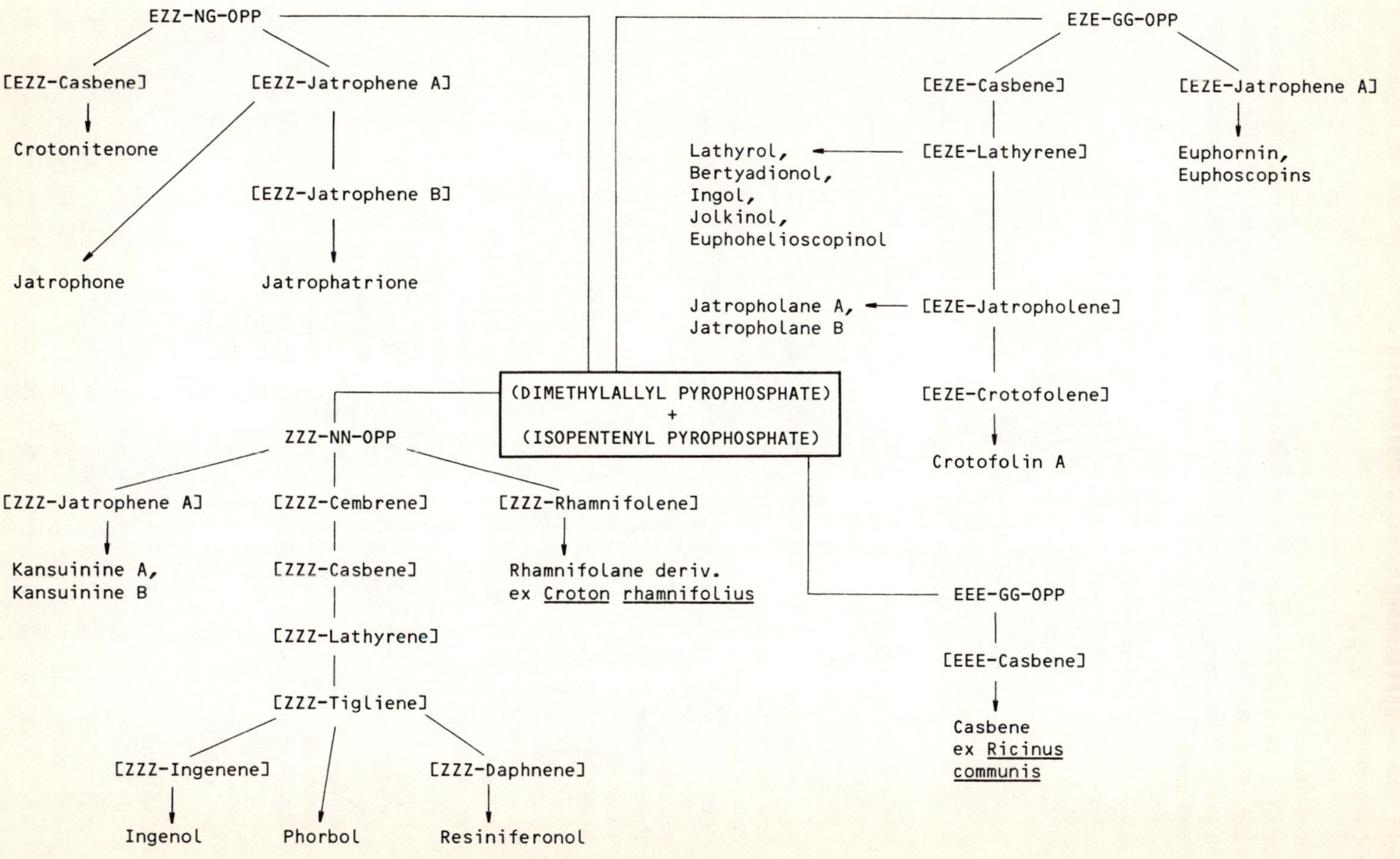

FIGURE 5. Proposed biosynthetic routes to head-to-tail cyclized tetraprenyl pyrophosphate-derived diterpenoids of the Euphorbiaceae and Thymelaeaceae.

II. CHEMOSYSTEMATICS

Currently, the families Euphorbiaceae and Thymelaeaceae are understood to have the following structures.

A. Euphorbiaceae

The family is remarkably varied. Almost 300 genera have been described, with perhaps 7650 species. Most are shrubs and trees, with a few herbaceous species.[27] The most recent taxonomic study of the family is that of Webster (see Chapter 3).[28] The family was divided into five subfamilies, and 52 tribes:

Subfamilies	
Phyllanthoideae	(Tribes 1—13)
Oldfieldioideae	(Tribes 14—17)
Acalyphoideae	(Tribes 18—36)
Crotonoideae	(Tribes 37—47)
Euphorbioideae	(Tribes 48—52)

Tribes		No. of genera	Tribes		No. of genera
1.	Wielandieae	10	27.	Pycnocomeae	6
2.	Amanoeae	2	28.	Bernardieae	4
3.	Bridelieae	2	29.	Epiprineae	7
4.	Dicoelieae	1	30.	Adelieae	4
5.	Poraanthereae	3	31.	Alchornieae	8
6.	Spondiantheae	1	32.	Acalypheae	32
7.	Antidesmeae	6	33.	Plukenetieae	13
8.	Aporuseae	7	34.	Dalechampieae	1
9.	Drypeteae	3	35.	Omphaleae	1
10.	Phyllantheae	18	36.	Pereae	1
11.	Hymenocardieae	1			
12.	Uapaceae	1	37.	Micrandreae	3
13.	Bischofieae	1	38.	Manihoteae	4
			39.	Adenoclineae	6
14.	Hyaenancheae	13	40.	Gelonieae	1
15.	Petalostigmateae	2	41.	Elateriospermae	1
16.	Caletieae	4	42.	Joannesieae	10
17.	Picrodendreae	1	43.	Codiaeae	19
			44.	Ricinocarpeae	6
18.	Clutieae	1	45.	Trigonostemoneae	3
19.	Pogonophoreae	1	46.	Aleuritideae	16
20.	Chaetocarpeae	2	47.	Crotoneae	3
21.	Cheiloseae	2			
22.	Erismantheae	3	48.	Stomatocalyceae	4
23.	Ampereae	2	49.	Hippomaneae	20
24.	Chrozophoreae	11	50.	Pachystromateae	1
25.	Agrostistachydeae	4	51.	Hureae	3
26.	Caryodendreae	3	52.	Euphorbieae	11

B. Thymelaeaceae

This family of some 500 species of mainly erect shrubs in about 50 genera is distributed through temperate and tropical regions.[29] There has been relatively little specific taxonomic interest in the family. The systematic treatment of Domke,[30] in which the family is divided into four subfamilies, is implicitly accepted by Airy Shaw;[29] the tribes are described by Waggenitz:[31]

Subfamilies	Tribes
Gonystyloideae	
Aquilarioideae	
	Microsemmateae
	Solmsieae
	Octolepideae
	Aquilarieae
Gilgiodaphnoideae	
Thymelaeoideae	
	Dicranolepideae
	Phalerieae
	Daphneae
	Thymelaeae

Takhtajan[32] classifies both the Thymelaeaceae and the Euphorbiaceae in the subclass Dilleniidae of the class Magnoliopsida in the division Magnoliophyta (or Angiospermae). The Thymelaeaceae is considered the only family in the order Thymelaeales; the Euphorbiaceae is considered the largest of six families — Euphorbiaceae (including Picrodendraceae), Buxaceae, Pandaceae, Simmondsiaceae, Daphniphyllaceae, and Dichapetalaceae — in the order Euphorbiales. There are other opinions as to the structure of the Thymelaeales. Hutchinson[33] considers the Thymelaeaceae to be one of six families (Gonystylaceae, Aquilariaceae, Geissolomataceae, Penaeaceae, Thymelaeaceae, and Nyctaginaceae) in the order Thymelaeales. Waggenitz[31] envisages the Thymelaeales as comprising five families (Geissolomataceae, Penaeaceae, Dichapetalaceae, Thymelaeaceae, and Elaeagnaceae). Willis[29] states that the family Thymelaeaceae is a very natural one with no close affinities, but probably connected through the subfamily Gonystyloideae with the Sphaerosepalaceae of Madagascar (Malagasy Republic). This implies an acceptance that the families Gonystylaceae and Aquilariaceae should be reduced to and included in the Thymelaeaceae. With regards to the Euphorbiaceae, an alternative classification proposed by Hutchinson[33] involves the retention of the family Euphorbiaceae in the order Euphorbiales, but the transference of the Dichapetalaceae to the Rosales; the Daphniphyllaceae and Buxaceae (including the Simmondsiaceae) to the Hamamelidales; the Pandaceae to the Celastrales; and the Picrodendraceae to the Juglandales.

A further alternative classification places the families Euphorbiaceae, Daphniphyllaceae, and Dichapetalaceae in the order Geraniales, a movement to another subclass, the Rosidae.[27]

The Thymelaeales are considered to share a common origin with the Euphorbiales and the Malvales, while the Euphorbiales are thought to have arisen from some ancient group intermediate between the Malvales and the family Flacourtiaceae of the Violales.[32] The order Euphorbiales, although apparently highly specialized in evolutionary terms, is nonhomogeneous, suggesting that it is polyphyletic in origin. It is thought, in fact, that the Euphorbiales is a synthetic group made up of derivatives from several ancestors brought together merely because of superficial resemblance.[27] Hutchinson[33] provides a diagrammatic overview of the presumed evolutionary relationships between the Euphorbiales and Thymelaeales showing the polyphyletic nature of the Euphorbiales (see Figure 6). It should be noted that Hutchinson[33] perceives the Euphorbiales as comprising the Euphorbiaceae only, with the Thymelaeales comprising six families as noted above. It is evident that there is much disagreement between taxonomists as to the affinities of the Euphorbiaceae and Thymelaeaceae, and the evolutionary relationships between the various families and orders discussed above. There is certainly reason to investigate the possibility of utilizing phytochemical information in the taxonomic discussion.

Although little attempt has been made to relate the results of phytochemical research to the systematics of the two families Euphorbiaceae and Thymelaeaceae, it is evident that the presence of similar diterpenoids in the two families provides further reason to suppose that

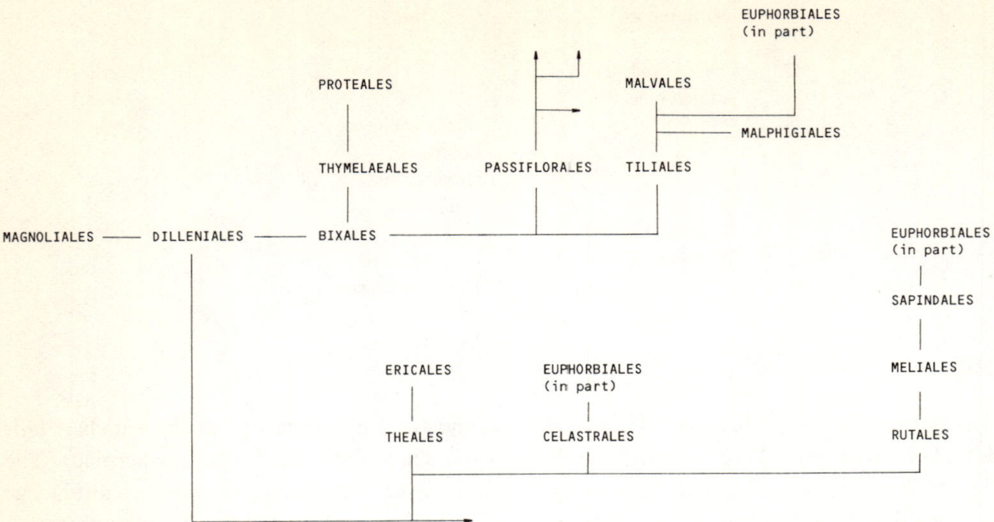

FIGURE 6. Dendrogram showing the supposed evolutionary relationships between the orders Euphorbiales and Thymelaeales (adapted from Hutchinson[33]).

a close relationship exists between them. It is perhaps surprising that similar diterpenoids are found in the Euphorbiaceae and the Thymelaeaceae but not in other families within the Euphorbiales as perceived by Takhtajan.[32] This would support the alternative classification systems described by Hutchinson[33] in which the order Euphorbiales is retained only for the family Euphorbiaceae.

Reference to the work of Mitchell and Rook[34] provides literature references to the occurrence of skin irritancy in certain members of other families thought to be related to the Euphorbiaceae and Thymelaeaceae. Caution should be exercised in interpreting the literature references cited, since plants can cause skin trauma by several mechanisms other than that by which the irritant members of the Euphorbiaceae and Thymelaeaceae exert their effects.[35] There would appear to be little or no relevant phytochemical information supporting the observations, but the discovery in the future of tiglianes, ingenanes, and/or daphnanes in these families would not be entirely unexpected and would, of course, provide useful data for further chemosystematic consideration.

Within the Euphorbiaceae and Thymelaeaceae, reported occurrences of tigliane, ingenane, or daphnane polyol esters and/or reports of skin irritancy that might be ascribed to the presence of these compounds, are summarized in Table 1. In the Euphorbiaceae, the subfamily Oldfieldioideae appears to be almost devoid of such compounds, although the relative obscurity of the 20 genera included in this subfamily may be a significant factor. The Phyllanthoideae (56 genera) and Acalyphoideae (107 genera) appear to be more or less equivalently constant sources of irritants. However, it must be said that reports of irritancy and/or piscicidal activity are rather less dramatic than those associated with plants from the subfamilies Crotonoideae and Euphorbioideae. The reports associated with the genus *Phyllanthus,* for instance, are indicative of piscicidal activity, but early research has suggested that the piscicidal activity is associated with the presence of saponins.[36] Thus, the Phyllanthoideae, Acalyphoideae, and Oldfieldioideae may well be phytochemically distinct from the Crotonoideae and Euphorbioideae. This is consistent with the belief that the Euphorbiaceae is of polyphyletic origin.[27,33] In contrast, 17 of the 72 genera of the Crotonoideae and 15 of the 39 genera in the Euphorbioideae have reportedly irritant constituent species, the irritancy being associated in many species with the presence of tigliane, daphnane, and

Table 1
DISTRIBUTION OF DITERPENE SKELETONS IN THE EUPHORBIACEAE AND THYMELAEACEAE WITH REFERENCE TO SYSTEMATICS AND REPORTED IRRITANCY

		Diterpene hydrocarbon skeletons[a]										
Tribes	Genera	T	I	D	R	C	L	Jo	JaA	JaB	Ca	Irritancy[b]
Euphorbiaceae												
Phyllanthoideae												
3. Bridelieae	*Cleistanthus* Hook. f.	:	:	:	:	:	:	:	:	:	:	+[36,38,39]
6. Spondiantheae	*Spondianthus* Engl.	:	:	:	:	:	:	:	:	:	:	+[40]
7. Antidesmeae	*Antidesma* L.	:	:	:	:	:	:	:	:	:	:	+[41]
10. Phyllantheae	*Phyllanthus* L.	:	:	:	:	:	:	:	:	:	:	+[36,37,42,43]
	Flueggea Willd.	:	:	:	:	:	:	:	:	:	:	+[40]
	Glochidion Forster & Forster f.	:	:	:	:	:	:	:	:	:	:	+[41]
Oldfieldioideae												
17. Picrodendreae	*Picrodendron* Planchon	:	:	:	:	:	:	:	:	:	:	+[42]
Acalyphoideae												
18. Clutieae	*Clutia* L.	:	:	:	:	:	:	:	:	:	:	+[44]
24. Chrozophoreae	*Chiropetalum* Adr. Juss.	:	:	:	:	:	:	:	:	:	:	+[43]
	Chrozophora Necker ex A.L. Juss.	:	:	:	:	:	:	:	:	:	:	+[38]
	Melanolepis Reichb. F. & Zoll.	:	:	:	:	:	:	:	:	:	:	+[41]
27. Pycnocomeae	*Pycnocoma* Benth.	:	:	:	:	:	:	:	:	:	:	+[40]
32. Acalypheae	*Acalypha* L.	:	:	:	:	:	:	:	:	:	:	+[41]
	Claoxylon Adr. Juss.	:	:	:	:	:	:	:	:	:	:	+[37,41]
	Macaranga Thouars	:	:	:	:	:	*[45]	:	:	:	:	+[41]
	Mareya Baillon	:	:	:	:	:	:	:	:	:	:	+[40]
	Mercurialis L.	:	:	:	:	:	:	:	:	:	:	+[46,c]
	Ricinus L.	:	:	:	:	:	:	:	:	:	:	+[41]
36. Pereae	*Pera* Mutis	:	:	:	:	:	:	:	:	:	:	+[47,d]
Crotonoideae												
37. Micrandreae	*Cunuria* Baillon	:	:	*[48]	:	:	:	:	:	:	:	(−)[c]
	Hevea Aublet	:	:	:	:	:	:	:	:	:	:	+[41]

Table 1 (continued)
DISTRIBUTION OF DITERPENE SKELETONS IN THE EUPHORBIACEAE AND THYMELAEACEAE WITH REFERENCE TO SYSTEMATICS AND REPORTED IRRITANCY

Tribes	Genera	Diterpene hydrocarbon skeletons[a]										
		T	I	D	R	C	L	Jo	JaA	JaB	Ca	Irritancy[b]
38. Manihoteae	*Cnidoscolus* Pohl	:	:	:	:	:	:	:	:	:	:	+[37,43,f]
	Manihot Miller	:	:	:	:	:	:	:	:	:	:	+[41]
	Victorinia Leon	:	:	:	:	:	:	:	:	:	:	+[37]
39. Adenoclineae	*Endospermum* Benth.	:	:	:	:	:	:	:	:	:	:	+[41]
42. Joannesieae	*Jatropha* L.	*[49]	:	:	:	:	:	*[50]	*[22]	*[24]	:	+[34,37-41]
43. Codiaeae	*Baliospermum* Blume	*[51]	:	*[51]	:	:	:	:	:	:	:	+[38,39,52]
	Codiaeum Adr. Juss.	:	:	:	:	:	:	:	:	:	:	+[41,43]
	Ostodes Blume	*[54]	:	:	:	:	:	:	:	:	:	(−)[c]
44. Ricinocarpeae	*Bertya* Planchon	:	:	:	:	:	*[55]	:	:	:	:	(−)[c]
46. Aleuritideae	*Aleurites* Forster & Forster f.[56]	:	:	:	:	:	:	:	:	:	:	+[36,40,41]
	Reutealis Airy Shaw[56]	:	:	:	:	:	:	:	:	:	:	+[36,37,41]
	Vernicia Lour.[56]	*[57]	:	:	:	:	:	:	:	:	:	+[34]
47. Crotoneae	*Croton* L.	*[58]	:	.	*[59]	*[60]	:	:	:	:	*[23]	+[34,37,38]
Euphorbioideae												
49. Hippomaneae	*Carumbium* Reinw.	:	:	:	:	:	:	:	:	:	:	(−)[c]
	(syns *Omalanthus* Adr. Juss.,	:	:	:	:	:	:	:	:	:	:	(−)[c]
	Homalanthus Adr. Juss. corr. Reichb.)	:	:	:	:	:	:	:	:	:	:	+[36,61]
	Colliguaja Molina	:	:	:	:	:	:	:	:	:	:	+[62]
	Excoecaria L.	:	:	*[63]	:	:	:	:	:	:	:	+[34,41]
	Grimmeodendron Urb.	:	:	:	:	:	:	:	:	:	:	+[37,64]
	Hippomane L.	*[26]	:	*[26]	:	:	:	:	:	:	:	+[35]
	Sapium P. Br.	*[65]	:	:	:	:	:	:	:	:	:	+[34,38]
	Sebastiania Sprengel	:	:	:	:	:	:	:	:	:	:	+[34,37]
	Spirostachys Sonder	:	:	:	:	:	:	:	:	:	:	+[34,37]
	Stillingia Garden ex L.	*[66]	:	*[66]	:	:	:	:	:	:	:	+[43,66]
51. Hureae	*Hura* L.	:	:	*[67]	:	:	:	:	:	:	:	+[34,37,38]

52.	Euphorbieae	*Anthostema* Adr. Juss.	:	:	:	:	:	:	:	:	:	:	+[40]
		Chamaesyce Gray	:	:	:	:	:	:	:	:	:	:	+[42]
		Euphorbia L.	*[68]	*[69]	*[70]	:	:	*[71]	:	*[19]	:	:	+[34,41,72]
		(incl. *Elaeophorbia* Stapf)	:	*[73]	:	:	:	:	:	:	:	:	+[40]
		Pedilanthus Necker ex Poit.	:	:	:	:	:	:	:	:	:	:	+[40,41]
		Synadenium Boiss.	:	:	:	:	:	:	:	:	:	:	+[34]
Thymelaeaceae													
Gonystyloideae													
		Gonystylus Teijsm. & Binnend.	:	:	:	:	:	:	:	:	:	:	+[37]
Aquilarioideae													
	Aquilarieae	*Aquilaria* Lam.	*[74]	:	:	:	:	:	:	:	:	:	(−)[e]
Thymelaeoideae													
	Dicranolepideae	*Synaptolepis* Oliver	:	:	*[75]	:	:	:	:	:	:	:	(−)[e]
	Daphneae	*Daphne* L.	:	:	*[76]	:	:	:	:	:	:	:	+[34]
		Daphnopsis Martius & Zucc.	*[77]	:	*[75]	:	:	:	:	:	:	:	+[37,46]
		Dirca L.	:	:	*[78]	:	:	:	:	:	:	:	+[34]
		Wikstroemia Endl.	:	:	*[79]	:	:	:	:	:	:	:	+[37,80]
	Thymelaeceae	*Gnidia* L.	:	:	*[81]	:	:	:	:	:	:	:	+[44]
		Lasiosiphon Fresen.	:	:	*[82]	:	:	:	:	:	:	:	+[39,44,83]
		Pimelea Banks & Sol.	*[84]	:	*[75]	:	:	:	:	:	:	:	+[44,85]
		Thymelaea Endl.	:	:	*[86]	:	:	:	:	:	:	:	+[46]

Note: Literature references given are intended to be representative but not exhaustive; see also Chapters 6 through 9.

[a] T = Tigliane, I = ingenane, D = daphnane, R = rhamnifolane, C = crotofolane, L = lathyrane, Jo = jatropholane, JaA = jatrophane A, JaB = jatrophane B, Ca = casbane.

[b] Irritancy to skin, but drastic purgative or piscicidal properties also taken as indicative.

[c] Probably not caused by diterpenoid esters.[87]

[d] Probably caused by a quinone, plumbagin.[34]

[e] No positive or negative reports found.

[f] Members of this genus have urticating hairs.[88]

ingenane polyol esters. These figures are summarized below, and related to the currently accepted 294 genera in the family Euphorbiaceae:

Phyllanthoideae	6/56	11%
Oldfieldioideae	1/20	5%
Acalyphoideae	12/107	11%
Crotonoideae	17/72	24%
Euphorbioideae	15/39	38%
Total	51/294	17%

The figures should be put into perspective by the observation that the subfamily Euphorbioideae comprises only 13% of the genera in the Euphorbiaceae, but contributes 59% of the species.

The occurrence of tiglianes and/or daphnanes in two of the four subfamilies of the Thymelaeaceae, and the reported piscicidal activity of an unidentified *Gonystylus* species[37] in another, provides support for the reduction of the families Gonystylaceae and Aquilariaceae to subfamilies of the Thymelaeaceae. The remaining subfamily, the Gilgiodaphnoideae, comprises just one species which has yet to be investigated.

It is only too evident from Table 1 that, considering the vast size of the family Euphorbiaceae, very few genera (and species) have been chemically investigated for their diterpenoid content. It is probably true that most of the common and best-known species have been investigated, but it is unfortunate that many investigators have been satisfied with identifying only one or just a few of the constituents of any particular species. However, it must be said in mitigation that research into this area has been driven not by an academic interest in chemosystematics, but by a desire to study the compounds responsible for the toxicity of the plants.

The vast majority of reports concern members of the subfamilies Crotonoideae and Euphorbioideae. Tiglianes and daphnanes have been found to co-occur regularly but not invariably within the same genus (and often within the same species). The fact that no tiglianes nor daphnanes have yet been isolated from the other three subfamilies (Phyllanthoideae, Oldfieldioideae, and Acalyphoideae) of the Euphorbiaceae may be of chemosystematic significance, but the reports of skin irritancy from members of these three subfamilies suggest that it may be premature to speculate further.

The occurrence of a particular type of diterpene skeleton within a genus should not be taken to imply that all members of that genus will contain such compounds. Currently available information concerning the families Euphorbiaceae and Thymelaeaceae generally suggests that the diterpenoids are not of universal occurrence. The most detailed study to date has been carried out by Evans and Kinghorn[89] who investigated the genus *Euphorbia*. Using a screening procedure involving hydrolysis of the diterpenoid esters to their parent diterpene polyols, followed by acetylation and gas-liquid chromatographic (GLC) analysis of the products, they found that although the most common type of diterpene skeleton present in the genus as a whole was ingenane, the three species they studied from the sections Anisophyllum and Poinsettia did not appear to yield any diterpene polyols. Some species from the section Euphorbium also yielded no recognizable diterpenes while others afforded phorbol derivatives (tiglianes) but no ingenanes. Several were recognized as containing "unknown ingenanes". However, caution is required in interpreting these results. *Euphorbia hirta* L. in the section Anisophyllum was found to contain no diterpenes, yet Uhe,[61] Mitchell and Rook,[34] and Morton[42] provide evidence, albeit anecdotal, that the species can cause dermatitis. (Although of no significance to the discussion here, it should be noted that authorities on the family Euphorbiaceae now consider *Euphorbia hirta* L. to belong to the genus *Chamaesyce* Gray.) *Euphorbia tirucalli* L. from three different geographical locations (Malagasy Republic, South Africa, and Colombia) has been found to contain different diterpene profiles, comprising esters of phorbol, 4-deoxyphorbol, and ingenol.[90] Perhaps more remarkably, *E. tirucalli* has also been reported by Baslas and Gupta[91-93] to contain a selection of 12-deoxyphorbol and resiniferonol esters that have previously been found in *Euphorbia poissonii* Pax. Further, *Euphorbia triangularis* Desf. is nonirritant according to Watt and Breyer-Brandwijk,[44] but has been reported to contain irritant 12-deoxyphorbol esters by Gschwendt and Hecker.[68] These observations are strongly suggestive of the existence of chemical races, thus rendering the results of single sample screening studies (such as those of Kinghorn and Evans[89] and Upadhyay et al.[94]) difficult to interpret. One conclusion about which there can be little doubt is that the genus *Euphorbia* will yet yield many new compounds.

Perhaps the most important observation to be made concerning the data in Table 1 is the occurrence of ingenanes only in the genus *Euphorbia*. The presumed formation of ingenanes from tiglianes, as discussed above, would appear to require a specific enzyme site, the existence of which would be determined genetically. Thus, ingenanes may be regarded as valuable chemosystematic marker compounds. The discovery of ingenol esters in two species of *Elaeophorbia* by Evans and Kinghorn[73,89] provides support for the reduction of *Elaeophorbia* to a section of the genus *Euphorbia*, as has already been proposed on morphological grounds.[95]

It is interesting that the Thymelaeaceae has yielded to date principally daphnane-type esters, although tigliane-type esters are being described with increasing frequency.[74,77,85] This suggests that the individual classes of compounds are biosynthesized *in toto* on a single enzyme system; if tiglianes are indeed precursors of daphnanes, then they are not necessarily released from the enzyme site before being transformed into daphnanes. Similarly, the cooccurrence of mainly tiglianes and fewer daphnanes in members of the Euphorbiaceae is best explained by the existence of two separate enzyme systems, rather than just one from which both types of compound may arise. A similar situation may well exist for each type of diterpene. These observations need to be borne in mind when considering putative biosynthetic relationships between the various classes of diterpenoids under discussion.

III. CONCLUDING REMARKS

Several questions concerning the fine details of biosynthesis and their implications to the chemosystematic discussion remain unanswered. Is the diterpene nucleus biosynthesized on a single enzyme system *in toto,* or on several in sequence? To what extent are levels of hydroxylation, epoxidation, unsaturation, etc., determined genetically, and at what stage are such functional groups introduced? Is esterification carried out in a genetically predetermined manner or are the compositions of fatty acid pools the determining factors? What is the significance of the existence of chemical races? Until answers to such questions are forthcoming, speculation as to the significance of particular diterpene polyols in plant species is likely to be difficult to interpret. Having said that, patterns are beginning to emerge. The distribution of diterpenes among individual species is discussed in Chapters 6 through 9.

REFERENCES

1. **Banthorpe, D. V. and Charlwood, B. V.,** The terpenoids, in *Encyclopedia of Plant Physiology, New Series,* Vol. 8, Pirson, A. and Zimmermann, M. H., Eds., *Secondary Plant Products,* Bell, E. A. and Charlwood, B. V., Eds., Springer-Verlag, Berlin, 1980, 185.
2. **Cane, D. E.,** The stereochemistry of allylic pyrophosphate metabolism, *Tetrahedron,* 36, 1109, 1980.
3. **Geissman, T. A. and Crout, D. G. H.,** *Organic Chemistry of Secondary Plant Metabolism,* Freeman, Cooper & Company, San Francisco, Calif., 1969, chap. 9.

4. **Cordell, G. A.,** Biosynthesis of sesquiterpenes, *Chem. Rev.*, 76, 425, 1976.
5. **Herz, W.,** Biogenetic aspects of sesquiterpene lactone chemistry, *Isr. J. Chem.*, 16, 32, 1977.
6. **Geissman, T. A. and Crout, D. G. H.,** *Organic Chemistry of Secondary Plant Metabolism*, Freeman, Cooper & Company, San Francisco, Calif., 1969, chap. 10.
7. **Dauben, W. G., Thiessen, W. E., and Resnick, P. R.,** Cembrene, a 14-membered ring diterpene hydrocarbon, *J. Am. Chem. Soc.*, 84, 2015, 1962.
8. **Kobayashi, H. and Akiyoshi, S.,** Thunbergene, a macrocyclic diterpene, *Bull. Chem. Soc. Jpn.*, 35, 1044, 1962.
9. **Roberts, D. L. and Rowland, R. L.,** Macrocyclic diterpenes. α- and β-4,8,13-Duvatriene-1,3-diols from tobacco, *J. Org. Chem.*, 27, 3989, 1962.
10. **Erdtman, H., Norin, T., Sumimoto, M., and Morrison, A.,** Verticillol, a novel type of conifer diterpene, *Tetrahedron Lett.*, (51), 3879, 1964.
11. **Robinson, D. R. and West, C. A.,** Biosynthesis of cyclic diterpenes in extracts from seedlings of *Ricinus communis* L. I. Identification of diterpene hydrocarbons formed from mevalonate, *Biochemistry*, 9, 70, 1970.
12. **Robinson, D. R. and West, C. A.,** Biosynthesis of cyclic diterpenes in extracts from seedlings of *Ricinus communis* L. II. Conversion of geranylgeranyl pyrophosphate into diterpene hydrocarbons and partial purification of thc cyclization enzymes, *Biochemistry*, 9, 80, 1970.
13. **Adolf, W. and Hecker, E.,** Diterpenoid irritants and cocarcinogens in Euphorbiaceae and Thymelaeaceae: structural relationships in view of their biogenesis, *Isr. J. Chem.*, 16, 75, 1977.
14. **Crombie, L., Kneen, G., Pattenden, G., and Whybrow, D.,** Total synthesis of the macrocyclic diterpene (−)-casbene, the putative biogenetic precursor of lathyrane, tigliane, ingenane, and related terpenoid structures, *J. Chem. Soc. Perkin Trans.* 1, 1711, 1980.
15. **Drew, M. G. B., Templeton, D. H., and Zalkin, A.,** The crystal and molecular structure of cembrene, *Acta Crystallogr.*, 25B, 261, 1969.
16. **Patil, V. D., Nayak, U. R., and Dev, S.,** Chemistry of Ayurvedic crude drugs. II. Guggulu (resin from *Commiphora mukul)* -2: diterpenoid constituents, *Tetrahedron*, 29, 341, 1973.
17. **Kodama, M., Matsuki, Y., and Itô, S.,** Synthesis of macrocyclic diterpenoids by intramolecular cyclisation. I. (±)-Cembrene A, a termite trail pheromone, and (±)-nephthenol, *Tetrahedron Lett.*, 35, 3065, 1975.
18. **Uemura, D. and Hirata, Y.,** Stereochemistry of kansuinine A, *Tetrahedron Lett.*, (21), 1701, 1975.
19. **Uemura, D., Katayama, C., Uno, E., Sasaki, K., Hirata, Y., Chen, Y.-P., and Hsu, H.-Y., Kansuine B,** a novel multi-oxygenated diterpene from *Euphorbia kansui* Liou., *Tetrahedron Lett.*, (21), 1703, 1975.
20. **Sahai, R., Rastogi, R. P., Jakupovic, J., and Bohlmann, F.,** A diterpene from *Euphorbia maddeni, Phytochemistry*, 20, 1665, 1981.
21. **Shizuri, Y., Kosemura, S., Ohtsuka, J., Terada, Y., and Yamamura, S.,** Structural and conformational studies on euphohelioscopins A and B and related diterpenes, *Tetrahedron Lett.*, (24), 2577, 1983.
22. **Kupchan, S. M., Sigel, C. W., Matz, M. J., Gilmore, C. J., and Bryan, R. F.,** Structure and stereochemistry of jatrophone, a novel macrocyclic diterpenoid tumor inhibitor, *J. Am. Chem. Soc.*, 98, 2295, 1976.
23. **Burke, B. A., Chan, W. R., Pascoe, K. O., Blount, J. F., and Manchand, P. S.,** The structure of crotonitenone, a novel casbane diterpene from *Croton nitens* Sw. (Euphorbiaceae), *J. Chem. Soc. Perkin Trans.* 1, 2666, 1981.
24. **Torrance, S. J., Wiedhopf, R. M., Cole, J. R., Arora, S. K., Bates, R. B., Beavers, W. A., and Cutler, R. S.,** Antitumor agents from *Jatropha macrorhiza* (Euphorbiaceae). II. Isolation and characterization of jatrophatrione, *J. Org. Chem.*, 41, 1855, 1976.
25. **Hecker, E.,** New toxic, irritant and cocarcinogenic diterpene esters from Euphorbiaceae and from Thymelaeaceae, *Pure Appl. Chem.*, 49, 1423, 1977.
26. **Adolf, W. and Hecker, E.,** On the irritant and cocarcinogenic principles of *Hippomane mancinella, Tetrahedron Lett.*, (19), 1587, 1975.
27. **Nair, N. C.,** Euphorbiales, in *The New Encyclopedia Britannica, Macropaedia*, Vol. 6, 15th ed., Goetz, P. W., et al., Eds., Helen Hemingway Benton, Chicago, 1974, 1027.
28. **Webster, G. L.,** Conspectus of a new classification of the Euphorbiaceae, *Taxon*, 24, 593, 1975.
29. **Willis, J. C.,** *A Dictionary of the Flowering Plants and Ferns*, 8th ed. (revised by H. K. Airy Shaw), Cambridge University Press, Cambridge, 1973.
30. **Domke, W.,** Untersuchungen über die systematische und geographische Gliederung der Thymelaeaceen nebst einer Neubeschreibung ihrer Gattungen, *Bibl. Bot.*, 27(111), 1, 1934.
31. **Waggenitz, G.,** Thymelaeales, in *A. Engler's Syllabus der Pflanzenfamilien*, Vol. 2, 12th ed., Melchior, H., Ed., Gebrüder Borntraeger, Berlin-Nikolassee, 1964, 316.
32. **Takhtajan, A.,** Angiosperm, in *The New Encyclopedia Britannica, Macropaedia*, Vol. 1, 15th ed., Goetz, P. W., et al., Eds., Helen Hemingway Benton, Chicago, 1974, 876.

33. **Hutchinson, J.,** *Evolution and Phylogeny of Flowering Plants. Dicotyledons: Fact and Theory,* Academic Press, London, 1969.
34. **Mitchell, J. C. and Rook, A.,** *Botanical Dermatology,* Greengrass, Vancouver, Canada, 1979.
35. **Evans, F. J. and Schmidt, R. J.,** Plants and plant products that induce contact dermatitis, *Planta Med.,* 38, 289, 1980.
36. **Burkill, I. H.,** *A Dictionary of Economic Products of the Malay Peninsula,* Vol. 1 and 2, Crown Agents for the Colonies, London, 1935.
37. **Von Reis, S. and Lipp, F. J.,** *New Plant Sources for Drugs and Foods from The New York Botanical Garden Herbarium,* Harvard University Press, Cambridge, Mass., 1982.
38. **Chopra, R. N. and Badhwar, R. L.,** Poisonous plants of India, *Indian J. Agric. Sci.,* 10, 1, 1940.
39. **Uphof, J. C. Th.,** *Dictionary of Economic Plants,* H. R. Engelmann (J. Cramer), Weinheim, 1959.
40. **Irvine, F. R.,** *Woody Plants of Ghana,* Oxford University Press, London, 1961.
41. **Souder, P.,** Poisonous plants on Guam, in *Venomous and Poisonous Animals and Noxious Plants of the Pacific Region,* Keegan, H. L. and Macfarlane, W. V., Eds., Pergamon Press, Oxford, 1963, 15.
42. **Morton, J. F.,** *Atlas of Medicinal Plants of Middle America, Bahamas to Yucatan,* Charles C Thomas, Springfield, Ill., 1981.
43. **Altschul, S. von R.,** *Drugs and Foods from Little-Known Plants,* Harvard University Press, Cambridge, Mass., 1973.
44. **Watt, J. M. and Breyer-Brandwijk, M. G.,** *The Medicinal and Poisonous Plants of Southern and Eastern Africa,* 2nd ed., E & S Livingstone, Edinburgh, 1962.
45. **Hui, W.-H., Li, M.-M., and Ng, K.-K.,** Terpenoids and steroids from *Macaranga tanarius, Phytochemistry,* 14, 816, 1975.
46. **Pammel, L. H.,** *A Manual and Poisonous Plants,* Torch Press, Cedar Rapids, Iowa, 1911.
47. **Thomson, R. H.,** *Naturally Occurring Quinones,* 2nd ed., Academic Press, London, 1971.
48. **Gunasekera, S. P., Cordell, G. A., and Farnsworth, N. R.,** Potential anticancer agents. XIV. Isolation of spruceanol and montanin from *Cunuria spruceana* (Euphorbiaceae), *J. Nat. Prod.,* 42, 658, 1979.
49. **Adolf, W., Opferkuch, H. J., and Hecker, E.,** Irritant phorbol derivatives from four *Jatropha* species, *Phytochemistry,* 23, 129, 1984.
50. **Purushothaman, K. K., Chandrasekharan, S., Cameron, A. F., Connolly, J. D., Labbé, C., Maltz, A., and Rycroft, D. S.,** Jatropholones A and B, new diterpenoids from the roots of *Jatropha gossypiifolia* (Euphorbiaceae) — crystal structure analysis of jatropholone B, *Tetrahedron Lett.,* (11), 979, 1979.
51. **Ogura, M., Koike, K., Cordell, G. A., and Farnsworth, N. R.,** Potential anticancer agents. VIII. Constituents of *Baliospermum montanum* (Euphorbiaceae), *Planta Med.,* 33, 128, 1978.
52. **Behl, P. N., Captain, R. M., Bedi, B. M., and Gupta, S.,** *Skin-Irritant and Sensitizing Plants Found in India,* P. N. Behl, New Delhi, 1966.
53. **Airy Shaw, H. K.,** The Euphorbiaceae of Borneo, *Kew Bulletin Additional Series* IV, Her Majesty's Stationery Office, London, 1975.
54. **Handa, S. S., Kinghorn, A. D., Cordell, G. A., and Farnsworth, N. R.,** Plant anticancer agents. XXII. Isolation of a phorbol diester and its $\Delta^{5,6}$-7β-hydroperoxide derivative from *Ostodes paniculata, J. Nat. Prod.,* 46, 123, 1983.
55. **Ghisalberti, E. L., Jeffries, P. R., Toia, R. F., and Worth, G. K.,** Stereochemistry of bertyadionol and related compounds, *Tetrahedron,* 30, 3269, 1974.
56. **Airy Shaw, H. K.,** Notes on Malaysian and other Asiatic Euphorbiaceae, *Kew Bull.,* 20, 379, 1966.
57. **Okuda, T., Yoshida, T., Koike, S., and Toh, N.,** New diterpene esters from *Aleurites fordii* fruits, *Phytochemistry,* 14, 509, 1975.
58. **Hecker, E. and Schmidt, R.,** Phorbolesters — the irritants and cocarcinogens of *Croton tiglium* L., *Fortschr. Chem. Org. Naturstoffe,* 31, 377, 1974.
59. **Stuart, K. L. and Barrett, M.,** A phorbol derivative from *Croton rhamnifolius, Tetrahedron Lett.,* (28), 2399, 1969.
60. **Chan, W. R., Prince, E. C., Manchand, P. S., Springer, J. P., and Clardy, J.,** The structure of crotofolin A, a diterpene with a new skeleton, *J. Am. Chem. Soc.,* 97, 4437, 1975.
61. **Uhe, G.,** Medicinal plants of Samoa, *Econ. Bot.,* 28, 1, 1974.
62. **Schwartz, L., Tulipan, L., and Birmingham, D. J.,** *Occupational Diseases of the Skin,* 3rd ed., Lea & Febiger, Philadelphia, 1957.
63. **Ohigashi, H., Katsumata, H., Kawazu, K., Koshimizu, K., and Mitsui, T.,** A piscicidal constituent of *Excoecaria agallocha, Agric. Biol. Chem., Jpn.,* 38, 1093, 1974.
64. **Kinghorn, A. D.,** Cocarcinogenic irritant Euphorbiaceae, in *Toxic Plants,* Kinghorn, A. D., Ed., Columbia University Press, New York, 1979, 137.
65. **Ohigashi, H., Kawazu, K., Koshimizu, K., and Mitsui, T.,** A piscicidal constituent of *Sapium japonicum, Agric. Biol. Chem., Jpn.,* 36, 2529, 1972.
66. **Adolf, W. and Hecker, E.,** New irritant diterpene-esters from roots of *Stillingia sylvatica* L. (Euphorbiaceae), *Tetrahedron Lett.,* (21), 2887, 1980.

67. **Sakata, K., Kawazu, K., Mitsui, T., and Masaki, N.,** Structure and stereochemistry of huratoxin, a piscicidal constituent of *Hura crepitans, Tetrahedron Lett* , (16), 1141, 1971.
68. **Gschwendt, M. and Hecker, E.,** Tumor promoting compounds from *Euphorbia triangularis:* mono- and diesters of 12-desoxy-phorbol, *Tetrahedron Lett.,* (40), 3509, 1969.
69. **Opferkuch, H. J. and Hecker, E.,** New diterpenoid irritants from *Euphorbia ingens, Tetrahedron Lett.,* 3, 261, 1974.
70. **Hergenhahn, M., Adolf, W., and Hecker, E.,** Resiniferatoxin and other esters of novel polyfunctional diterpenes from *Euphorbia resinifera* and *unispina, Tetrahedron Lett.,* (19), 1595, 1975.
71. **Zechmeister, K., Röhrl, M., Brandl, F., Hechtfischer, S., Hoppe, W., Hecker, E., Adolf, W., and Kubinyi, H.,** Röntgenstrukturanalyse eines neuen Makrozyklischen Diterpen-Esters aus der Springwolfsmilch *(Euphorbia lathyris* L.), *Tetrahedron Lett.,* (35), 3071, 1970.
72. **Kinghorn, A. D. and Evans, F. J.,** A biological screen of selected species of the genus *Euphorbia* for skin irritant effects, *Planta Med.,* 28, 325, 1975.
73. **Evans, F. J. and Kinghorn, A. D.,** The succulent Euphorbias of Nigeria. I., *Lloydia,* 38, 363, 1975.
74. **Gunasekera, S. P., Kinghorn, A. D., Cordell, G. A., and Farnsworth, N. R.,** Plant anticancer agents. XIX. Constituents of *Aquilaria malaccensis, J. Nat. Prod.*, 44, 569, 1981.
75. **Zayed, S., Adolf, W., Hafez, A., and Hecker, E.,** New highly irritant 1-alkyldaphnane derivatives from several species of Thymelaeaceae, *Tetrahedron Lett.,* (39), 3481, 1977.
76. **Nyborg, J. and La Cour, T.,** The structure of mezerein, a major toxic principle of *Daphne mezereum* L., *Nature (London),* 257, 824, 1975.
77. **Adolf, W. and Hecker, E.,** On the active principles of the Thymelaeaceae. II. Skin irritant and cocarcinogenic diterpenoid factors from *Daphnopsis racemosa, J. Med. Plant Res.,* 45, 177, 1982.
78. **Badawi, M. M., Handa, S. S., Kinghorn, A. D., Cordell, G. A., and Farnsworth, N. R.,** Plant anticancer agents. XXVII. Antileukemic and cytotoxic constituents of *Dirca occidentalis* (Thymelaeaceae), *J. Pharm. Sci.,* 72, 1285, 1983.
79. **Jolad, S. D., Hoffmann, J. J., Timmermann, B. N., Schram, K. H., Cole, J. R., Bates, R. B., Klenck, R. E., and Tempesta, M. S.,** Daphnane diterpenes from *Wikstroemia monticola:* wikstrotoxins A - D, huratoxin, and excoecariatoxin, *J. Nat. Prod.,* 46, 675, 1983.
80. **Arnold, H. L.,** *Poisonous Plants of Hawaii,* Charles E. Tuttle, Rutland, Vt., 1968.
81. **Kupchan, S. M., Sweeny, J. G., Baxter, R. L., Murae, T., Zimmerly, V. A., and Sickles, B. R.,** Gnididin, gniditrin, and gnidicin, novel potent antileukemic diterpenoid esters from *Gnidia lamprantha, J. Am. Chem. Soc.,* 97, 672, 1975.
82. **Coetzer, J. and Pieterse, M. J.,** The isolation of 12-hydroxy-daphnetoxin, a degradation product of a constituent of *Lasiosiphon burchellii, J. Suid-Afrik. Chem. Inst.,* 24, 241, 1974.
83. **Nadkarni, A. D.,** *Dr. K. M. Nadkarni's Indian Materia Medica,* Vol. 1, Popular Prakashan, Bombay, 1976.
84. **Cashmore, A. R., Seelye, R. N., Cain, B. F., Mack, H., Schmidt, R., and Hecker, E.,** The structure of prostratin: a toxic tetracyclic diterpene ester from *Pimelea prostrata, Tetrahedron Lett.,* (20), 1737, 1976.
85. **Zayed, S., Adolf, W., and Hecker, E.,** On the active principles of the Thymelaeaceae. I. The irritants and cocarcinogens of *Pimelea prostrata, J. Med. Plant Res.,* 45, 67, 1982.
86. **Rizk, A. M., Hammouda, F. M., Ismail, S. E., El-Missiry, M. M., and Evans, F. J.,** Irritant resiniferonol derivatives from Egyptian *Thymelaea hirsuta* L., *Experientia,* 40, 808, 1984.
87. **Kinghorn, A. D.,** Some Biologically Active Constituents of the Genus *Euphorbia,* Ph.D. thesis, University of London, London, 1975.
88. **Thurston, E. L. and Lersten, N. R.,** The morphology and toxicology of plant stinging hairs, *Bot. Rev.,* 35, 393, 1969.
89. **Evans, F. J. and Kinghorn, A. D.,** A comparative phytochemical study of the diterpenes of some species of the genera *Euphorbia* and *Elaeophorbia* (Euphorbiaceae), *J. Linn. Soc. Bot.,* 74, 23, 1977.
90. **Kinghorn, A. D.,** Characterisation of an irritant 4-deoxyphorbol diester from *Euphorbia tirucalli, J. Nat. Prod.,* 42, 112, 1979; *Chem. Abstr.,* 98, 68891.
91. **Baslas, R. K. and Gupta, N. C.,** Chemical constituents of the bark of *Euphorbia tirucalli, Indian J. Pharm. Sci.,* 44, 113, 1982.
92. **Baslas, R. K. and Gupta, N. C.,** Chemical investigation of Indian medicinal plants possessing anticancer activity: roots of *Euphorbia tirucalli* Linn., *J. Indian Chem. Soc.,* 60, 506, 1983; *Chem. Abstr.,* 99, 191694.
93. **Baslas, R. K. and Gupta, N. C.,** Chemical investigation of the root of *Euphorbia tirucalli* L., *Herba Hung.,* 22, 35, 1983; *Chem. Abstr.,* 100, 64928.
94. **Upadhyay, R. R., Bakhtavar, F., Mohseni, H., Sater, A. M., Saleh, N., Tafazuli, A., Disaji, F. N., and Mohaddes, G.,** Screening of *Euphorbia* from Azarbaijan for skin irritant activity and for diterpenes, *Planta Med.,* 38, 151, 1980.
95. **Webster, G. L.,** The genera of Euphorbiaceae in the southeastern United States, *J. Arnold Arbor.,* 48, 363, 1967.

Chapter 5

NON-DITERPENOID CONSTITUENTS OF EUPHORBIACEAE AND THYMELAEACEAE

Abdel-Fattah M. Rizk and Mustafa M. El-Missiry

TABLE OF CONTENTS

I. EUPHORBIACEAE

The Euphorbiaceae, or spurge family, comprises some 263 genera and about 7300 species of almost cosmopolitan distribution, mainly of the tropics, but extending also into the temperate regions of the northern and southern hemispheres.[1] Fifteen genera of this family have more than 100 species each: *Euphorbia* (about 2000 species), *Croton* (700 species), *Phyllanthus* (480 species), *Acalypha* (430 species), *Glochidion* (280 species), *Macaranga* (240 species), *Manihot* (160 species), *Jatropha* (150 species), and *Tragia* (140 species).[2,3]

Plants of this family have acrid, milky, or colorless juice. They are of considerable economic importance, and products obtained from this family include castor oil *(Ricinus)*, tung oil *(Aleurites)*, cassava and tapioca *(Manihot)*, and rubber *(Hevea)*.[1]

In addition to the co-carcinogenic properties and the antitumor activity of certain plants of this family, particularly *Euphorbia* species, a number of alleged folkloric uses have been ascribed to several species of this family, including *Croton* and *Euphorbia*.[3,4] It has been reported that plants exhibit antimicrobial properties, antimalarial, antiatherogenic, androgenic, and insecticidal activities.[5-7]

A. Terpenoids and Related Substances

Several triterpenoids have been reported as constituents of the Euphorbiaceae. Reviews on the triterpenoid investigations, particularly of *Euphorbia* species, were written by several authors.[8-11] A review on these investigations in connection with plant taxonomy was given early on by Ponsinet and Ourisson.[12] In the triterpene field, several pentacyclic (e.g., bauerenol, germanicol, and phyllanthol) and tetracyclic (e.g., euphol, euphorbol., and tirucallol) alcohols were isolated from plants of this family. In *Euphorbia* several groups of species can be defined according to their tetracyclic terpenes. These groups were found to correspond partially with the established chemotaxonomic divisions.[13] Hundreds of species of Euphorbiaceae have been investigated and more than 50 triterpenes have been identified (examples are shown in Table 1 and Figure 1).

On the basis of the triterpene components (15 tetra- and pentacyclic compounds) identified from the latex of 75 *Euphorbia* species, Ponsinet and Ourisson[14] suggested that the species fell into five groups: group A with euphol and euphorbol in a 2:1 ratio; group B with euphol and tirucallol in a 2:1 ratio; group C with cycloartenol and 24-methylene-cycloartenol as major products; group D with cycloartenol, 24-methylene cycloartenol, lupeol, and euphol in equal amounts; and group E with α-amyrin, but lacking tetracyclic triterpenes. Species in group C have been further subdivided on the basis of minor constituents such as lanosterol, lanostenol, lupeol, β-amyrin, butyrospermol, euphol, and tirucallol.[14] Further studies showed that herbaceous euphorbias all contained cycloartenol; cactus-like species contained euphol and euphorbol, while coral-like species contained euphol and tirucallol.[15]

Investigations of various *Glochidion* species[108—115] have shown that this genus is characteristic in yielding lupene derivatives which include some new compounds: glochidone (lup-1,20(29)-diene-3-one), glochidiol (lup-20(29)-en-1β,3α-diol), glochidonal (1β-hydroxy-lup-20(29)-ene-3-one), lup-20(29)-en-,1β-3β-diol, lup-20(29)-en-3α,23-diol, glochilocudiol (lup-20(29)-en-1α,3β-diol), and glochidol. The number of new compounds isolated from this species is seven and all those possessing two oxygen functions have these existing in a 1,3 relationship which might be another significant features of this genus.[110]

β-Sitosterol has been identified in the sterol fractions of the different studied Euphorbiaceae species. However, the presence of other sterols; e.g., campesterol, stigmasterol, euphorbosterol (a distinct stereoisomer of stigmasterol) and others have also been reported (Table 2 and Figure 2).

Though β-sitosterol represents the major constituent of the sterol fractions of the different *Euphorbia* species, in *E. pulcherimma* the most abundant component was cholesterol, con-

Table 1
TRITERPENES OF SOME EUPHORBIACEOUS PLANTS

Species	Plant part	Triterpenoids	Ref.
1. *Aleurites moluccana* L.	l	Moretenone, moretenol, α-amyrin	16
	s	Moretenone	
2. *Antidesma bunias* Spreng.	l, s	Epifriedelinol, dammara-20,24-diene-3-β-ol	17
3. *Apurosa cardisperma* Merr.	f	Friedelan deriv.	18
4. *Baccaurea sapida* Muell.	l	Epifriedelinol, β-amyrin, taraxerol	17
5. *Beyeria brevifolia*	l, s	Lup-20-ene-3β,16β-diol	19
6. *Beyeria leschenaultii*	l	Lup-20-ene-3β,16β-diol	19
7. *Bischofia trifoliata* (Roxb.) HK.f.	l, s	Friedelan-3α-yl acetate, friedelin, friedelan-3β-ol, friedelan-3α-ol, epifriedelinol, friedelinol	16,17
8. *Bridelia micrantha* Baill.	b	Epifriedelinol, taraxerone, taraxerol	17
9. *B. moonii*		Glochidone	18
10. *B. stipularis* Blume.	b	Epifriedelinol	17
11. *Croton sparsiflorus* Morong. *(C. bonplandianum* Bail.)	l, s	Taraxerol	20
12. *Euphorbia abyssinica* Raeus	t	Euphol, euphorbol	1
13. *E. adenochlora* Morr.	h	Taraxerol, taraxerone, euphol, cycloartenol	21
14. *E. ammak* Schw.	t	Euphol, euphorbol	14
15. *E. amygdaloides* L.	t	Lanosterol	14
16. *E. antiquorum* Forsk.	t	Euphol, euphorbol, β-amyrin, cycloartenol	22, 23
	r	Taraxerol	24
	s, b	Taraxerol, epifriedelanol	25
	s	Taraxerol, taraxerone, friedelan-3α-ol, epifriedelinol-3β-ol, friedelinol, euphol	21, 24—26
17. *E. aphylla* Brouss.	t	Handianol (= cycloartenol), lanosterol, lanostenol	14, 27
18. *E. atropurpurea* Brouss.	t	Lanosterol	14
19. *E. balsamifera* Ait.	t	Germanicol, epigermanicol, cycloartenol, lanosterol, lanostadienol	14, 28—30
20. *E. brachiata* E. mey.	t	Lanosterol	14
21. *E. bravoona* Soveut.	t	Cycloartenol, lanosterol, lanostenol	31
22. *E. canariensis* L.	t	Euphol, euphorbol	14, 32
23. *E. candelabrum* Trém.	t	Euphol, euphorbol	14
24. *E. capuroni* Ursch et Léandri	t	β-Amyrin, cycloartenol, 24-methylenecycloartenol, butyrospermol	14
25. *E. caput-meduras* L.	t	Euphol, tirucallol	14
26. *E. carascana*	t	Euphol, euphol acetate	33
27. *E. cattimandoo*	t	Euphol, nerifoliol, euphorbol	34
	s	Taraxerol	
28. *E. caudicifolia* Mans.	t	Cyclolaudenol, 3-epi-cyclolaudenol	35
	s	Cycloartenol, 3-ketomethyluroslate	
29. *E. condylocarpa*	r	β-Euphorbol, euphadienol	36
30. *E. coralloides*	t	Lanosterol	14
31. *E. corollata*	h	Corollatadiol	37
32. *E. cristata* Lem.	t	Euphol, euphorbol	14
33. *E. cyparissias* L.	h	Euphol, 24-methylenecycloartenol, cycloart-23-en-3,25-diol, glut-5(6)-en-3-one, glut-5(6)-en-3β-ol, β-amyrin, hopenone-β	38, 39
	t	Euphol, lanosterol	14
34. *E. deightoni* L. Ct.	t	Euphol, euphorbol	14
35. *E. delphinsis* Ursch et Léandri	t	β-Amyrin, cycloartenol, 24-methylenecycloartenol, butyrospermol	14
36. *E. dendroides* L.	t	Lanosterol	14
37. *E. didieroides* M. Den.	t	β-Amyrin, cycloartenol, 24-methylenecycloartenol, butyrospermol	14
38. *E. dracuncloides* Lam.	h	Euphorbol	14, 40

Table 1 (continued) TRITERPENES OF SOME EUPHORBIACEOUS PLANTS

Species	Plant part	Triterpenoids	Ref.
39. *E. durani* Ursch et Leandri	t	β-Amyrin, cycloartenol, 24-methylenecycloartenol, butyrospermol	14
40. *E. ebracteolata* Hayata	t	β-Amyrin acetate, 24-methylenecycloartenol	41
41. *E. echinus* Hook.	t	Cycloartenol, 24-methylenecycloartenol, euphol, lupeol, obtusifoldienol, lanosterol, lanostenol	42, 43
42. *E. enopla* Bois.	t	Euphol, tirucallol	14
43. *E. enterophora*	t	Euphol, tirucallol	14
44. *E. epithymoides* Jacq.	t	Lanosterol	14
45. *E. erythraea* Hemsl.	t	Euphol, euphorbol	14
46. *E. esula* L.	h	24-methylenecycloartenol, cycloartenol, lupeol	44, 45
47. *E. exigua* L.	t	Lanosterol	14
48. *E. fianarantsoe* Ursch et Léandri	t	β-Amyrin, cycloartenol, 24-methylenecycloartenol, butyrospermol	14
49. *E. gatbergensis* N.E. Br.	t	Euphol, tirucallol	14
50. *E. geniculata* Ortega	t, h	β-Amyrin acetate	46, 47
	h	Geniculatin (saponin)	48
51. *E. genoudiana*	t	β-Amyrin, butyrospermol, cycloartenol, 24-methylene-cycloartenol	14
52. *E. grandidens* Haw.	t	Euphol, euphorbol	14
53. *E. handiensis* Burch	t	Cycloartenol, euphol	42
54. *E. helioscopia* L.	t	Lanosterol	14
55. *E. hermentiana* Lem.	t	Euphol, euphorbol	14
56. *E. hernandez*	t	Cycloartenol	42
57. *E. hernandez-pachecoi* Cab.	t	Cycloartenol, lanostenol, lanosterol	42
58. *E. heterophylla* L.	t	β-Amyrin	14
	l	Euphyl acetate, moretenone	49
59. *E. hirta* L.	h	Cycloartenol, 24-methylenecycloartenol, β-amyrin, taraxerol, taraxerone	50—52
	t	β-Amyrin, lupeol	14
	s	β-amyrin, taraxerol, friedelin	53, 54
60. *E. hirta* var. *procumbens*	h	Friedelin, β-amyrin	51
61. *E. indica* Lam. = *E. hypercifolia* L.	h	Taraxerol, 24-methylenecycloartenol	55
62. *E. ingens* E. Mey.	t	Euphol, euphorbol	14,56
63. *E. intysi* Drake	t	Euphol, tirucallol	14
64. *E. jaxartica*	h	Euphol, euphorbol	57, 58
	r	Euphol, euphorbol, euphorbol-hexacozonate	
65. *E. jolkini* Boiss.	h	Taraxerol	51
66. *E. kamerunica* Pax.	t	Euphol, euphorbol	14
67. *E. lactea* Haw.	t	Euphol, euphorbol	14
68. *E. laro* L.	t	Euphol, tirucallol	14
69. *E. latazi* var. *glabra* Steyermark	t	Lanosterol, cycloartenol, 24-methylenecycloartenol	59
70. *E. lateriflora* Schum. and Thonn.	h	Cholest-5-one, moretenone, moretenol, lupeol, 24-methylenecycloartenol	60
71. *E. lathyris* L.	t	Lanosterol	14
	s	Taraxerol, taraxerone, betulin	61
72. *E. leuconeura* Boiss.	t	24-methylenecycloartenol, cycloartenol, butyrospermol, β-amyrin	14
73. *E. longana*	l	Epifriedelinol	62

Table 1 (continued)
TRITERPENES OF SOME EUPHORBIACEOUS PLANTS

Species	Plant part	Triterpenoids	Ref.
74. *E. lophogona* Lam.	t	β-Amyrin, cycloartenol, 24-methylenecycloartenol, butyrospermol	14
75. *E. maculata*	h	Maculatol	63
76. *E. maddani*		*epi*-Lupeol acetate, lupenone, lupeol acetate	64
77. *E. mamillaris* L.	t	Euphol, tirucallol	14
78. *E. marignata* Pursh.	t	24-methylenecycloartenol, cycloartenol, β-amyrin, lanosterol, butyrospermol	14
79. *E. mauritanica* L.	t	Lanosterol	14
80. *E. millii* Desm.	h	Cycloartenol, β-amyrin, β-amyrin acetate	65
81. *E. mitsembinensis* Ursch et Léandri	t	24-Methylenecycloartenol, β-amyrin, cycloartenol, butyrospermol	14
82. *E. monteirii* Hook.	t	Cycloartenol, 24-methylenecycloartenol	14
83. *E. myrsinites* Linn.	h	β-Amyrin, taraxerol	66
	t	Lanosterol, cycloartenol, 24-methylenecycloartenol	14
84. *E. nerifolia* L.	t	Euphol, nerifoliol	67
	l	Friedelan-3α-ol, taraxerol	68, 69
	s	Friedelan-3α-ol, taraxerol	68
	b	Euphol, 24-methylenecycloartenol	70
85. *E. neohumberti* Boit.	t	β-Amyrin, 24-methylenecycloartenol, cycloartenol, butyrospermol	14
86. *E. obtusifolia* Poir.	t	Cycloartenol, obtusifoliol, lanosterol, obtusifoldienol	14, 71
87. *E. officinarum* L.	t	Euphol, lupeol, cycloartenol, 24-methylenecycloartenol	14
88. *E. oncoclada* Drake	t	Euphol, tirucallol	14
89. *E. pallassii*	r	Lupeol, cycloartenol acetate	72
90. *E. paralias* L.	h	α-Amyrin, β-amyrin, uvaol, betulin, oleanolic, ursolic, cycloartenol, 24-methylenecycloartenol	73, 74
	t	Lanosterol, cycloeucalenol, cycloartenol, 24-methylenecycloartenol	
91. *E. parviflora*	r	β-Amyrin, moretenone, δ-amyrenone, taraxerol	49
92. *E. pauliani* Ursch et Léandri	t	24-Methylenecycloartenol, β-amyrin, cycloartenol, butyrospermol	14
93. *E. pedilanthoides* D. Men.	t	24-Methylenecycloartenol, β-amyrin, cycloartenol, butyrospermol	14
94. *E. peplus* L.	t	Lanosterol	14
	h	β-Amyrin acetate	75
95. *E. perrieri* Drake	t	24-Methylenecycloartenol, cycloartenol, butyrospermol, β-amyrin	14
96. *E. petiolata*	h	Cycloartenol, 24-methylenecycloartenol	76
97. *E. phosphorea*	c	Lupeol, lupenone, taraxenone, taraxerol acetate, olean-13(18)-en-3-one	77
98. *E. pilulifera*	h	Taraxerol, taraxerone, β-amyrin, euphol, euphorbol, tirucallol	78, 79
99. *E. polygonifolia* L.	h	β-Amyrin, cycloartenol, lupeol, cycloart-23-en-3-β-ol	80
100. *E. pugniformis* Boiss.	t	Euphol, tirucallol	14
101. *E. pulcherrima* Wild.	f, s	Germanicol, β-amyrin, pseudotaraxasterol	81
	h	Germanicol acetate, α-amyrin	82
	t	Germanicol, pseudotaraxasterol, β-amyrin, α-amyrin	14, 81

Table 1 (continued)
TRITERPENES OF SOME EUPHORBIACIOUS PLANTS

Species	Plant part	Triterpenoids	Ref.
	s	Germanicyl acetate	83
	l	Germanicol, germanicol acetate	84
102. *E. regis-jubae* Webb.	t	Lanosterol, lanostenol, taraxerol	14, 31, 51
103. *E. resinifera* Boiss.	t	Euphol, euphorbol, α-euphol, β-euphol, taraxerol, β-amyrin, obtusifoldienol	14, 85, 86
104. *E. royleana* Boiss.	h	Taraxerol	87
	t	Euphol, cycloeucalenol, euphorbol, cycloroylenol	14, 88, 89
	s	Taraxerol, *epi*-taraxerol-glut-5-en-3-β-ol	90, 91
105. *E. segetalis* L.	h	Lupeol, cycloartenol, 24-methylenecycloartenol	92
106. *E. serrata* L.	t	Lanosterol	14
107. *E. shimperi* Presl.	t	Lanosterol	14
108. *E. sikkimensis*	t	Glut-5-en-3-on, butyrospermol	93
109. *E. sipolisii* N.E. Br.	t	Cycloartenol, 24-methylenecycloartenol	14
110. *E. stenoclada* Baill.	t	Euphol, tirucallol	14
111. *E. sudanica* N.E. Br.	t	Euphol, euphorbol	14
112. *E. terracina* L.	h	β-Amyrin	94
113. *E. thymifolia* L.	r	Taraxerol, tirucallol, euphorbol, 24-methylenecycloartenol	95, 96
114. *E. tinctoria*	h	γ-Euphorbol	97, 98
115. *E. tirucalli* L.	t	Euphol, isoeuphorbol, taraxasterol, tirucallol, euphorbinol	99—101
	h, s	Euphol, tirucallol, taraxerol	14, 102, 103
	b	Euphorbol, euphorbol hexacosonate, taraxerone	104
	l	β-Amyrin, cycloartenol	105
116. *E. trigona* Haw.	l	Euphol, euphorbol	14
117. *E. verrucosa* Lam.	l	Lanosterol	14
118. *E. veguieri* M. Den.	l	24-Methylenecycloartenol, β-amyrin, cycloartenol, butyrospermol	14
119. *E. virose* Willd.	l	Euphol, euphorbol	14
120. *E. wallichii* Hr.f.	r	Euphol, euphorbol	106
121. *E. watanabei* Mark.	h	Alnusenone, taraxerol acetate, taraxerone, taraxasterol, lupenone, lupeol, lupeol acetate	107
122. *Glochidion accuminatum*		Glochidonol, glochidone	108
123. *G. dasyphyllum*	s	Lupenone, glochidone, glochidiol	109
124. *G. eriocarpum*	s	Lupenone, lupeol, glochidone, glochidol, glochidinol, lup-20(29)-en-1β,3β-diol	110
125. *G. hohenackeri*	b, r	3-epi-Lupeol, glochidone, glochidiol	111
126. *G. honkongense*	s	Epilupeol, lupeol, glochidinol, glochidone, glochidiol	109
127. *G. macrophyllum*	s	Friedelin, friedelan-3β-ol, lup-20(29)-ene-1β,3β-diol	109
	l	Friedelin, frieden-3β-ol	109
	s	Lup-20(29)-en-3α-23-diol	112
128. *G. multiloculare*		Glochilocudiol, glochidiol, glochidione	113
129. *G. thomsoni*		Lup-20(29)-ene-1β,3β-diol	108
130. *G. venulatum*	l	Glochidone	114
131. *G. wrightii* Benth.	s	Glochidinol	115
132. *Hura crepitans* L.	t	Butyrospermol, cycloartenol, 24-methylene cycloartenol	116
133. *Phyllanthus acidus* Skeel.	b	β-Amyrin, phyllanthol	117
134. *P. discoides* Muell.	r	Betulinic acid	118

Table 1 (continued)
TRITERPENES OF SOME EUPHORBIACEOUS PLANTS

Species	Plant part	Triterpenoids	Ref.
135. *P. distichus*	b, r	Lupeol	119
136. *P. emblica*	b	Lupeol, lupenone, epitaraxerol	120, 121
137. *P. engleri*		Phyllanthol	122
138. *P. muellerianus*	b	Friedelane deriv.	123
139. *P. reticulatus* Poir.	l	Friedelin, glochidonol, friedelan-3β-ol, 21α-hydroxy-friedelan-3-one, 21α-hydroxy-friedel-4(23)-en-3-one	124
	s	Friedelin, betulinic acid	
140. *Poddadenia thwaitesii*	h	Aleuritolic acid	9
141. *Putranjiva roxburghii*	h	Friedelinol, 3α-hydroxy-friedelan-7-one	17
142. *Sapium discolor* Muell.-Arg.	l	Eipfriedelinol	17
143. *S. sebiferum* Roxb.	l	Friedelin	125
	s	Moretenone, moretenol	

Note: b, Bark; c, cortex; f, flowers; h, herb; l, leaves; r, roots; s, stems; t, latex.

stituting half the steroid mixture, followed in order of decreasing concentration by 24-α-ethylcholesterol (sitosterol), 24-α-methylcholesterol (campesterol), and 24β-methyl-cholesterol (22-dihydrobrassicasterol).[133] The relative amount of cholesterol in *E. pulcherimma* has been reported to be the highest found so far in a tracheophyte.[133]

In the course of investigations on the terpenes and sterols in various *Euphorbia* species, several workers[92-96,135-139] have encountered long-chain fatty alcohols and hydrocarbons. *n*-Octacosanol and/or *n*-hexacosanol are the alcohols commonly isolated from *Euphorbia* species. They have been identified in the 15 *Euphorbia* species studied for their alcohol content. The presence of other homologues, e.g., tetracosanol,[138] *n*-docosanol,[135] and *n*-triacontanol[139] was also reported.[139] Helioscopiol[131] and pulcherrol[81] were identified from *E. helioscopia* and *E. pulcherrima,* respectively. The hydrocarbons isolated from the different *Euphorbia* species have been shown by GLC and/or MS to be mixtures of *n*-alkanes. All the hydrocarbons isolated before applying the GLC/MS techniques and thought to be single pure hydrocarbons by m.p. and elemental analysis were shown to be complex mixtures of *n*-alkanes.[74,75,140] The leaves of *E. heterophylla* yielded 10,10-dimethyl hexacosan-7-one.[141]

Euphorbia species have been found to yield a considerable amount of waxy material (hydrocarbons), alcohols, and terpenes, and a number of these species have been suggested as potential hydrocarbon-producing crops (renewable sources for petroleum-derived products and petrochemicals) for production of oil low in isoprene units.[142-146] Research with several latex-rich species showed that low molecular weight hydrocarbons suitable for cracking to fuels occur in reasonable percentages (10% or more of total dry weight) in many of the Euphorbiaceae; the gopher plant *(E. lathyris),* ranked among the highest species analyzed, containing up to 10% reduced hydrocarbons (acetone-extractable materials). *E. lathyris* and *E. tirucalli* have been estimated to yield approximately ten barrels of hydrocarbon material per acre per year.[147] The product was a mixture of hydrocarbons consisting mainly of open-chain and cyclic terpenoid isoprenes. According to Calvin,[148] 95% of the oil from *E. lathyris* can be cracked to useful products, making it a very valuable petrochemical feedstock from which gasoline itself can be made.

B. Flavonoids

Species of the Euphorbiaceae, and in particular *Euphorbia* species, have been reported as plants bearing flavonoids. The different classes of the flavonoids, viz., flavanone (e.g.,

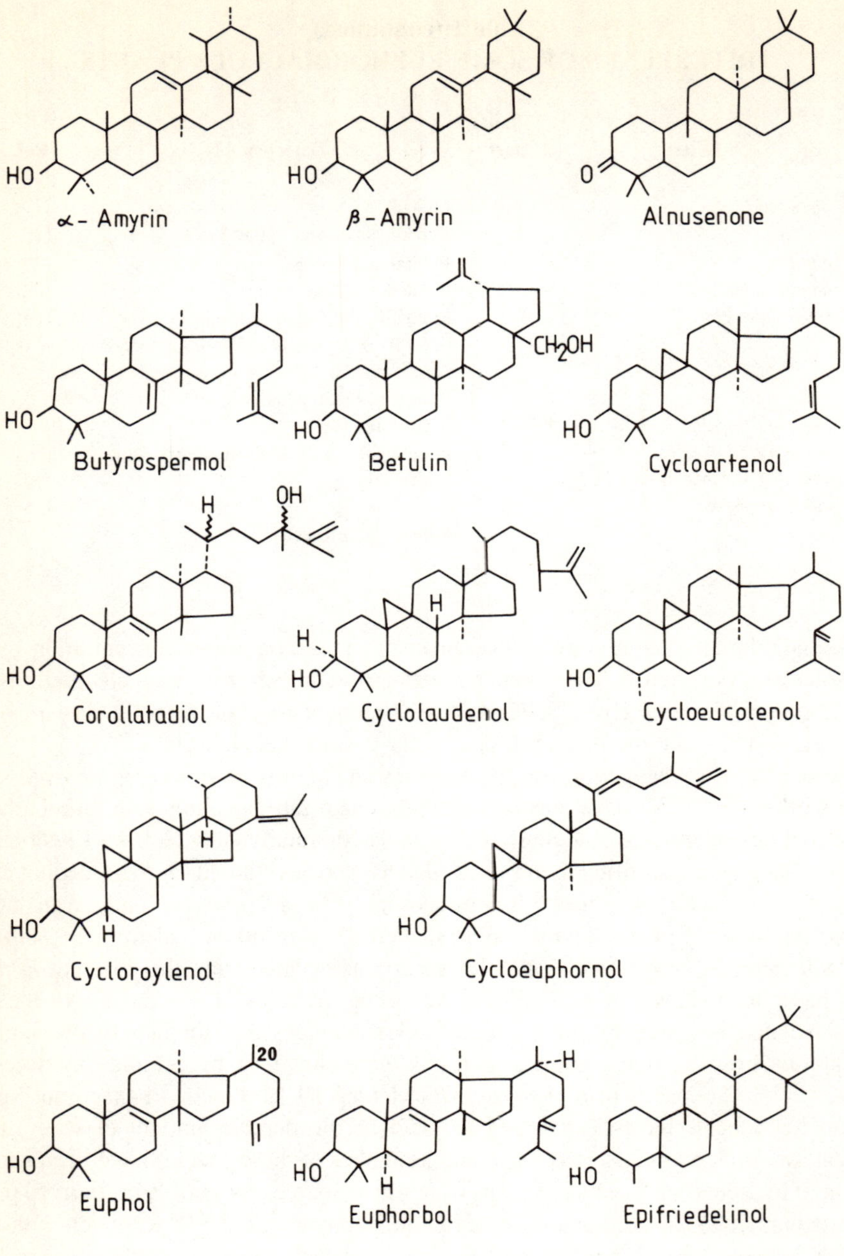
HO
α- Amyrin
HO
β-Amyrin
O
Alnusenone
HO
Butyrospermol
CH2OH
HO
Betulin
HO
Cycloartenol
H
OH
HO
Corollatadiol
H
H
HO
Cyclolaudenol
HO
Cycloeucolenol
H
HO
H
Cycloroylenol
HO
Cycloeuphornol
20
HO
Euphol
H
HO
H
Euphorbol
HO
Epifriedelinol

FIGURE 1. Triterpenes of the family Euphorbiaceae.

naringenin), flavonol (e.g., quercetin, rutin, kaempferol, tithymalin, myricetin, etc.) and flavone (vitexin, saponaretin, orientin, apigenin, etc.) have been represented in the family (Table 3 and Figure 3). Both O- and C-glycosides have been identified in the family.

C. Coumarins

Although qualitative detection of coumarins in several plants of the family Euphorbiaceae has been reported, yet only few have been isolated (Table 4). The seeds of *Euphorbia lathyris* L. yielded, in addition to esculetin, two bicoumarins (euphorbetin and isoeuphorbetin).[205-206] A coumarino-lignan was identified from the roots of *Jatropha glandulifera*.[207]

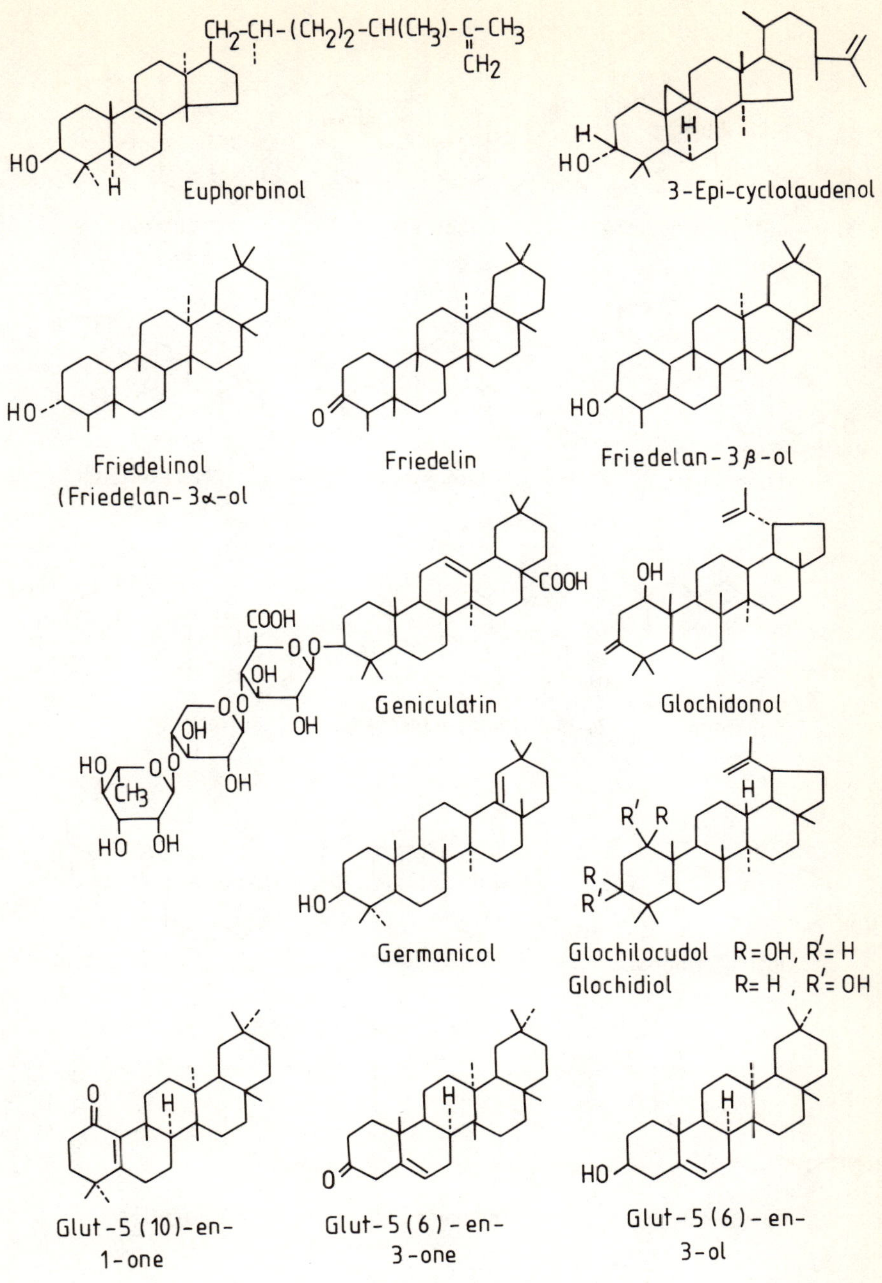

Euphorbinol

3-Epi-cyclolaudenol

Friedelinol (Friedelan-3α-ol

Friedelin

Friedelan-3β-ol

Geniculatin

Glochidonol

Germanicol

Glochilocudol R=OH, R'=H
Glochidiol R=H, R'=OH

Glut-5(10)-en-1-one

Glut-5(6)-en-3-one

Glut-5(6)-en-3-ol

FIGURE 1 (continued)

D. Tannins, Phenanthrenes, and Other Related Phenolic Compounds

The distribution of tannins in several species of the Euphorbiaceae, e.g., *Euphorbia maculata, Gleditsia japonica, Phyllanthus emblica, Sapium sebiferum,* and *Mallotus japonicus* has been reported (Figure 5).[184,212-215] The tannin of the bark of *Phyllanthus emblica* amounts to 16.64% and is almost exclusively catechuic, with only traces of pyrogallic tannins.[212] *Euphorbia maculata* contains 15.7 to 17.1% of tannin substances, predominantly of the pyrogallol group (46%) and pyrocatechols (39.4%).[184] Okuda et al.[215] have recently found that geraniin[216] is the main tannin of *Triadica sebifera* (= *Sapium sebiferum*), and that geraniin and mallotusinic acid[216] are the main tannins in *Mallotus japonicus*.

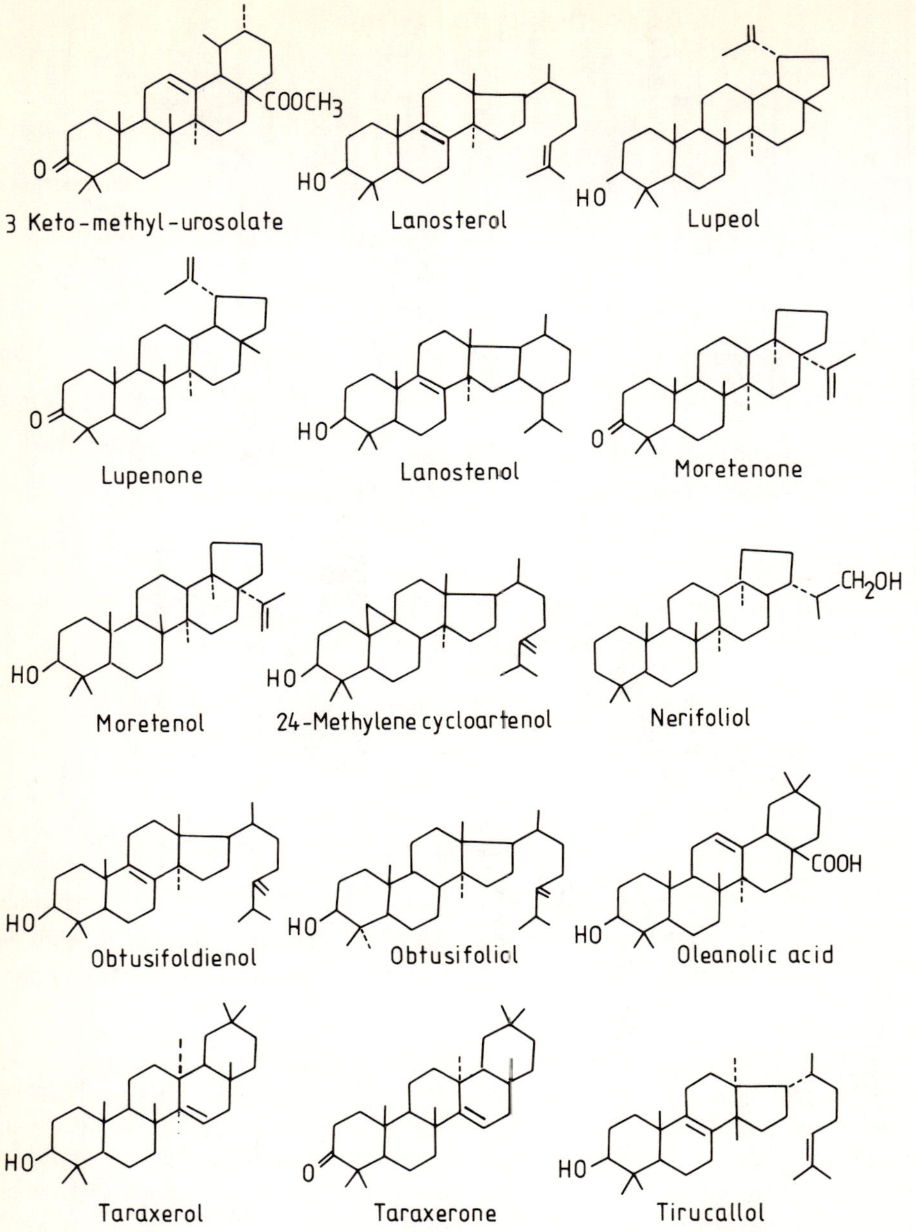

COOCH3

3 Keto-methyl-urosolate Lanosterol Lupeol

Lupenone Lanostenol Moretenone

CH2OH

Moretenol 24-Methylene cycloartenol Nerifoliol

COOH

Obtusifoldienol Obtusifoliol Oleanolic acid

Taraxerol Taraxerone Tirucallol

FIGURE 1 (continued)

Ellagic acid (a constituent of some tannins) has been found in several species, e.g., *Bischofia trifoliata,*[16] *Euphorbia cornigera,*[217] *E. granulata,*[164] *E. hypericifolia* (= *E. indica),*[174] *E. iberica,*[218] *E. macroceras,*[180] *E. paralias,*[185] *E. petrophila,*[218] and *E. wallichii.*[217] The presence of other phenolic acids was also reported, e.g., veratric, and vanillic acids in *E. resinifera.*[219] Quinic and shikimic acids (nonphenolic acids) were identified in *E. resinefera*[219] and *E. pilulifera,*[220] respectively.

The trunk wood of *Sagotia racemosa* Baill. contains two micrandrols, E(6-hydroxy-7-methoxy-1,2-dimethylphenanthrene) and F(6-hydroxy-7-methoxy-1,2-dimethyl-9,10-dihydrophenanthrene).[221] On the other hand, the trunk of *Jatropha glandulifera* (Roxb.) has been found to contain 3,3-dimethylacryl shikonin as the major pigment responsible for the trunk and branch color together with acetylshikonin.[222]

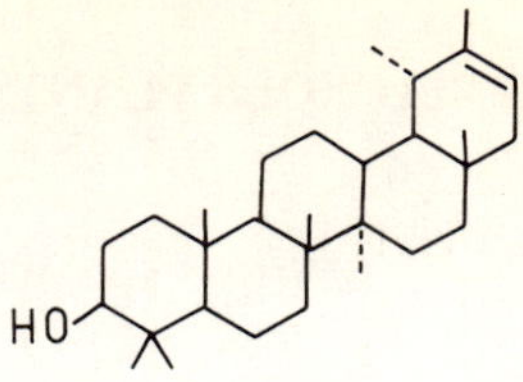

Pseudo - Taraxasterol

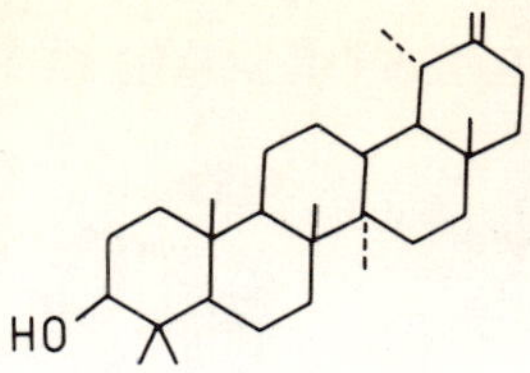

Taraxasterol

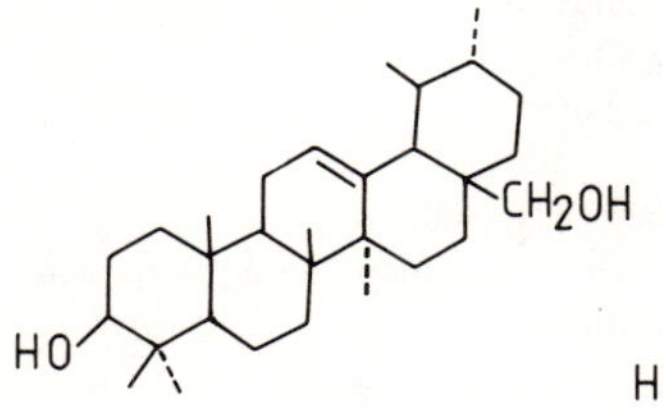

Uvaol

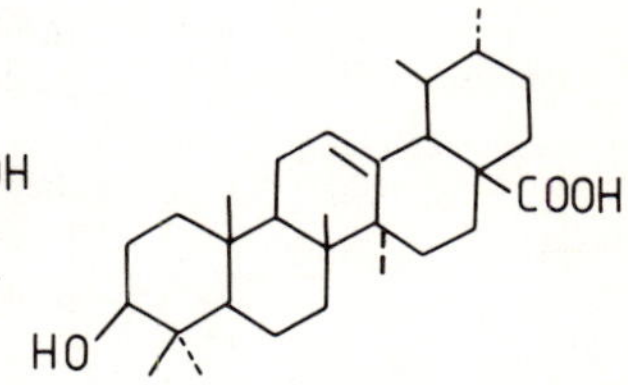

Ursolic acid

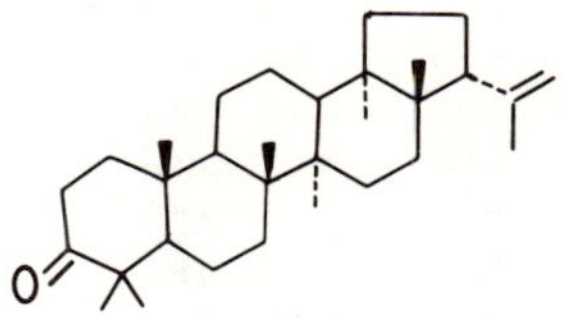

Hopenone - B

FIGURE 1 (continued)

The leaves of *Phyllanthus niruri* contain two lignans identified as phyllanthin and hypophyllanthin.[223,224] Brevifolin (2-hydroxy-4,6-dimethoxyacetophenone) and α-carotene were isolated from the leaves of *Hippomane mancinella.*[225]

E. Alkaloids

Alkaloids representing different classes have been identified from a number of euphorbiacious plants (Table 5). Aporphine alkaloids (thaliporphine and glaucine) as well as a proaporphine, crotsparine, and its dihydroderivative (crotosparinine) were isolated from *Croton* species. Securinine alkaloids, e.g., securinine and phyllchrysine, were isolated from *Phyllanthus* species. *Glochidion* species yielded several imidazole alkaloids (e.g, glochidine and glochidicine). The presence of other classes, such as ricinine (a pyridone alkaloid) from *Ricinus communis,* alochronine (a pyrimidine alkaloid) and guanidine alkaloids from *Alchornea javanesis,* as well as others including thapsine (an alkaloid of unusual structure), was also reported (Table 5, Figure 6).

F. Cyanogenic Glucosides

Cyanogensis in the Euphorbiaceae has been recently reviewed by Valen.[251] Both subfamilies, i.e., the Hyllanthoideae and Euphorbioideae sensustricta[252] comprise taxa which are able to produce hydrocyanic acid. The distribution of cyanogensis in most of the genera has been analyzed by Hegnauer.[253] Several species belonging to the following genera have been reported as cyanopheric: *Andrachne, Bridelia, Cnidoscolus, Euphorbia, Gymnanthes, Hevea, Jatropha, Manihot, Mercurialis, Poranthera,* and *Securinega.*[251]

Figure 7 illustrates and Table 6 summarizes the cyanogenic glucosides identified from the Euphorbiaceae.

Table 2
STEROLS OF SOME EUPHORBIACEOUS PLANTS

	Species	Plant part	Sterols	Ref.
1.	*Acalypha indica*	h	ν-Sitosterol	126
2.	*Antidesma bunias* Spreng.	l, s	β-Sitosterol	17
3.	*Breynia rhamnoides*	r	β-Sitosterol	127
4.	*Bridelia stipularis*	b	β-Sitosterol	17
5.	*Croton sparsiflorus*	h, l, s	β-Sitosterol	20
6.	*Euphorbia acanthothamnos* Heldr. et Sort.	h	β-Sitosterol	128
7.	*E. acualis*	r	Sterol glycoside	129
8.	*E. ebracteolata* Hayata	r	β-Sitosterol, stigmasterol, campesterol	41
9.	*E. escula* L.	h	β-Sitosterol	44
10.	*E. fischeriana* Steud.	r	7-Oxosterols (campesterol deriv.), 7α-hydroxysterols (stigmasterol deriv.), 7β-hydroxysterols (sitosterol deriv.)	130
11.	*E. geniculata* Jacq.	h	β-Sitosterol, stigmasterol, campesterol, cholesterol	47
12.	*E. helioscopia* L.	h	β-Dihydrofucosterol	131
13.	*E. hirta* L.	s	β-Sitosterol	52
14.	*E. hirta* var. *procumbens*	h	β-Sitosterol	50
15.	*E. jaxartica*	h, r	β-Sitosterol	57, 58
16.	*E. lateriflora* Schum. & Thonn.	h	β-Sitosterol	60
17.	*E. lathyris* L.	l	β-Sitosterol	61
18.	*E. longana*	s	β-Sitosterol	62
19.	*E. millii* Desm.	h	β-Sitosterol	65
20.	*E. pallisii*	r	β-Sitosterol	72
21.	*E. paralias* L.	h	β-Sitosterol stigmasterol, campesterol, cholesterol	74
22.	*E. peplus* L.	h	Campesterol, cholesterol, stigmasterol, β-sitosterol, 28-isofucosterol, Δ^7-isofucostenol	75, 132
23.	*E. pilulifera*	h	Campesterol, cholesterol, β-sitosterol	79
24.	*E. polygonifolia*	h	β-Sitosterol	80
25.	*E. pulcherimma*	s, h	β-Sitosterol, cholesterol, campesterol, dihydrobrassicasterol	133
26.	*E. pulcherrima*	s, h	β-Sitosterol	81—83
27.	*E. segetalis*	h	β-Sitosterol	92
28.	*E. tirucalli*	t	Euphorbosterol	105
29.	*E. watanabei*	h	β-Sitosterol	107
30.	*Jatropha glandulifera*	d	α-Sitosterol	134

b, Bark; d, seeds; h, herb; l, leaves; r, roots; s, stems; t, latex.

G. Glucosinolates

According to Ettlinger and Kjaer,[259] the glucosinolates and myrosinase may be confined in the Euphorbiaceae to a small minority of the genera. However, within the Euphorbiaceae, glucosinolates and myrosinase occur in *Putranjiva roxburghii* Wall and in *Drypetes gossweileri* S. Moore (= *D. armoracia,* Pax and R. Hoffm.).[259,260] Examples of the glucosinolates identified from the Euphorbiaceae are glucoputranjivin, glucocochlearin, and glucojaputin, from kernels of *Putranjiva roxburghii* Wall, and benzyl isothiocyanate from the latex of *Jatropha multifida* L.[261]

II. THYMELAEACEAE

Plants of the family Thymelaeceae are highly poisoncus shrubs or sometimes trees, seldom

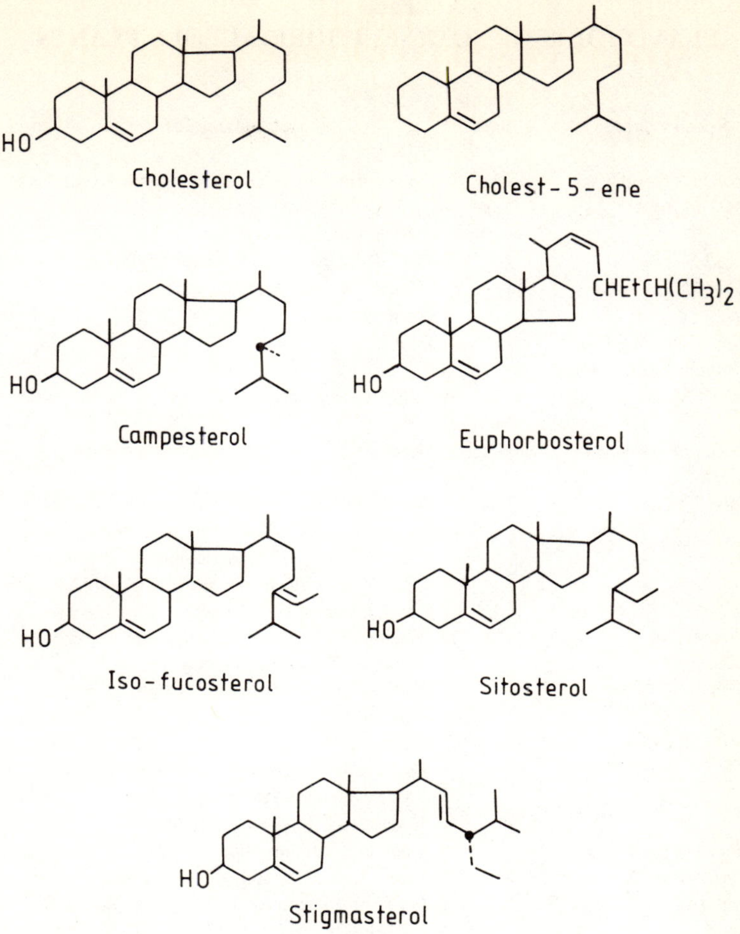

HO Cholesterol

Cholest-5-ene

HO Campesterol

HO Euphorbosterol CHEtCH(CH3)2

HO Iso-fucosterol

HO Sitosterol

HO Stigmasterol

FIGURE 2. Sterols of the family Euphorbiaceae.

herbs or lianas, mostly with unicellular hairs, not strongly tanniferous, seldom producing proanthocyanins, lacking ellagic acid and iridoid compounds, rarely cyanogenic, but sometimes producing glycosides, and characteristically accumulating the simple coumarin dephnin (or allied compounds).[262]

A. Coumarins

Several species of the Thymelaeaceae have been reported to contain coumarins (Figure 8). Daphnetin (7,8-dihydroxy coumarin) occurs as daphnin (7-β-glucosyloxy-8-hydroxy coumarin) in *Daphne alpina* L., *D. kiusiana, D. laureola* L., *D. odora* Thunb., and *D. mezereum* L., as well as in *Edgeworthia papyrifera, Gnidia polycephala, Thymelaea hirsuta* (L.) Endl. and *Wikstroemia gnapi.*[263-267] Umbelliferone was isolated from *Daphne mezereum*[268] and *Thymelaea hirsuta,*[267] while daphnoretin (6-methoxy-7-hydroxy,3,7-dicoumarin ether) was separated by Tschesche et al.[269,270] from both *Daphnopsis racemosa* Griseb. and *Daphne mezereum* L., as well as from *Daphne cannabina,*[271] *Thymelaea hirsuta,*[272] and *T. tarton-raira* L.[273]

The glucoside daphnorin (daphnoretin-7-β-glucoside was isolated from *Daphne mezereum*[269,270] and *Thymelaea hirsuta.*[267]

The latter plant contains, in addition to the above coumarins, esculetin, daphnetin-8-glucoside, and scopoletin.[267] Kosheleva and Nikonov[274] isolated from *Daphne mezereum* L. a coumarin glucoside, identified as 7-hydroxycoumarin-8-β-D-glucopyranoside and a phenol

Table 3
FLAVONOIDS OF SOME EUPHORBIACEOUS PLANTS

	Species	Plant part	Flavonoids	Ref.
1.	*Aleuritis cordata* Stand.	h	Quercerin-3-rhamnoside, quercetin-3-rutinoside	149
2.	*Beyeria brevifolia*	l, s	5,7,3′-Trihydroxy-3,8,4′,5′-tetramethoxyflavone	150
3.	*B. leschenaultii*	h	5,4′-Dihydroxy-3,7,8-trimethoxyflavone	151
4.	*Croton oblongifolius* Roxb.	l	Quercetin, isorhamnetin, quercetin-3-galactoside	152
5.	*C. sparsiflorus* Morong *(C. bonplandianum* Baill. Haines)	l	Rutin	152, 153
6.	*C. zambezicus* Muell.	l	Vitexin, saponaretin, orientin, iso-orientin, vicenin-1	154
7.	*Euphorbia acanthothamnos* Heldr. et Start.	l	Kaempferol, quercetin, quercetin-3-glucoside	155
8.	*E. amygdaloides* L.	h, r	Kaempferol, rhamnetin, rhamnetin-3α-arabinoarabinoside, rhamnetin-3α-arabopyranoside, rhamnetin-3β-arabinoarabinoside, rhamnetin-3β-rhamnosido-β-rhamnoside, rhamnetin-3α-arabofuranoside, rhamnetin-3-diarabinoside	156
9.	*E. chamaesyce*	l	Kaempferol-3-glucoside, quercetin-3-glucoside	157
10.	*E. condylocarpa*	f	Naringen-7-*0*-β-glucofuranoside	158
11.	*E. cyparissias* L.	h	Kaempferol-3-glucuronide, quercetin-3-glucuronide	159
12.	*E. dracunculoides* Lam.	h	Kaempferol	40, 160
13.	*E. dulcis*		Quercetin 3-glucoside, hyperoside, quercetin-3-galactoside-6″-gallate	161
14.	*E. esula*	h	Kaempferol-3β-D-glucuronide	162
15.	*E. geniculata* Ortega	h	Quercetin-3-rhamnoside	163
16.	*E. granulata* Forssk.	l	Rutin, quercetin, apigenin 7-glucoside	164
17.	*E. helioscopia* L.	h	Quercetin, kaempferol, tithymalin (quercetin-5,3-digalactoside), kaempferol-3-glucoside, hyperoside	165, 169
			Quercetin 3-glucoside, hyperoside, quercetin-3-galactoside-2″-gallate	161
18.	*E. hirta*	h	Quercetin, quercitrin, leucocyanidol, derivative containing rhamnose and xanthorhamnetin	170, 171
19.	*E. humifuse* Willd.	h	Quercetin	172
20.	*E. iberica*	l	Quercetin-3-*O*-galactopyranoside	173
21.	*E. indica* L. (= *E. hypericifolia)*	h	Quercetin, quercitrin	174
22.	*E. jaxartica*	h	Rutin, kaempferol	175
		r	Quercetin	
23.	*E. kalenicznkii*	h	Quercetin hyperoside, myricetin, isomyricitrin, steptoside	176
24.	*E. lamprocarpa*	h	Rutin, kaempferol	175
		r	Quercetin	
25.	*E. larica*	l	Kaempferol-3-*O*-glucoside, quercetin-3-*O*-glucoside, kaempferol-3-rutinoside, rutin	157
26.	*E. lathyris* L.	l	Kaempferol-3-mono-D-β-glucuronide, quercetin-3-mono-β-D-glucuronide	177
27.	*E. longana*	l	Quercetin	178
28.	*E. lunulata* Bge.	h	Kaempferol, quercetin, kaempferol-3-mono-L-rhamnoside, quercetin-3-mono-L-rhamnoside	179
29.	*E. macroceras*	f	Hyperoside	180
30.	*E. maculate*	f	Quercetin	181
31.	*E. maddeni*	h	Kaempferol-4′-*O*-glucoside, hyperin	64

Table 3 (continued)
FLAVONOIDS OF SOME EUPHORBIACEOUS PLANTS

Species	Plant part	Flavonoids	Ref.
32. *E. magalanta*	l	Kaempferol-3-*O*-glucoside, quercetin-3-*O*-glucoside, kaempferol-3-rutinoside, rutin	157
33. *E. myrsinitis* L.		Kaempferol-3α-rhamnopyranoside, kaempferol-3β-glucoside, kaempferol-3β-galactoside and three rhamnetin glycosides	182
34. *E. oblongifolia*	f	Hyperoside	180
35. *E. palustris*	h	Kaempferol, quercetin, hyperoside, myricetin, steptogenin, steptoside, robidanol, robidanol-3-gallate, isomyricitrin, kaempferol-3-rhamnoglucoside (quercetin-3-rhamnoglucoside)	183, 184
36. *E. paralias* L.	h	Quercetin 3′-xyloside, quercetin, quercetin-3-galactoside, guaijaverin	168, 185, 186
37. *E. peplus* L.	h	Quercetin, hyperoside, kaempferol-3-*O*-glucoside, rhamnetin-3-*O*-galactoside, rhamnetin	168, 187
38. *E. petrophila*	h	Hyperoside	173
39. *E. pilulifera*		Quercitrin	188
40. *E. platiphyllos*		Quercetin 3-D-galactopyranoside-6″-gallate	189
41. *E. prostrata* Ait.	h	Apigenin-7-glucoside, rhamnetin-3-galactoside	163
42. *E. seguieriana*	h	Quercetin-3-glucoside, hyperoside, quercetin-3β-galactoside-2-gallate, iso-quercitrin, rutin	190, 191
43. *E. semivilosa*		Isoquercitin	191
44. *E. stepposa*	h	Steppogenin, stepposide, quercetin, kaempferol, myricetin, robidanol, robidanol-3-gallate, rutin, hyperoside, isomyricitrin, kaempferol-3-rhamnoglucoside	183, 192—194
45. *E. stricta*		Hyperoside, quercetin-3-arabinoside	161
46. *E. subina* Refin		Cosmosiin	195
47. *E. tirucalli*	s	Kaempferol	99
48. *E. thymifolia*	h	Cosmosiin, hyperoside	96, 195
49. *E. verrucosa*		Hyperoside, quercetin-3-galactoside-2″-gallate	161, 189
50. *E. vigata*	l	Kaempferol-3-*O*-glucoside, quercetin-3-*O*-glucoside, kaempferol-3-rutinoside, rutin	157
51. *E. virgultosa*	h	Astragalin	191
52. *Hevea brasiliensis* H.B.K.	l	Vitexin, isovitexin	151
53. *Jatropha curcas* L. (= *J. urens* L.)	l	Apigenin, vitexin, isovitexin	151
54. *J. gossypifolia* L.	l	Apigenin, vitexin, isovitexin	196
55. *J. heynii* Bal. nom. nov.	l	Quercetin, hyperoside, vitexin, isovitexin	151
	s	Vitexin, isovitexin	151
56. *Mobea caudata*	f, r	Naringenin, naringenin-3,6-dicoumaroylglucosyl, naringenin-3-*p*-coumaroyl glucosyl	197
57. *Manihot utilissima* Pohl. (= *M. esculenta* Crantz.)	l	Rutin	151
58. *Mercurialis annua* L.		Rutin, narcissin (isorhamnetin-3-rutinoside)	198
59. *M. perennis* L.	h	Rutin, kaempferol-3-rutinoside	199
60. *Phyllanthus emblica*	l	Kaempferol, kaempferol-3-glucoside	151
61. *Ricinus communis*	l	Quercetin, rutin, hyperoside	149, 200, 201
62. *Ricinocarpos muricatus*	h	3′,4′,5,7-Tetrahydroxy-3,8-dimethoxy flavone	202
63. *R. stylosus*		5,4′-Dihydroxy-3,7,8-trimethoxyflavone, gossypetin hexamethylether, myricetin hexamethylether	203
64. *Sapium sebiferum* Roxb.	l	Quercetin-3-glucoside	204

f, Flower; h, herb; l, leaves; s, stem; r, root.

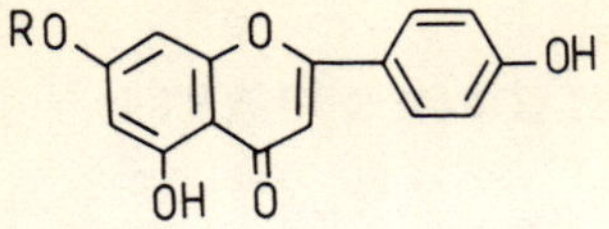

Apigenin : R = H
Cosmosiin : R = Glucosyl

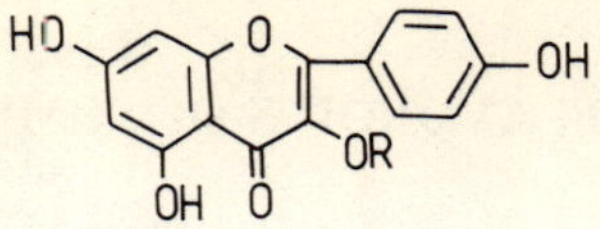

Kaempferol : R = H
Kaempferol-3-glucuroride : R = Glucuronic acid

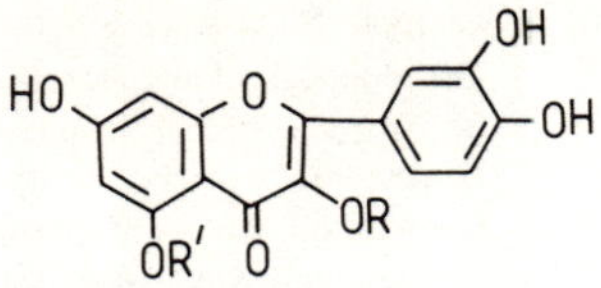

Quercetin : R = R′ = H
Quercetrin : R = Rhamnosyl ; R′ = H
Hyperoside : R = Galactosyl ; R′ = H
Tithymallin : R = R′ = Galactosyl
Rutin : R = Rhamnoglucosyl ; R′ = H

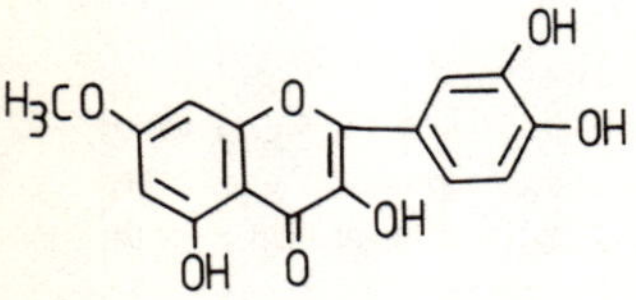

Rhamnetin

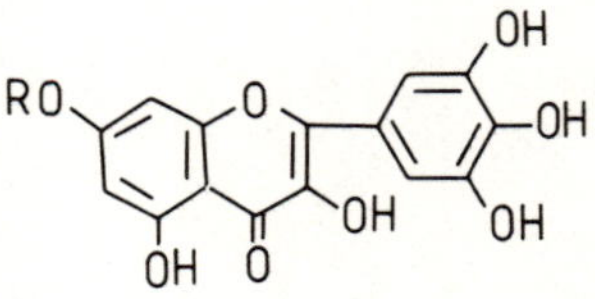

Myricetin : R = H
Isomyricitrin : R = Glucosyl

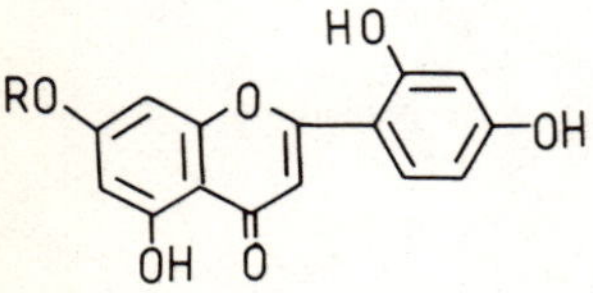

Steppogenin : R = H
Steposide R = Glucosyl

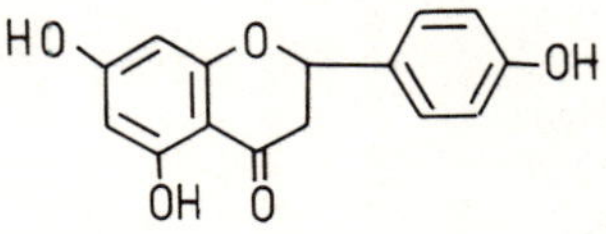

Naringenin

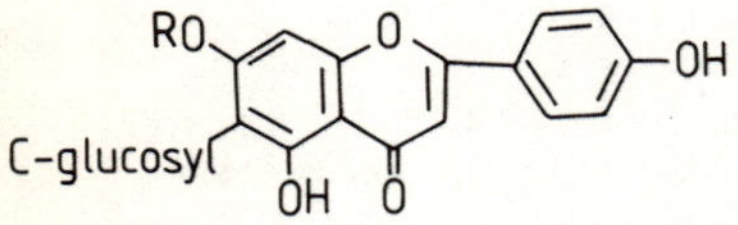

Isovitexin : R = H (saponaretin)
Saponarin : R = Glucosyl

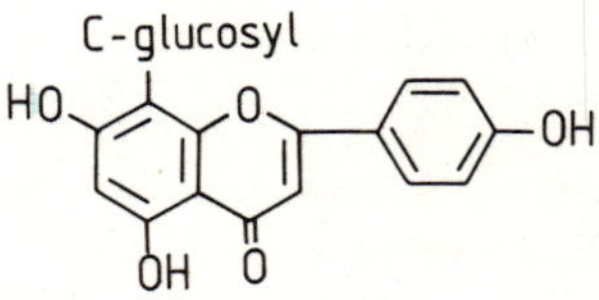

Vitexin

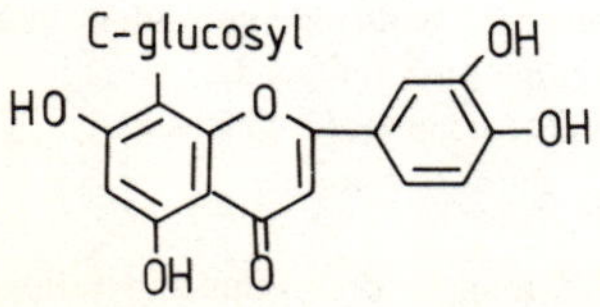

Orientin

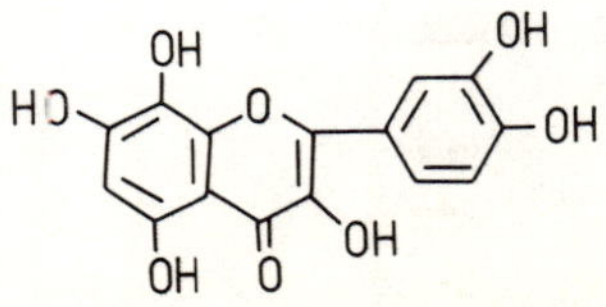

Gossypetin

FIGURE 3. Flavonoids of the family Euphorbiaceae.

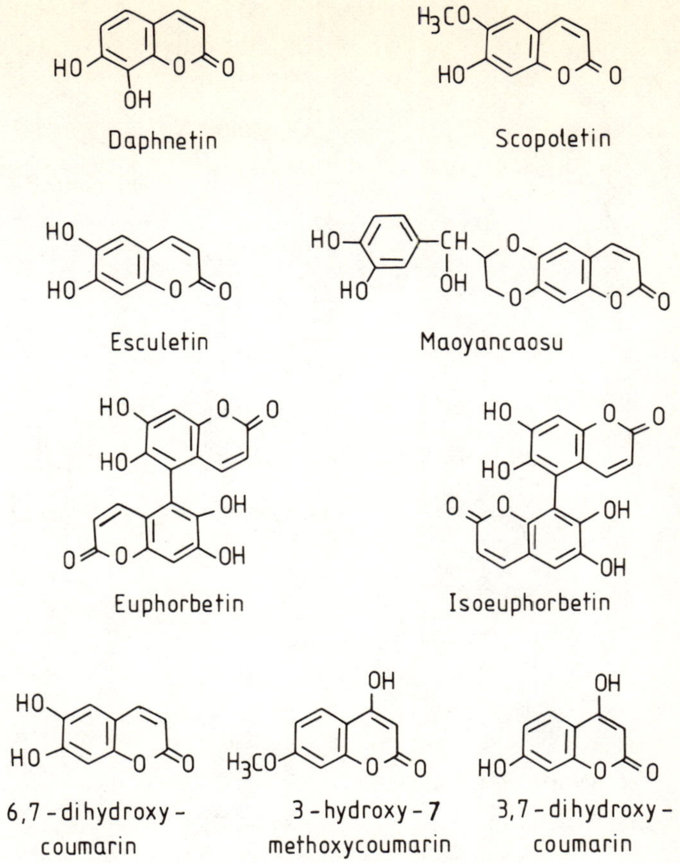

Daphnetin

Scopoletin

Esculetin

Maoyancaosu

Euphorbetin

Isoeuphorbetin

6,7-dihydroxy-coumarin

3-hydroxy-7 methoxycoumarin

3,7-dihydroxy-coumarin

FIGURE 4. Coumarins of the family Euphorbiaceae.

glycoside named daphnoside (3-methoxy-4-hydroxybenzyl-4-α-glucopyranoside), in addition to daphnin, daphnoretin, and umbelliferone. Umbelliferone and scopoletin were also identified in *Thymelaea passerina*.[275]

B. Flavonoids

Certain species of the family Thymelaeaceae (mostly *Daphne*) have been reported to contain flavonoids (Table 6). The most widely distributed acylated flavonol glycoside, tiliroside (Kaempherol-3-*p*-coumaroyl glucoside), was isolated from *Thymelea hirsuta*.[276] Both O- and C-glycosides were identified from the family (Table 7, Figure 9).

C. Triterpenoids, Steroids, and Related Compounds

Relatively scanty information is available about the distribution of triterpenoids in the family Thymelaeaceae. Table 8 summarizes the triterpenes, sterols, and related substances identified from the family (Table 8).

Two sesquiterpenes (noroxoagarofuran and 4-hydroxy dihydroagarofuran) (Figure 10) were isolated from the genus *Aquilari*.[291,292] GC of the essential oil obtained from the flowers of *Daphne odora* Thunb. revealed the presence of β-phellandrene, *n*-hexanol, Ψ-hexanol, nonanol, *trans*-linalool oxide, *cis*-linalool oxide, (−)-linalol, citronellyl acetate, nerol, geranial (−)-citronellol, and geraniol.[293] Watanabe et al.[294] recently identified by GC/MS 145 compounds, including 30 hydrocarbons, 30 alcohols, 7 phenols, 10 acids, and 10 miscellaneous compounds, from the same oil.

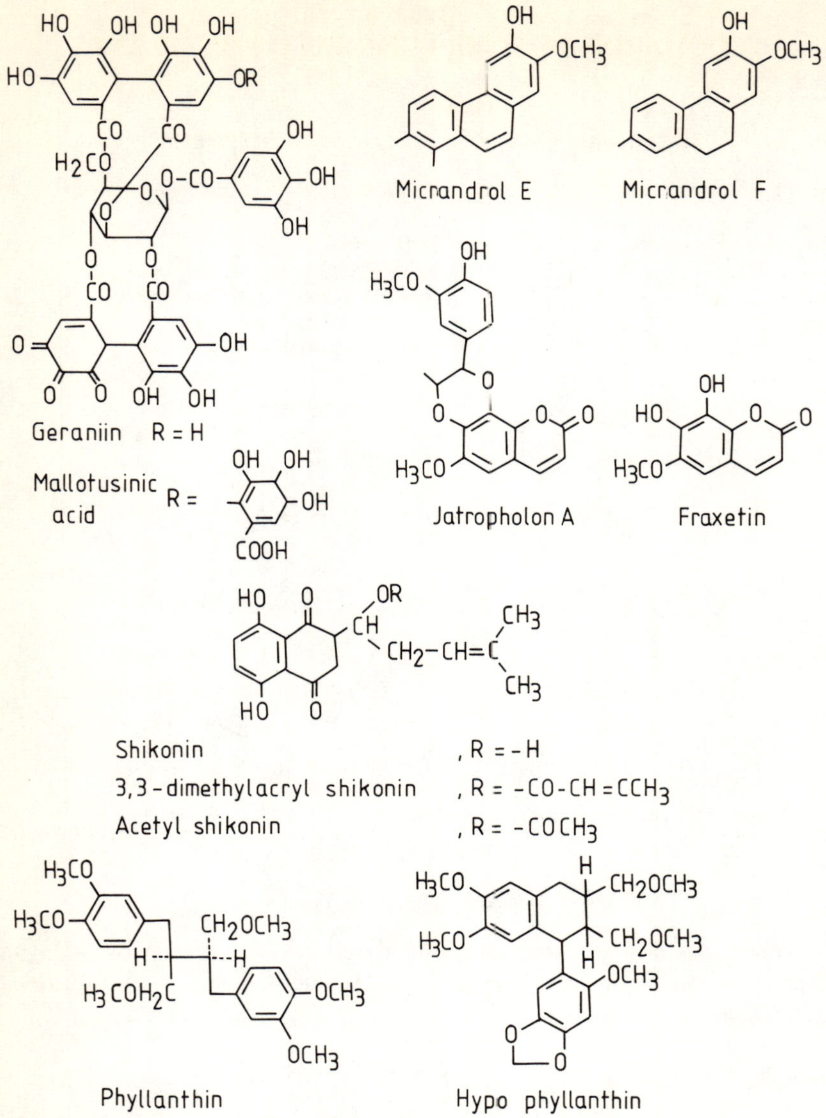

Geraniin R = H

Mallotusinic acid R =

Micrandrol E

Micrandrol F

Jatropholon A

Fraxetin

Shikonin , R = -H

3,3-dimethylacryl shikonin , R = $-CO-CH=CCH_3$

Acetyl shikonin , R = $-COCH_3$

Phyllanthin

Hypo phyllanthin

FIGURE 5. Tannins and Phenanthrenes of the family Euphorbiaceae.

Table 4
COUMARINS OF SOME EUPHORBIACEOUS PLANTS

	Species	Plant part	Coumarins	Ref.
1.	*Euphorbia acanthothamnos* Heldr. et Sart.	h	Esculine, scopoline, esculetine, scopoletin	154
2.	*E. dracunculoides* Lam.	f, r	Daphnetin	208
3.	*E. lunulata* Bge.	h	6,7-Dihydroxycoumarin, maoyancaosu	178
4.	*E. lathyris* L.	s	Esculetin, euphorbetin, isoeuphorbetin	206, 207
5.	*E. paralias* L.	h	3-Hydroxy-7-methoxycoumarin	209
6.	*E. seguieriana*	h	Scopoletin	190
7.	*E. terracina*	r	3-Hydroxy-7-methoxycoumarin, 3,7-dihydroxy-coumarin, esculetin	209
8.	*Fluegga microcarpa* Bl.	l	Bergenin	210
9.	*Nealchornea gapurenis* Huber	r	Scopoletin	211
10.	*Ricinus communis* L.	f	6,7-Dihydroxy-8-methoxy (or 6,8-dihydroxy-7-methoxy) coumarin, 3,4-dimethoxy-6,8-dihydroxycoumarin	201

f, Flowers; fr, fruits; h, herb; l, leaves; r, roots; s, stems.

Table 5
ALKALOIDS OF SOME EUPHORBIACEOUS PLANTS

	Species	Plant part	Alkaloids	Ref.
1.	*Alchornea javanensis*	h	Alchronine, alchornidine, N',N'-diisopentenyl-guanidine, N^1,N^2,N^3-triisopentenylguanidine, 2,2-dimethylacrylamide	226, 227
2.	*Croton draconoides* Muell.-Arg.	l	Thaliporphine, glaucine, thaspine	228
3.	*C. lechlery*	t	Thaspine	229
4.	*C. sparsiflorus* Morong	h	Crotosparine, crotsparnine, sparsiflorine, pronuciferine	230—232
5.	*Euphorbia atoto* Fort.	h	(+)-9-Aza-1-methylbicyclo-(3,3,1)-nonan-3-one	233, 234
6.	*E. millii* Ch. de Moulins	r	Milliamine A, milliamine B	235
7.	*Flueggea virosa* Baill.	h	Flueggeine	236
8.	*Glochidion philippicum* (Cav.) C.B. Rox.	l	Glochidine, glochidicine, N^{α}-4'-oxodecanoy-lhistamine, N^{α}-cinnamoylhistamine	237, 238
9.	*Jatropha basiacantha*	h	Purine nuclei alkaloids	239
10.	*Julocroton camporum*	h	Julocrotine	240
11.	*J. montevidensis*	h	Julocrotine	240
12.	*J. subpannosus*	h	Julocrotine	240
13.	*Phyllanthus discoides* Müll.	r	Phyllochrysine, securinine	241
		l	Securinine (?)	242
		r, b	Securinine, methoxysecurinine, phyllantine, phyllantidine	243
		s, b	Securinine (?)	242
		h	Norsecurinine (?) and other unidentified alkaloids	244, 245
14.	*Ricinus communis* L.	l, f, s	Ricinine	201, 246—249
15.	*Securinega suffruticosa*	r	Securinine	241, 250
16.	*S. virosa*	r	Securinine	241

b, Bark; f, flowers; h, herb; l, leaves; r, roots; s, stems; t, latex.

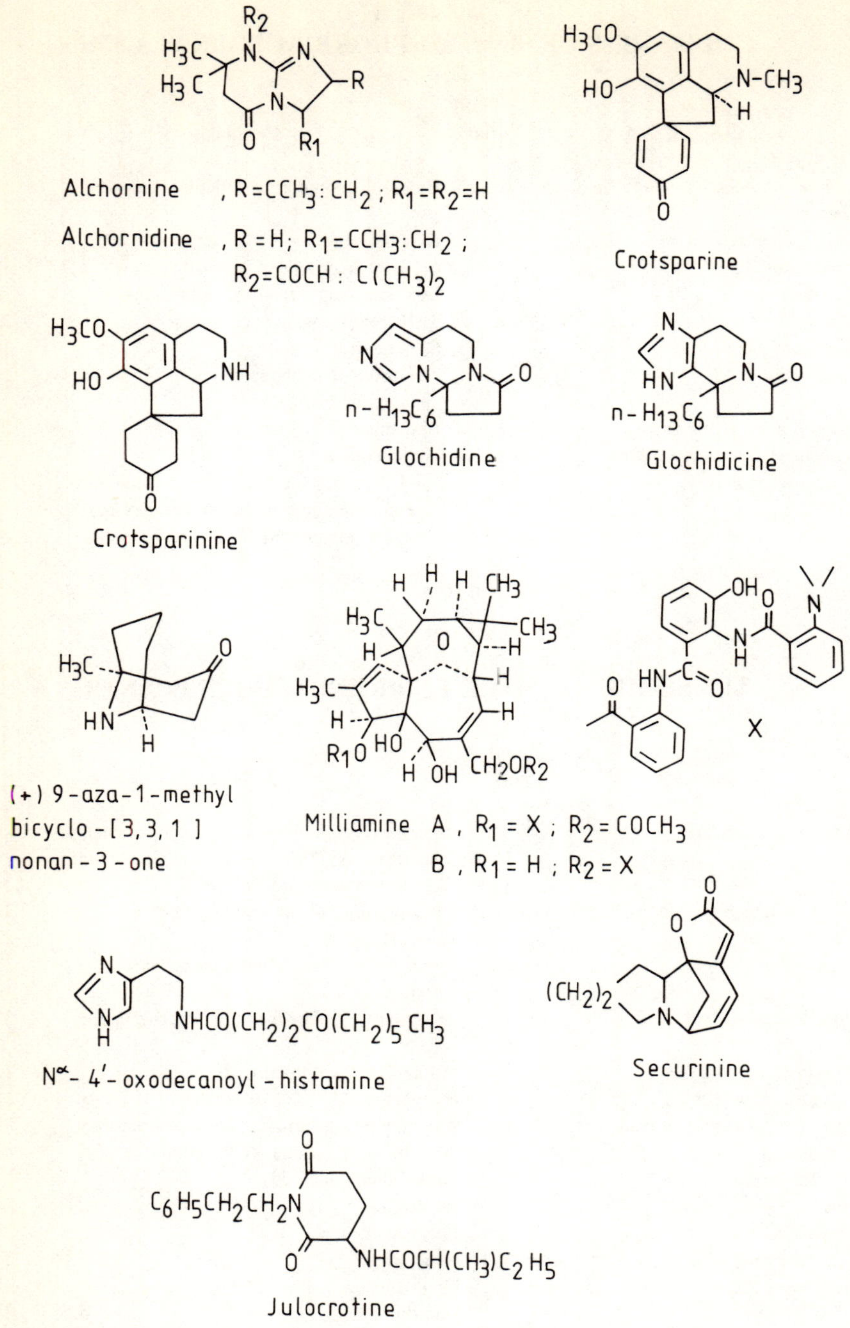

Alchornine, $R = CCH_3{:}CH_2$; $R_1 = R_2 = H$

Alchornidine, $R = H$; $R_1 = CCH_3{:}CH_2$; $R_2 = COCH:C(CH_3)_2$

Crotsparine

Crotsparinine

Glochidine

Glochidicine

(+) 9-aza-1-methyl bicyclo-[3,3,1] nonan-3-one

Milliamine A, $R_1 = X$; $R_2 = COCH_3$

B, $R_1 = H$; $R_2 = X$

N^{α}-4′-oxodecanoyl-histamine

Securinine

Julocrotine

FIGURE 6. Some alkaloids of the family Euphorbiaceae.

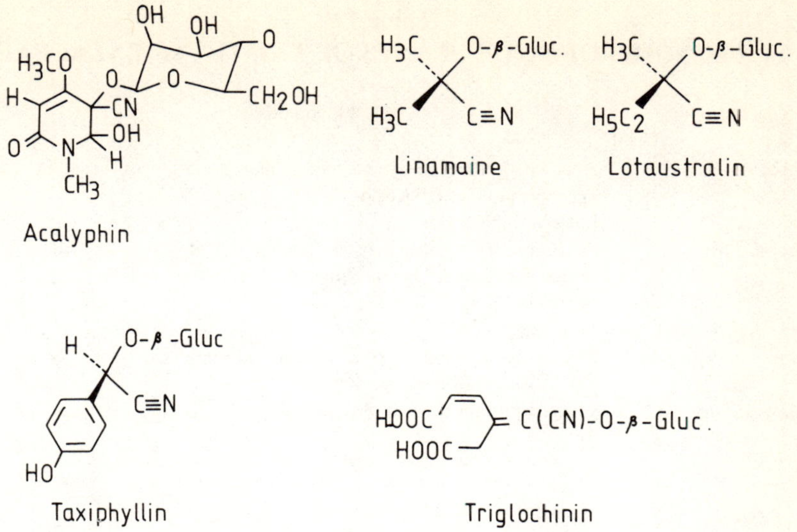

FIGURE 7. Cyanogenic glycosides of the family Euphorbiaceae.

Table 6
CYANOGENIC GLUCOSIDES OF SOME EUPHORBIACEAE PLANTS

	Species	Plant part	Cyanogenic glucosides	Ref.
1.	*Acalypha indica* L.	h	Acalyphin	254
2.	*Andrachne colchia* Fisch. et Mey.	l, f	Triglochinin	251
3.	*Cnidoscolus texanus* (Muell. Arg.) Small	r	Linamarin	255
4.	*Hevea brasiliensis* Muell. Arg.	s	Linamarin	256
5.	*Manihot aipi* Pohl *(M. palmata* Muell. Arg.)	r	Linamarin	251, 257
6.	*M. carthaginensis* Muell. Agr.	r	Linamarin, lotaustralin	256
7.	*M. utilissima* Pohl (= *M. esculenta* Crantz)	r	Linamarin (phaseolunatin = manihotoxin)	251, 257
8.	*Phyllanthus gasstroemii* Muell. Arg.	l	Taxiphyllin (= phyllanthin)	251, 258
9.	*Šeucrinega suffruticosa* (Pall.) Rehder	l, f	Triglochinin	251

f, Flowers; h, herb; l, leaves; r, root; s, stems.

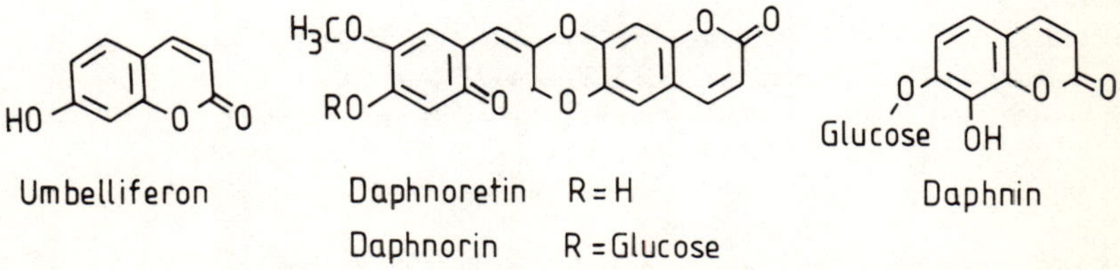

FIGURE 8. Some coumarins of the family Thymelaeaceae.

Table 7
FLAVONOIDS OF SOME PLANTS OF THE THYMELAEACEAE

Species	Flavonoids	Ref.
Daphne genkwa	Apigenin, genkwanin, yuenkanin (genkwanin-5-glucoxyloside or genkwanin-5-xyloglucoside)	277—279
D. oleoides	Luteolin-7-glucoside	280
Ovidia pillo-pillo Meisner	Apigenin-5-*O*-xylosylglucoside, luteolin-7,4′-dimethylether-5-*O*-xylosylglucoside, luteolin-7-methylether-5-*O*-xylosylglucoside, luteolin-7,4′-dimethylether	281, 282
Thymelaea hirsuta L. Endl.	Tiliroside	276
T. tartonraira	Orientin, isoorientin, vitexin, vicenin-2, kaempferol, genkwanin, 5-*O*-β-D-primeverosyl genkwanin	273
Wikstroemia virdiflora	Wikstroemin (genkwanin-5-*O*-α-D-glucosyl-α-D-glucose)	283

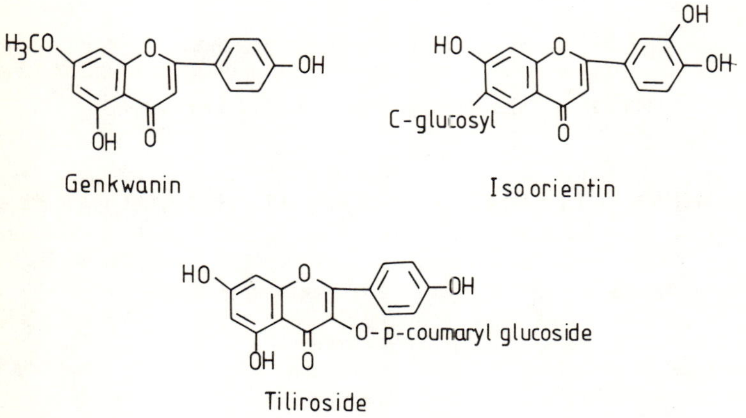

FIGURE 9. Some Flavonoids of the family Thymelaeaceae.

Table 8
TRITERPENOIDS, STEROIDS, AND RELATED SUBSTANCES OF SOME THYMELAEACEAE SPECIES

	Species	Components	Ref.
1.	*Daphne cannabina*	Taraxerol, taraxerone	284
2.	*D. genkwa*	β-Sitosterol	278, 279
3.	*D. mezereum*	β-Sitosterol	285
4.	*D. oleoides*	β-Sitosterol	286
5.	*Edgeworthia papyrifera*	Ergosta-5,8,17-(20)-trien-3β-ol	287
6.	*Thymelaea hirsuta*	Lupeol, α-amyrin, betulin, lanosterol, β-sitosterol, cholesterol, stigmasterol,[a] campesterol, β-sitosterol-β-D-glucoside, octacosanol, hexacosanol, tetracosanol, docosanol	272, 288—290
7.	*T. passerina* (Linn.) Cass.	β-Amyrin, sitosterol, stigmasterol, triacontanol	275

[a] Stigmasterol was reported by Gharbo et al.,[288] but could not be detected by Rizk et al.[290] from the plant.

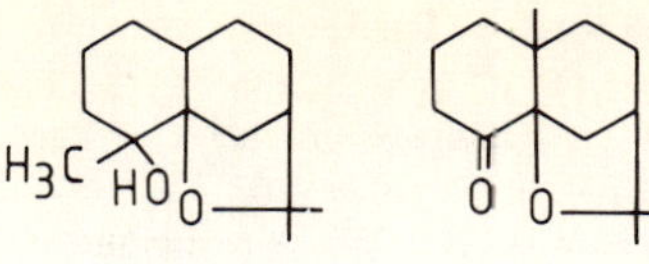

FIGURE 10. Some sesquiterpenes of the family Thymelaeaceae.

REFERENCES

1. **Lawrence, G. H. M.,** *Taxonomy of Vascular Plants,* Macmillan, New York, 1951.
2. **Hecker, E.,** Carcinogenic substances from Euphorbiaceae, *Planta Med. Suppl.,* p. 24, 1968.
3. **Farnsworth, N. R., Blomster, R. N., Messme, W. M., King, J. C., Persinos, G. J., and Wilkes, J. D.,** A phytochemical and biological review of the genus *Croton, Lloydia,* 32, 1, 1969.
4. **Watt, J. M. and Breyer-Brandwijk, M. G.,** *The Medicinal and Poisonous Plants of Southern and Eastern Africa,* 2nd ed., E. & S. Livingston, Edinburgh, 1962.
5. **Jiu, J.,** A survey of some medicinal plants of Mexico for selected biological activities, *Lloydia,* 29, 250, 1966.
6. **Spencer, C. F., Koniuszy, F. R., Rogers, E. F., Shavel, J., Jr., Easton, N. R., Kacska, E. A., Kuehl, F. A., Jr., Phillips, R. F., Walti, A., Folkers, K., Malanga, C., and Seeler, A. O.,** Survey of plants for anti-malarial study, *Lloydia,* 10, 145, 1947.
7. **Meal, R. E., Rogers, E. F., Wallace, R. T., and Starnes, O.,** A survey of plants for insecticidal activity, *Lloydia,* 13, 89, 1950.
8. **El-Missiry, M. M.,** Chemical Investigation of Certain *Euphorbia* Species, M. Pharm. thesis, Cairo University, Cairo, 1976.
9. **El-Missiry, M. M.,** Contribution to the Study of the Chemistry of Certain *Euphorbia* Species Growing in Egypt, Ph.D. thesis, Cairo University, Cairo, 1979.
10. **Baslas, R. K.,** Phytochemical study of plants of the genus *Euphorbia.* II., *Herba Hung.,* 21, 115, 1982.
11. **Radwan, H. M. A.,** Chemical Investigation of Certain *Euphorbia* Species, M.Sc. thesis, Cairo University, Cairo, 1983.
12. **Ponsinet, G. and Ourisson, G.,** Chemotaxonomic studies of the Euphorbiaceae. I. General introduction and seperation and identification of the natural tetracyclic triterpenes, *Phytochemistry,* 4, 799, 1965.
13. **Ourisson, G., Bisset, N. G., Diaz, M. A., Ehret, Ch., Palmade, M., Pesnelle, P., and Sreith, J.,** Chemotaxonomic study of the Euphorbiaceae. Chemotaxonomic study of the Dipterocarpaceae, *Mem. Soc. Bot. Fr.,* 14, 1965.
14. **Ponsinet, G. and Ourisson, G.,** Chemotaxonomy of the *Euphorbiaceae.* III. Distribution of triterpenes in *Euphorbia* latex, *Phytochemistry,* 7, 89, 1968.
15. **Ponsinet, G., Ourisson, G., and Ochlschlager, A. C.,** Systematic aspects of distribution of di- and triterpenes, *Re. Adv. Phytochem.,* 1, 271, 1966.
16. **Hui, W. H. and Ho, C. T.,** An examination of the Euphorbiaceae of Hong Kong. I. The occurrence of triterpenoids, *Aust. J. Chem.,* 21, 1675, 1968.
17. **Sainsbury, M.,** Friedelin and epifridelinol from the bark of *Prunus turfosa* and a review on their natural distribution, *Phytochemisty,* 9, 2209, 1970.
18. **Carpenter, R. C., Sotheeswaran, S., Suluanbawa, M. U. S., and Balasubramaniam, S.,** Chemical investigation of Ceylonese plants. XL. Triterpenes of five Euphorbiaceae species of Sri Lanka, *Phytochemistry,* 19, 1171, 1980.
19. **Baddeley, G. V., Bealing, A. J., Jefferies, P. R., and Retallack, R. W.,** The chemistry of Euphorbiaceae. VI. A triterpene from *Beyeria leschenaultii, Aust. J. Chem.,* 17, 908, 1964.
20. **Bhakuni, D. S., Gupte, N. C., Satish, S., Sharma, S. C., Shukla, Y. N., and Tandon, J. S.,** Chemical constituents of *Actinodaphne kaugustifolia, Croton sparsiflorus, Duabanga sonneratiodes, Glycosmis mauritiana, Hedyotis auricularia, Lyonia ovalifolia, Micromelum pubescens, Pyrus pashia* and *Rhododendron niveum, Phytochemistry,* 10, 2247, 1971.
21. **Takemoto, T., Kondo, Y., and Ishiguro, T.,** Constituents of *Euphorbia odenochlora.* I., *Yakugaku Zasshi,* 86, 528, 1966.

22. **Anjaneyulu, V., Rao, D. N., and Row, L. R.,** The triterpenoids of *Euphorbia antiquorum, Curr. Sci.,* 33, 583, 1966.
23. **Anjaneyulu, V. and Row, L. R.,** Crystalline constituents of Euphorbiaceae. VIII. Triterpenes of *E. antiquorum* Latex, *Curr. Sci.,* 36, 204, 1967.
24. **Anjaneyulu, V., Rao, D. N., and Row, L. R.,** Crystalline constituents of Euphorbiaceae. VII. Triterpenes of *Euphorbia antiquorum, J. Indian Chem. Soc.,* 44, 123, 1967.
25. **Sengupta, P. and Ghosh, S.,** Isolation of taraxerol and epifriedelanol from *Euphorbia antiquorum, Indian J. Chem.,* 2, 298, 1964.
26. **Rao, D. N. and Row, L. R.,** Crystalline constituents of Euphorbiaceae. VII. The triterpenes of *Euphorbia antiquorum* Linn., *J. Ind. Chem. Soc.,* 44, 123, 1967.
27. **Gonzalez, A. G. and Toste, A. H.,** Latex of *Euphorbia aphyla, An. R. Soc. Exp. Fis. Quim. Ser. B:,* 50, 597, 1954.
28. **Chapon, S. and David, S.,** A secondary constituents of the latex of *Euphorbia balsamifera, Bull. Soc. Chim. Fr.,* 456, 1952.
29. **Gonzalez, A. G. and Mora, M. L. G.,** Latex of Canary Island Euphorbias. Germanicol and lanosterol from the latex of *Euphorbia balsamifera, An. R. Soc. Esp. Fis. Quim. Ser. B:,* 48, 483, 1952.
30. **Gonzalez, A. G.,** Trimethylsteroids of the latex of *Euphorbia balsamifera, An. Estud. Atlanticos,* 6, 61, 1960.
31. **Gonzalez, A. G. and Padron, A. G.,** Latex of the Canary Island Euphorbias. XVII. *Euphorbia regis-jubae* and *E. bravoona, An. R. Soc. Esp. Fis. Quim. Ser B:,* 54, 695, 1958.
32. **Gonzalez, A. G. and Calero, A.,** The latex of the Canary Island Euphorbias. I. Latex of *Euphorbia canariensis, An. Soc. Esp. Fis. Quim. Ser. B:,* 45, 269, 1949.
33. **Morales-Mendez, A.,** *Euphorbia carascana* triterpenes, *Rev. Soc. Quim. Mex.,* 13(3), 116A, 1969.
34. **Anjaneyulu, V., Rao, C. S., Srinivasulu, C., and Row, L. R.,** The crystalline constituents of Euphorbiaceae. IX. Triterpenes of *Euphorbia cattimandoo, J. Indian Chem. Soc.,* 45, 404, 1968.
35. **Govardhan, Ch., Reddy, P. R., and Sundrararamaiah, T.,** 3-Epicyclolaudenol and known triterpenes from *Euphorbia caudicifolia, Phytochemistry,* 23, 411, 1984.
36. **Roshchin, Yu. V. and Kiryalov, N. P.,** Tetracycyclic triterpenoid compounds from *Euphorbia concylocarpa, Khim. Prir. Soedin.,* 6, 483, 1970.
37. **Piatak, D. M. and Reimann, K. A.,** Plant investigation. IV. Corollatadiol, a new triterpene from *Euphorbia corollata* (tetracyclic triterpene), *Tetrahedron Lett.,* 44, 4525, 1972.
38. **Starrat, A. N.,** Triterpenoid constituents of *Euphorbia cyparissias, Phytochemistry,* 5, 1341, 1966.
39. **Starrat, A. N.,** Isolation of Hopenone-B-from *Euphorbia cyparissias, Phytochemistry,* 8, 1831, 1969.
40. **Singh, A. and Srivastava, S. N.,** Chemical examination of *Euphorbia dracunculoides* Lam., *Indian J. Chem.,* 4, 520, 1966.
41. **Itokawa, H., Kushida, K., and Fujita, M.,** Constituents of *Euphorbia ebracteolate, Chem. Pharm. Bull.,* 18, 1276, 1970.
42. **Gonzalez, A. G. and Berrera, R.,** Latex of Canary Island Euphorbias. XIV. Sources of the triterpenes handianol, aphyldienol and obtusifoliol, *Publ. Inst. Quim A Barba,* 10, 199, 1956; *C.A.,* 51, 12428, 1957.
43. **Gonzalez, A. G. and Breten, J. L.,** Latex of *Euphorbia* species from Canary Island. XVIII. Structure of the new triterpene obtusifoldienol *An. R. Soc. Esp. Fis. Quim.,* 55B, 93, 1959.
44. **Fransworth, N. R., Wagner, H., Hörhammer, L., Horhammer, H. P., and Fong, H. S.,** *Euphorbia esula.* I. Preliminary phytochemical and biological evaluation, *J. Pharm. Sci.,* 57, 933, 1968.
45. **Starrat, A. N.,** Cycloartenol and Lupeol from *Euphorbia esula, Phytochemistry,* 12, 231, 1973.
46. **Rizk, A. M., Hammouda, F. M., El-Shamy, A. M., and El-Missiry, M. M.,** Triterpenoids from latices of *Euphorbia paralias* and *Euphorbia geniculata, Fitoterapia,* 51, 313, 1980.
47. **Rizk, A. M., Hammouda, F. M., Seif El-Nasr, M. M., and El-Missiry, M. M.,** Constituents of Egyptian Euphorbiaceae. VI. phytochemical investigation of *Euphorbia geniculata.* Jacq. and *E. prostrata* Ait, *Pharmazie,* 33, 540, 1978.
48. **Tripathi, R. D. and Tiwari, K. P.,** Geniculatin, a triterpenoid saponin from *Euphorbia geniculata, Phytochemistry,* 19, 2163, 1980.
49. **Tiwari, K. P., Kumar, P., Minocha, P. K., and Masood, M.,** Chemical constituents of *Euphorbia heterophylla* and *Euphorbia parviflora, Proc. Natl. Acad. Sci. India Sect. A.,* 51, 213, 1981.
50. **Estrada, H.,** Study of *Euphorbia hirta* var. *procumbens, Pedilanthus calcaratus* and *P. tehaucanus, Bol. Inst. Quim. Univ. Nac. Auton. Mex.,* 11, 15, 1959; C.A., 56, 7706, 1962.
51. **Takemoto, T. and Ingaki, M.,** Constitutents of *Euphorbia* R., *Yakugaku Zasshi,* 78, 289, 1958.
52. **Baslas, R. K. and Agarwal, R.,** Isolation and characterization of different constituents of *Euphorbia hirta* Linn., *Curr. Sci.,* 49, 311, 1980.
53. **Rangaswami, S. and Sambamurthy, K.,** Chemical examination of the leaves of *Rhododendron campanulatum., Proc. Indian Acad. Sci.,* 53, 98, 1961.
54. **Gupta, D. R. and Garg, S. K.,** A chemical examination of *Euphorbia hirta, Bull. Chem. Soc. Jpn.,* 39, 2532, 1966.

55. **Rizk, A. M. and Rimpler, H.,** Constituents of Egyptian Euphorbiaceae. VI. Investigation of *Euphorbia indica, Planta Med.*, 32, 177, 1978.
56. **Barbour, J. B., Warren, F. L., and Wood, D. A.,** The characterization of the groups in euphorbol. *J. Chem. Soc.*, 2537, 1951.
57. **Azimov, M. A. and Nazirov, Z. N.,** *Euphorbia jaxartica* triterpenes, *Khim. Prir. Soedin.*, 5, 599, 1969.
58. **Azimov, M. A.,** Euphorbol hexacozonate from *Euphorbia jaxartica, Khim. Prir. Soedin.*, 6, 136, 1970.
59. **Morales-Mendes, A.,** *Euphorbia latazi* var. *glabra* triterpenes, *An. Quim.*, 67, 1239, 1971.
60. **Levi, D., Jain, M. K., and Orebamjo, T. O.,** Terpenoids. VII. Constitutents of *Euphorbia lateriflora, Phytochemistry*, 7, 657, 1968.
61. **Dutta, P. K. and Chakravarti, R. N.,** Constituents of *Euphorbia lathyris, Phytochemistry*, 10, 2550, 1971.
62. **Mahato, S. B., Sahu, N. P., and Chakrovarti, R. V.,** Chemical investigation of *Euphorbia longena, Phytochemistry*, 10, 2847, 1971.
63. **Takemoto, T. and Ingaki, M.,** Constituents of *Euphorbia maculata, Yakugaku Zasshi*, 78, 292, 1958.
64. **Sahai, R., Dube, M. P., and Rastogi, R. P.,** Chemical and pharmacological study of *Euphorbia maddeni, Indian J. Pharm. Sci.*, 43, 216, 1981.
65. **Pancorbo, S. and Hammer, R. H.,** Preliminary phytochemical investigation of *Euphorbia millii, J. Pharm. Sci.*, 61, 954, 1972.
66. **Aynehchi, Y., Mojtabaii, M., and Yazdizadeh, K.,** Chemical examination of *Euphorbia myrsinites* Lim, *J. Pharm. Sci.*, 61, 292, 1972.
67. **Rao, D. N. and Row, L. R.,** Crystalline constituents of Euphorbiaceae. III. The triterpenes of *Euphorbia neriifolia, Curr. Sci.*, 34, 432, 1966.
68. **Anjaneyulu, V. and Row, L. R.,** Crystallization principles of Euphorbiaceae. VI. Triterpenes from the stems and leaves of *E. nerifolia, Curr. Sci.*, 34, 608, 1965.
69. **Anjeneyulu, A. S. R., Row, L. R., Subrahmanyam, C., and Murty, K. S.,** Crystalline constituents from Euphorbiaceae. XIII. The structure of a new triterpene from *Euphorbia nerifolia* L., *Tetrahedron*, 29, 3909, 1973.
70. **Baslas, R. K. and Agarwal, R.,** Chemical examination of the bark of *Euphorbia neriifolia* Linn., *Indian J. Pharm. Sci.*, 42, 66, 1980.
71. **Gonzalez, A. G. and Breton, J. L.,** The Canary Island of *Euphorbia* latex. V. *Euphorbia obtusifolia, An. R. Soc. Esp. Fis. Quim.*, 47B, 363, 1951; *C.A.*, 46, 5140, 1952.
72. **Tarmaeva, Z. V. and Belova, N. V.,** Components of *Euphorbia pallassii roots, Khim. Prir. Soedin.*, 855, 1980.
73. **Breton, J. L., Martin, J. D., Fraga, B. M., and Gonzalez, A. F.,** Latex de las *Euphorbia* canaris. XX. Triterpenes de la *E. paralias L., An. Quim.*, 65, 61, 1969; *C.A.*, 70, 112331, 1969.
74. **Rizk, A. M., Youssef, A. A., Diab, M. A., and Salem, H. M.,** Triterpenoids and related substances of *Euphorbia paralias, Z. Naturforsch.*, 29C, 529, 1974.
75. **Rizk, A. M. Hammouda, F. M., Seif El-Nasr, M. M., and Abou-Youssef, A. A.,** Phytochemical investigation of *Euphorbia peplus, Fitoterapia*, 51, 223, 1980.
76. **Rustaiyan, A., Niknejad, A., Sharif, Z., and Izaddoost, M.,** Triterpenes from *Euphorbia petiolata, Fitoterapia*, 53, 143, 1982.
77. **Carrazzoni, P.,** Chemical study of Euphorbiaceae. I. Triterpenes from *Euphorbia phosphorea, An. Acad. Brasil Cienc.*, 38, 431, 1966, *C.A.*, 68, 36705.
78. **Takemoto, T. and Ingaki, M.,** Constituents of *Euphorbia pilulifera, Yakugaku Zasshi*, 78, 294, 1958.
79. **Atallah, A. M. and Nicholas, H. J.,** Triterpenoids and steroids of *Euphorbia pilulifera, Phytochemistry*, 11, 1860, 1972.
80. **Starrat, A. N.,** Triterpenoids of *Euphorbia polygonifolia, Phytochemistry*, 8, 795, 1969.
81. **Dominguez, X. A., Delgado, J. G., Maffey, Ma de L., Mares, J. G., and Rombold, C.,** Chemical study of the latex, stems, bracts and flowers of christmas flower *(Euphorbia pulcherrima), J. Pharm. Sci.*, 56, 1184, 1967.
82. **Khastgir, H. N. and Pradhan, B. P.,** Terpenoids and related compounds. IV. Chemical investigation of *Euphorbia pulcherrima, J. Indian Chem. Soc.*, 44, 159, 1967.
83. **Gibbs, R. D., Edward, J. T., and Ferland, J. M.,** A novel color reaction of some *Euphorbia* and *Oxyanthus* species, *Phytochemistry*, 6, 253, 1967.
84. **Khafagy, S., Nazmi, N., Abdel-Salam, N., and Seif Eldin, A.,** Steroid triterpenoid, and flavonoid constituents of *Euphorbia pulcherrima* Willd. leaves, *Acta Pharm. Jugosl.*, 30, 103, 1980.
85. **Dupont, G., Kopaczewski, W., and Broadki,** Study of *Euphorbia* resins. II. Latex of *Euphorbia resinifera, Bull. Soc. Chem. Fr.*, 1068, 1947; *C.A.*, 42, 2647, 1948.
86. **Dupont, G., Julia, M., and Wragg, W. R.,** Euphorbiaceae resins. VIII. The identification of taraxerol, of the β-amyrin portion and of a new triterpene alcohol, resiniferol, as the less abundant constituents of the latex of *Euphorbia* species, *Bull. Soc. Chim. Fr.*, 852, 1953.

87. **Sharma, R. C., Zaman, A., and Kidwai, A. R.,** Chemical examination of *Euphorbia royleana, Indian J. Chem.*, 2, 254, 1964.
88. **Nazir, M., Naeemuddin, I. A., Khan, S. A., Bhatty, M. K., and Karimullah,** Chemical constituents of *Euphorbia royleana, Pak. J. Sci. Ind. Res.*, 8(3), 80, 1965; C.A., 64, 13089, 1966.
89. **Bhat, V. S., Joshi, V. S., and Nanavati, D. D.,** Cycloroylenol, A cyclopropane containing euphoid from *Euphorbia royleana, Tetrahedron Lett.*, 5207, 1980.
90. **Anjaneyulu, A. S. R., Row, L. R., Subrahmanyam, C., and Murty, K. S.,** Isolation of epitaraxerol from *Euphorbia royleana, Curr. Sci.*, 43, 10, 1974.
91. **Sengupta, P. and Ghosh, S.,** Triterpenoids and related compounds. VI. Triterpenoids of *Euphorbia royleana., J. Indian Chem. Soc.*, 42, 543, 1965.
92. **Breton, J. L., Castaneda, J. P., Fraga, B. M., and Gonzalez, A. G.,** Latex de las *Euphorbia* canarias. XII. Triterpenos de la *E. segetalis* L., *An. Quim.*, 66, 213, 1970.
93. **Pradhan, B. P. and Khastgir, H. N.,** Triterpenoids and related compounds. IX. Chemical investigation of *Euphorbia sikkimensis, J. Indian Chem. Soc.*, 46, 331, 1969.
94. **Khafagy, S. M., Abdel-Salam, N. A., Mohamed, Y. A., and Mahmoud, Z. F.,** Crystalline principles of *Euphorbia terracina* L., *Pharmazie*, 32, 82, 1977.
95. **Gupta, D. R. and Garg, S. K.,** Chemical examination of *Euphorbia thymifolia, Indian J. Appl. Chem.*, 29, 39, 1966.
96. **Agarwal, R. and Baslas, R. K.,** Chemical examination of the aerial parts of *Euphorbia thymifolia, Indian J. Pharm. Sci.*, 43, 182, 1981.
97. **Aynehchi, Y. and Kiumehr, N.,** Constituents of *Euphorbia tinctoria, Phytochemistry*, 11, 2887, 1972.
98. **Aynehchi, Y. and Kiumehr, N.,** Chemical examination of *Euphorbia tinctoria* Boiss., *Acta Pharm. Suec.*, 11, 185, 1974.
99. **Gopalachari, R. and Siddiqui, S.,** Chemical examination of the latex from *Euphorbia tirucalli, J. Sci. Ind. Res. India*, 8B, 129, 1949.
100. **Haines, D. W. and Warren, F. L.,** *Euphorbia* resins. II. The isolation of taraxasterol and a new triterpene, tirucallol, from *Euphorbia tirucalli, J. Chem. Soc.*, 2554, 1949.
101. **Afza, N., Malik, A., and Siddiqui, S.,** A new triterpenoid from *Euphorbia tirucalli, Pak. J. Sci. Ind. Res.*, 22, 124, 1979.
102. **Gupta, P. K. and Mahadevan, V.,** Chemical examination of the stems of *Euphorbia tirucalli, Indian J. Pharm.*, 29, 152, 1967.
103. **Afza, N., Malik, A., and Siddiqui, S.,** Isolation and structure of cycloeuphorbol, a new triterpene from *Euphorbia tirucalli, Pak. J. Sci. Ind. Res.*, 22, 173, 1979.
104. **Baslas, R. K. and Gupta, N. C.,** Chemical constituents of the bark of *Euphorbia tirucalli, Indian J. Pharm. Sci.*, 44, 113, 1982.
105. **Malik, A., Afza, N., and Siddiqui, S.,** Further studies in the fresh latex of *Euphorbia tirucalli, Pak. J. Sci. Ind. Res.*, 24, 1, 1981.
106. **Rishi, A. K., George, V., and Kapoor, R.,** Chemical investigation of local plants. II. Triterpenoids of *Euphorbia wallichii, Planta Med.*, 35, 193, 1979.
107. **Takemoto, T. and Ishiguro, T.,** Consituents of *Euphorbia watanabei*. I. *Yakugaku Zasshi*, 86, 530, 1966.
108. **Talapatra, B., Dutta, S., Maiti, B. C., Pradhan, D. K., and Talapatra, S. K.,** Terpenoid constituents of Indian *Glochidion* species *(G. accuminatum* and G. *thomsoni)*. Partial synthesis of glochidone *Aust. J. Chem.*, 27, 2711, 1974.
109. **Hui, W. H., Lee, W. K., Ng, K. K., and Chan, C. K.,** The occurrence of triterpenoids and steroids in three *Glochidion* species, *Phytochemistry*, 9, 1099, 1970.
110. **Hui, W. H. and Li, M. M.,** Lupene triterpenoids from *Glochidion eriocarpum, Phytochemistry*, 15, 561, 1976.
111. **Ganguly, A. K., Govindachari, T. R., Mohamed, P. A., Rahimtulla, A. D., and Viswanthan, N.,** Constituents of *Glochidion hohenackeri, Tetrahedron*, 22, 1513, 1966.
112. **Hui, W. H. and Lee, W. K.,** Euphorbiaceae of Hong Kong. VII. Lup-20(29)-ene-3α,23-diol, a new triterpene from *Glochidion marophyllum, J. Chem. Soc. C:*, 1004, 1971.
113. **Talapatra, S. K., Bhattacharya, S., Maiti, B. C., and Talapatra, B.,** Structure of glochicudiol. New triterpenoid from *Glochidion multioculare*. Natural occurrence of dimedon, *Chem. Ind.*, 1033, 1973.
114. **Ahmad, S. A. and Zaman, A.,** Chemical constituents of *Glochidon venulatum, Maesa indica* and *Rhamnus triquetra, Phytochemistry*, 12, 1826, 1973.
115. **Hui, W. H. and Fung, M. L.,** An examination of the Euphorbiaceae of Hong Kong. VI. Isolation and structure of glochidinol, a new triterpene ketol from *Glochidion wrightii* Benth., *J. Chem. Soc. C:*, 1710, 1969.
116. **Ponsinet, G. and Ourisson, G.,** Etudes chimico-taxonomique dans la famille des Euphorbiacées. II. Triterpénes de *Hura crepitans* L., *Phytochemistry*, 4, 813, 1965.
117. **Sengupta, P. and Mukhopadhyay, J.,** Triterpenoids and related compounds. VII. Triterpenoids of *Phyllanthus acidus, Phytochemistry*, 5, 531, 1966.

118. **Savior, R.,** Betulic acid in *Phyllanthus discoides, Bull. Soc. Chim. Belges,* 74, 52, 1965.
119. **Dekker, S.,** Phytochemical notes, *Pharm. Weekbl.,* 45, 1156, 1909; C.A., 3, 659.
120. **Seshadri, T. R. and Laumas, L.,** Components of the bark of *Phyllanthus emblica, J. Sci. Ind. Res. India,* 17B, 167, 1958.
121. **Hui, W. H. and Sung, M. L.,** An examination of the Euphorbiaceae of Hong Kong. II. The occurrence of epitaraxerol and other triterpenoids, *Aust. J. Chem.,* 21, 2137, 1968.
122. **Simonsen, J. and Ross, W. C. J.,** *The Terpenes,* Vol. 4, Cambridge University Press, London, 1957, 369.
123. **Adesida, G. A., Girgis, P., and Taylor, D. A. H.,** Friedeline derivatives from *Phyllanthus muellerianus, Phytochemistry,* 11, 851, 1972.
124. **Hui, W. H., Li, M. M., and Wong, K. M.,** A new compound, 21 α-hydroxy-friedel-4-(23)-en-3-one and other triterpenoids from *Phyllanthus reticulatus, Phytochemistry,* 15, 797, 1976.
125. **Hui, W. H., Fung, T. L., and Ng, K. K.,** An examination of the Euphorbiaceae of Hong Kong. V., *Phytochemistry,* 8, 331, 1969.
126. **Mukherjee, J.,** Chemical examination of *Acalypha indica, J. Indian Chem. Soc.,* 44, 292, 1967.
127. **Sengupta, P. and Ghosh, S.,** Chemical examination of *Breynia rhamnoides, Indian J. Chem.,* 2, 83, 1964.
128. **Kallimanis, G. and Philianos, S.,** Constituents chimiques des feuilles d'*Euphorbia acanthothamnos, Plant. Med. Phytother.,* 14, 230, 1980.
129. **Khanna, N. M.,** Examination of *Euphorbia acaulis, Indian J. Pharm.,* 16, 110, 1954.
130. **Schroeder, G., Rohmer, M., Beck, J. P., and Anton, R.,** 7-Oxo-7α-hydroxy-and 7β-hydroxysterols from *Euphorbia fischeriana, Phytochemistry,* 19, 2213, 1980.
131. **Durrani, A. A., Rafiullah, M., and Ikram, M.,** Studies on *Euphorbia helioscopia* Linn., *Pak. J. Sci. Ind. Res.,* 10, 167, 1967.
132. **Baisted, D. J.,** Sterols of *Euphorbia peplus:* the fate of 28-isofucosteol in phytosterol biosynthesis, *Phytochemistry,* 8, 1697, 1969.
133. **Sekula, B. and Nes, W.,** The identification of cholesterol and other steroids in *Euphorbia pulcherimma, Phytochemistry,* 19, 1509, 1980.
134. **Alimchandani, R. L., Badami, R. C., and Katti, M. C. T.,** Chemical examination of the seeds of *Jatropha glandulifera, J. Indian Chem. Soc.,* 26, 523, 1949.
135. **Estrada, H.,** *n*-Docosanol from *Euphorbia calyculata, Bol. Inst. Quim. Univ. Nac. Auton. Mex.,* 18, 85, 1966.
136. **Piatak, D. M. and Reimann, K. A.,** Isolation of 1-octacosanol from *Euphorbia corollata, Phytochemistry,* 2, 2585, 1970.
137. **Starrat, A. N.,** The identification of long-chain alcohols from *Euphorbia species, Phytochemistry,* 11, 293, 1972.
138. **Khamidova, Kh. A. and Nazirov, Z. N.,** Hydrocarbons and alcohols of *Euphorbia severtzovii* and *E. falcata, Khim. Prir. Soedin.,* 8(1), 112, 1972.
139. **Azimov, M. A. and Nazirov, Z. N.,** Hydrocarbons and alcohols of *Euphorbia lamprocarpa, Khim. Prir. Soedin.,* 5, 432, 1969.
140. **Farnsworth, N. R., Wagner, H., Hoerhammer, L., and Hoerhammer, H.-P.,** Nomenclature correction of certain plant *n*-alkanes, *J. Pharm. Sci.,* 56, 1369, 1967.
141. **Tiwari, K. P., Kumar, P., and Masood, M.,** 10, 10-Dimethyl hexcosan-7-one from *Euphorbia heterophylla, J. Indian Chem. Soc.,* 57, 530, 1980.
142. **Hinman, C. W., Hoffmann, J. P., McLaughlin, S. P., and Peoples, T. R.,** Hydrocarbon production from arid land plant species, *Proc. Annu. Meet. Am. Sect. Int. Sol. Energy Soc.,* 3(Sect. 1), 110, 1980.
143. **Nemethy, E. K., Otovos, J. W., and Calvin, M.,** Natural production of high-energy liquid fuels from plants, in *Fuels from Biomass and Wastes,* Klass, D. L. and Emert, G. H., Eds., Ann Arbor Scientific Publications, Ann Arbor, Mich., 1981, 405.
144. **Nemethy, E. K., Otvos, J. W., and Calvin, M.,** Hyrdrocarbons from *Euphorbia lathyris, Pure Appl. Chem.,* 53, 1101, 1981.
145. **Calvin, M., Nemethy, E. K., Redenbaugh, K., and Otvos, J. W.,** Plants as a direct source of fuel, *Experientia,* 38, 18, 1982.
146. **Ward, R. F.,** *Euphorbia* — is it the source of hydrocarbons in the future?, *Sol. Energy,* 29, 83, 1982.
147. **Calvin, M.,** Hydrocarbons from plants: analytical method and observations, *Naturwissenschaften,* 67, 525, 1980.
148. **Calvin, M.,** Energy agriculture. Symposium on controversial topics in agricultural and food chemistry, American Chemical Society, Kansas City, Mo., Sept. 12 to 17, 1982.
149. **Nakaoki, T. and Morita, N.,** *J. Pharm. Soc. Jpn.,* 77, 108, 1957; cited in Ref. 156.
150. **Chow, P. W. and Jefferies, P. R.,** The chemistry of Euphorbiaceae. XXI. Compounds from *Beyeria brevifolia, Aust. J. Chem.,* 21, 2529, 1968.

151. **Jefferies, P. R. and Retallack, R. W.,** the chemistry of the Euphorbiaceae. XX. Further compounds from *Beyeria leschemoultii, Aust. J. Chem.*, 21, 2085, 1968.
152. **Wagner, H., Iyengar, M. A., Michahelles, E., and Herz, W.,** Quercetin-3-(*O*-acetyl)-β-D-glucopyranoside in *Plummera floribunda* and *Helenium hoopensii, Phytochemistry,* 10, 2547, 1971.
153. **Subramanian, S. S., Nagarajan, S., and Sulochana, N.,** Flavonoids of some Euphorbiaceous plants, *Phytochemistry,* 10, 2548, 1971.
154. **Wagner, H., Hoerhammer, L., and Kiraly, I. C.,** Flavon-glykoside in *Croton zambezicus, Phytochemistry,* 9, 897, 1970.
155. **Kallimanis, G. and Philianos, S.,** Chemical constituents of *Euphorbia acanthothamnos* Heldr. and Sart. leaves. II., *Plant. Med. Phytother.*, 14, 233, 1980.
156. **Mueller, R. and Pohl, R.,** Die Flavonolglykoside von *Euphorbia amygdaloides* und ihre Quantitative Bestimmung in verschiedenen Entwicklungsstadien der pflanze, *Planta Med.*, 18, 114, 1970.
157. **Ulubelen, A., Oksuz, S. H. B., Aynehchi, Y., and Mabry, T. J.,** Flavonoids from *Euphorbia larica, E. virgata, E. Chamaesyce* and *E. magalanta, J. Nat. Prod.*, 46, 598, 1983.
158. **Roshchin, Yu. V., Shinkarenko, A. L., and Oganesyan, E. T.,** Flavon-7-glucoside from *Euphorbia condylocarpa, Khim. Prir. Soedin.*, 6, 472, 1970.
159. **Stadmann, H. and Pohl, R.,** Quercetin-3-glucuronide and kaempferol-3-glucuronoide, the main flavonoids in *Euphorbia cyparissias* L., *Naturwissenschaften,* 53, 362, 1966.
160. **Pal, S. K. and Dutta, S. P.,** Flavonoid constituents of *Euphorbia dracunculoides, J. Indian Chem. Soc.*, 46, 1066, 1969.
161. **Pohl, R., Janistyn, B., and Nahrstedt, A.,** Die Flavonolygycoside von *Euphorbia helioscopia, E. stricta, E. verrucosa* and E. *dulcis.* IX. Mitteilung über die Flavonoide einheimisher Euphorbiaceen, Planta Med., 27, 310, 1975.
162. **Wagner, H., Danninger, H., Seligmann, O., Noraday, M., Farkas, L., and Farnsoworth, N.,** Synthesis of glucosiduronic acid in the flavonoid series. II. Isolation of kaempferol 3-β-D glucuronide from *Euphorbia esula, Chem. Ber.*, 103, 3678, 1970.
163. **Ismail, S., El-Missiry, M. M., Hammouda, F. M., and Rizk, A. M.,** Flavonoids of *Euphorbia geniculata* and *Euphorbia prostrata, Pharmazie,* 32, 538, 1977.
164. **Rizk, A. M., Al-Nagdy, S. A., and El-Missiry, M. M.,** Contituents of plants growing in Qatar. II. Investigation of *Euphorbia granulata, Pharmazie,* 37, 737, 1982.
165. **Kutani, N., Kawase, A., and Gunji, S.,** The isolation and identification of flavonoids occurring in the herb of *Tithymalus helioscopius* Hill. I., *Women's Univ.*, 14, 77, 1962.
166. **Kawase, A. and Kutani, N.,** Some properties of a new flavonoid, tithmalin, isolated from the herb of *Euphorbia helioscopia, Agric. Biol. Chem. (Tokyo),* 32, 121, 1968.
167. **Vololueva, M. A.,** Phytochemical study of *Euphorbia helioscopia, Tr. Alam. At. Med. Inst.*, 26, 451, 1979; *C.A.*, 77, 72542, 1972.
168. **Abdel-Salam, N. A., Mahmoud, A., El-Sayed, S., and Khafagy, S. M.,** spectrophotometric estimation of individual flavone glycosides in three *Euphorbia* species, *Pharmazie,* 30, 402, 1975.
169. **Chen, Y., Tang, Z. -J., Jiang, F. -X., Zhang, X. -X., and Lao, A. -N.,** Studies on the active principles of Ze-Qi *(Euphorbia helioscopia* L.): a drug used for chronic bronchitis. I., *Yoo H. Sueh Hsuehpao,* 14, 91, 1979.
170. **Blanc, P. and De Saqui-Sannes, G.,** Les flavonoides d'*Euphorbia hirta* L. (Euphorbiaceés), *Plant. Med. Phytother.*, 6, 106, 1972.
171. **Ueda, H. and Hsu, Ch.,** A chemical study of *Euphorbia, J. Taiwan Pharm. Assoc.*, 1, 40, 1949.
172. **Husii, K., Huzikawa, H., and Kudo, Y.,** Constituents of *Euphorbia humifusa, J. Pharm. Soc. Jpn.*, 57, 140, 1937.
173. **Roshchin, Yu., Shinkarenko, A. L., and Oganesyan, E. T.,** Hyperoside from *Euphorbia petrophila* and *Euphorbia iberica, Khim. Prir. Soedin.*, 5, 442, 1969.
174. **Rizk, A. M., Ripler, H., and Ismail, S. I.,** Flavonoids and ellagic acid from *Euphorbia hypericifolia* L. (=*E. indica* Lam.), *Fitoterapia,* 48, 99, 1977.
175. **Azimov, M. A. and Nazirov, Z. N.,** Flavonoids from *Euphorbia jaxartica* and *Euphorbia lamprocarpa, Khim. Prir. Soedin.*, 6, 271, 1970.
176. **Soboleva, V. A. and Chagovets, R. K.,** Polyphenolic compounds from *Euphorbia kaleniczenkii, Khim. Prir. Soedin.*, 7, 528, 1971.
177. **Dumkow, K.,** Kaempferol-3-glucuronide and quercetin-3-glucuronide, main flavonoids of *Euphorbia lathyris* L. and their separation on acetylated polyamide, *Z. Naturforsch.*, 24, 358, 1969.
178. **Karrer, W.,** *Konstitution und Verkommen der organischen Pflanzenstoffe,* Birkhauser Verlag, Basel, 1958.
179. **Shang, T. -M., Wang, L., Liang, H. -T., Lin, X. Y., Xia, J. -M., and Niu, S. L.,** Studies on the constituents of Mao-Yan-Cao *(Euphorbia lunulata* Bge.), *Hua Hsueh Hsueh Pao,* 37, 119, 1979.
180. **Roshchin, Yu. V.,** Polyphenolic compounds from *Euphorbia oblongifolia* and *Euphorbia macroceras, Khim. Prir. Soedin.*, 6, 280, 1970.

181. **Zhuk, L. M. and Mudzhiri, K. S.,** *Euphorbia maculata* studied for tannin content, *Tr. Inst. Farmakokhim. Akad. Nauk Gruz. SSR,* 11, 69, 1969; *C.A.,* 73, 127776, 1970.
182. **Dumkow, K.,** Die Flavonolglykoside von *Euphorbia myrisinitis,* ihre Isolierung und Identifizerung-6-Mitteilung uber die Flavonoide einheimischer Euphorbiaceen, *Planta Med.,* 19, 197, 1971.
183. **Bondarendko, O. M., Chagovets, R. K., Litvinenko, V. I., Obolentseva, G. V., Sila, V. I., and Kigel, T. B.,** *Euphorbia palustris* and *Euphorbia stepposa* flavonoids and their pharmacological properties, *Farm. Zh. (Kiev),* 26, 46, 1971; *C.A.,* 76, 121697, 1972.
184. **Sotnikova, O. M. and Chagovets, R. K.,** Phytochemical study of *Euphorbia palustris.* II. Separation and chemical study of compound A, *Farm. Zh. (Kiev),* 21, 49, 1966; *C.A.,* 65, 9006, 1966.
185. **Rizk, A. M., Youssef, A. M., Diab, M. A., and Salem, H. M.,** Constituents of Egyptian Euphorbiaceae. II. Flavonoids of *Euphorbia paralias* L., *Pharmazie,* 31, 405, 1976.
186. **Rizk, A. M., Ahmed, S. S., and Diab, M. A.,** Constituents of Egyptian Euphorbiaceae. VII. Further investigation of the flavonoids of *Euphorbia paralias* L., *Planta Med.,* 36, 189, 1979.
187. **Khafagy, S. M., Gharbo, S. A., and Abdel-Salam, N. A.,** Phytochemical study of *Euphorbia peplus, Planta Med.,* 27, 387, 1975.
188. **Hallett, F. P. and Parks, M.,** A note on the isolation of quercitrin from *Euphorbia pilulifera, J. Am. Pharm. Assoc.,* 40, 56, 1951.
189. **Nahrstedt, A., Dumkow, K., Janistyn, B., and Pohl, R.,** Quercetin-Galaktosid-Gallate in Euphorbiaceen, *Tetrahedron Lett.,* 559, 1874.
190. **Pohl, R. and Janistyn, B.,** Die Flavonolglycoside von *Euphorbia sequieriana, Planta Med.,* 26, 90, 1974.
191. **Soboleva, V. A.,** Flavonoids of some species of *Euphorbia, Khim. Prir. Soedin.,* 855, 1979.
192. **Sotnikova, O. M. and Litvinenko, V. I.,** Isomyricitrin from *Euphorbia stepposa, Khim. Prir. Soedin.,* 4, 50, 1968.
193. **Sotnikova, O. M., Chagovets, R. K., and Litvinenko, V. I.,** New flavanes of *Euphorbia stepposa, Khim. Prir. Soedin.,* 4, 82, 1968.
194. **Sotnikova, O. M. and Chagovets, R. K.,** *Euphorbia stepposa* flavonoids. I. Separation and chemical behaviour of flavonoid compounds from grass, *Farm. Zh. (Kiev),* 21, 60, 1968; *C.A.,* 68, 93487, 1968.
195. **Nagase, M.,** Flavonol glycoside from *Euphorbia thymifolia* L., *J. Agric. Chem. Soc. Jpn.,* 17, 483, 1941.
196. **Subramanian, S. S., Nagarajan, S., and Sulochana, N.,** Flavonoids of the leaves of *Jatropha gossypifolia, Phytochemistry,* 10, 1690, 1971.
197. **Barros, D. A. D., De Alvarenga, M. A., Gottlieb, O. R., and Gottlieb, M. E.,** The chemistry of Brazilian Euphorbiaceae. IV. Naringenin coumaroylglycosides from *Mabea caudata, Phytochemistry,* 21, 2107, 1982.
198. **Dumkow, K.,** Die Flavonoide einheimischer Euphorbiaceen. IV. Die Flavonolglykoside von *Mercurialis annua,* L., *Z. Naturforsch.,* 24b, 1203, 1969.
199. **Dumkow, K.,** Die Flavonoide einheimischer Euphorbiaceen. III. Isolierung und Identifizierung der Flavonolglykoside von *Merucrialis perennis* L., *Planta Med.,* 17, 391, 1969.
200. **Nakaoki, T. and Morita, N.,** Components of the leaves of *Cornus controversa, Ailanthus altissima* and *Ricinus communis, J. Pharm. Soc. Japn.,* 78, 558, 1958.
201. **Khafagy, S. M., Mahmoud, Z. F., and Abdel-Salam, N. A.,** Coumarins and flavonoids of *Ricinus communis* growing in Eygpt, *Planta Med.,* 37, 191, 1979.
202. **Henrick, C. A. and Jeffries, P. R.,** Chemistry of Euphorbiaceae. XIII. Flavones and minor terpenes from *Ricinocarpos muricatus, Tetrahedron,* 21, 3219, 1965.
203. **Henrick, C. A. and Jeffries, P. R.,** New flavones from *Ricinocarpos stylosus, Aust. J. Chem.,* 17, 934, 1964.
204. **Shimokoriyama, M.,** The flavone glycosides isoquercitrin from the leaves of *Sapium sebiferum, J. Chem. Soc. Jpn.,* 68, 1, 1947; *C.A.,* 44, 3983, 1950.
205. **Dutta, P. K., Banerjee, D., and Dutta, N. L.,** Euphorbetin: a new bicoumarin from *Euphorbia lathyris* L., *Tetrahedron Lett.,* 601, 1972.
206. **Dutta, P. K., Banerjee, D., and Dutta, N. L.,** Euphorbetin, a new bicoumarin from *Euphorbia lathyris, Indian J. Chem.,* 11, 831, 1973.
207. **Parthasarathy, M. R. and Saradhi, K. P.,** A coumarimolignan from *Jatropha glandulifera, Phytochemistry,* 23, 867, 1984.
208. **Chawla, H. M., Chakrabrty, K., Chibber, S. S., Kalia, A. N., and Chaudhury, N. C.,** Daphnetin from *Euphorbia dracunculoides* fruits, *Indian J. Pharm. Sci.,* 42, 138, 1980.
209. **Mahmoud, Z. F. and Abdel-Salam, N. A.,** Coumarins of *Euphorbia terracina* and *Euphorbia paralias, Pharmazie,* 34, 446, 1979.
210. **Ahmad, S. A., Kapoor, S. K., and Zaman, A.,** Bergenin in *Flueggea microcarpa, Phytochemistry,* 11, 452, 1972.
211. **Gunasekera, S. P., Cordell, G. A., and Farnsworth, N. R.,** Constituents of *Nealchornea yapurensis* Euphorbiaceae), *J. Nat. Prod.,* 43, 285, 1980.

212. **Heim DeBalsac, F., Maheu, J., Léfevre, L., and Parveaud, A.,** The value of the barks of "Cay-Xu" *(Aegiceras majus* Goertin-Myrsin) and of "Chu-Me" *(Phyllanthus emblica* L. Euphorb.) from Tonquin as tanning materials, *Bull. Agence Gen. Colon. Fr.,* 23, 710, 1930; *C.A.,* 25, 230, 1931.
213. **Damodaran, M. and Nair, K. R.,** A tannin from the Indian gooseberry *(Phyllanthus emblica)* with a protective action on ascorbic acid, *Biochem. J.,* 30, 1014, 1936.
214. **Hsu, S. H., Chang, H. L., and Tien, Y. W.,** A survey of common materials containing vegetable tannin found in Szechwan, *J. Chem. Eng. China,* 7, 26, 1940; *C.A.,* 36, 2438, 1942.
215. **Okuda, T., Mori, K., and Hatano, T.,** The distribution of geraniin and mallotusinic acid in the order Geraniales, *Phytochemistry,* 19, 547, 1980.
216. **Okuda, T., Yoshida, T., and Hatano, T.,** Equilibrated stereostructures of hydrated geraniin and mallotusinic acid, *Tetrahedron Lett.,* 2561, 1980.
217. **Hussein, S. P.,** Ellagic acids from *Euphorbia cornigera* and *Euphorbia wallichi, Phytochemistry,* 13, 867, 1974.
218. **Roshchin, Yu. V. and Dzhumyrko, S. F.,** Ellagic acid from *Euphorbia petrophila* and *E. iberica, Khim. Prir. Soedin.,* 5, 610, 1969.
219. **Boe, J. E., Winsnes, R., Nordal, A., and Bernatek, E.,** New constituents of *Euphorbia resinifera* Berg., *Acta Chem. Scand.,* 23, 3609, 1969.
220. **El-Naggar, L., Beal, J. L., Parks, M., Salman, K. N., Patil, P., and Schwarting, A. E.,** A note on the isolation and identification of two pharmacologically active constituents of *Euphorbia pilulifera, Lloydia,* 41, 73, 1978.
221. **De Alvarenga, M. A., Gottlieb, O. R., and Magalhàes, M. T.,** Methylphenanthrenes from *Sagotia racemosa, Phytochemistry,* 15, 844, 1976.
222. **Balantine, J. A.,** The isolation of two esters of the naphthaquinone alcohol, shikonin, from the shrub *Jatropha glandulifera, Phytochemistry,* 8, 1587, 1969.
223. **Krishnamurti, G. V. and Seshardi, T. R.,** Bitter principle of *Phyllanthus niruri, Proc. Indian Acad. Sci.,* 24A, 357, 1946.
224. **Row, L. R., Srinivasulu, C., Smith, M., and Subba Rao, G. S. R.,** New lignans from *Phyllanthus niruri, Tetrahedron Lett.,* 1557, 1964.
225. **Schaeffier, H. J., Lauter, W. M., and Foote, P. A.,** A preliminary phytochemical study of *Hippomane mancinella, J. Am. Pharm. Assoc.,* 43, 43, 1954.
226. **Hart, N. K., Johns, S. R., and Lamberton, J. A.,** Hexahydroimidazopyrimidines, a new class of alkaloids from *Alchornea javanensis, J. Chem. Soc.,* 1484, 1969.
227. **Hart, N. K., Johns, S. R., Lamberton, J. A., and Willing, R. I.,** Alkaloids of *Alchornea javanensis.* The isolation of hexahydroimidazo (1,2-a) pyrimidines and guanidines, *J. Chem.,* 1679, 1970.
228. **Bettolo, R. M. and Scarpati, M. L.,** Alkaloids of *Croton draconoides, Phytochemistry,* 18, 520, 1979.
229. **Persinos, J. G.,** Antiinflammation composition containing thaspine or its salts, *C.A.,* 78, 7827, 1973.
230. **Bhakuni, D. S. and Dhar, M. M.,** Crotsparinine, a Dihydroproaporphine alkaloid from *Croton sparsiflorus, Experientia,* 25, 354, 1969.
231. **Bhakuni, D. S. and Dhar, M. M.,** Crotosparine, a new proaporphine alkaloid from *Croton sparsiflorus* Morong., *Experientia,* 24, 10, 1968.
232. **Srish, K. S.,** Chemical investigation of *Croton sparsiflorus, Sci. Cult.,* 24, 572, 1959.
233. **Hart, N. K., Johns, S. R., and Lamberton, J.,** (+)-9-Aza-1-methyl-bicyclo (3,3,1) nonan-3-one, a new alkaloid from *Euphorbia atoto, Aust. J. Chem.,* 20, 561, 1967.
234. **Beecham, A. F., Johns, S. R., and Lamberton, J. A.,** The absolute configuration of (+)-9-Aza-1-methylbicyclo [3,3,1] nonan-3-one, an alkaloid from *Euphorbia atoto, Aust. J. Chem.,* 20, 2291, 1967.
235. **Uemura, D. and Hirata, Y.,** Isolation and structures of two new alkaloids, milliamines A and B, obtained from *Euphorbia millii, Tetrahedron Lett.,* 3673, 1971.
236. **Paris, R. A., Moyse Mignon, H., and Le Men, J.,** Alkaloids of *Flueggea virosa, Ann. Pharm. Fr.,* 13, 245, 1955.
237. **Johns, S. R. and Lamberton, J. A.,** New histamine alkaloids from a *Glochidon* species, *Chem. Commun.,* 312, 1966.
238. **Johns, S. R. and Lamberton, J. A.,** New imidazole alkaloids from a *Glochidion* species, *Aust. J. Chem.,* 20, 555, 1967.
239. **Ramos, C. E. M.,** The active principles of *Coidoscolus basiacantha (Jatropha basiacantha), An. Fac. Farm. Bioquim. Univ. Nac. Mayor San Marcos,* 7, 126, 1956.
240. **Nakano, T., Djerassi, C., Corral, R. A., and Orazi, O. O.,** Structure of julocrotine, *Tetrahedron Lett.,* 8, 1959.
241. **Parello, J., Melera, A., and Goutarel, R.,** Phyllochrysine and securinine, alkaloids of *Phyllanthus discoides, Bull. Soc. Chem.,* 898, 1963.
242. **Bevan, C. W. L., Patel, M. B., Rees, A. H., and Taylor, D. A.,** Alkaloid from *Phyllanthus discoides, Chem. Ind. (London),* 838, 1964.

243. **Parello, J. and Munavalli, S.,** Phyllantine and phyllantidine, alkaloids of *Phyllanthus discoides, Compt. Rend.* 260, 337, 1965.
244. **Oletta, S. A.,** Therapeutic alkoloid from *Phyllanthus discoides, C.A.*, 57, 9963.
245. **Bevan, C. W. L., Patel, M. B., and Rees, A. H.,** Minor alkaloid of *Phyllanthus discoides, Chem. Ind.*, 2054, 1964.
246. **Boettcher, B.,** *Chem. Ber.*, 51, 673, 1981.
247. **Spaeth, E. and Koller, G.,** Constitution of ricinin., *Chem. Ber.*, 56B, 880, 1923.
248. **Massart, S. A. and Massart, A.,** Feed for farm Stock "Centtrexico", Central d'explotations industrielles et commerciale Belg., 438, 744, 1940; *C.A.*, 36, 2950, 1942.
249. **Rao, N. V. S.,** Chemical composition of caster leaves, *Proc. Indian Acad. Sci.*, 21A, 123, 1944.
250. **Satoda, M., Murayama, J., Tsji, and Yoshii., E.,** Securinine and allosecurinine *Tetrahedron Lett.*, 1199, 1962 *C.A.*, 59, 687, 1963.
251. **Valen, F.,** Contribution to the knowledge of cyanogensis in angiosperms. X. Cyanogenis in Euphorbiaceae, *Planta Med.*, 34, 408, 1978.
252. **Scholz, H.,** Geraniales, in *Syllabus der Pflanzen-familien*, Vol. 2, 12th ed., Engler, A., Ed., Gerbrüder Bornträger, Berlin, 1964, 246.
253. **Hegnauer, R.,** *Chemotaxonomie der Pflanzen*, Vol. 4, *Dicotyledoneae Daphniphyllaceae-Lathraceae*, Birkhäuser Verlag, Basel, 1966, 124.
254. **Nahrstedt, A., Kant, J.-D., and Wray, V.,** Acalyphin, a cyanogenic glucoside from *Acalyphy indica, Phytochemistry*, 21, 101, 1982.
255. **Seigler, D. S. and Bloomfield, J. J.,** Constituents of the genus *Cnidoscolus, Phytochemistry*, 8, 935, 1969.
256. **Butler, G. W.,** The distribution of the cyanoglucosides linamarin and lotaustralin in higher plants, *Phytochemistry*, 4, 127, 1965.
257. **Dunstan, W. R., Henry, T. A., and Auld, S. J. M.,** Cyanogenesis in plants. V. The occurrence of phaseolunation in *Cassava (Manihot aipi* and *Manihot utilissima), Proc. R. Soc. London Ser. B:*, 78, 1906.
258. **Finnemore, H., Reichard, S. K., and Large, D. K.,** Cyanogenetic glucosides in Australian plants. V. *Phyllanthus gasstroemii, J. Proc. R. Soc. N.S. Wales*, 70, 257, 1936.
259. **Ettlinger, M. G. and Kjaer, A.,** Sulfur compounds in plants, in *Recent Advances in Phytochemistry*, Vol. 1, Mabry, T. J., Alston, R. E., and Runeckles, V. C., Eds., North-Holland, Amsterdam, 1968, 50.
260. **Kjaer, A.,** The natural distribution of glucosinolates: a uniform group of sulfur-containing glucosides, in *Chemistry in Botanical Classification*, Bendz, G. and Santesson, J., Eds., Academic Press, London, 1974, 229.
261. **Kjaer, A. and Friis, P.,** Isothiocyanates from *Putranjiva roxburghii* Wall. including (*S*)-2-methylbutyl isothiocyanate, a new mustard oil of natural derivation, *Acta Chem. Scand.*, 16, 936, 1962.
262. **Cronquist, A.,** *An Integrated System of Classification of Flowering Plants*, Columbia University Press, New York, 1981, 634.
263. **Leone, G.,** Constitution and synthesis of daphnin, *Gazz. Chim. Ital.*, 55, 673, 1925.
264. **Asai, T.,** Daphnins of *Daphne odora, Acta Phytochim. Jpn.*, 5, 9, 1030.
265. **Mindl, R.,** *S. Afr. J. Sci.*, 30, 455, 1933 (cited in Ref. 272).
266. **Nakabayashi,** *J. Pharm. Soc. Jpn.*, 74, 192, 1924 (cited in Ref. 272).
267. **Rizk, A. M., Hammouda, F. M., and Ismail, S. I.,** Phytochemical investigation of *Thymelaea hirsuta*. III. Comaris, *Acta Chem. Acad. Sci. Hung.*, 85, 107, 1975.
268. **Zwenger, C.,** *Liebigs Ann. Chem.*, 115, 1, 1860 (cited in Ref. 272).
269. **Tschesche, R., Schacht, U., and Legler, G.,** Daphnorin, a new naturally occuring derivative of 3,7-dicoumaryl ether, *Liebigs Ann. Chem.*, 662, 113, 1963.
270. **Tschesche, R., Schact, U., and Legler, G.,** Daphnoretin, a coumarin glycoside from *Daphne mezereum. Naturwissenschaften*, 15, 521, 1963.
271. **Majumder, P. L. and Sengupta, C. C.,** Chemical investigation of *Daphne cannabina, J. Indian Chem. Soc.*, 45, 1058, 1968.
272. **Rizk, A. M. and Rimpler, H.,** Isolation of daphnoretin and β-sitosterol-β-D-glucoside from *Thymelaea hirsuta, Phytochemistry*, 11, 473, 1970.
273. **Garcia-Granados, A. and Saenz de Buruagea, J. M.,** Thymeleacea phytochemistry. II. Flavone and coumarin components of *Thymelea tartonraira* L., *An. Quim. Soc. C:*, 76, 96, 1980.
274. **Kosheleva, L. I. and Nikonov, G. K.,** Phytochemical study of *Daphne mezereum, Farmatsiya (Moscow)*, 17, 40, 1968; *C.A.*, 70, 103683, 1969.
275. **Varughese, G. and Rishi, A. K.,** Consituents of *Thymelaea passerina, Fitoteropia*, 53, 191, 1982.
276. **Ismail, S. I.,** Tiliroside (Kaempferol-3-*p*-coumaroyl-glucoside) from *Thymelea hirsuta*. IV., *Fitoterapia*, 49, 156, 1978.
277. **Chen, C.-L. and Tseng, K.-F.,** Flavonoids present in Chinese drugs. XI. Components of the root bark of *Daphne genkwa, Yao Hsuch Poa*, 12, 119, 1965; *C.A.*, 63, 1654, 1965.
278. **Nakano, M. and Tseng, K. F.,** Constituents of *Daphne genkawa*. I., *J. Pharm. Soc. Jpn.*, 52, 903, 1932.

279. **Nakano, M. and Tseng, K. F.,** Components of *Daphne genkawa,* II. Constitution of genkwanin, *J. Shanghai Sci. Inst.,* 1, 1, 1933.
280. **Kapoor, S. K., Kohli, J. M., and Zaman, A.,** Chemical investigation of *Daphne oleoides, Indian J. Appl. Chem.,* 32, 105, 1969.
281. **Nùnez-Alarcon, J., Rodriguez, E., Schmid, R. D., and Mabry, T. J.,** 5-*O*-xylosyl-glucosides of apigenin and luteolin-7-and 7,4′-methylethers from *Ovida pillopillo, J. Org. Chem.,* 36, 3829, 1971.
282. **Nùnez-Alacron, J., Rodriguez, E., Schmid, R. D., and Mabry, T. J.,** 5-*O*-xylosylgucosides of apipenin and luteolin 7-and 7,4′-methylethers from *Ovidia pillo-pillo, Phytochemistry,* 12, 1451, 1973.
283. **Tseng, K. F. and Choa, C. Y.,** Flavonoids in Chinese Drugs. A new flavone glycoside isolated from *Wikstroemia viridiflora, Yao Husch Pao,* 10, 286, 1963, *C.A.,* 59, 15372, 1963.
284. **Maiti, P. C. and Toy, S.,** Triterpenes from *Daphne cannabina, Curr. Sci.,* 36, 99, 1967.
285. **Kosheleva, L. I. and Nikonov, G. K.,** Seasonal dynamics of the hydroxycoumarins content in *Daphne mezerum.,* from *Ref. Zh. Biol. Khim.,* 1970; *C.A.,* 75, 126577, 1971.
286. **Das, A. K. and Mutherji, K.,** Stearic acid from the leaves of *Daphne oleoides, Curr. Sci.,* 35, 178, 1966.
287. **Sygiyama, N., Yamamoto, M., and Yamada, K.,** Studies on the components of *Edgeworthia papyrifera, Nippon Kagaku Zasshi,* 89, 710, 1968; *C.A.,* 70, 862, 1969.
288. **Gharbo, S. A., Khafagy, S. M., and Sarg, T. M.,** Phytochemical investigation of *Thymelaea hirsuta, U.A.R. J. Pharm. Sci.,* 11, 101, 1970.
289. **Carcia-Granados, A. and Saenz de Buruaga, A.,** Thymelaeaceae phytochemistry. I. Diterpenes, triterpenes and sterols of *Thymelea hirsuta* L. leaves, *An. Quim. Soc. C:,* 76, 94, 1980.
290. **Rizk, A. M., Hammouda, F. M., and Ismail, S. I.,** Phytochemical investigation of *Thymelaea hirsuta.* II. Lipid fraction, *Planta. Med.,* 26, 346, 1974.
291. **Mentzer, C.,** in *Comparative Phytochemistry,* Swain, T., Ed., Academic Press, London, 1966.
292. **Moheshwari, M. L., Varman, K. R., and Bhattacharya, S. C.,** Terpenoids. XLVII. Structure and absolute configuration of noroxoagarofuran, 4, hydroxy-dihydroagarofuran,3,4-dihydroxy-dihydroagarofuran and conversion of β-agarofuran to α-agarofuran, *Tetrahedron Lett.,* 19, 1519, 1963.
293. **Sisido, K., Kurozumi, S., and Utimoto, K.,** Fragrant flower constituents of *Daphne odora, Perfum, Essential Oil Rec.,* 58, 528, 1967.
294. **Watanabe, I., Yanai, T., Awano, K., Kogami, K., and Hayashi, K.,** Volatile components of zinchoge flower *(Daphne odora,* Thunb., *Agric. Biol. Chem.,* 47, 483, 1983.

Chapter 6

MACROCYCLIC DITERPENES OF THE FAMILY EUPHORBIACEAE

Fred J. Evans

TABLE OF CONTENTS

I. INTRODUCTION

The isolation of phorbol and its related esterified diterpenes from plants of the family Euphorbiaceae[104] seemed to explain the diverse toxicological and medical uses of these plants.[5-8] Irritant and tumor-promoting diterpenes of the tigliane, daphnane, and ingenane groups, when shown to be present in plant species, would be expected to be responsible for therapeutic indications of plants of the Euphorbiaceae including their use as purgatives and abortifacients internally, and for external use such as counterirritants and for skin complaints including fungal, bacterial, and parasite infestations (see Chapter 1). In traditional medicine, species of this plant family have several uses which may not be attributed to the presence of toxic phorbol type derivatives. These indications include the use of Euphorbiaceae species for the treatment of asthma,[9] their reputed analgesic activity,[10] and the reported antipeptic ulcer actions[11,12] of some plants. Clearly these activities were due to different chemical groups of compounds (see Chapter 5). Considering the involvement of phorbol esters in tumor promotion in mammalian[13] systems, perhaps the most surprising traditional use of these plants is their repeated worldwide use in the treatment of human tumors.[14,15] According to traditional literature from geographical locations including China, South America, the Far East, Africa, the U.S., and Europe,[15] extracts of plant species of the Euphorbiaceae are well known for their anticancer activities. Certain phorbol-related compounds including phorbol esters,[16] ingenol esters,[17] and daphnane derivatives[18] have demonstrated antileukemic actions in mice under laboratory conditions. Such observations may open new avenues for cancer chemotherapy as an approximation to Alexander Haddows'[19] suggestion that cytostatic alkylating agents also exhibit carcinogenic actions. However as Hecker[13] has pointed out, the validity of this paradox for cocarcinogens is uncertain. The main problem when assessing potential antileukemic activities of tumor-promoting agents is associated with the lack of complete data. These experiments involved survival times in mice, no postmortems were performed, and complete experimental data including statistical evaluations were absent. Whether certain tumor-promoting agents possess anticancer activity and the contribution that such compounds make to the documented anticancer actions of plant extracts is unclear at the present time. Plants produce a number of secondary metabolites during growth[20] and whole plants or extracts from them used for the treatment of disease will contain a variety of compounds, many of which may be responsible for the therapeutic effects of the drug. Related or even chemically unrelated secondary plant metabolites may be antagonists, agonists, potentiators, or synergistic agents when used in medicine, all of which contribute to the activity of the crude drug itself. In terms of the use of plants of the family Euphorbiaceae in the treatment of human cancers it was accordingly significant that certain groups of macrocyclic and related diterpenes which are formed in these plants together with phorbol-ester tumor-promoting agents, have been shown to demonstrate antitumor[21-23] and cytotoxic[24,25] actions. Groups of macrocyclic compounds which have currently exhibited antitumor actions include those of the jatrophane[21] and jatropholane[22] types, while lathyrane macrocyclics have reported cytotoxic actions against tumor cells in culture.[24,25] In general, the macrocyclic compounds are not irritating to skin, and do not possess tumor-promoting activities[26] related to the phorbol esters. There is a possibility that these compounds may be relatively nontoxic antitumor drugs. Related compounds have also been shown to exhibit analgesic[10] and antiulcer activities.[11,12]

The first macrocyclic diterpene to be isolated from a *Euphorbia* species was "euphorbiasteroid", isolated by Dubylanska[27] in 1937, but it was not until 1970[28] that the chemical structure of this novel lathyrane diterpene was elucidated. In recent years the macrocyclic and related diterpenes of the family Euphorbiaceae have been shown to be a large and biologically as well as chemically interesting group of natural products.

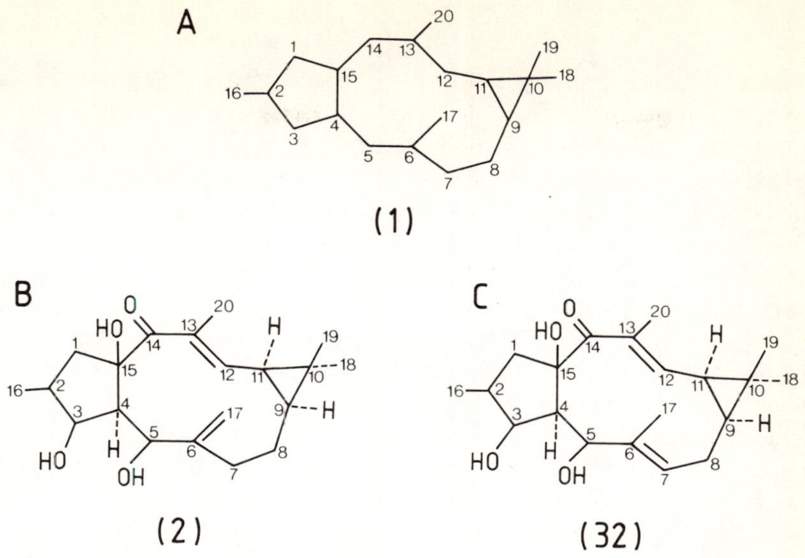

FIGURE 1.
A. (**1**) Lathyrane
B. (**2**) Lathyrol
C. (**32**) Isolathyrol

II. THE LATHYRANE DITERPENES

A. Lathyrols

These diterpenes are based upon the lathyrane (**1**) (Figure 1) hydrocarbon nucleus. Lathyrane (**1**) occurs in plants of the genus *Euphorbia* in the oxygenated form, typically exhibiting a number of secondary and tertiary hydroxy groups. Lathyrol (**2**), a polyhydroxylated macrocyclic diterpene is typical of the group. This compound (Figure 1) consists of a cyclopropane ring *cis*-fused to C-9 and C-11 of the macrocyclic undecane ring. The protons at C-9 and C-11 have the α-configuration. The cyclopentane and the cycloundecane rings are *trans*-linked with the H-4α. Accordingly lathyrol (**2**) possesses the 4α-lathyrane skeleton. Lathyrol (**2**) and closely related lathyrane derivatives are obtained from plants of the genus *Euphorbia* as a series of esters. Euphorbiasteroid[27] was shown to consist of one such ester known as L_1[28,29] (**3**), three esters originally known as L_1 (**3**), L_2 (**4**), and L_3 (**5**) were isolated from the seed oil of *Euphorbia lathyris* by crystallization after dilution of the oil with acetone.[29] They therefore readily separated from the tumor-promoting ingenane diterpenes of this plant species.

Compound L_1[29] (**3**) was the diacetate-phenylacetate of 6,17-epoxylathyrol (**6**) (Figure 2). It had a molecular formula of $C_{32}H_{40}O_8$ as determined by mass spectrometry (MS) and the parent polyol (**6**) could be prepared from it by alkaline hydrolysis with 0.5% KOH in methanol. On acetylation with acetic anhydride/pyridine, 6,17-epoxylathyrol produced a mixture of mono-, di-, and triacetates. The structure of 6,17-epoxylathyrol (**6**) was deduced from its ^{1}H-NMR spectrum utilizing extensive decoupling experiments. Additional evidence was produced from the UV maximum at 273 nm (ϵ = 15,000), which on borohydride reduction of the carbonyl group gave the vinyl-cyclopropane (**7**). Hydrogenation of 6,17-epoxylathyrol (**6**) over 10% Pd/C gave a dihydro-derivative (**8**), which on prolonged treatment produced a tertiary alcohol (**9**). Acetylation of the tertiary alcohol produced a diacetate (**10**) the ^{1}H-NMR of which clearly demonstrated the additional methyl resonance, providing evidence for the presence of the epoxide moiety in 6,17-epoxylathyrol (**6**). $LiAlH_4$ reduction

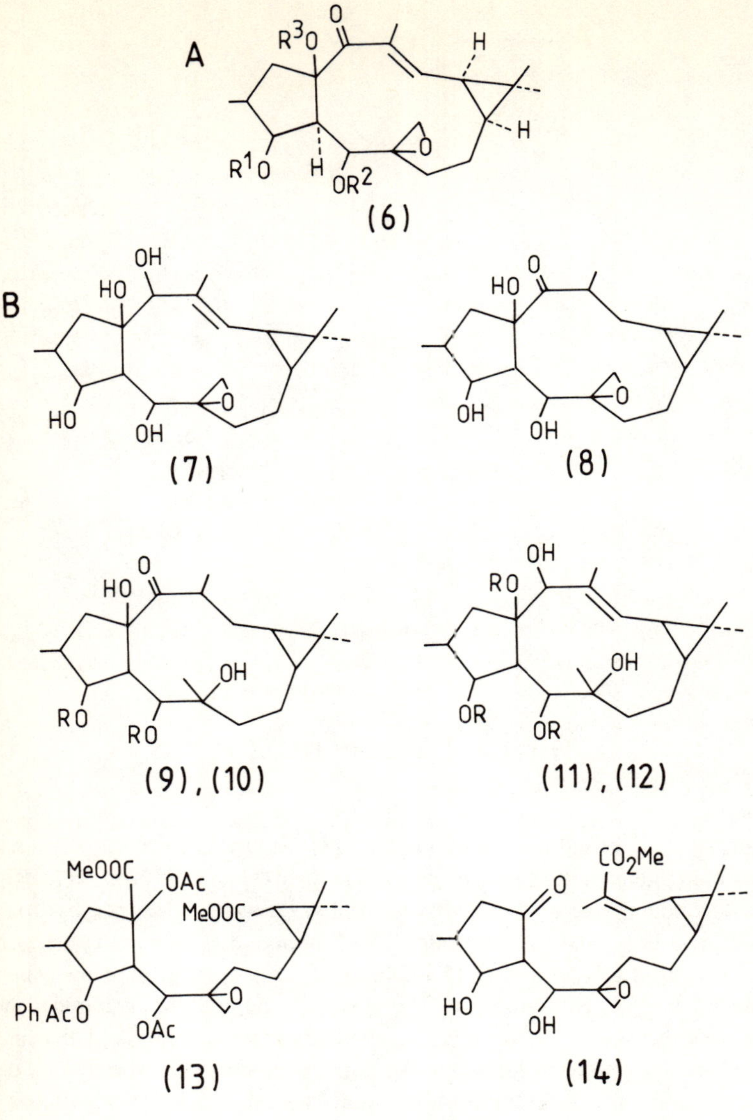

FIGURE 2.
A. (**6**) 6,17-Epoxylathyrol
B. (**9**), (**10**), (**11**), (**12**) R = H
R = acetate
(**7—14**) reaction products of lathyrol-6,17-epoxide

of 6,17-epoxylathyrol (**6**) produced a pentol (**11**) which was able to afford the 3,5,14-triacetate (**12**) on acetylation. The macrocyclic ring of 6,17-epoxylathyrol (**6**) was cleaved by ozonolysis, followed by methylation to produce a diester (**13**); whereas periodate oxidation followed by methylation gave the cyclopentanone monoester (**14**). These reactions lead to the structure of 6,17-epoxylathyrol (**6**) as shown in Figure 2, but the ester group positions of the natural product L_1 (**3**) were obtained from the study of the acid-catalyzed rearrangement products. The structure of 6,17-epoxylathyrol (**6**) was later[28] confirmed by X-ray crystallography of its triester and on x-y-projection of L produced. This compound was later isolated from *Macaranga tanarius* by Hui and Ng[30] in 1975. The genus *Macaranga* is also of the plant family Euphorbiaceae.

Compound L_2 (**4**) (Figure 3) was purified from the seed oil of *Euphorbia lathyris* by

FIGURE 3.

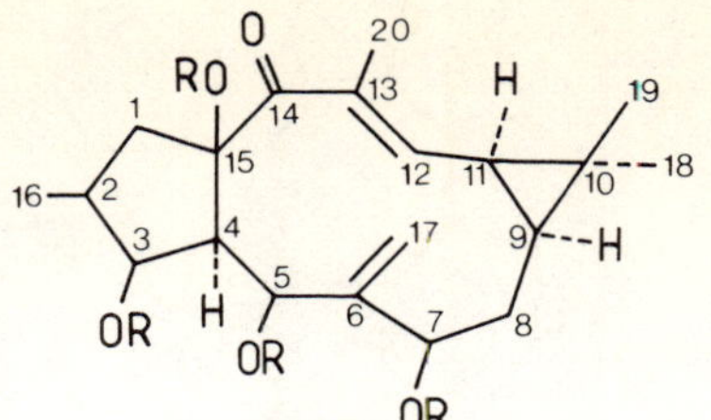

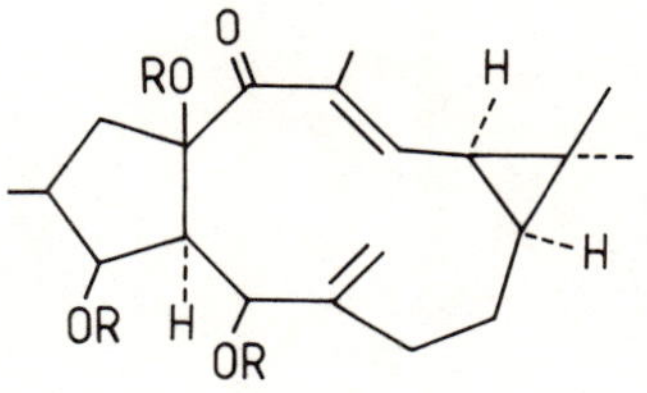

A. (**16**) 7-Hydroxylathyrol; (**17**) R = 3 × acetate, 1 × H; (**4**) R = 2 × benzoate, 2 × acetate
Empirical formula $C_{38}H_{42}O_9$
m.p. 205—207°C
UV (MeOH) λ_{max} (ε) nm 229 (29,800), 273 (14,200)
IR (KBr) $\nu_{max}^{cm^{-1}}$, 1735, 1713, 1650, 1623, 907, 712
^{1}H-NMR δ 1.02 (3H-16), 1.27 (3H-19), 1.34 (3H-18), 1.55 (1H-11), 1.88 (3H-20), 2.40 (1H-2), 2.99 (1H-4), 3.47 (2H-1), 5.25, 5.53 (2H-17), 5.55 (1H-7), 5.80 (1H-3), 6.39 (1H-5), 6.53 (1H-12), ppm. (**17**) R = 3 × acetate, 1H
According to Reference 31

B. (**2**) Lathyrol
(**5**) R = 2 × acetate, 1 × benzoate
Empirical formula $C_{31}H_{38}O_7$
mp 156—158°C
UV (MeOH), λ_{max} (ε) nm, 229 (16,400), 275 (15,300)
IR (KBr)$\nu_{max}^{cm^{-1}}$, 1730, 1705, 1640, 1613, 897, 705
^{1}H-NMR δ 0.98 (3H-16), 1.22 (6H-18, 19), 1.40 (1H-11), 1.76 (3H-20), 2.30 (1H-2), 2.92 (1H-4), 3.60 (2H-1), 4.78, 5.0 (2H-17), 5.81 (1H-3), 6.20 (1H-5), 6.53 (1H-12) ppm
According to Reference 26

means of Craig multistage distribution of the hydrophobic fraction.[31] Compound L_2 (**4**) was the diacetate-dibenzoate tetraester of the new lathyrane diterpene 7-hydroxylathyrol (**16**), recognized by comparison of its UV, IR, and ^{1}H-NMR spectra to that of compound (**6**). No assignments for the positions of acylation in compound (**4**) were made.[31] Comparison of the ^{1}H-NMR for the parent lathyrane, 7-hydroxylathyrol (**16**), with those of (**6**) revealed the absence of the signal for the protons of the epoxide group and the presence of two singlets of protons indicative of an exocyclic methylene group, together with the signal of an extra geminal ester proton. Hydrolysis of L_2 (**4**) with 0.5 *M* KOH in methanol produced the polyol (**16**). The structure of (**16**) was confirmed by means of X-ray diffraction analysis of its triacetate, and finally solved by the application of triple-product releations and Sayre's equation. The absolute configuration of 7-hydroxylathyrol-2,5,7-triacetate (**17**) was obtained using the anomalous scattering of CuK α radiation by the oxygen atoms. Accurate intensity measurements were made on Bijvoet pairs for which the calculated dispersion effects were strongest. Thus, it was found that 7-hydroxylathyrol (**16**) has an absolute configuration which is identical to that of phorbol (see Chapter 7). The polyol, lathyrol (**2**), was later found to

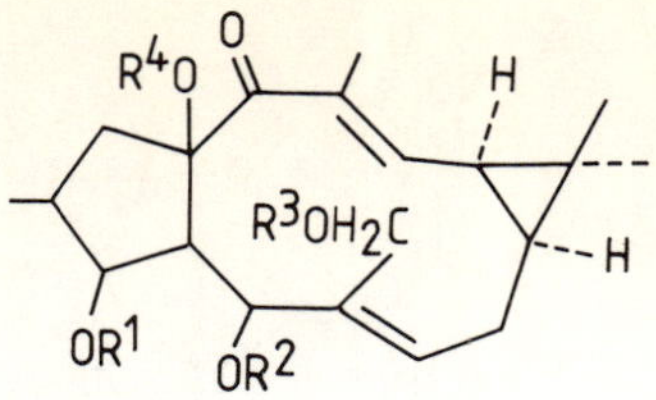

(18)

FIGURE 4.

(**18**) 17-Hydroxylathyrol
(**19**) R^1 = benzoyl, R^2 = R^3 = R_4 = acetate

MS	$M^{+\cdot}$ m/z 580
UV	(MeOH), λ_{max} (ϵ) nm 273 (12,900), 229 (14,400)
IR	(KBr), $\nu^{cm^{-1}}_{max}$, 1735, 1713, 1648, 1625
^{1}H-NMR	δ 0.96 (3H-16), 1.20, 1.37 (6H-18, 19), 1.80 (3H-20), 2.81 (1H-4), 3.43 (1H-1), 4.50 (2H-17), 5.69 (1H-7), 5.90 (1H-3), 6.40 (1H-5), 6.70 (1H-12)

According to Reference 32
(**20**) R^1 = R^2 = R^3 = R^4 = H
(**21**) R^1 = benzoate, R^2 = R^4 = acetate, R^3 = H
According to Reference 32

be the parent diterpene of compound L_3 (**5**) also isolated from *Euphorbia lathyris* seed oil by means of multistage Craig distribution techniques.[26] Ester (**5**) was the diacetate-benzoate of lathyrol (**2**). Its structure was obtained by comparison of UV, IR, and ^{1}H-NMR to that of 7-hydroxylathyrol (**16**) and 6,17-epoxylathyrol (**6**). The ^{1}H-NMR data of L_3 (**5**) corresponded to that for L_2 (**4**) but a signal of the geminal ester in position 7 was not apparent. The configuration of lathyrol (**2**) was adopted from that of 7-hydroxylathyrol (**16**).

A new diterpene parent polyol was isolated from the same plant in 1984 by Adolf et al.[32] This compound, known as 17-hydroxyisolathyrol (**18**), occurred in the neutral fraction produced by Craig distribution of the seed oil.[32] It was isolated as its tetraester (**19**) of three acetates and a benzoate moiety by means of thinlayer chromatography (TLC). The presence of the β-cyclopropylenone system in (**19**) was supported by its UV adsorption at 273 nm (α = 12,900) and its ^{1}H-NMR spectrum together with decoupling experiments. In this sepctrum an additional doublet of doublets at 5.69 ppm suggested the presence of a further double bond in the lathyrane nucleus, and an extra acyloxy function at C-17 was indicated by the presence of an AB quartet at 4.28 ppm. The positions of the acyl groups in 17-hydroxyisolathyrol (**18**) were deduced from a transesterification reaction which produced products (**20** and (**21**) (Figure 4).

B. Jolkinols

Closely related lathyrane diterpenes known as the jolkinols (**22**) were isolated in 1976[33] from extracts of *Euphorbia jolkini*. Four compounds were obtained from this plant, known as the jolkinols A (**23**), B (**24**), C (**25**), and D (**26**). Jolkinols A (**23**) and B (**24**) were characterized by the presence of an α-orientated 5,6-epoxide (Figure 5), but differed in that jolkinol A (**23**) exhibited a C-18 primary hydroxy group while jolkinol B (**24**) exhibited a methyl group at this position. Jolkinol C (**25**) was a new parent diterpene characterized by the presence of a $\Delta^{5,6}$ double bond but was otherwise identical to the parent nucleus of jolkinol B (**24**), while jolkinol D (**26**) was the C-3 acetate of (**24**) (Figure 5).

Jolkinol B (**24**) was chemically transformed to lathyrol (**2**) by reaction on alumina without solvent at 60°C. Lathyrol (**2**) was identified by comparison of its ^{1}H-NMR, melting point, and IR to an authentic sample. The cinnamic acid moiety at C-3 of jolkinol B was deduced

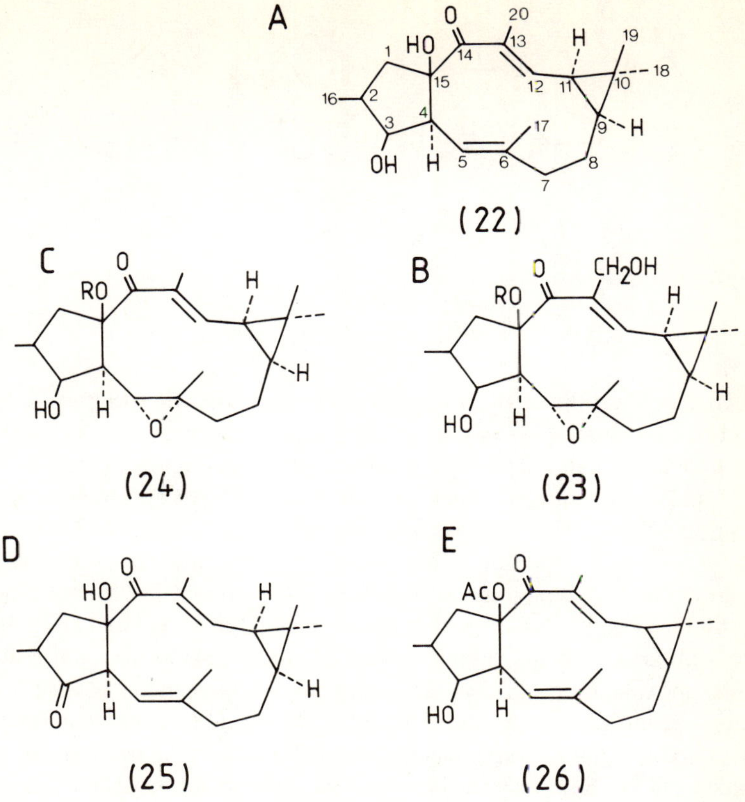

FIGURE 5.

A. (**22**) Jolkinol

B. (**23**) Jolkinol-A, R = cinnamate
Empirical formula $C_{29}H_{36}O_6$
MS M$^{+\cdot}$, m/z 480
IR ($CHCl_3$), $\vee^{cm^{-1}}_{max}$, 3520, 1720, 1640, 1590
^{1}H-NMR δ 1.12 (3H-16), 1.74 (1H-11), 3.68 (1H-4), 4.16 (1H-3), 4.34 (2H-20) ppm
According to Reference 33

C. (**24**) Jolkinol-B, R = cinnamate
Empirical formula $C_{29}H_{36}O_5$
MS M$^{+\cdot}$, m/z 464
IR ($CHCl_3$), $\vee^{cm^{-1}}_{max}$, 3520, 1720, 1630, 1580
^{1}H-NMR δ 1.13 (3H-16), 1.88 (3H-20), 3.65 (1H-5), 4.14 (1H-3), 7.0 (1H-12) ppm

D. (**25**) Jolkinol-C
Empirical formula $C_{20}H_{28}O_3$
MS M$^{+\cdot}$, m/z 316
IR ($CHCl_3$), $\vee^{cm^{-1}}_{max}$, 3460, 1740, 1630, 1610
^{1}H-NMR δ 1.29 (3H-16), 2.87 (1H-4), 5.50 (1H-5), 7.40 (1H-12) ppm
According to Reference 33

E. (**26**) Jolkinol-D
Empirical formula $C_{22}H_{32}O_4$
MS M$^{+\cdot}$, m/z 360
IR ($CHCl_3$), $\vee^{cm^{-1}}_{max}$, 3580, 1740, 1640, 1610
^{1}H-NMR δ 1.46 (3H-16), 1.83 (3H-30), 3.90 (1H-3), 5.68 (1H-5), 6.64 (1H-12) ppm
According to Reference 33

F. (**27**), (**28**) Jolkinol reaction products

F

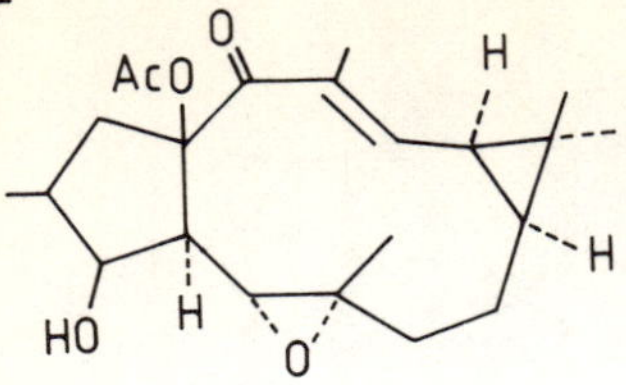

(27) 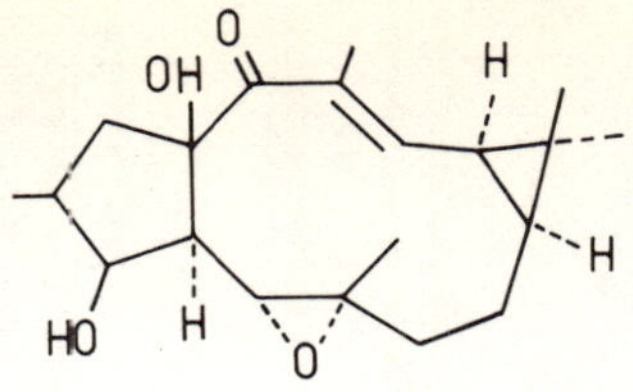 (28)

FIGURE 5F

from the characteristic shift of H-3 of 1.31 ppm in the ^{1}H-NMR spectrum of the acetate of jolkinol B (**24**). Compound (**24**) was accordingly a lathyrane diterpene ester with an identical configuration to that of lathyrol (**2**) except concerning the configuration of C-6. The structure of jolkinol A (**23**) was obtained by comparison of its ^{1}H-NMR spectrum to that of (**24**). In this spectrum the 2H signal of the primary hydroxy group at C-18 of (**23**) was exhibited at 4.34 and 4.54 ppm, and the allylic methyl signal at 1.88 ppm of (**24**) was absent. The *trans* configuration of the $\Delta^{5,6}$ double bond of jolkinols C (**25**) and D (**26**) were unequivocally established by the observation of NOE between H-5 and hydroxyl and H-12 and hydroxyl. Finally, the configuration at C-6 in jolkinols A (**23**) and B (**24**) were established by chemical methods. The epoxidation product of (**26**) using *m*-chloroperbenzoic acid (**27**) produced the diol (**28**) after saponification. Because jolkinol B (**24**) could be converted to lathyrol (**2**) the stereochemistry corresponds to epoxidation from the α-side of the *trans* double bond in (**26**). It was pointed out by Seip and Hecker[34] in 1983 that in the structure elucidation of the jolkinols the α-position of the OH-5 lathyrol (**2**) was erroneously assumed from which the α-position of the epoxide was derived for jolkinol B. The β-position of the OH-5 was arrived at by X-ray structure analysis for 7-hydroxylathyrol (**16**)[31] and 6,20-epoxylathyrol (**6**)[28] unequivocally. Accordingly the epoxide oxygen of jolkinol B (**24**) should be in the β-configuration.

Jolkinol-B (**24**) was recently[33] isolated from the roots of *Euphorbia lathyris* and jolkinol A (**23**) from the callus culture of the same plant. Also isolated from the seed oil of *Euphorbia lathyris*[32] was the diacetyl, cinnamoyl triester (**58**) of the new lathyrane diterpene 17-hydroxyjolkinol (**15**). In the ^{1}H-NMR spectrum of (**15**) in contrast to that of jolkinol D (**26**), the 3H signal at 1.46 ppm for the vinylic methyl group of C-17 was missing and an AB system of two protons at 4.28 ppm was present. This indicated that hydroxylation of the C-17 methyl had occurred in (**15**) (Figure 6).

In 1983 further diterpenes of the jolkinol type were isolated from the latex of *Euphorbia characias* by Seip and Hecker.[34] Five of these diesters were esterified froms of jolkinol-5β,6β-epoxide (**29**) and differed from the parent nucleus of jolkinol A (**23**) in the configuration of the 5,6-epoxy group (Figure 7). These compounds were isolated from the acetone extract of the latex of *E. characias* after Craig distribution followed by preparative TLC. The structure of these five new diterpenes was obtained from their UV, IR, and ^{1}H-NMR spectral comparisons to lathyrol (**2**), extensive decoupling experiments, and base-catalyzed transesterification to jolkinol-5β,6β-epoxide (**29**). The nature and position of acylation in these diesters was obtained from the fragmentation pattern exhibited in the high resolution mass spectra and from partial transesterification reactions of one of them. Furthermore, treatment of an ester of (**29**) with acid furnished two esters of isolathyrol (**30, 31**) from which isolathyrol (**32**) was obtained by hydrolysis.

Also from the same fraction of the latex of *Euphorbia characias* a triester of the novel

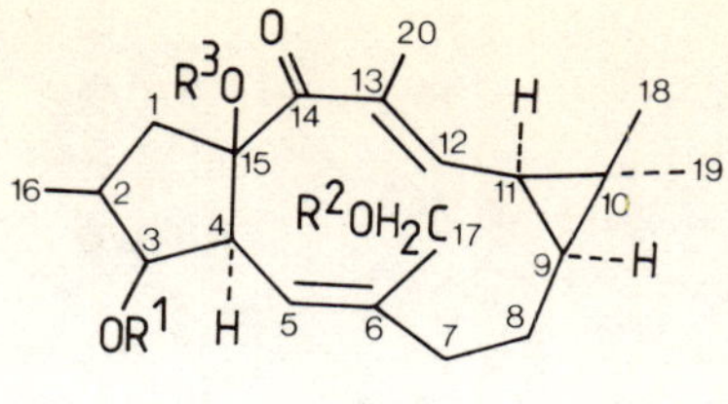

(15)

FIGURE 6.

(**15**) 17-Hydroxyjolkinol

R^1 = cinnamate, R_2 = R^2 = acetate

Empirical formula $C_{33}H_{40}O_7$

MS	$M^{+\cdot}$, m/z 548.
UV	(MeOH), λ_{max} (ϵ), nm, 277 (33,500), 223 (17,000), 217 (19,800)
IR	(KBr), ν_{max}^{cm-1}, 1735, 1708, 1640, 1610
^{1}H-NMR	δ 0.98 (3H-16), 1.05 (6H-18, 19), 1.87 (3H-20), 2.80 (1H-4), 3.60 (1H-1), 4.28 (2H-17), 5.44 (1H-3), 5.69 (1H-5), 6.60 (1H-12) ppm

According to Reference 32

A

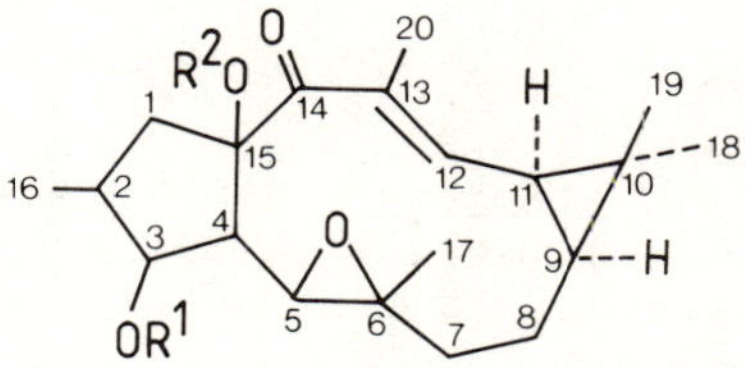

(29)

B

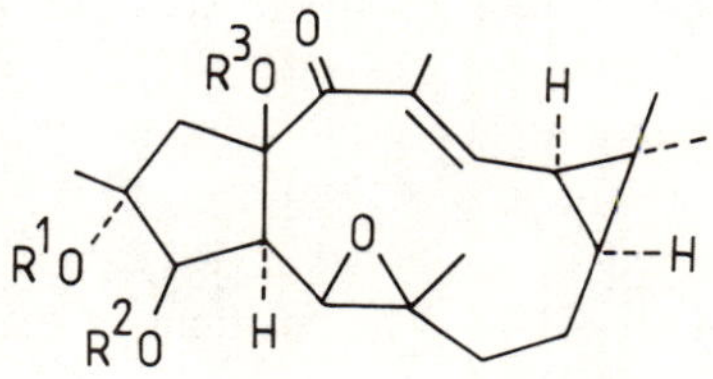

(33)

FIGURE 7.

A. (**29**) Jolkinol-5β,6β-epoxide

	R^1	R^2
(1)	$COCH_2Me$	COMe
(2)	$COCHMe_2$	COMe
(3)	COC(Me) = CHMe	COMe
(4)	COC_6H_5	COMe
(5)	COC_5H_4N	COMe

(**30**) R^1 = $COCHMe_2$, R^2 = H, R^3 = COMe

(**31**) R^1 = H, R^2 = $COCHMe_2$, R^3 = COMe

B. (**33**) 2α-Hydroxyjolkinol-5β-6β-epoxide

(**34**) R^1 = R^2 = R^2 = COMe, $COCHMe_2$, COC_5H_4N

A

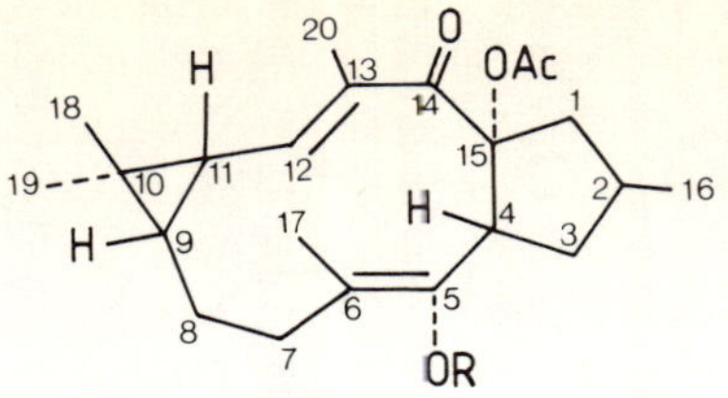

(93) (94)

B

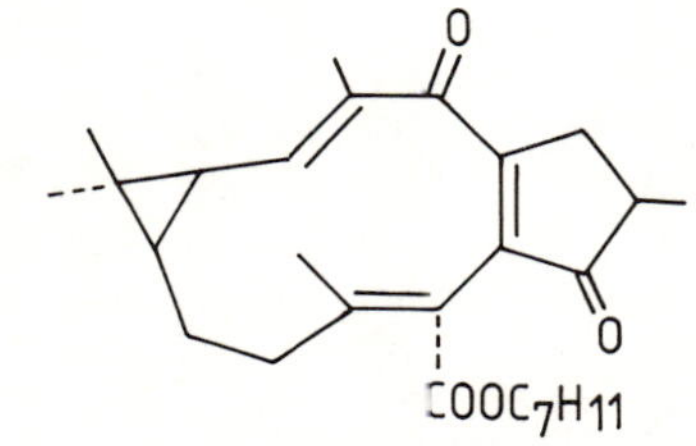

(95)

FIGURE 8.

A. (**93**) Euphohelioscopin-A, R =
Empirical formula, $C_{30}H_{42}O_6$
Colorless oil
IR (film), $\nu_{max}^{cm^{-1}}$, 3500, 1735, 1710, 1620
^{1}H-NMR δ 0.72 (3H,t), 0.85 (3H,s), 0.96 (3H,s), 1.00 (3H,d), 1.58 (3H,b.s), 1.65 (3H,s), 1.93 (3H,s), 2.61 (1H,d.d), 2.93 (1H,d.d), 3.57 (1H,d.d), 5.0 (1H,d.d), 5.56 (1H,d.t), 5.98 (1H,d), 6.00 (1H,b.t), 6.31 (1H,b.d), 6.62 (1H,b.d), 7.87 (1H,d.d) ppm
(**94**) Euphohelioscopin-B, R =

B. (**95**) Oxidation product of (**93**)

lathyrane diterpene 2α-hydroxyjolkinol-5β,6β-epoxide (**33**) was isolated by means of preparative TLC. The molecular formula of this ester (**34**) was derived by high resolution MS as $C_{32}H_{41}NO_8$ and its fragmentation pattern indicated the presence of acetic, isobutyric and nicotinic acids (Figure 7). The structure of the polyol (**33**) was obtained from the ^{1}N-NMR spectrum of (**34**) which indicated an additional acyloxy group at C-2 of the nucleus because of the absence of the doublet for 3H-16. The positions of esterification in (**34**) could not be established due to a paucity of isolated material.

A closely related group of lathyrane diterpenes were isolated in 1983[56] from the leaves and roots of *Euphorbia helioscopia* L. Two compounds known as euphohelioscopin-A (**93**) and euphohelioscopin-B (**94**) (Figure 8) were purified from the methanol extracts of the plant by solvent partition into ether followed by column chromatography on silica gel. Compound (**93**) by the action of potassium carbonate in methanol was converted to its parent diterpene triol, which acetylated to give the diacetate under normal conditions. Oxidation of (**93**) with PCC in dichloromethane afforded the conjugated dione (**93**). The structure and stereochemistry of euphohelioscopins A (**93**) and B (**94**) were assigned from NOE experiments in the ^{1}H-NMR spectra together with NOE experiments on their derivatives using related compounds of known stereochemistry for comparison purposes.

C. Bertyadionol

The publication of the structure 6,17-epoxylathyrol[29] (**6**) prompted the rapid publication of an unusual new lathyrane diterpene from *Bertya, sp. nov.*[35] This plant species was later classified by Airy Shaw[36] as *Bertya cuppressoidea* of the family Euphorbiaceae. Bertyadionol

(**35**) (Figure 9) was a diketo alcohol, empirical formula $C_{20}H_{26}O_3$, the oxygen functions of which were characterized by its IR and UV spectra together with reactivities including the formation of a monoacetate and a dioxime. The structure of (**35**) without stereochemical assignments was obtained from its ^{1}H-NMR spectrum and degradative chemical methods.[35] Bertyadionol (**35**) was characterized by the presence of a double bond at $\Delta^{4,15}$, possibly as a result of dehydration of a 15-hydroxy-3-ketone. Base-catalyzed isomerization of (**35**) produced the 1,5-ene-dione isomer (**36**). Double irridation experiments[37] in the ^{1}H-NMR of (**36**) allowed assignments as shown in Figure 9 and the chemical shifts and coupling constants of the cyclopentenone moiety of (**36**) were as for the tigliane diterpene phorbol. Intramolecular NOE measurements were additionally carried out for (**36**). Saturation of the signal for H-15 in the ^{1}H-NMR spectrum of (**36**) resulted in an increase in the intensity of the signals due to H-1, H-12, and H-5, confirming that these protons were on the same side of the molecule, and that the $\Delta^{12,3}$ olefinic link was in the *trans* configuration. Saturation of the 3H-6 resonance signal increased the intensity of the signals for H-4 but not for H-5, confirming the *trans* configuration for the $\Delta^{5,6}$ double bond. The cyclopentenone ring was accordingly in *trans* fusion. The configuration of the OH-7 was obtained as β from application of the INDOR technique to the monoacetate of (**36**) (Figure 9). The absolute stereochemistry of (**36**) was deduced from these results as shown in Figure 9, but extrapolation of these results to bertyadionol was only possible if it was assumed that *cis-trans* isomerization of the double bonds did not occur during the formation of (**36**) or its monoacetate. Further bertyadionol derivatives were also isolated from *Bertya* species.[37] 5-Hydroxy-4-hydrobertyadionol (**37**) and 1,2-dehydro-4,15-dihydrobertyadionol (**38**) also exhibited the typical *trans* function of the 5- and 11-membered rings of lathyrol (**2**).

D. Ingols

Plant species of the genus *Euphorbia* have also yielded a variety of esters of a further lathyrane diterpene known as ingol (**39**) (Figure 10). The isolation of ingol (**39**) was first described by Opferkuch and Hecker[38] in 1971 and later in 1973.[39] Ingol (**39**) was obtained by a combination of adsorption chromatography and Craig distribution as its triacetate-nicotinate tetraester (**40**) from the acetone-soluble extract of the latex of *Euphorbia ingens*. Base-catalyzed transesterification produced 12-acetylingol from its mass and ^{1}H-NMR spectra and a partial structure for ingol (**39**) was obtained by extensive decoupling experiments of its acetate tribenzoate. The functional groups of ingol were ascertained from their chemical reactivity. Reduction of ingol tetraacetate with $NaBH_4$ followed by methanolysis of the product produced the pentol (**41**), whereas oxidation with OSO_4 produced the diol (**42**), and treatment with 3-Cl-perbenzoic acid led to the epoxide (**43**), confirming the presence of a single double bond in the structure of ingol. 12-Acetylingol in the presence of 1 *M* KOH/CH_3OH lost the elements of acetic acid to furnish the $\Delta^{11,12}$ derivative (**44**). Ring opening of the 11-carbon ring was achieved by means of $NaIO_4$ oxidation of the pentol (**41**) to the α,β-unsaturated aldehyde (**45**) together with the semiacetate confirming the C-8, C-12 diol arrangement. $LiAlH_4$ reduction of ingol monoacetate produced a hexol (**46**) suggesting the existence of a ditertiary expoxide in (**39**). From this data the structure of ingol was proposed, but no stereochemical assignments were implied. The crystal structure and stereochemistry of ingol (**39**) were obtained on the basis of the X-ray crystallographic analysis of ingol-3,7,8,12-tetraacetate after crystallization from diisopropylether as white prisms (m.p. 172 to 175°C).[40] Independent reflections were measured on an automated diffractometer with CuKα radiation and the structure was solved with MULTAN. An E Map and Subsequent Fourier Maps revealed the presence of 38 nonhydrogen atoms. In contrast to lathyrol (**2**), ingol (**39**) has the 4β-lathyrane skeleton. The three ring systems are cyclopentane, cycloundecane, and cyclopropane (Figure 10). The junction between the 5- and 11-membered rings is bridged by a 4,15β orientated epoxide group, thereby fixing the ring junction in the *cis*

A

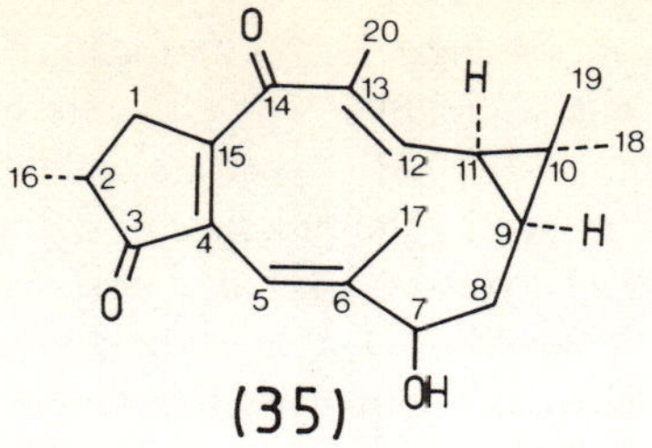

(35)

B

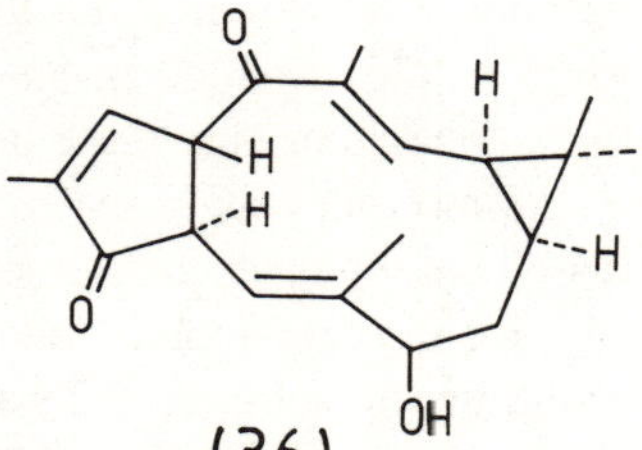

(36)

C

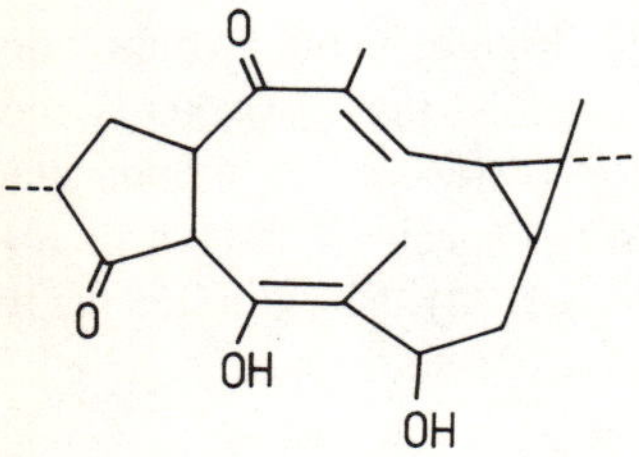

(37)

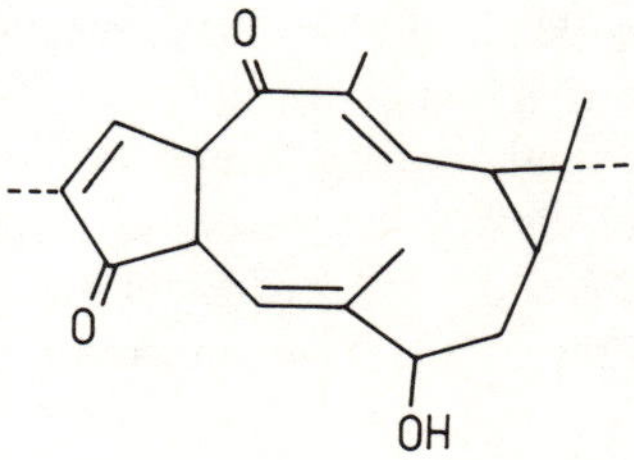

(38)

FIGURE 9.

A. (**35**) Bertaydional

Empirical formula $C_{20}H_{26}O_3$

m.p.	159—160°C
UV	λ_{max} (ϵ), nm, 220 (10,800), 263 (9400), 332 (3400)
IR	$\nu_{max}^{cm^{-1}}$, 3610, 1710, 1665
^{1}H-NMR	δ 1.11 (6H-18, 19), 1.22 (3H-16), 1,47 (3H-17), 1.81 (3H-20), 4.25 (1H-7), 5.95 (1H-5), 5.98 (1H-12)

According to Reference 35

B. (**36**)

Proton(s)	$\alpha(C_6D_6)$	J(Hz)
1-H	7.30	$J_{1\text{-}15} = 1.5$, $J_{1\text{-}19} = 2$
2-Me	1.69	$J_{1\text{-}19} = 2$, $J_{15\text{-}19} = 1.5$
4-H	2.79	$J_{4\text{-}15} = 4.9$, $J_{4\text{-}5} = 11.2$
5-H	5.58	$J_{4\text{-}5} = 11.2$, $J_{5\text{-}20} = 1$
6-Me	1.43	
7-H	4.07	$J_{7\text{-}8a} = 9$, $J_{7\text{-}8b} = 2$
8-Ha, Hb	2.5,1.1	$J_{8a\text{-}8b} = 14$
12-H	6.05	$J_{11\text{-}12} = 11$, $J_{12\text{-}18} = 1$
13-Me	1.89	
15-H	3.58	$J_{1\text{-}15} = 1.5$, $J_{15\text{-}19} = 1.5$, $J_{4\text{-}5} = 4.9$

C. (**37**) 5-Hydroxy-4-hydrobertyandionol

(**38**) 1,2-Dihydro-4,15-dihydrobertyadionol

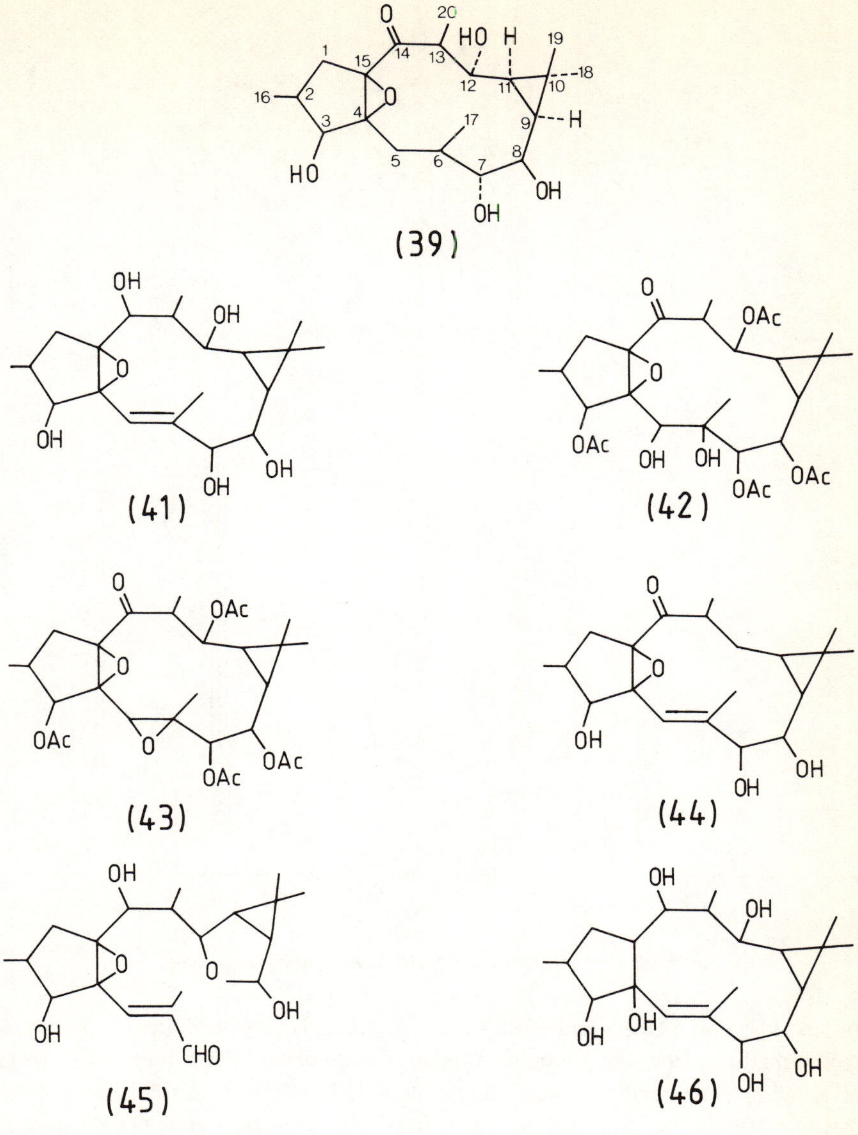

FIGURE 10. (**39**), Ingol; (**41**)—(**46**), ingol reaction products.

configuration. The junction between the 11- and the 3-membered rings is also in *cis* configuration as for lathyrol (**2**), and the $\Delta^{5,6}$ olefinic bond is in the *trans* configuration. The absolute configuration of ingol (**39**) was assigned by analogy to other lathyrane diterpenes.

Besides ingol-3,7,12-triacetate-8-nicotinate an unseparable mixture of further ingol esters was obtained from *Euphorbia ingens.*[39] Mass-spectral analysis of this mixture indicated the presence of three tetraesters of ingol. Base-catalyzed transesterification of the mixture produced the reaction product 12-acetyl-ingol-8-tigliate. It was suggested that the 3,7,12-triacetate-8-tigliate of ingol (**47**) was one of the constituents of this mixture. Two tetraesters of ingol (**48, 49**) were later isolated from euphorbium.[46] This material is a dried latex resin produced from *Euphorbia resinifera.* The structures of these compounds were ascertained by comparison of their ^{1}H-NMR and other date with previous ingol esters. In a similar manner 3,12-diacetylingol-8-tigliate (**50**) a new ester of ingol (**39**) was isolated by chromatographic and partition methods from the latex of *Euphorbia lactea.*[22] More recently a

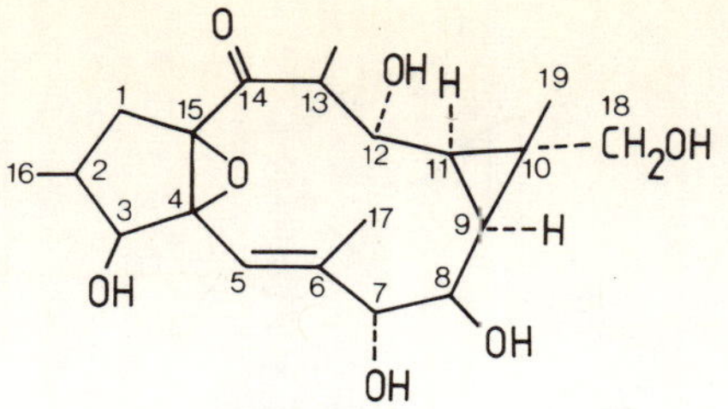

FIGURE 11. (**63**) 18-Hydroxyingol

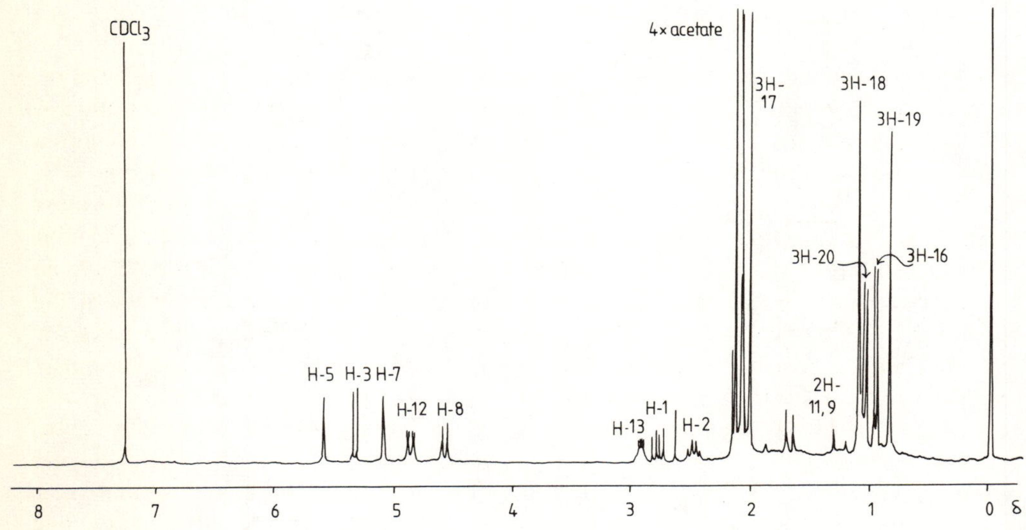

FIGURE 12. ^{1}H-NMR (250 MHz, $CDCl_3$, ingol-tetraacetate)

series of seven (**51** to **57**) tri- and tetraesters of ingol (**39**) were isolated by partition chromatographic methods from the nonpolar fraction of *Euphorbia kamerunica*.[25,43,44] In 1983,[48] Lin and Kinghorn obtained two new ingol esters (**59, 60**) from *Euphorbia hermentiana* acetone-soluble fraction of the latex and identifications were based upon comparison of their spectroscopic data to known compounds. Two further ingol (**39**) esters 3,12-diacetyl-7-angiloyl-8-methoxyingol (**61**) and 3,7,12-triacetyl-8-benzoyl-18-hydroxyingol (**62**) were tentatively identified in the same plant fraction after attempts at selective transesterification reactions. 18-Hydroxyingol (**63**) (Figure 11) was originally isolated from the latex of *Euphorbia marginata* and *Euphorbia unispina*.[46] The compounds from *Euphorbia hermentiana*[45] were isolated using droplet countercurrent chromatographic (DCC) methods. A mixture of four tetraesters of ingol (**39**) and four triesters of the same lathyrane diterpene were separated from the latex of *Euphorbia royleana*.[47] These mixtures failed to separate further using conventional chromatographic methods. By means of spectroscopic analysis and hydrolysis reactions the four tetraesters and the four triesters were found to be approximately equally mixtures of esters differing only in the nature of the C-8 acyl function.

The separation of mixtures of ingol-esters has proven difficult even using Craig distribution[39] and partition chromatographic techniques,[25] but the utilization of DCC methods may prove of use in the future. However, because the absolute configuration of ingol (**39**) was obtained at an early stage[40] and its ^{1}H-NMR and mass spectra (Figure 12) are known, the elucidation of the structure and assignment of positions of acylation are well understood from a com-

Table 1
¹H-NMR OF INGOL-ESTERS

Ingol-ester	Chemical shifts (δ ppm) of acyl protons			
	C-3	C-7	C-8	C-12
3,7,8-Triacetylingol-12-tigliate	5.25	5.23	4.60	4.88
12-Tigloylingol-7,8-diacetate	4.29	5.37	4.60	4.88
3,7-Diacetylingol-12-tigliate	5.27	5.10	3.54	4.88
7-Tigloylingol-3,12-diacetate	5.26	5.10	3.55	4.90
8-Tigloylingol-12-acetate	4.37	4.30	4.58	4.87
8-Acetylingol-12-tigliate	4.29	4.30	4.60	4.88
Ingol-12-acetate	4.36	4.34	2.87	4.84

bination of transesterification reactions and spectroscopic information.[25,43,44] Furthermore, comparisons of ¹H-NMR spectra and mass spectra have made possible the identification of new ingol (**39**) polyols such as 18-hydroxyingol (**63**); further derivatives may be identified in the near future. Ester functions are present on the ingol (**39**) nucleus at positions 3, 7, 8, or 12 for known derivatives of this lathyrane diterpene. It is not possible from the mass spectrum alone to assign the positions of acylation. Electron-impact mass spectra[38,47] exhibit $M^{+\cdot}$ ions sufficiently intense for accurate mass measurements and molecular formulas to be ascertained. The main feature of the fragmentation patterns of these compounds is the sequential loss of the acyl moieties in the carboxyl form, thereby determining their molecular weight. Chemical ionization mass spectra[48] are essentially similar to the electron-impact spectra, with the exception that a quasi $M^{+\cdot} + 1$ ion of relative abundance up to 95% may be produced, depending upon the nature of the reagent gas used and the inlet temperature. Fast atom bombardment MS (FAB) is also a useful technique in that further quasi-molecular ions involving M^+M^+ and higher ions provide useful information concerning the molecular weight of these unstable compounds. The assignment of the various acyl substituents to one or another of the positions of acylation requires the application of stepwise hydrolysis techniques in combination with ¹H-NMR spectroscopy of the isolated reaction products.[48] Mild alkaline hydrolysis of esters using 0.1 *M* KOH in methanol produces a series of products consisting of tetra-, tri-, di-, and monoesters. The ultimate product of the reaction is 12-acetylingol together with smaller quantities of the parent diterpene ingol (**39**) both as baseline TLC products. These reactions should be carried out under a variety of conditions, as it is possible for cyclic transesterifications to take place and artifacts to be produced. In the ¹H-NMR spectrum of ingol-tetraacetate (**58**) the protons geminal to the acyl functions are exhibited downfield (Figure 12). The assignments of these protons were confirmed by extensive decoupling experiments[25,39,48] using an FT 250 MHz instrument. The absence of one or more of the acyl moieties at these positions resulted in an upfield shift of the geminal proton of the order of 1 ppm and distinguished an acylated from a free secondary hydroxy in the ingol (**39**) nucleus (Table 1).

Toxicological properties have in general not been reported for lathyrane diterpenes. There is a possibility, however, that lathyrane derivatives may possess antitumor action, even if weak when compared to other macrocyclic compounds. Cytotoxic activity has been attributed to 6,17-epoxylathyrol and its aliphatic esters by Schroeder and others[24] in 1979. One of these esters exhibited cytotoxic activity at 132 μg/mℓ in vitro against 7288 Morris rat hepatoma cells. More recently, several ingol esters exhibited cytotoxic actions against mouse TLX/5 lymphoma cells in vitro,[48] and Marsh et al.[49] demonstrated that ingol-ester containing extracts of the spurge hawkmoth (*Hyles euphorbiae* L.) exhibited pronounced cytotoxic activity in vitro and antitumor activity in vivo. The distribution of known lathyrane diterpenes in plants of the family Euphorbiaceae is shown in Table 2.

Table 2
DISTRIBUTION OF LATHYRANE DITERPENES IN PLANTS OF THE FAMILY EUPHORBIACEAE

Compound	Species	Ref.
6,17-Epoxylathyrol-3-phenyl-acetate-5,15-diacetate (**3**)	*Euphorbia lathyris* *Macaranga tanarias*	29,30
7-Hydroxylathyrol-dibenzoate-diacetate (**4**)	*E. lathyris*	31
Lathyrol-diacetate-benzoate (**5**)		26
Lathyrol-diacetate-nicotinate		26
17-Hydroxyisolathyrol-5,15,17-triacetate-3-benzoate (**19**)		32
Jolkinols A (**23**), B (**24**), C (**25**), D (**26**)	*E. jolkini*	33
17-Hydroxyjolkinol-5,17-diacetate-3-cinnamate (**58**)	*E. lathyris*	32
5β,6β-Jolkinolepoxide-3-propionate-15-acetate	*E. charachias*	34
5β,6β-Jolkinolepoxide-3-isobutyrate-15-acetate		34
5β,6β-Jolkinolepoxide-3-tigliate-15-acetate		34
5.β,6β-Jolkinolepoxide-3-benzoate-15-acetate		34
5β,6β-Jolkinolepoxide-3-nicotinoate-15-acetate		34
2-Hydroxy-5β,6β-Jolkinolepoxide-acetate-isobutyrate-nicotinate		34
Bertyadional (**35**)	*Bertya cuppressoidea*	35
5-Hydroxy-4-hydro-bertyadional (**37**)		37
1,2-Dehydro-4,15-dihydrobertyadionol (**38**)		37
Ingol-3,7,12-triacetate-8-nicotinate (**40**)	*E. ingens*	39
Ingol-3,7,12-triacetate-8-tigliate (**47**)		39
Ingol-3,8,12-triacetate-7-phenylacetate (**48**)	*Euphorbium*	41
Ingol-3,8,12-triacetate-7-*p*-methoxyphenylacetate (**49**)		41
Ingol-3,12-diacetate-8-tigliate (**50**)	*E. lactea* *E. royleana*	22,47
Ingol-3,7,8-triacetate-12-tigliate (**51**)	*E. kamerunica*	25
Ingol-3,7-diacetate-12-tigliate (**52**)		25
Ingol-tetraacetate (**53**)		25
Ingol-3,7,12-triacetate-8-benzoate (**54**)	*E. royleana* *E. kamerunica*	43,47
Ingol-3,7,12-triacetate-8-tigliate (**55**)	*E. royleana* *E. kamerunica*	43,47
Ingol-3,7,12-triacetate-8-angelate (**56**)	*E. royaleana* *E. kamerunica*	43,47
Ingol-3,12-diacetate-7-tigliate (**57**)		44
8-Methoxy-ingol-3,12-diacetate-7-benzoate (**59**)	*E. hermentiana*	45
8-Methoxy-ingol-3,12-diacetate-7-tigliate (**60**)		45
8-Methoxy-ingol-3,12-diacetate-7-angelate (**61**)		45
18-Hydroxy-ingol-3,7,12-triacetate-8-benzoate (**62**)		45
Ingol-3,7,12-triacetate-8α-methylbutyrate	*E. royleana*	47
Ingol-3,12-diacetate-8-angelate		47
Ingol-3,12-diacetate-8-benzoate		47
Ingol-3,12-diacetate-8-α-methylbutyrate		47
Euphohelioscopin A (**93**)	*E. helioscopia*	56
Euphohelioscopin B (**94**)		56

III. THE JATROPHANE DITERPENES

Jatrophane macrocyclic diterpenes are characterized by the absence of a cyclopropane ring. These compounds have been isolated from species of the family Euphorbiaceae and are potent antileukemic agents in vivo.[21,22]

The first member of this group of natural products to be isolated was jatrophone (**64**), obtained from the roots of *Jatropha gossypiifolia*. In a brief communication in 1970 Kupchan

A

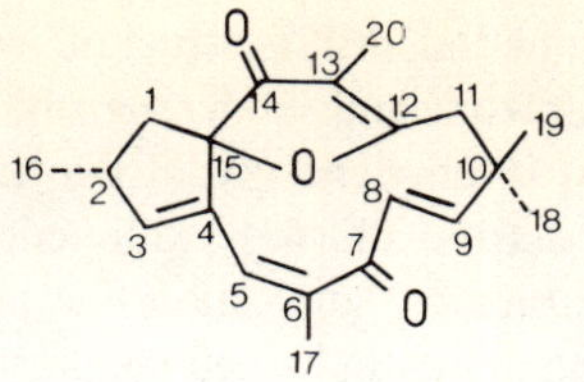

(64)

B

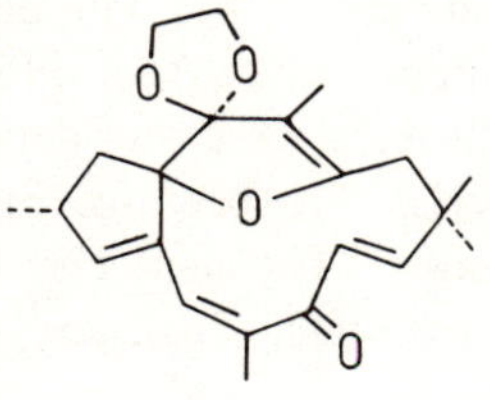

(65)

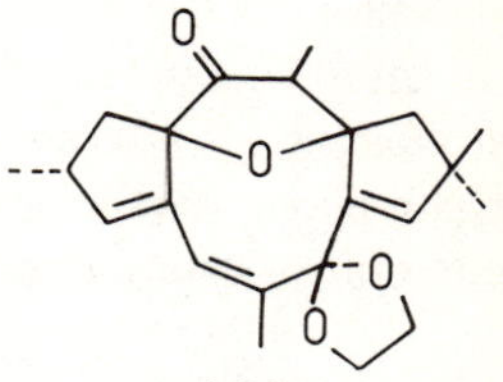

(66)

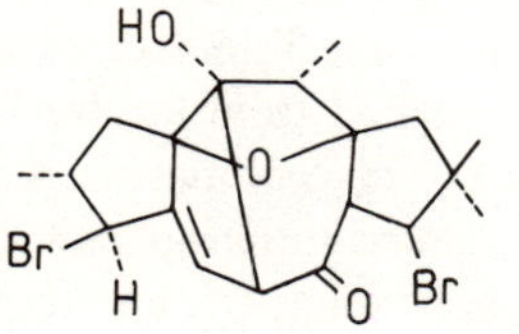

(67)

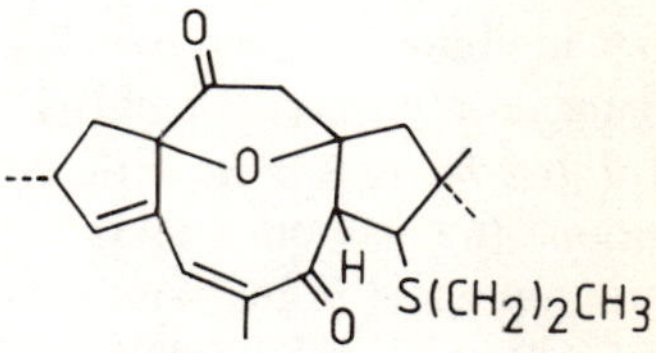

(68)

FIGURE 13.

A. (**64**) Jatrophone

Empirical formula $C_{20}H_{24}O_3$

m.p.	152—153°C
UV	(EtOH), λ_{max} (ϵ) nm 285 (10,200)
IR	(KBr), ν_{max}, 3.35, 3.43, 3.46, 5.96, 6.06, 6.20, 7.10, 7.35, 8.07, 8.17, 10.10
MS	$M^{+\cdot}$, m/z 312
^{1}H-NMR	δ 0.93 (3H-16), 1.20, 1.33 (6H-18, 19), 1.82 (3H-20), 1.83 (3H-17), 2.01 (1H-1), 2.13 (1H-1), 2.39 (1H-11), 2.88 (1H-11), 2.91 (1H-2), 5.86 (1H-3), 6.04 (1H-5), 6.26 (1H-8), 6.75 (1H-9) ppm

B. Reaction products of jatrophone (**64**)

According to References 21 and 50

and associates[50] proposed the structure shown in Figure 13 for jatrophone (**64**). The structure and stereochemistry of jatrophone was ascertained at a later date.[21] This compound is a dicyclic macrolide which exhibits a 15β,12-epoxide bridge across the macrocyclic ring B. The $\Delta^{5,6}$ double bond is in the *cis* configuration and the two remaining double bonds in the structure are in *trans* configuration. Jatrophone (**64**) was extracted from ground roots of the plants with ethanol. After evaporation of the ethanolic extract, the benzene-soluble fraction was obtained by trituration, and from the residue left by evaporation of this fraction a hexane-soluble portion was produced in a similar manner. Following decolorization of the hexane portion and column chromatogrpahy, jatrophone (**64**) was obtained as colorless needles (m.p. 152 to 153°C) by crystallization from hexane.[50] On the basis of elemental analysis and high resolution MS jatrophone (**64**) was assigned the empirical formula $C_{20}H_{24}O_3$. Both the IR and UV spectra of this compound were indicative of a conjugated carbonyl system.[50] Information concerning the oxygen function in this structure was obtained by chemical reactions. Attempts to produce the hydrazone and semicarbazone were unsuccessful, but when reacted with *p*-toluene sulfonic acid in ethylene glycol two ketal analogues (**65**) and (**66**) were produced.[21] Since the UV spectrum of (**65**) exhibited an absorbance at 285 nm a conjugated ketone remained in its structure, suggesting that jatrophone (**64**) possessed two ketone functions. The data for (**66**) no logner indicated the presence of a conjugated carbonyl system, but the IR band at 5.70 μm suggested a ketone in a 5-membered ring. Compound (**66**) would have arisen from an αβ-unsaturated ketone in the 5-membered ring of (**64**) which became saturated during an acid-catalyzed ring closure. Treatment of (**64**) with dry HBr in glacial acetic acid produced jatrophone-bis (hydrogen bromide) adduct (**67**), which could be reconverted to jatrophone (**64**) by the action of sodium hydroxide. From this data together with spectral information of (**64**) through (**67**) including the ^{1}H-NMR spectrum (Figure 13), the structure of jatrophone (**64**) with *cis* $\Delta^{5,6}$ bond was proposed. The stereochemistry and absolute configuration of jatrophone (**64**) was established by direct single crystal X-ray analysis.[21]

Further investigations[21] of the reactivity of jatrophone (**64**) were undertaken with a view to understanding its biological activity as a tumor inhibitor. Jatrophone (**64**) rapidly reacted with *n*-propylthiol in buffer solution to produce (**68**), and as would be expected, jatrophone (**64**) reacted with thiol groups on protein including DNA-dependent RNA polymerase from *Escherichia coli,*[51] in a covalent manner. It was concluded[21,51] that jatrophone (**64**) may act as a tumor inhibitor by selective alkylation of growth regulatory macromolecules in vivo.

At a later date[22] a jatrophone (**64**)-related diterpene was isolated from the fresh ground roots of *Jatropha macrorhiza* of the family Euphorbiaceae. This compound, known as jatrophatrione (**69**) is both structurally and biosynthetically related to jatrophone (**64**), and is an antitumor agent in the same way. It is, however, a tricyclic diterpene and the epoxide bridge of (**64**) is absent in (**69**) as is the $\Delta^{8,9}$ olefinic bond which is responsible for the nucleophilic attack by jatrophone (**64**) on thiol groups of proteins. The structure of jatrophatrione (**69**) was elucidated by comparison of its spectral data including ^{13}C and ^{1}H-NMR to (**64**) (Figure 14). In the ^{1}H-NMR spectrum the downfield position of the signal for the quaternary methyl group at C-20 suggested that is was flanked by two carbonyl groups, and the presence of the 5-membered ring was suggested by the characteristic IR absorbance at 1746 cm^{-1}. Jatrophatrione (**69**) was extracted from the roots in ethanol, purified by partition with chloroform, and then with ether. The ether soluble fraction was separated into fractions by column chromatogrpahy over silica gel and jatrophatrione (**69**) obtained from one of these fractions as colorless spears, m.p. 148 to 150°C by crystallization from ether-hexane. An X-ray study[22] confirmed the constitution and relative configuration of jatrophatrione (**69**). The absolute configuration was not determined crystallographically, but was based upon the assumption that (**69**) and (**64**) possessed the same configuration at C-2. Jatrophatrione (**69**) was active at a dose of 0.5 mg/kg in the 3PS antitumor test system, based upon the measurement of survival of treated animals over that of a control group. *n*-Butyl mercaptan does not react with jatrophatrione (**69**) in the way it does with (**64**) due to the absence of the $\Delta^{8,9}$ double bond, and the $\Delta^{3,5}$ bond is not conjugated between the diene system and the carbonyl group. It was suggested[22] that the observed activity of (**69**) in the test system used was due to the ability of jatrophatrione (**69**) to undergo reverse Michael addition to give compound (**70**), which can then react in a manner similar to jatrophone (**64**).

Two novel jatrophane diterpenes were isolated from *Euphorbia kansui* Liou in 1975 by Uemura and Hirata.[10] These compounds were the first of the jatrophane class shown to possess analgesic and antiwrithing properties in mice. The *E. kansui* diterpenes were known as the kansuinines A (**71**) and B (**72**) They were isolated from the ethanol extract of the plant by means of column and preparative TLC followed by high pressure liquid chromatography (HPLC) on a Micropak® NH_2 column. Kansuinine A (**71**), crystallized from ether-

A

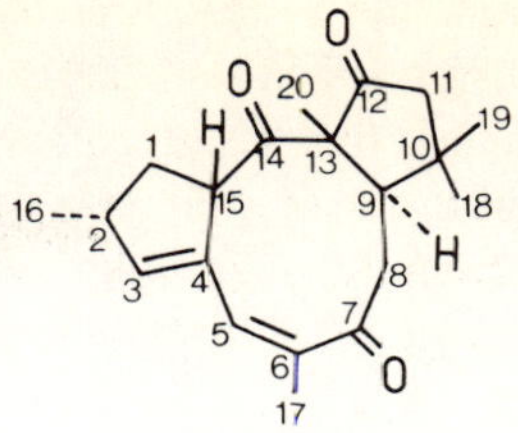

(69)

B

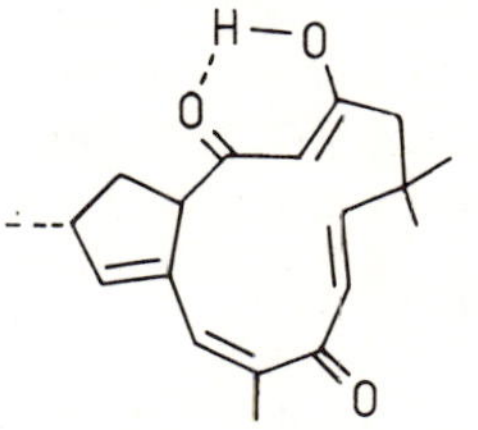

(70)

FIGURE 14.

A. (**69**) Jatrophatrione

Empirical formula $C_{20}H_{26}O_3$

m.p.	148—150°C
UV	(EtOH), λ_{max}(log ϵ) nm, 280 (3.49), 244 (3.95), 223 (4.04)
IR	($CHCl_3$), $\nu_{max}^{cm^{-1}}$, 1746, 1690, 1640, 1632
MS	$M^{+\cdot}$ m/z 314
^{1}H-NMR	δ 0.91 (3H,s), 1.11 (3H,d), 1.23 (3H,s), 1.47 (3H,s), 1.93 (3H,b.s), 2.2—3.0 (8H), 4.08 (1H,b.t), 5.82 (1H,m), 6.08 (1H,m) ppm
^{13}C-NMR	δ 13.0 (q), 19.2 (q), 21.6 (q), 25.6 (q), 34.1 (t), 35.5 (t), 35.2 (s), 36.6 (d), 46.8 (d), 47.2 (d), 51.0 (t), 59.1 (s), 116.7 (d), 124.5 (s), 126.8 (s), 131.8 (d), 193.8 (s), 196.2 (s), 198.8 (s)

B. (**70**) Hypothetical biologically active species of jatrophatrione (**69**)

petroleum ether, m.p. 218 to 220°C (Figure 15), was the naturally occurring diacetate of the jatrophane diterpene. The structure of kansuinine A (**71**) was established by means of spectroscopic data (Figure 15) together with its chemical reactivties. Compound (**71**) upon oxidation with osmium tetroxide yielded the diol (**73**) which was converted to the *p*-bromobenzoate (**74**) and the keto derivative (**75**). Kansuinine A (**71**) failed to acetylate under normal conditions, but gave the methyl ether with silver oxide and methyl iodide. Reduction of (**71**) with sodium borohydride produced compound (**76**). These reactions suggested the presence of a carbonyl moiety in the structure of kansuinine A (**71**). Furthermore, compound (**76**) afforded (**77**) by hydrolysis. This reaction product contained one acetoxy group as determined from its spectroscopic data. This information demonstrated that of the nine oxygen functions seven were present as hydroxyl, one was keto, and the other an etheral oxygen.[10] Further derivatives of kansuinine A (**71**) were produced as being more suitable for comprehensive decoupling experiments in their ^{1}H-NMR,[10] and the structure of (**71**) proposed as shown in Figure 15 without stereochemical assignments. The stereochemistry of kansuinine A (**71**) was obtained from NOE measurements.[52] Compound (**71**) yielded a lactone (**78**) with sodium hydroxide which was converted via its methyl ester to (**79**) by oxidation with periodic acid in acetone-water. The coupling constants H2-H3, H3-H4, and H4-H5, were all 3Hz, which suggested an all-*cis* relationship. Thus the ^{1}H-NMR spectrum established the necessary stereochemistry at C-7 and C-15 of kansuinine A (**71**). In the NOE experiments of (**80**) which was derived from (**71**) by transannular reaction, it was established that the protons

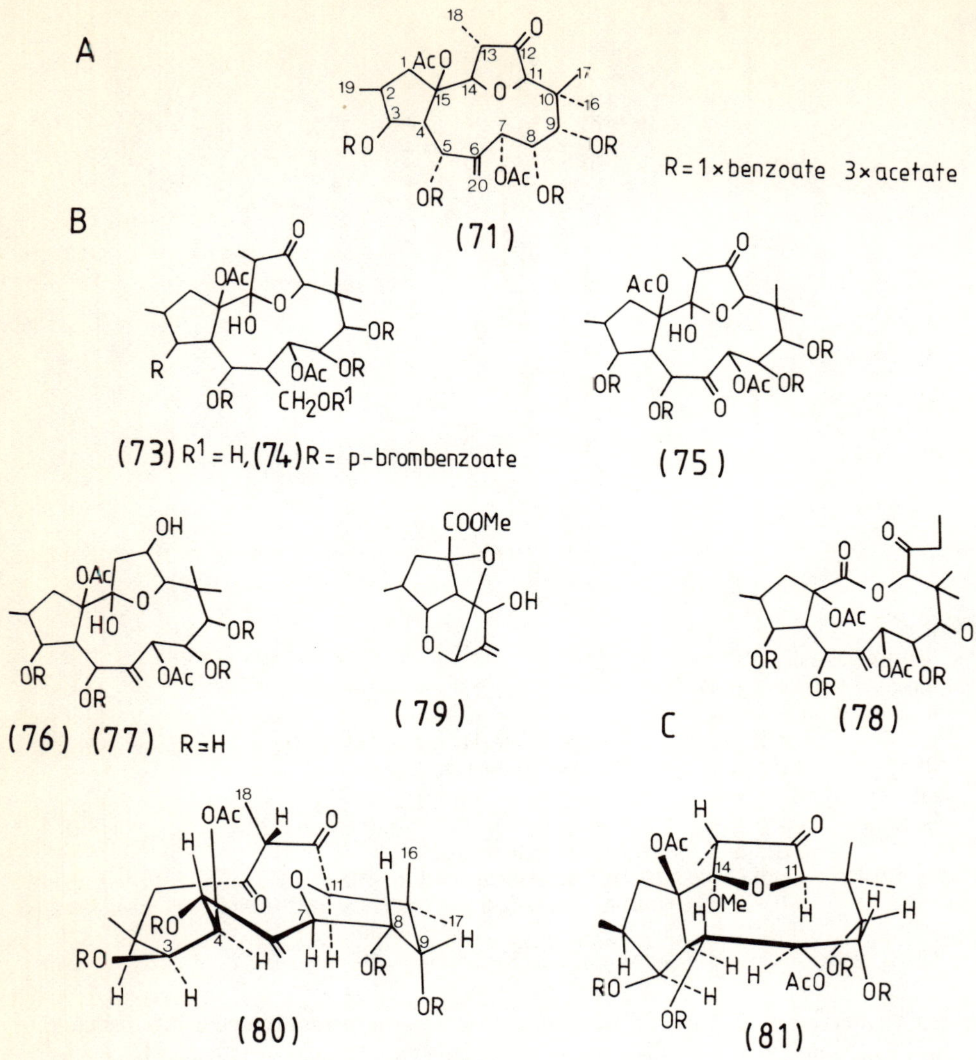

A

(71)

R = 1 × benzoate 3 × acetate

B

(73) R1 = H, (74) R = p-brombenzoate

(75)

(76) (77) R = H

(79)

C

(78)

(80)

(81)

FIGURE 15.

A. (**71**) Kansuinine-A

Empirical formula $C_{37}H_{46}O_{15}$		
m.p.	218—220°C	
UV	(MeOH) λ_{max} (ϵ), nm, 230 (12,000)	
IR	(KBr), $\nu_{max}^{cm^{-1}}$, 3350, 1745, 1680, 1605, 1590, 1220—1280	

B. (**73—79**) Reaction products of kansuinine-A

C.

	Irradiated	Observed	Increase (%)
(**80**)	16-CH_3	H-9	13
	17-CH_3	H-9	13
		H-11	14
	H-4	H-7	10
	H-11	H-7	10
	H-7	H-4	9
		H-11	8
(**81**)	OCH_3	H-11	19
	H-4	H-3	9
		H-7	16

According to Reference 52

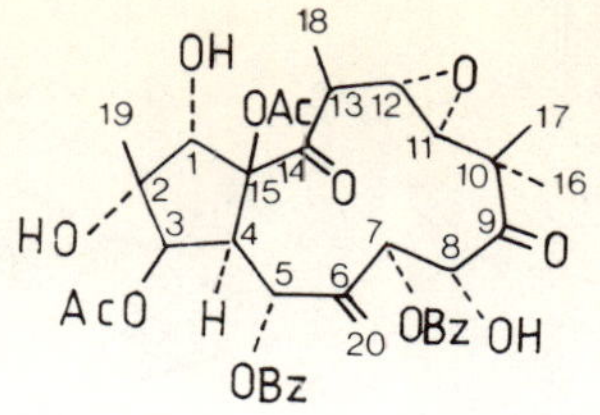

(72)

FIGURE 16.

(**72**) Kansuinine-B

Empirical formula $C_{38}H_{42}O_{14}$

m.p.	160—162°C
MS	$M^{+\cdot}$, m/z 722
IR	(KBr), $\nu_{max}^{cm^{-1}}$, 3500, 1720, 1640, 1609, 1590

of C-7, C-8, and C-9 were α, β, and β configurations. Furthermore, NOE experiments carried out on compound (**81**) disclosed the *cis* relationship between H-11 and OCH_3-14 (Figure 15).

Kansuinine B (**72**) was obtained from the same plant by a further recycling operation in high speed liquid chromatography. A crystalline compound was obtained, m.p. 160 to 162°C, $[\alpha]_D^{23}$ + 37 (C, 0.27 MeOH) (Figure 16). The structure of (**72**) was unambiguously established by X-ray analysis.[53] Kansuinine B (**72**) is the dibenzoate-diacetate of a novel jatrophane polyol.

Euphornin (**82**) is a jatrophane diterpene isolated from *Euphorbia maddeni*.[23] This compound was obtained from the chloroform extract of the whole plant following defatting with hexane. Euphornin (**82**) was purified by means of column chromatography on silica gel and recrystallized from ether as colorless needles, m.p. 180°C, $[\alpha]_D$ − 25° (C, 1.3m MeOH). This compound exhibited marked antitumor activity. The structure of euphornin (**82**) was obtained from its ^{1}H-NMR and chemical ionization mass spectrum together with that of its tetraacetate. Inspection of models indicated that the couplings obtained in the ^{1}H-NMR spectrum of (**82**) were in agreement with the stereochemistry at C-1 to C-12 as shown in Figure 17. The absolute configuration was not, however, determined.[23]

Euphornin (**82**) was later isolated from the leaves and roots of *Euphorbia helioscopia* L.,[54] together with three new jatrophane derivatives euphornins A (**83**), B (**84**), and C (**85**) (Figure 18). These compounds were separated from the methanol extract of the plant by partition into ether followed by column chromatography and preparative TLC. Euphornin (**82**) was converted to its acetonide (**86**) and to its *p*-bromobenzoate (**87**). Compound (**87**) was subjected to an X-ray crystallographic analysis and its correct absolute configuration obtained. The acetate of euphornin (**82**), when treated with *p*-TSOH in acetone, produced the tricyclic compound (**88**). This was the first example of a jatrophane diterpene being chemically converted to a *trans*-lathyrane compound. From this same plant species this group of workers also isolated a new group of jatrophane diterpenes which they refer to as the euphoscopins. Four derivatives, euphoscopin A (**89**), B (**90**), C (**91**), and D (**92**)[56] were obtained (Figure 19). These derivatives were purified from the methanol extract of the leaves and roots and separated by means of solvent-solvent partition followed by chromatogrpahy on silica gel. The euphoscopins are similar in structure to the euphornins but differ from them in the presence of a keto function at C-9 of the nucleus. The stereochemistry of euphoscopin A (**89**) was obtained from an X-ray crystallographic analysis of its *p*-bromobenzoate.[55] The botanical sources of the jatrophane macrocyclic diterpenes are summarized in Table 3.

A

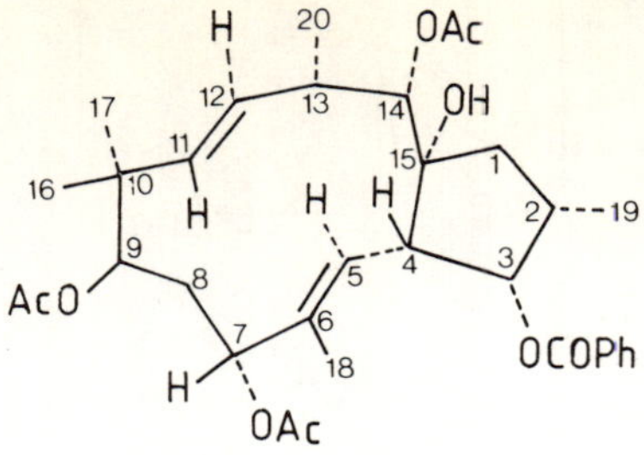

(82)

B

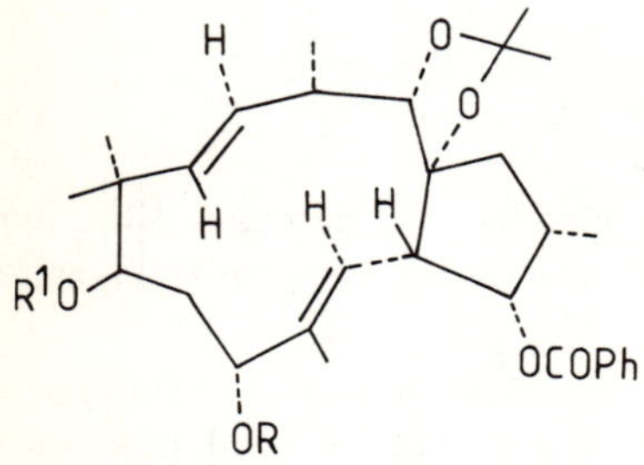

(86), (87)

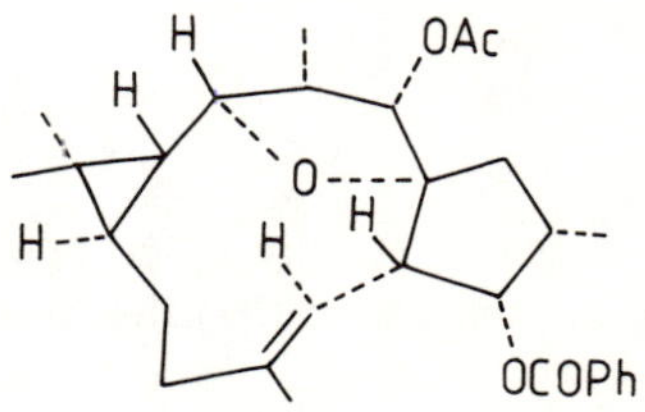

(88)

FIGURE 17.

A. (**82**) Euphornin

Empirical formula $C_{33}H_{44}O_9$

m.p.	180°C
UV	(MeOH), λ_{max} (log ϵ), nm, 228 (4.16)
IR	$\nu_{max}^{cm^{-1}}$, 3500, 2980, 2880, 1740, 1705, 1600, 1500, 1370, 1280, 1245, 1138, 1040, 965, 720
MS	(CI-butane), $M^+ + 1 - H_2O$, m/z 567
^{1}-H-NMR[23]	δ 0.93 (3H-17), 1.01 (3H-19), 1.06 (3H-16), 1.07 (3H-20), 1.49 (3H-18), 1.85 (1H-8), 1.8 (1H-1), 1.85 (1H-2), 2.06 (1H-8), 2.25 (1H-1), 2.52 (1H-13), 2.69 (1H-4), 4.98 (1H-14), 5.06 (1H-9), 5.12 (1H-3), 5.53 (1H-7), 5.58 (1H-11), 5.87 (1H-12), 6.06 (1H-5) ppm

B. (**86**) R = R^1 = Ac

(**87**) R = COC_6H_4Br-*p*, R^1 = H

(**88**) *trans*-Lathyrane derivative

IV. JATROPHOLANE AND CROTOFOLANE DITERPENES

The jatropholane and crotofolane diterpenes are related compounds which have been isolated from *Jatropha* and *Croton* species. They differ in the fact that in the crotofolane nucleus the dimethyl-cyclopropane ring of jatropholane has opened out to form an isopropylidene side chain.

The jatropholanes A (**96**) and B (**97**) were isolated from the roots of *Jatropha gossypiifolia*.[57] Compound (**97**) was the C-2 epimer of compound (**96**). The structure of jatropholane B (**97**) was obtained from its spectroscopic data, including the ^{13}C-NMR and the ^{1}H-NMR (Figure 20), which suggested that this compound was a cyclized casbene derivative. This structure was confirmed by X-ray crystallographic analysis, but the absolute configuration was not established, although it was reasonable to assume that the absolute configuration of (**97**) was similar to that of bertyadionol (**35**). Jatropholane A (**96**) was similar to (**97**) on the basis of its spectral data including the CD curves.[57] their epimeric relationship followed

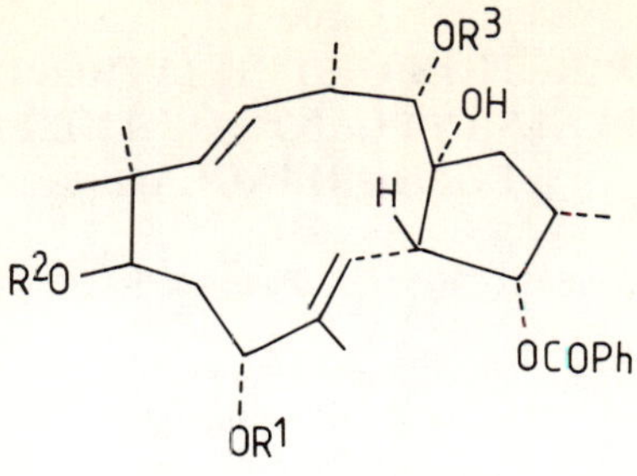

(83),(84),(85)

FIGURE 18.
(**83**) Euphornin-A, R[1] = OH, R[2] = R[3] = acetate
m.p. 98—102°C
IR (film) $\nu_{max}^{cm^{-1}}$, 3470, 1710, 1600, 1580
^{1}H-NMR[54] δ 0.94 (6H,s), 0.96 (3H,d), 0.99 (3H,d), 1.68 (3H,b.s), 2.05 (3H,s), 2.22 (3H,s), 2.97 (1H,d.d) 4.09 (1H,d), 4.37 (1H,d.d), 4.95 (1H,d), 5.05 (1H,d), 5.41 (1H,d.d), 5.63 (1H.d.d), 5.82 (1H,b.d) 7.4—7.6 (3H,m), 8.07 (2H,m) ppm
(**84**) Euphornin-B, R[1] = R[3] = acetate, R[2] = OH
(**85**) Euphornin-C, R[2] = R[3] = acetate, R[1] = keto

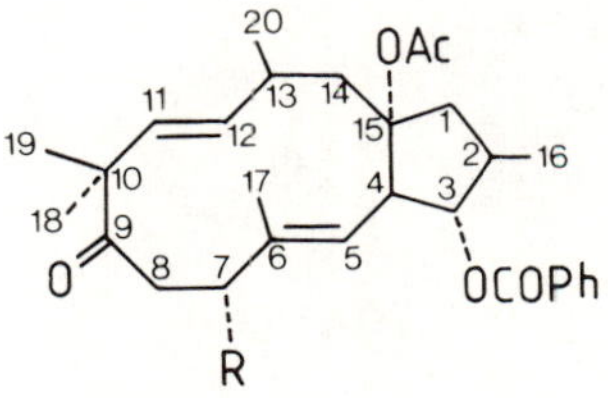

(89),(90),(91),(92)

FIGURE 19.
(**89**) Euphoscopin-A, R = H
Empirical formula $C_{31}H_{40}O_8$
Colorless oil
IR (film) $\nu_{max}^{cm^{-1}}$, 3500, 1740, 1710, 1600, 1580
^{1}H-NMR[55] δ 0.91 (3H-16), 1.10 (3H-20), 1.08 (3H-18), 1.22 (3H-19), 1.44 (1H-1), 1.82 (3H-17), 2.15 (3H-acetate), 2.18 (3H-acetate), 2.2—2.6 (1H-2, 1H-13), 2.63 (1H-8), 2.99 (1H-8), 3.01 (1H-1), 3.29 (1H-4), 4.44 (1H-7), 5.12 (1H-12), 5.23 (1H-3), 5.36 (1H-11), 5.68 (1H-5), 5.93 (1H-14), 7.35—7.65 (3H,m), 7.99 (2H,m) ppm
(**90**) Euphoscopin-B, R = acetate
(**91**) Euphoscopin-C, R = ocoph
(**92**) Euphoscopin-D, R = keto

from the fact that mild base treatment of either compound produced an equimolar mixture of both. The jatropholanes were inactive in antitumor tests.

The tricyclic crotofolane derivatives were isolated from the Jamaican plant *Croton corylifolius* L. of the family Euphorbiaceae. Crotofolin A (**98**)[58] was the first compound of this type to be isolated in the crystalline form, m.p. 277 to 279°C. Its structure was deduced from spectroscopic data (Figure 21) and from a single crystal X-ray analysis.[58] The structures of crotofolin B (**99**) and C (**100**) were published at a later date without chemical evidences,[59] and the presence of a related compound crotofolin D was also claimed to co-occur in this plant, but its structure is unknown. The fifth member of this group, crotofolin E (**101**) was isolated from *Croton corylifolius* in 1979 by Burke et al.[59] Crotofolin E (**101**) contained the

Table 3
DISTRIBUTION OF THE JATROPHANE DITERPENES IN PLANTS OF THE FAMILY EUPHORBIACEAE

Compound	Plant species	Ref.
Jatrophone (**64**)	*Jatropha gossypiifolia*	21,50
Jatrophatrione (**69**)	*J. macrorhiza*	22
Kansuinine A (**71**)	*Euphorbia kansui*	10
Kansuinine B (**72**)		10
Euphornin (**82**)	*E. maddeni*	23
	E. helioscopia	54
Euphornin A (**83**)	*E. helioscopia*	54
Euphornin B (**84**)	*E. helioscopia*	54
Euphornin C (**85**)	*E. helioscopia*	54
Euphoscopin A (**89**)	*E. helioscopia*	55
Euphoscopin B (**90**)	*E. helioscopia*	55
Euphoscopin C (**91**)	*E. helioscopia*	56
Euphoscopin D (**92**)	*E. helioscopia*	56

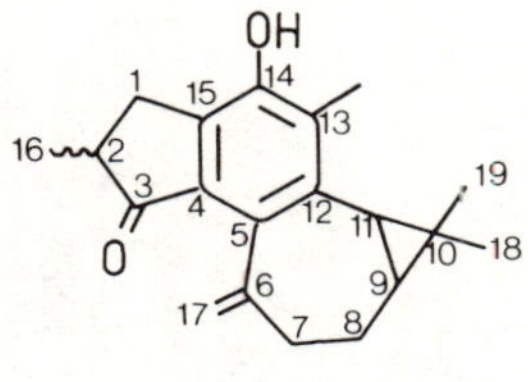

(96),(97)

FIGURE 20.
(**96**) 2β-CH_3 Jatropholone-A
(**97**) 2α-CH_3 Jatropholone-B
Empirical formula $C_{20}H_{24}O_2$

UV	(EtOH), λ_{max}, nm, 225, 238, 275, 335
IR	(CCl_4), $\nu_{max}^{cm^{-1}}$, 3603, 1715
^{1}H-NMR[57]	δ 0.36 (3H-19), 1.29 (3H-18), 1.34 (3H-16), 4.66, 5.19 (2H-17), 5.42 (OH) ppm
^{13}C-NMR[57]	Cresol ring δ 150.2, 137.5, 137.2, 131.8, 131.1, 134.4, 28.2
	Methyls 17.1, 16.1, 13.3
	Exomethylene 145.6, 115.2
	Ketonic carbonyl 208.2 ppm

rare functional group, a pseudo-acid moiety.[59] Compound (**101**), empirical formula $C_{20}H_{22}O_5$, m.p. 208 to 211°C, $[\alpha]_D^{25}$ − 65°C ($CHCl_3$, C-1.00), was isolated as a crystalline compound. Its spectral data (Figure 21) did not differentiate crotofolin E (**101**) from diterpenes of the daphnetoxin-type (see Chapter 8), and accordingly its structure and relative stereochemistry were obtained by single-crystal X-ray analysis using direct methods.

The distribution in plant species of the crotofolane and jatropholane diterpenes is recorded in Table 4.

V. CASBANE DITERPENES

The casbane diterpenes have been considered to be the possible biosynthetic precursors of several groups of diterpenes of the families Euphorbiaceae and Thymelaeaceae[60] (see Chapter 4). Casbane diterpenes (Figure 22) are bicyclic macrolides and were initially isolated

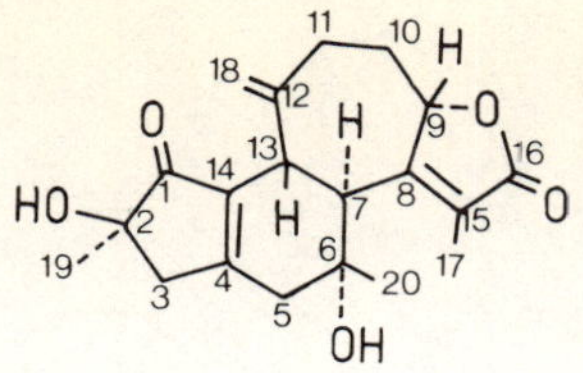

(98)

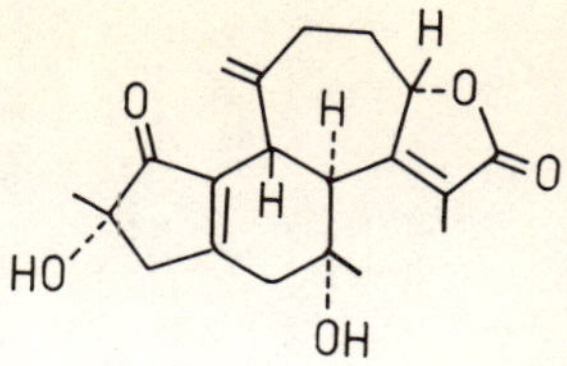

(99)

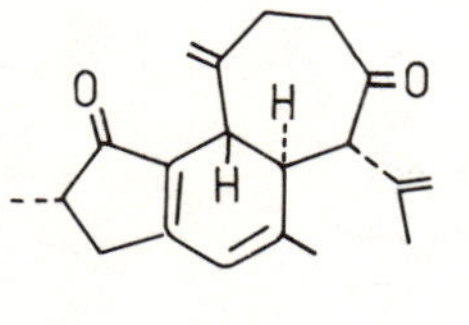

(100)

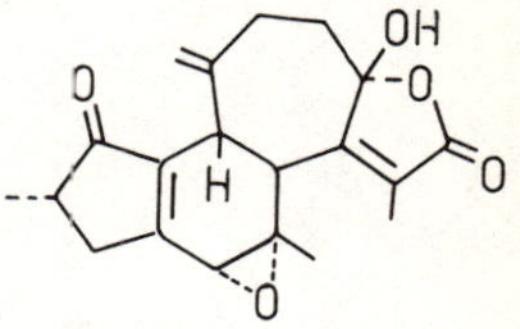

(101)

FIGURE 21.
(**98**) Crotofolin-A
Empirical formula $C_{20}H_{24}O_5$

m.p.	277—279°C
MS	$M^{+\cdot}$, m/z 344
UV	(MeOH), λ_{max}, (ε), nm, 227 (23,800)
IR	ν_{max}^{cm-1}, (nujol), 3465, 1740, 1705, 1650, 900
^{1}H-NMR[58]	δ 1.92 (3H-17), 4.20, 4.84 (2H-18), 5.15 (1H-9) ppm

(**99**) Crotofolin-B
(**100**) Crotofolin-C
(**101**) Crotofolin-E
Empirical formula $C_{20}H_{22}O_5$

m.p.	208—211°C
MS	$M^{+\cdot}$, m/z 342
UV	(EtOH), λ_{max} (ε), nm, 215 (15,700)
IR	(KBr), ν_{max}^{cm-1}, 3515, 3417, 1698, 1670, 905
^{1}H-NMR[59]	δ 1.21 (3H-19), 1.31 (3H-20), 1.93 (3H-17) 4.46, 4.91 (2H-18) ppm
^{13}C-NMR[59]	δ 9.5 (C-20), 16.8 (C-19), 19.6 (C-17), 34.3 (C-10), 35.7 (C-11), 39.4 (C-7), 40.0 (C-13), 41.2 (C-3), 52.2 (C-2), 53.7 (C-5), 59.2 (C-6), 107.9 (C-9), 111.7 (C-18), 128.4 (C-12), 143.6 (C-15), 146.1 (C-14), 159.1 (C-8), 165.8 (C-4), 169.7 (C-16), 206.6 (C-1) ppm

Table 4
DISTRIBUTION OF THE JATROPHOLANE AND CROTOPHOLANE DITERPENES IN PLANTS OF THE FAMILY EUPHORBIACEAE

Compound	Plant species	Ref.
Jatropholone A (**96**)	*Jatropha gossypiifolia*	57
Jatropholone B (**97**)		
Crotofolin-A (**98**)		58
Crotofolin-B (**99**)		59
Crotofolin-C (**100**)	*Croton corylifolius*	59
Crotofolin-D		59
Crotofolin-E (**101**)		59

A

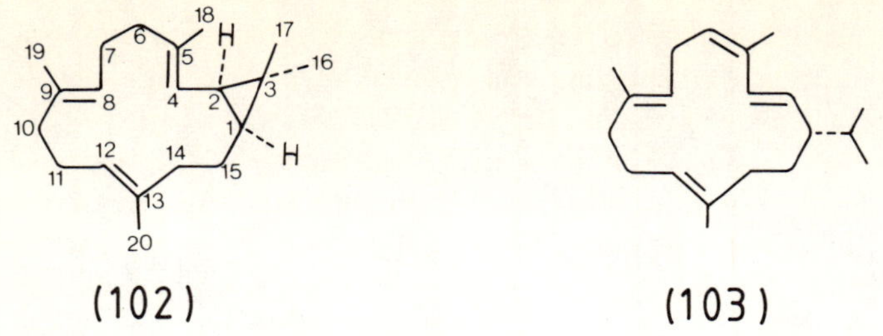

(102) (103)

B

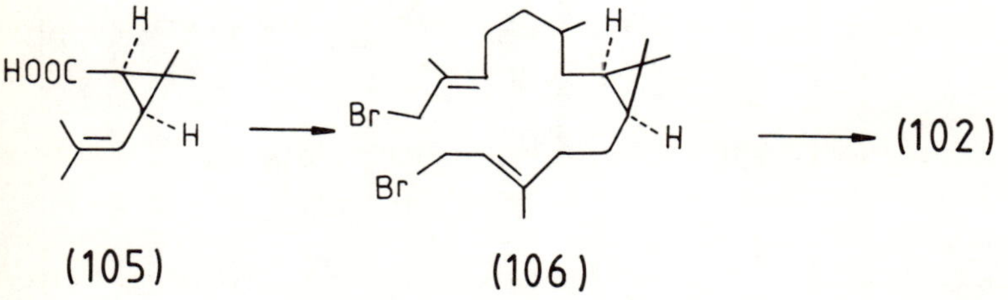

(105) (106) (102)

C

(104)

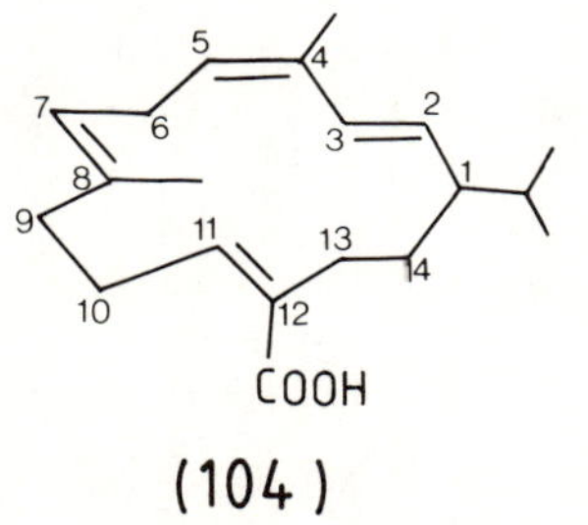

FIGURE 22.
A. (**102**) Casbene
(**103**) Cembrene
B. Synthesis of casbene

C. Poilaneic acid
Empirical formula $C_{20}H_{30}O_2$

m.p.	94—95°C
UV	(EtOH), λ_{max} (log ϵ), nm, 223 (4.1), 245, 255
IR	(Nujol), $\nu_{max}^{cm^{-1}}$, 3000, 1680, 1620
^{1}H-NMR[65]	δ 0.80 (3H,d), 0.83 (3H,d), 1.66 (3H,t), 1.82 (3H,t), 1.5—2.5 (11H,m), 3.05 (1H,d.d.d), 5.21 (1H,d.d), 5.56 (1H,d.d), 6.04 (1H,d.d), 6.05 (1H,d) ppm
^{13}C-NMR	δ 14.51 (q), 19.38 (q), 19.98 (q), 20.98 (q), 25.92 (t), 26.27 (t), 29.52 (t), 32.15 (d), 32.78 (t), 47.96 (d), 125.73 (d), 128.04 (d), 128.88 (s), 130.54 (d), 131.00 (s), 131.30 (d), 135.17 (s), 147.81 (d), 173.69 (s) ppm

from seedlings of *Ricinus communis*[61] in the form of the unsaturated macrolide casbene (**102**) by Robinson and West in 1970. Soluble enzyme preparations of the seedling of the castor oil bean converted mevalonic acid in the presence of ATP into a mixture of hydrocarbons. Casbene (**012**) was separated from this mixture of five derivatives and tentatively identified on the basis of its spectral data. The similarity of (**102**) to the naturally occurring macrolide cembrene (**013**) was noted. Compound (**103**) was earlier structurally elucidated as a 14-membered ring hydrocarbon in 1962,[62] and later (±)-cembrene (**103**) was synthesized using nickel carbonyl-catalyzed coupling of a terminal allylic bromide as the key step.[63] In a later paper an alternative synthesis of (**103**) was described which utilized the facile formation of the 1,3-diene and 1-ene-1-3-ol systems via the sulfoxide.[64] A cembrene (**103**) derivative has been isolated from a species of the family Euphorbiaceae.[65] The cembranoid diterpene known as poilaneic acid (**104**) was isolated from *Croton poilanei* growing in Thailand. This compound was (IR,* 2E, 4Z, 7E, 11Z)-12-carboxyl-1-isopropyl-4,8-dimethylcyclotetradecatetrene as determined from its spectral data and $LiAlH_4$ reduction to (−)-cembrene (**103**). Casbene (**102**), the bicyclic derivative of cembrene (**103**), was obtained by total synthesis in 1980 by Crombie and others.[66] Spectroscopic data supported the structure of (**102**) as possessing a 1S,3R-*cis*-cyclopropane ring and a 4,8,12-all-E-triene system (Figure 22). Compound (**102**) was elaborated from the cyclopropane *cis*-chrysanthemic acid (**105**) to the *bis*-allylic bromide (**106**), followed by closure of the 14-membered ring at C-10, C-11 with tetracarbonylnickel (Figure 22).

Crotonitenone (**107**) was the first naturally occurring oxygenated derivative of casbene (**102**) to be isolated. It was separated from the dried leaves and twigs of *Croton nitens*[67] during an investigation aimed at producing precursors of the crotofolane groups of diterpenes. Crotonitenone (**107**) was obtained from the benzene extract of the plant by means of column chromatographic purification followed by preparative TLC. Compound (**107**) crystallized from methanol as needles, m.p. 128 to 129°C, $[\alpha]_D^{25}$ + 176° (C, 1 in $CHCl_3$). The constitution of its functional groups was obtained from spectroscopic data including its ^1H-NMR and ^{13}C-NMR spectra (Figure 23) and by means of Jones oxidation to (**108**). The structure and relative stereochemistry of (**107**) was eventually established by a single-crystal X-ray analysis using direct methods.[67]

VI. RHAMNOFOLANE AND MYRSINOL DITERPENES

A diterpene of the rhamnofolane type was reported[68] to have been isolated from *Croton rhamnifolius*. This compound was of particular interest because of its close structural relationship to phorbol (see Chapter 7). 13,15-Seco-4,12-dideoxy-14-phorbol-20-acetate (**109**) was identified by comparison of its spectral data to that of phorbol-triacetate (Figure 24). In the ^1H-NMR spectrum, features could be identified corresponding to the protons of a β,β-dimethyl *cisoid* α, β-unsaturated ketone structure. It was suggested that this moiety could easily be derivable from the *gem*-dimethylcyclopropane system of phorbol. Phorbol has been converted by chemical methods to (**109**) and this reaction product was identified in all respects with the diterpene isolated from *Croton rhamnifolius*.[68]

The name myrsinol was given in 1982[69] to a new group of tetracyclic diterpenes which were isolated from *Euphorbia myrsinites* L. These compounds were nonirritant substances obtained in the same plant extract as the well-known pro-inflammatory ingenol esters (see Chapter 9). A series of six mixtures of esters of myrsinol (**110 to 115**) were separated by conventional methods; three of them were crystalline derivatives, while the other three were resinous in nature. The crystalline compound (**114**), $C_{30}H_{42}O_9$, m.p. 147°C was a mixture of the diacetate-hexanoate and the acetate-propionate-valerate of myrsinol (Figure 25). The positions of acylation were not assigned in the myrsinol nucleus. The structure and nature of the functional groups of (**114**) were obtained from its spectroscopic data (Figure 25), including extensive decoupling in the ^1H-NMR spectrum and the ^{13}C-NMR spectrum together with its chemical reactivity. Reduction of (**114**) with $LiAlH_4$ produced a 2/1 mixture of 14-deoxo-14-hydroxymyrsinol (14β, 14α) (**116**). Hydrolysis of (**114**) with KOH in MeOH produced isomyrsinol (**117**)[70] which by means of $LiAlH_4$ reduction was converted to compound (**118**), and by the action of TSOH in acetone to (**119**) (Figure 26). The stereochemistry

A

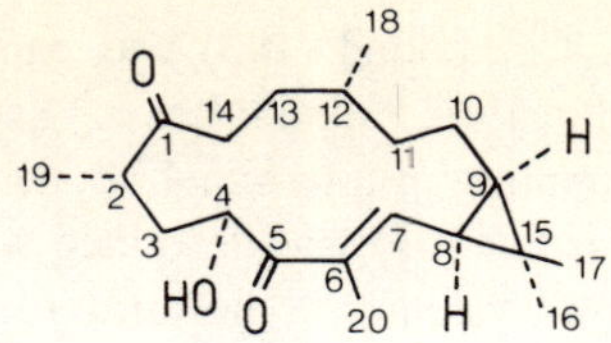

(107)

B

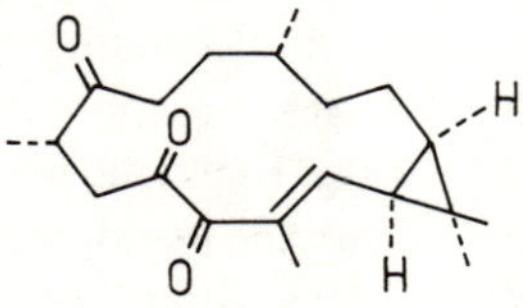

(108)

FIGURE 23.
A. (**107**) Crotonitenone
Empirical formula $C_{20}H_{32}O_3$
m.p. 128—129°C
UV λ_{max} (ϵ), nm, 267 (16,000)
IR $\nu_{max}^{cm^{-1}}$, 3430, 1710, 1650, 1620
Raman (neat) 1650, 1620, 1045 cm^{-1}
^{1}H-NMR[67] δ 0.92 (3H-18), 0.94 (3H-17), 1.15 (3H-19), 1.17 (3H-16), 1.91 (3H-20), 3.57 (1H, OH), 4.54 (1H-7, 1H-4), 6.11 (1H-7), ppm
^{13}C-NMR δ 12.23 (C-17), 16.25 (C-16), 17.83 (CH_3), 20.83 (CH_3), 23.25 (CH_2), 26.51 (CH_2), 27.10 (C-15), 27.60 (C-12), 29.23 (C-20), 30.18 (CH_2), 31.55 (C-9), 33.50 (CH_2), 34.00 (C-8), 42.39 (CH_2), 45.19 (C-2), 69.77 (C-4), 134.85 (C-6), 142.76 (C-7), 201.74 (C-5), 213.63 (C-1) ppm
B. (**108**) Jones oxidation product of **107**.

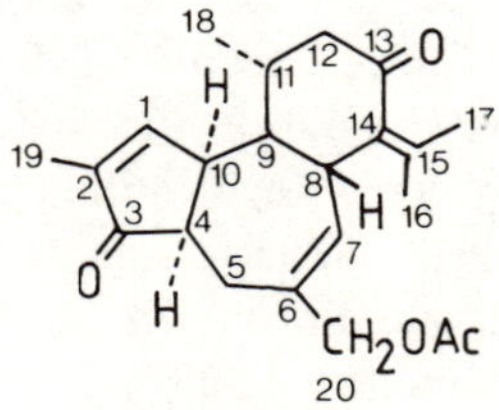

(109)

FIGURE 24.
(**109**) 13,15-Seco-4,12-dideoxy-4α-phorbol-20-acetate
Empirical formula $C_{22}H_{28}O_5$
m.p. 171°C
UV (EtOH), λ_{max} (ϵ), nm, 205 (14,180), 235 (12,210)
IR (Nujol), $\nu_{max}^{cm^{-1}}$, 3410, 1736, 1689, 1661, 1626
^{1}H-NMR[68] δ 1.15 (3H-18), 1.58, 1.87 (6H-16, 17), 1.83 (3H-19), 3.40 (1H-8), 4.46 (2H-20), 5.10 (1H-7), 7.15 (1H-1) ppm

of 14-desoxo-14-hydroxymyrsinol (**116**) was obtained by X-ray analysis following crystallization from methanol/water (4/1).[69]

The distribution of the casbene, rhamnofolane, and myrsinol diterpenes in plants of the family Euphorbiaceae are shown in Table 5.

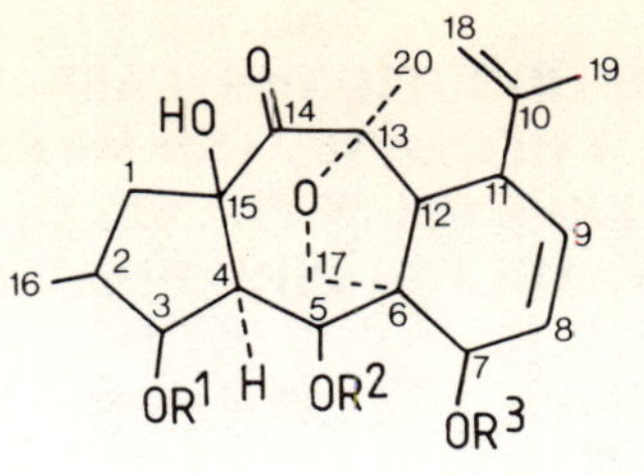

(110 - 115)

FIGURE 25.

(**110**) m.p. 96°C — R^1, R_2 = $COCH_3$, $CO(CH_2)_3CH_3$, R^3 = COC_5H_4N
R^1, R^2 = $COCH_2CH_3$, $CO(CH_2)_2CH_3$, R^3 = COC_5H_4N

(**111**) m.p. 135°C — R^1, R^2 = $COCH_3$, $CO(CH_2)_4CH_3$, R^3 = COC_5H_4N
R^1, R^2 = $COCH_2CH_3$, $CO(CH_2)_3CH_3$, R^3 = COC_5H_4N

(**112**) R^1, R^2 = $COCH_3$, $CO(CH_2)_5CH_3$, R^3 = COC_5H_4N
R^1, R^2 = $CO(CH_2)_2CH_3$, $CO(CH_2)_3CH_3$, R^3 = COC_5H_4N

(**113**) R^1, R^2, R^3 = $COCH_3$, $COCH_3$, $CO(CH_2)_3CH_3$
R^1, R^2, R^3 = $COCH_2CH_3$, $COCH_2CH_3$, $COCH_2CH_3$

(**114**) R^1, R^2, R^3 = $COCH_3$, $COCH_3$, $CO(CH_2)_4CH_3$
R^1, R^2, R^3 = $COCH_3$, $COCH_2CH_3$, $CO(CH_2)_3CH_3$

Empirical formula $C_{30}H_{42}O_9$

m.p. 147°C

UV (MeOH) λ_{max}, (ϵ), nm, 300 (13,500)

IR (KBr), $\nu_{max}^{cm^{-1}}$, 3084, 3010, 1740

^{1}H-NMR[69] δ 0.8 (3H-16), 1.3 (3H-20), 1.4, 3.7 (2H-1), 1.75 (1H-2), 1.81 (3H-19), 2.19 (1H-4), 3.18 (1H-12), 3.22 (1H-11), 3.44, 3.85 (2H-17), 4.8 (2H-18), 4.82 (1H-7), 5.24 (1H-3), 5.56 (1H-9), 6.04 (1H-8), 6.13 (1H-5) ppm

^{13}C-NMR δ 54.70, 89.76, 90.52, 113.07, 122.24, 134.21, 134.21, 146.94, 169.0, 173.0, 203.7 ppm

(**115**) R^1, R^2, R^3 = $COCH_3$, $COCH_3$, $CO(CH_2)_5CH_3$
R^1, R^2, R^3 = $COCH_3$, $CO(CH_2)_2CH_3$, $CO(CH_2)_3CH_3$

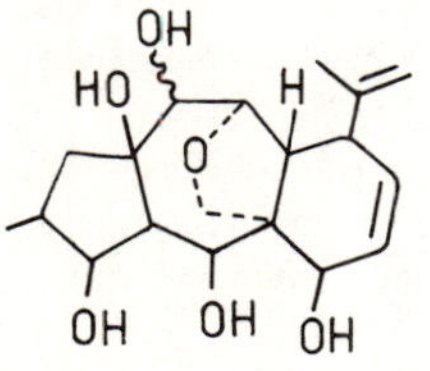

(116) 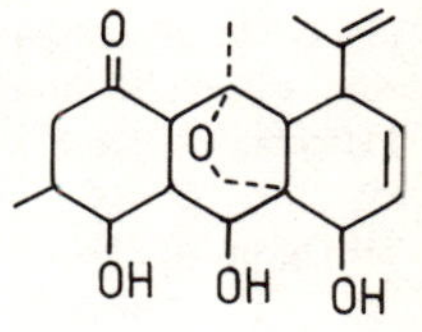 (117)

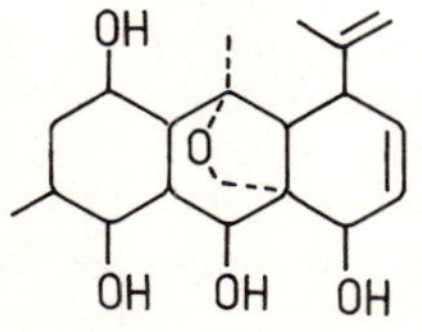

(118) 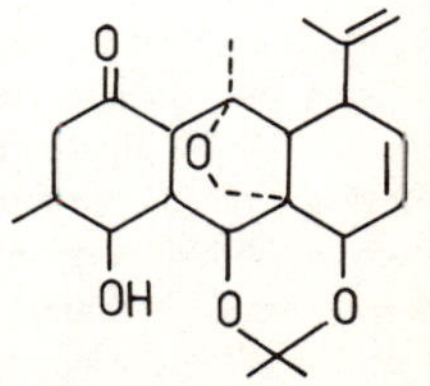 (119)

FIGURE 26.
(**116—119**) Reaction products of (**114**)

Table 5
DISTRIBUTION OF CASBANE, RHAMNIFOLANE, AND MYRSINOL DITERPENES IN PLANTS OF THE FAMILY EUPHORBIACEAE

Compound	Plant species	Ref.
Casbene (**102**)	*Ricinus communis*	61
Crotonitenone (**107**)	*Croton nitens*	67
Compound (**109**)	*C. rhamnifolius*	68
Myrsinol esters (**110—115**)	*Euphorbia myrsinites*	69

REFERENCES

1. **Hecker, E.,** Cocarcinogenic principles from the seed oil of *Croton tiglium* and from other Euphorbiaceae, *Cancer Res.*, 28, 2338, 1968.
2. **Hecker, E. and Schmidt, R.,** Phorbol esters, the irritants and co-carcinogens of *Croton tiglium, Fortschr. Chem. Org. Naturstoffe,* 31, 377, 1971.
3. **Evans, F. J. and Soper, C. J.,** The tigliane, daphnane and ingenane diterpenes, their chemistry distribution and biological activities, *J. Nat. Prod.*, 41, 193, 1978.
4. **Evans, F. J. and Taylor, S. E.,** Proinflammatory, tumor-promoting and anti-tumor diterpenes of the plant families Euphorbiaceae and thymeleaeceae, *Fortschr. Chem. Org. Naturstoffe,* 44, 1, 1983.
5. **Watt, J. M. and Breyer-Brandwijk, M. G.,** *The Medicinal and Poisonous Plants of Southern and Eastern Africa,* 2nd ed., E & S Livingstone, Edinburgh, 1962, 395.
6. **Von Reis Altschul, S.,** *Drugs and Foods from Little Known Plants,* Notes in Harvard University Herbaria, Harvard University Press, Boston, 1973, 143.
7. **Dymock, W., Warden, C. J. H., and Hooper, D.,** *Pharmacographic Indica* (reprinted 1972), Vol. 1, Hamdard National Foundation, Karachi, Pakistan, 1890, 247.
8. **Kerharo, J. and Adam, J. G.,** *La Pharmacopie Senegalaise Tradional,* Vergot Freres, Paris, 1974, 402.
9. **Evans, F. J.,** *Euphorbia.* Pharmacognosy section, *The British Herbal Pharmacopoeia,* Part I, British Herbal Medicine Association Scientific Committee, London, 1979, 201.
10. **Uemara, D. and Hirata, Y.,** The structure of kansuinine A, a new multi-oxygenated diterpene, *Tetrahedron Lett.*, 1697, 1975.
11. **Kitazawa, E. and Ogiso, A.,** Two diterpene alcohols from *Croton sublyratus, Phytochemistry,* 20, 287, 1981.
12. **Takahashi, S., Kurabayashi, M., Kitazawa, E., Haruyama, H., and Ogiso, A.,** Plaunolide, a furanoid diterpene from *Croton sublyratus, Phytochemistry,* 22, 302, 1983.
13. **Hecker, E.,** New toxic, irritant and cocarcinogenic esters from Euphorbiaceae and from Thymelaeaceae, *Pure and Appl. Chem.*, 49, 1423, 1977.
14. **Hartwell, J. L.,** Plants used against cancer. A survey, *Lloydia,* 32, 153, 1969.
15. **Fransworth, N. R.,** Biological and chemical screening of plants, *J. Pharm. Sci.*, 55, 225, 1966.
16. **Hecker, E.,** Isolation and characterization of the co-carcinogenic principles of *Croton* oil, in *Methods of Cancer Research,* Vol. 6, Busch, H., Ed., Academic Press, London, 1971.
17. **Adolf, W., Opferkuch, H. J., and Hecker, E.,** Über die diterpenoiden Inhaltsstoffe des Samenöls von *Euphorbia lathyris* und ihre tumorpronovierende Wirkung, *Fette. Siefen, Anstrichm.*, 70, 850, 1968.
18. **Hecker, E.,** cocarcinogensis and tumor-promoters of the diterpene ester type as possible carcinogenic risk factors, *J. Cancer Res. Clin. Oncol.*, 99, 103, 1981.
19. **Furst, A.,** *Chemistry of Chelation in Cancer,* Charles C Thomas, Springfield, Ill., 1963, 27.
20. **Stary, F. and Jirásek, V.,** in *Herbs,* Evans, F. J., Ed., Hamlyn, London, 1973, 14.
21. **Kupchan, S. M., Siegel, C. W., Matz, M. J., Gilmore, C. J., and Bryan, R. F.,** Structure and stereochemistry of jatrophone a novel macrocyclic diterpenoid tumor inhibitor, *J. Am. Chem. Soc.*, 98, 2295, 1976.
22. **Torrance, S. J., Wiedhopf, R. M., Cole, J. R., Arora, S. K., Bates, R. B., Beaver, W. A., and Cutler, R. S.,** Anti-tumor agents from *Jatropha macrorhiza.* II., Isolation and characterisation of jatrophatrione, *J. Org. Chem.*, 41, 1855, 1976.

23. **Sahai, R., Rastogi, R. P., Jakupovic, J., and Bohlmann, F.,** A diterpene from *Euphorbia maddeni, Phytochemistry,* 20, 1665, 1981.
24. **Schroeder, G., Rohmer, M., Beck, J. P., and Anton, R.,** Cytotoxic activity of 6,20-epoxylathyrol and its aliphatic esters, *Planta Med.,* 35, 235, 1979.
25. **Abo, K. and Evans, F. J.,** Macrocyclic diterpene esters of the cytotoxic fraction from *Euphorbia kamerunica, Phytochemistry,* 20, 2535, 1981.
26. **Adolf, W. and Hecker, E.,** Further new diterpenes from the irritant and co-carcinogenic seed oil and latex of the caper spurge *(Euphorbia lathyris* L.), *Experientia,* 27, 1393, 1971.
27. **Dublyanskaya, N. F.,** Über den biologischen Effekt der die Toxizität von *E. lathyris* bedingenden Fraktionen, *Pharm. Pharmacol.,* 11(12), 50, 1937.
28. **Zechmeister, K., Röhrl, M., Brandle, F., Hechtfischer, S., Hoppe, W., Hecker, E., and Kubinyi, H.,** Röntgenstrukturanalyse eines neuen makrozyklischen Diterpenester aus der Spring Wolfsmilch *(Euphorbia lathyris* L.), *Tetrahedron Lett.,* 3071, 1970.
29. **Adolf, W., Hecker, E., Balmain, A., Lhomme, M. F., Nakatani, Y., Ourisson, G., Ponsinet, G., Pryce, R. J., Santhanakrishnan, T. S., Matyukhira, L. G., and Saltikova, I. A.,** Euphorbiasteroid, epoxy-lathyrol, a new tricyclic diterpene from *Euphorbia lathyris* L., *Tetrahedron lett.,* 2241, 1970.
30. **Hui, W. H. and Ng, K. K.,** Terpenoids and steroids from *Macaranga tanarius, Phytochemistry,* 14, 816, 1975.
31. **Narayanan, P., Röhre, M., Zeichmeister, K., Engel, D. W., Hoppe, W., Hecker, E., and Adolf, W.,** Structure of 7-hydroxy-lathyrol, a further diterpene from *Euphorbia lathyris* L., *Tetrahedron Lett.,* 1325, 1971.
32. **Adolf, W., Köhler, J., and Hecker, E.,** Lathyrana type diterpene esters from *Euphorbia lathyris, Phytochemistry,* 23, 1461, 1984.
33. **Uemura, D., Nobuhara, K., Nakayama, Y., Shizari, Y., and Hirata, Y.,** The structure of new lathyrana diterpenes, jolkinols A, B. C, and D, from *Euphorbia jolkini* Boiss., *Tetrahedron Lett.,* 4593, 1976.
34. **Seip, E. H. and Hecker, E.,** Lathyrane type diterpenoid esters from *Euphorbia characias, Phytochemistry,* 22, 1791, 1983.
35. **Ghisalberti, E. L., Jefferies, P. R., Payne, T. G., and Worth, G. K.,** Bertyadionol — an unusual diterpene, *Tetrahedron Lett.,* 4599, 1970.
36. **Airy Shaw, H. K.,** *Kew Bull.,* 20(1), 67, 1971.
37. **Ghisalberti, E. L., Jefferies, P. R., Payne, T. G., and Worth, G. K.,** Structure and stereochemistry of bertyadionol, *Tetrahedron,* 29, 403, 1973.
38. **Opferkuch, H. J. and Hecker, E.,** Über ein neues Diterpen aus *Euphorbia ingens, in XVI Colloquium Spectroscopicum Internationale Heidelberg,* Hecker, E., Ed., Adam Hilger, London, 1971, 282.
39. **Opferkuch, H. J. and Hecker, E.,** Ingol, a new macrocyclic diterpene alcohol from *Euphorbia ingens, Tetrahedron Lett.,* 3611, 1973.
40. **Lotter, H., Opferkuch, H. J., and Hecker, E.,** Cyrstal structure and stereochemistry of ingol-3,7,8,12-tetra-acetate, *Tetrahedron Lett.,* 77, 1979.
41. **Hergenhahn, M., Kusumoto, S., and Hecker, E.,** Diterpene esters from euphorbium and their irritant and cocarcinogenic activity, *Experientia,* 30, 1438, 1974.
42. **Upadhyay, R. R. and Hecker, E.,** Diterpene esters of the irritant and cocarcinogenic latex of *Euphorbia lactea, Phytochemistry,* 14, 2514, 1975.
43. **Abo, K. and Evans, F. J.,** The composition of a mixture of ingol-esters from *Euphorbia kamerunica, Planta Med.,* 43, 392, 1981.
44. **Abo, K. and Evans. F. J.,** A triester of ingol from the latex of *Euphorbia kamerunica* Pax., *J. Nat. Prod.,* 45, 365, 1981.
45. **Lin, L. J. and Kinghorn, A. D.,** 8-Methoxyingol esters from the latex of *Euphorbia hermentiana, Phytochemistry,* 22, 2795, 1983.
46. **Adolf, W. and Hecker, E.,** Diterpenoid irritants and co-carcinogens in Euphorbiaceae and Thymelaeaceae. Structural relationships in view of their biogenesis, *Isr. J. Chem.,* 16, 75, 1977.
47. **Rizk, A. M., Hammouda, F. M., El-Missiry, M. M., Radwan, H. M., and Evans, F. J.,** Macrocyclic diterpene esters from *Euphorbia royleana, Phytochemistry,* 23, 2377, 1984.
48. **Abo, K. A.,** Cytotoxic and Pro-Inflammatory Diterpene Esters of Nigerian *Euphorbia kamerunica,* Ph.D. thesis, University of London, London, 1982.
49. **Marsh, N., Rothschild, M., Scutt, A., and Evans, F. J.,** A cytotoxin from the pupa of the spurge hawk moth *(Hyles euphorbiae* L.) inhibiting the growth of malignant cells *in vivo* and *in vitro,* unpublished results, 1982.
50. **Kupchan, S. M., Sigal, C. W., Matz, M. J., Saenz Renauld, J. A., Haltiwanger, R. C., and Bryan, R. F.,** Jatrophone, a novel macrocyclic diterpenoid tumor-inhibitor from *Jatropha gossypiifolia, J. Am. Chem. Soc.,* 92, 4476, 1970.
51. **Lillehaug, J. R., Kelppe, K., Siegel, C. W., and Kupchan, S. M.,** *Biochem. Biophys. Acta,* 372, 92, 1973.

52. **Uemura, D. and Hirata, Y.,** Stereochemistry of kansuinine A, *Tetrahedron Lett.,* 1701, 1975.
53. **Uemura, D., Katayama, C., Uno, E., Sasaki, K., and Hirata, Y.,** Kansuinine B., a novel multi-oxygenated diterpene from *Euphorbia kansui* Liou, *Tetrahedron Lett.,* 1703, 1975.
54. **Shizuri, Y., Kosemura, S., Ohtusuka, J., Terada, Y., Yamamura, S., Ohba, S., Ito, M., and Saito, Y.,** Structural and conformational studies on euphornin and related diterpenes, *Tetrahedron Lett.,* 1155, 1984.
55. **Yamamura, S., Kosemura, S., Ohba, S., Ito, M., and Saito, Y.,** The isolation and structures of euphoscopins A and B, *Tetrahedron Lett.,* 5315, 1981.
56. **Shizuri, Y., Kosemura, S., Ohtsuka, J., Terada, Y., and Yamamura, S.,** Structural and conformational studies on euphohelioscopins A and B and related diterpenes, *Tetrahedron Lett.,* 2577, 1983.
57. **Purushothaman, K. K. and Chandrasekharan, S.,** Jatropholones A and B. New diterpenoids from the roots of *Jatropha gossypiifolia* (Euphorbiaceae), crystal structure analysis of jatropholone B, *Tetrahedron Lett.,* 979, 1979.
58. **Chan, W. R., Prince, E. C., Manchard, P. S., Springer, J. P., and Clardy, J.,** The structure of crotofolin-A, a diterpene with a new skeleton, *J. Am. Chem. Soc.,* 97, 4437, 1975.
59. **Burke, B. A., Chan, W. R., and Pascoe, K. J.,** The structure of crotofin-E, a novel tricyclic diterpene from *Croton corylifolius, Tetrahedron Lett.,* 3345, 1979.
60. **Adolf, W. and Hecker, E.,** Diterpenoid irritants and co-carcinogens in Euphorbiaceae and Thymelaeaceae. Structural relationships in view of their biogenesis, *Isr. J. Chem.,* 16, 75, 1977.
61. **Robinson, D. R. and West, C. A.,** Biosynthesis of cyclic diterpenes in extracts from seedlings of *Ricinus communis* L. I. Identification of diterpene hydrocarbons formed from mevalonate, *Biochemistry,* 9, 70, 1970.
62. **Dauben, W. G., Thiessen, W. E., and Resnick, P. R.,** Cembrene, a 14-membered ring diterpene hydrocarbon, *J. Am. Chem. Soc.,* 84, 2015, 1962.
63. **Dauben, W. G., Beasley, G. H., Broadhurst, M. D., Muller, B., Peppard, D. J., Pesnelle, P., and Suter, C.,** A synthesis of (±)-cembrene, a 14-membered ring diterpene, *J. Am. Chem. Soc.,* 97, 4973, 1975.
64. **Kodama, H., Shimada, K., and Ito, S.,** Synthesis of macrocyclic terpenoids by intramolecular cyclisation. III. Model reactions for the synthesis of cembrene and thunbergol derivatives, *Tetrahedron Lett.,* 2763, 1977.
65. **Sato, A., Kurabayashi, M., Ogiso, A., and Kuwano, H.,** Poilaneikc acid, a cembranoid diterpene from *Croton polanai, Phytochemistry,* 20, 1915, 1981.
66. **Crombie, L., Kneen, G., Pattenden, G., and Whybrow, D.,** Total synthesis of the macrocyclic diterpene (−)-casbene, the putative biogenetic precursor of lathyrane, tigliane, ingenane and related terpenoid structures, *J. Chem. Soc.,* 8, 1711, 1980.
67. **Burke, B. A., Chan, W. R., Pascoe, K. D., Blount, J. F., and Manchand, P. S.,** The structure of crotonitenone, a novel casbene diterpene from *Croton rhamnifolius, Tetrahedron Lett.,* 2399, 1969.
69. **Rentzea, M., Hecker, E., and Latter, H.,** Neue tetrazyklische, polyfunktionelle diterpenoide aus *Euphorbia myrsinites* L. Röntgenstrukturanalyse und Stereochemie des 14-desoxo-14β-hydroxymyrsinols, *Tetrahedron Lett.,* 1781, 1982.
70. **Rentzea, M. and Hecker, E.,** α-Ketol-Umlagerung von Myrsinol zum iso-Myrsinol und mögliche Biogenese des Myrsinan-Gerüstes, *Tetrahedron Lett.,* 1785, 1982.

Chapter 7

PHORBOL: ITS ESTERS AND DERIVATIVES

Fred J. Evans

TABLE OF CONTENTS

I. INTRODUCTION

The phorbol esters belong to the group of natural products based upon the hydrocarbon nucleus known as tigliane.[1] These compounds were the first pure tumor-promoting agents[2] to be isolated from natural sources and remain the most potent compounds known to date.[3] The parent diterpene, phorbol (**1**) is a novel tetracyclic polyol which, in its esterified form, as a series of 11 closely related diesters of the C-12, 13 hydroxy groups, was originally isolated from *Croton tiglium* seed oil,[4] and later from several other species of the plant family Euphorbiaceae. Phorbol esters, mainly as tetradecanoyl phorbol acetate (TPA), have been extensively studied in recent years[5-8] for their role in the promotion of tumor formation in mammalian systems. Recently, attention has focused on the possibility that a specific receptor site exists for phorbol ester,[9] which may be located on a membrane cell surface.[10-13] Evidence for this conclusion originally derived from the structural specificity of phorbol derivatives for activity in a variety of biological systems. The 12,13-diesters of phorbol are amphiphilic in nature, the acyl functions increasing the lipid solubility of the inactive polar polyol[14] and conferring biological activity on the esterified forms. The incrreased lipid solubility of phorbol esters over that of phorbol (**1**) itself may allow penetration of membrane structures to the site of action, while the molecular shape or configuration of the molecule, together with the location of its functional groups, induce interaction with membrane proteins producing a disruption of normal membrane functions. Such an interaction may be envisaged as arising from a Michael addition in vivo of the αβ-unsaturated ketone of phorbol (**1**) rendering compounds such as TPA membrane-alkylating agents.[15] Further evidence for the phorbol-ester receptor was obtained from pharmacological experiments. Phorbol esters are active at nanomolar dose levels[16] and their activities may be inhibited by a variety of pharmacological inhibitors. Tumor promotion in vivo by TPA has been inhibited by corticosteroids[17] and by protease inhibitors.[18] Additionally, membrane-stabilizing agents including the imipramine group of drugs inhibit both in vivo[19] and in vitro[20] effects of phorbol esters, as do phospholipase A_2 inhibitors, calmodulin-Ca^{++} inhibitors,[19,20] and cyclosporin-A.[21]

The nature of the phorbol-ester receptor site has been made clearer due to the use of radioactive labeled phorbol compounds, mainly 20-^{3}H-phorbol-12,13-dibutyrate, a relatively simple derivative to synthesize from natural phorbol. The location of specific, saturatable phorbol-ester receptors on mammalian cells has been demonstrated on a number of cell types[23-28] from 1980 onwards, beginning with the work of Driedger and Blumberg.[22] In 1981 Salomon[29] demonstrated that epidermal growth factor binding to carcinoma cells of mice could be inhibited by phorbol esters mediated by a specific phorbol-ester receptor. The fact that Mori and others[30] in 1980 demonstrated that phospholipid-interacting drugs which had been shown to inhibit phorbol ester activity would also inhibit phospholipid-dependent protein kinase was a significant indication as to the mechanism of action of phorbol esters. Phorbol esters have been suggested as interacting with protein kinase-C in homogenous preparations of Ca^{++}-activated, phospholipid protein kinase.[31-34] Shortly afterward, phorbol esters and diacylglycerol were shown to induce protein phosphorylation at tyrosine,[35] suggesting that some of the effects of TPA in vivo may be mediated by protein phosphorylation at tyrosine residues. At the same time[36] phorbol-ester-induced activation of human platelets was shown to be associated with protein kinase-C phosphorylation of myrosin light chains. The mechanism of action of tumor-promoting agents which has remained largely mysterious now becomes clearer because of the initial results reported by Castagna and others,[31] in that phorbol-ester promoters act by binding to and activating a specific calcium- and lipid-dependent protein kinase. This suggestion may be of significance in the understanding of the actions of tumor-promoting agents as a group of health hazards, since Umezawa and others[37] have recorded a similarity in the effects of teleocidin-B and phorbol esters on

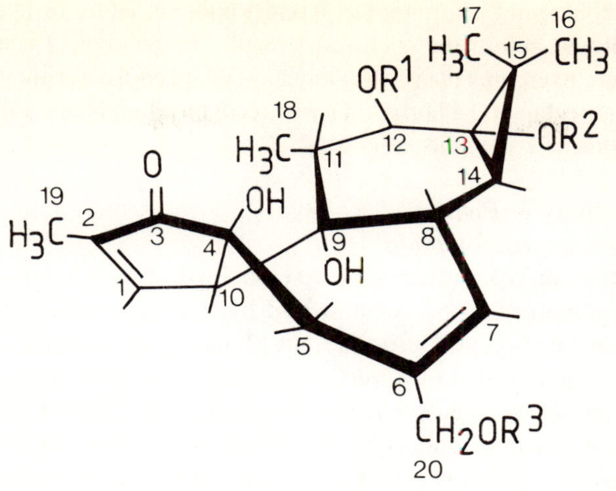

FIGURE 1

	R^1	R^2	R^3
Phorbol(**1**)	H	H	H
Phorbol-20-acetate (**32**)	OH	OH	Acetate
Phorbol-12,13-diacetate (**7**)	Acetate	Acetate	H
Phorbol-13,20 diacetate (**20**)	H	Acetate	Acetate
Phorbol-12,13,20-triacetate (**2**)	Acetate	Acetate	Acetate
Phorbol-12,13,20-tridecanoate (**10**)	Decanoate	Decanoate	Decanoate
Phorbol-13,20-dimethyl ether (**12**)	H	ME	ME
20-*O*-methyl-phorbol-12,13-diacetate (**13**)	Acetate	Acetate	ME

membrane receptors. Teleocidin-B has been shown to be as potent as TPA in several biochemical systems as well as for tumor-promotion in mouse skin.[38]

The phorbol esters, as the first pure group of tumor-promoting agents, have been shown to be of considerable consequence as biochemical tools in furthering the understanding of cancer proliferation in mammalian systems. The isolation and final structural elucidation of phorbol and its 12,13-diesters was acheived nearly 2 decades ago by Hecker and his group[4,39] at Heidelburg, and has since given rise to a new era of natural product research based upon the novel and diverse diterpenes of the plant families Euphorbiaceae and Thymelaeaceae.[40]

II. PHORBOL AND ITS ESTERS

Phorbol (**1**) was isolated as a hydrolysis product of *Croton tigilium* oil in 1935.[41] The correct empirical formula of phorbol (**1**) was obtained in 1966,[42] but its structure remained elusive in that several erroneous structures were initially proposed for this compound.[43-46] Determination of the structure of phorbol (**1**) was finally achieved by means of X-ray crystallography of neophorbol-13,20-diacetate-3*p*-bromobenzoate[47] and was later confirmed by X-ray analysis of phorbol-bromofuroate[48] and phorbol.[49] More recently, the geometry of phorbol (**1**) has been optimized and the change distribution calculated.[50] The structure and stereochemistry of phorbol was also derived by Crombie and others[51] by chemical methods.

Phorbol (**1**) is a tetracyclic diterpene; it consists of a 5-membered ring A connected in *trans* configuration to a 7-membered ring B. Ring C is 6-membered and the cyclopropane ring D is linked to it in the *cis*-configuration. An αβ-unsaturated ketogroup is present at C-3 of ring A, a glycol at C-12,13, and a primary hydroxyl at C-20. Tertiary hydroxyls are exhibited at C-4 and C-9 of the structure (Figure 1). The original structure elucidation of phorbol (**1**) as the hydrolysis product of the *Croton* factors proved difficult not only due to its instability but also due to the fact that phorbol (**1**) badly solvates in solvents. Problems were encountered concerning its empirical formula and the nature of its functional groups which required solving before the chemical strucutre of phorbol (**1**) could be proposed. These problems were overcome using a combination of spectroscopic methods together with traditional chemical technique. Phorbol (**1**) was recognized as being a diterpene by quantitative hydrogenation of its synthetic acetate.[45]

A. Chemical Reactivity of Phorbol

1. Reactivity of the Functional Groups

Phorbol (**1**) contains an αβ-unsaturated keto group in ring A of its structure; the presence of this functional group cannot be demonstrated by the normal reagents. This functional group as exhibited by phorbol-12,13,20-triacetate (**2**) undergoes reduction with either sodium borohydride or lithium aluminum hydride. In the latter case 3-deoxo-3β-hydroxyphorbol-triacetate (**3**) was produced,[52] but with the former reagent, some of the 3α-epimer was obtained as a minor reaction product.[53] The allylic hydroxyl of the 3β-epimer may be reconverted to the keto substituent by means of chromium trioxide/pyridine oxidation. Oxidation of the hydroxyl functions of phorbol (**1**) or its triacetate (**2**) may be achieved using a variety of reagents. Phorbol (**1**) exhibits primary, secondary, and tertiary hydroxyl functions and their reactivities are different. The tertiary hydroxyls at C-4 and C-9 are resistant to oxidation under moderate conditions, but the tertiary cyclopropanol moiety was readily oxidized as was demonstrated by the positive Fehling's and Tollen's test at an early stage in the elucidation of the structure.[55-58] At that time this was a new reaction and was independently confirmed by Schaefsma and others.[58] This oxidation is thought to commence with the abstraction of a proton from the C-13 tertiary hydroxyl, thereby generating the cation. The secondary hydroxyl function at C-12 of phorbol (**1**) was oxidized using chromium trioxide/pyridine to yield the 12-keto derivative (**4**) (Figure 2). This new carbonyl group exhibited low reactivity, thereby allowing the selective reduction of the C-3 keto group with sodium borohydride to give "neophorbol" (**5**). This compound (**5**), which exhibits a 3β-hydroxy and 12-keto in its structure, was instrumental in the final structure elucidation of (**1**) because its ester was used for X-ray analysis. The primary alcoholic group of 12,13-diesters of phorbol is the most reactive hydroxy function. With manganese dioxide or with chromium trioxide/pyridine the primary allylic group may be oxidized to the 20-aldehyde(6).[59-61] This reaction has since proved valuable for the introduction of tritium into the structure for the production of labeled analogues for receptor studies in biological systems.

The hydroxyl functions of phorbol may be substituted with halogen. Phorbol-12,13-diacetate (**7**) reacts with mesyl-chloride in pyridine to produce the 12-chloro analogue.[61,62] However, attempts to substitute the OH-12 by halogen, even under mild conditions, led to rearrangements of the cyclopropane ring. The substitution of the 20-acetyl moiety of phorbol-12,13,20-triacetate (**2**) with hydrogen was achieved using Raney nickle/ethanol. This reaction was carefully controlled to avoid saturation of the susceptible $\Delta^{6,7}$ bond, but the 20-chloro derivatrive more readily produced 20-deoxyphorbol (**8**).[61] Elimination of the 20-acetyl group of (**2**) was achieved using strongly basis conditions. This group eliminates with the simultaneous isomerization of $\Delta^{6,7}$ bond to give compound (**9**).[61] The tertiary OH at C-9 may be eliminated from 4-*O*-acetyl-phorbol-12,13,20-tridececanoate (**10**) by the use of thionyl chloride/pyridine to yield to $\Delta^{9,10}$ analogue (**11**).[61]

Phorbol ester hydroxyl groups may be selectively alkylated under mild reaction conditions.[57] These reactions have recently been of some importance for the synthesis of 4-*O*-methyl phorbol esters which are used as a negative control in biological experiments. Alkylation of the 4-OH may be achieved with methyl iodine/silver oxide in DMF either as solvent or in catalytic quantities, depending upon the nature of the reacting phorbol ester. In more basic conditions a simultaneous exchange of acetyl and alkyl groups may take place, but epimerization of the tertiary 4β-OH to the 4α-epimer does not occur.[57] The primary

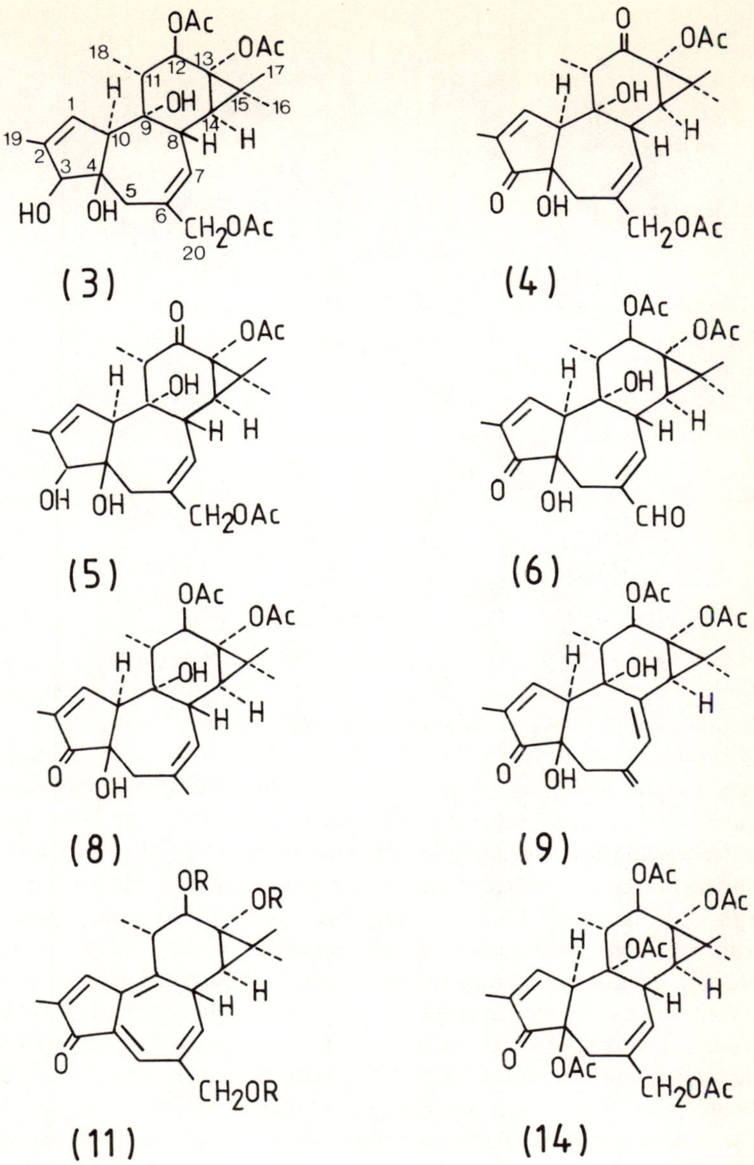

FIGURE 2. Reaction products of the functional groups of (**1**) and its esters.

allylic hydroxy of phorbol (**1**) is susceptible to alkylation and these reactions have proven useful for the production of tritiated derivatives. When phorbol (**1**) was reacted with tritylchloride/pyridine the 20-hydroxyl was selectively tritylated,[63] subsequent treatment with diazomethane produced the 13,20-dimethylether (**12**). The primary alcoholic group may be selectively alkylated by the use of diazoalkane/Al-i-propylate or methyl iodide with silver oxide to yield (**13**) using phorbol-12,13-diacetate (**7**) as starting material.[57]

Phorbol (**1**) acylates under appropriate reaction conditions, the three hydroxyl functions at C12, 13, and 20 being the most reactive. With acetic anhydride in pyridine at room temperature, phorbol-12,13,20-triacetate (**2**) was produced.[64] Reaction of this product with acetic anhydride/*p*-toluene sulfonic acid produced phorbol pentaacetate (**14**).[65] A combination of positional isomeric mono- and di-acetates was produced from (**2**) by transesterification reactions[66] in either acid or alkali and methanol. Similarly higher molecular weight tri-

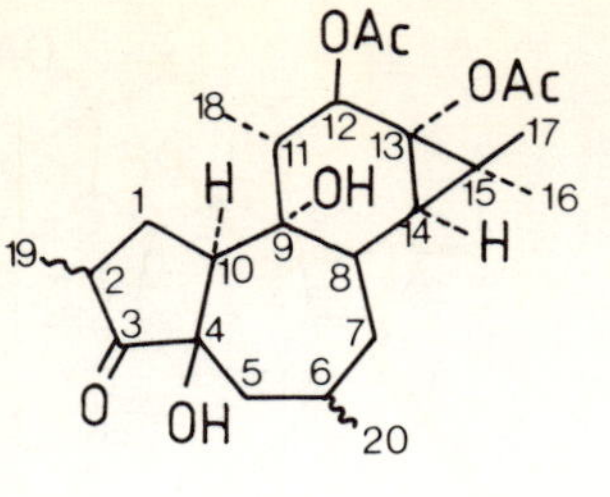

FIGURE 3. Reaction products of the double bonds of (**1**) and its esters.

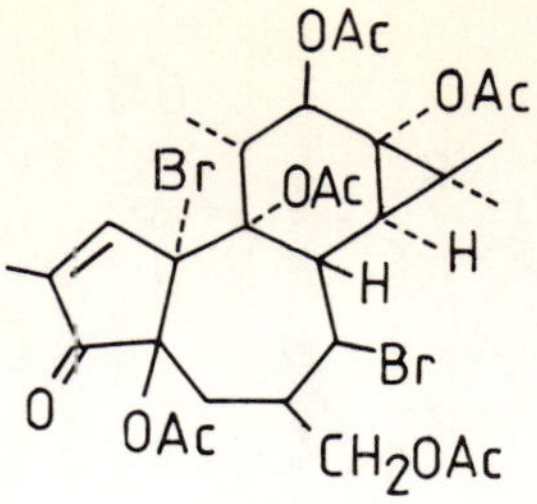

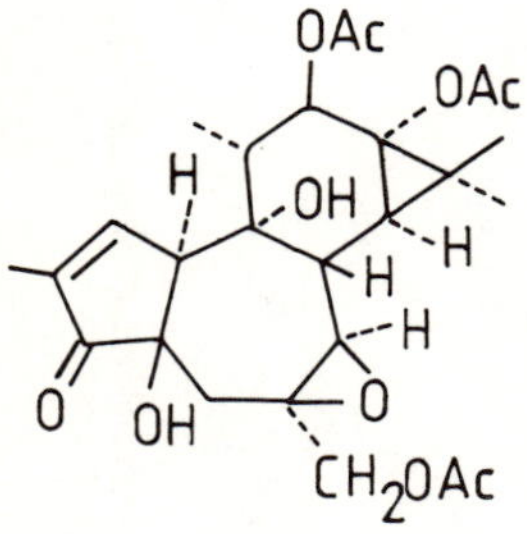

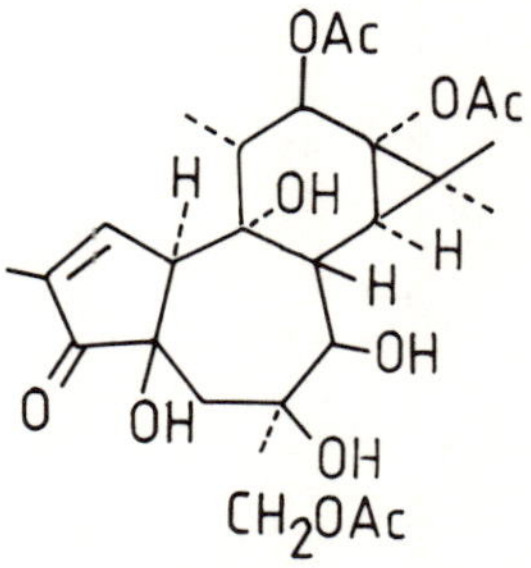

aceylates may be prepared from phorbol (**1**) by reaction with the respective acyl chlorides in pyridine and the 12,13-diesters subsequently prepared from them by base-catalyzed transesterification in methanol.[51,67-68] In general, the C-12 secondary hydroxyl is the least reactive function both to esterification and transesterification, while the tertiary hydroxyl at C-13 is considerably more reactive than might be expected. Using acetic anhydride/DMF the C-13 hydroxyl is even more rapidly acetylated than the C-20 primary hydroxyl group.[66]

Phorbol-12,13,20-triacetate (**2**) may be catalytically hydrogenated with palladium on charcoal. Three molecules of hydrogen were taken up as a result of this reaction by the $\Delta^{1,2}$, $\Delta^{6,7}$ double bonds and the allyl acetoxy at C-20 of the nucleus to yield compound (**15**).[45,65] Phorbol-pentaacetate (**14**) brominated with Br_2 in carbon tetrachloride at the C-20 position, but the primary product of this reaction was (**16**) due to saturation of the $\Delta^{6,7}$ bond by liberated HBr.[69] Although phorbol (**1**) reacts vigorously with Br_2, no products were isolated from this reaction.[52,70] Treatment of (**14**) with bromine in ether/glacial acetic acid produced the unstable tetrabromo derivative (**17**).[69] One of the more important reactions of the olifinic groups of phorbol (**1**) for the identification of tigliane derivatives which occur naturally was the oxidation of $\Delta^{6,7}$ bond with *m*-chloro-perbenzoic acid. As a result of this reaction, the 6β,7β-epoxide (**18**) was produced.[71] This reaction also occurs with phorbol-triacetate (**2**) and phorbol-pentaacetate (**14**). If perbenzoic acid was used as reagent then the 6α,7α-epoxide was obtained in low yield. Oxidation of phorbol-triacetate (**2**) with osmium tetroxide produced the 6β,7β-dihydroxy derivative (**19**)[72] (Figure 3).

2. Reactions of the Tigliane Nucleus

Several reactions of the phorbol (**1**) nucleus itself were instrumental in the assignment of the configuration to this novel tetracyclic diterpene. Phorbol (**1**) when treated with sodium methoxide in methanol, produced the AB-*cis* isomer 4α-phorbol (**20**) as a major product together with 10β-phorbol (**21**) and $\Delta^{1,10}$-isophorbol (**22**)[61,74-76] (Figure 4). Care has to be taken with this reaction to exclude oxygen and to adjust the strength of the base to 10^{-2} M

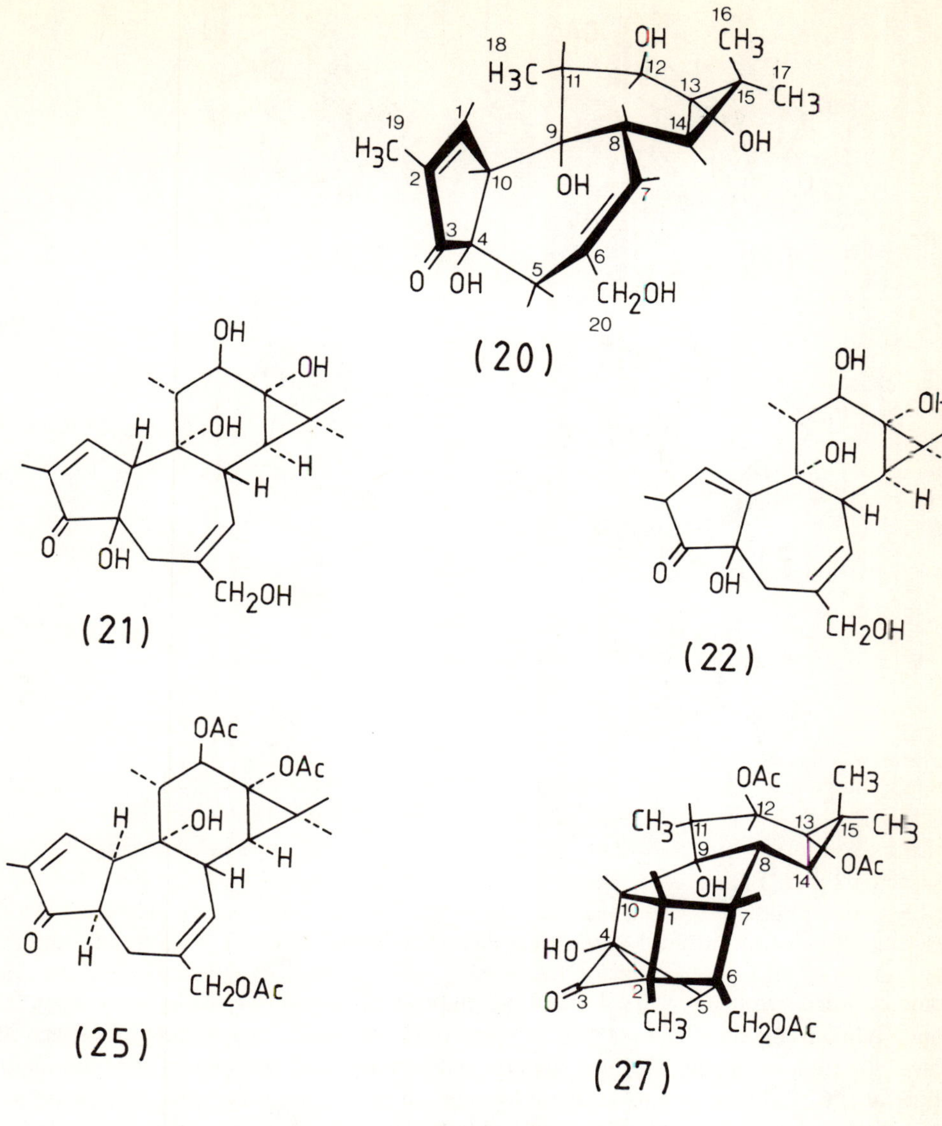

FIGURE 4. 4α-Phorbol (**20**) and its reaction products.

before epimerization will take place. The reactivity of the functional groups of (**20**) is similar to that of phorbol (**1**) (Figure 1) with the exception of the reactivity of the 4α-OH function. During acetylation the 4-*O*-acetyl analogue (**23**) was produced together with the triacetate[75-77] and methylation with methyl iodide/silver oxide/DMF produced the 4-*O*-methyl-4α-phorbol-triacetate (**24**). Furthermore, zinc/acetic acid on 4α-phorbol-triacetate (**26**) readily induced an acyloin reduction to 4-deoxy-4α-phorbol-triacetate (**25**). Irradiation of 4α-phorbol-triacetate (**26**) with UV light at 254 nm produced the isomeric lumiphorbol-12,13,20-triacetate (**27**). This derivative exhibits a cage structure of the cyclobutane type as was shown by means of X-ray analysis.[74,78] Such a structure could be envisaged as arising by intramolecular cycloaddition of the double bonds of (**27**).

Reaction of phorbol-13,20-diacetate (**28**) with mesyl chloride in pyridine at low temperature produced crotophorbolone-enol-13,20-diacetate (**29**) (Figure 5) and acetoxycrotophorbolone-20-acetate (**30**).[72] Crotophorbolone (**31**) was produced from (**29**) by means of base catalyzed transesterification. Treatment of phorbol triacetate (**2**) with acetic acid/boron tri-

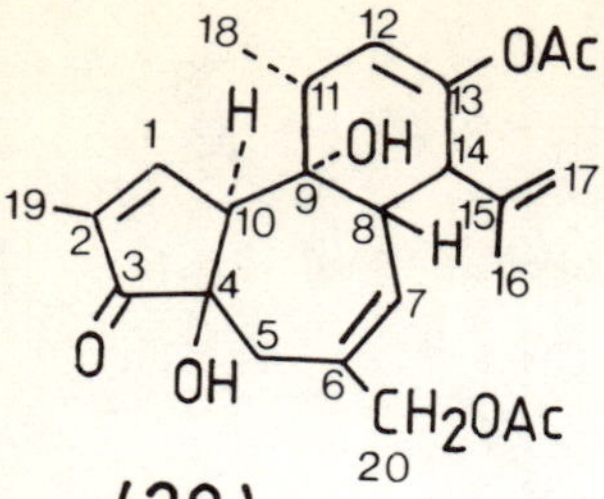

FIGURE 5. Rearrangement products of phorbol (**1**).

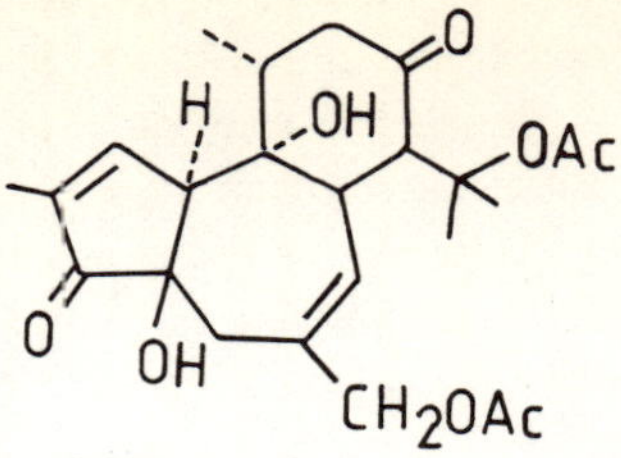

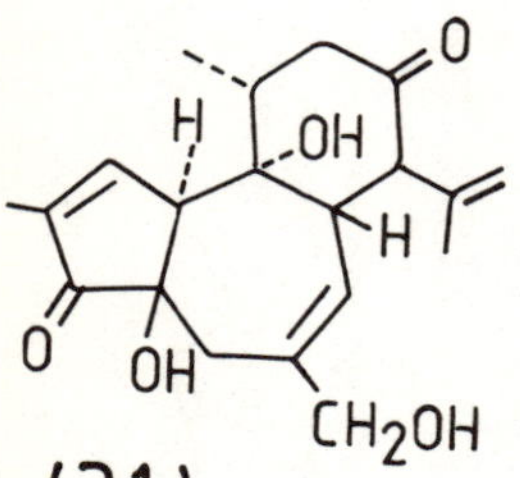

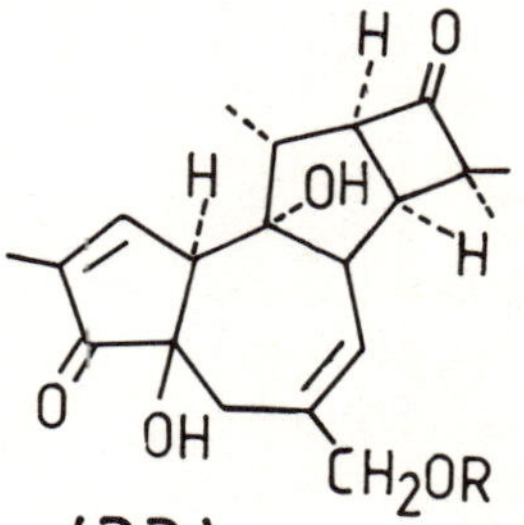

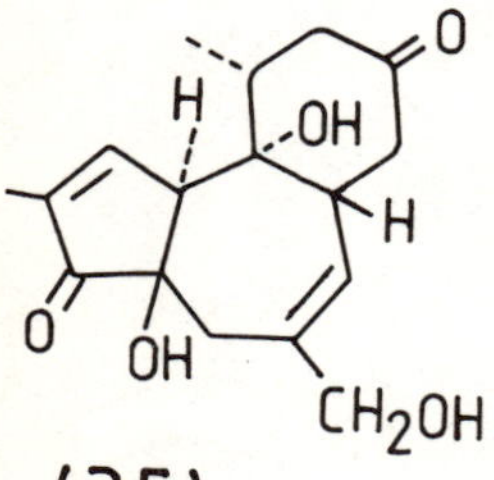

fluoride afforded compound (**30**). These reaction products are considered to be the result of a homoallyl rearrangement of [α-acetoxy-cyclopropyl]-carbinol group of phorbol (**1**) and its acetates. Phorbol-20-acetate (**32**) will undergo dehydration with phosphorus oxychloride in pyridine at 0°C, inducing a pinacol rearrangement of the [α-hydroxycyclopropyl]-carbinol group to give phorbol-butanone-20-acetate (**33**).[63] This product was converted to phorbobutanone (**34**) by means of phosphorus oxychloride/pyridine. Phorbobutanone (**34**) was one of the major products produced as a result of the Flaschenträger reaction.[70,79] Phorbol (**1**) was treated with hot dilute sulfuric acid and (**34**) together with crotophorbolone (**31**) and compound (**35**) were isolated.[62] This complex reaction was aimed at the structure elucidation of phorbol (**1**) and possibly arises as a result of acid catalyzed elimination of the C-12 hydroxyl.

Reactions altering the ring structure of phorbol (**1**) were also of importance for the structure elucidation of this compound. In ring A of (**3**) the *cis*-3β-4-gycol group was readily split with sodium periodate in dioxan and water to yield the 3,4-seco compound (**36**).[80] The carbonyl groups of (**36**) were reduced with sodium borohydride to produce the more stable tetrahydro derivative (**37**). Phorbol triacetate (**2**) was shown to be stable towards this reaction. Ring B of the tigliane nucleus may also be selectively cleared. Using the 6β,7β-dihydroxyl derivative (**18**) ring B cleavage with lead tetraacetate or sodium periodate produced the 4,7-semiacetal (**38**).[81] All of the hydroxyl functions of (**19**) were accessible to oxidation, and using lead tetraacetate, the corresponding 20-nor-6,7β-keto (**39**) was additionally produced.

This compound was reduced to the 3β,6ξ-diol (**40**) which yielded a pentaacetate on acetylation. These derivatives assisted in the location of the 2H-C5[51,83] in the structure of phorbol (**1**) (Figure 6). An aqueous solution of phorbol (**1**) has been shown to consume one molecule of lead tetraacetate to produce equimolar mixtures of bisdehydrophorbol (**41**) and tigliophorbol (**42**).[82] However, when reacted with sodium periodate, (**41**), (**42**) and the 9,13-semiketal (**43**)[83] were produced. If the oxidation was carried out with two molecules of sodium periodate, then phorbolactone-semiacetal (**44**) was mainly produced together with (**42**) and hydroxyphorbolactone-semiacetal (**45**). Crombie et al.[51] have also investigated ring D oxidation and additionally detected the 12-oxo-13-hydroxy-isomer (**46**) of (**41**) as a product of the reaction. The stereochemical assignments of these reactions products were of assistance in the determination of the configuration of phorbol (**1**) together with its spectroscopic and X-ray data.

B. Synthesis of Phorbol Esters

The synthesis of esters of phorbol (**1**) from the natural polyol, as isolated by crystallization from the hydrolysis of croton oil, was essential for the structure elucidation of these natural products.[51,84,85] More recently it has been found that supplies of phorbol esters for biological evaluation and mechanism of action studies are more readily available using the techniques of partial synthesis previously described by Hecker and co-workers.[51,84,85] Additionally, ^{14}C-labeled fatty acids may be used as starting materials for the ^{14}C-labeling of phorbol esters in the side chains. Considering that phorbol may be labeled with tritium at the C-20 allylic position,[59-61] it is feasible for double labeled phorbol esters to be synthesized by a series of relatively simple steps.

The synthetic routes for the production of the 12,13-diesters of phorbol initially involve the production of the 12,13,20-triacylated derivatives. The diesters of phorbol (**1**) may be produced as two 12,13-isomeric ester forms. In the first case a long-chain fatty acid residue is present at C-12 of the nucleus giving rise to the "A" series of esters, of which the well-known TPA of PMA is a member. In the second case the longer chain ester is present at C-13, giving rise to the less biochemically investigated "B" series of esters. The synthetic route to the "A" series esters from phorbol (**1**) involves acylation at the three reactive positions of the nucleus with an acyl chloride in pyridine. The acyl substituents at the tertiary position C-13 and the primary position C-20 are susceptible to transesterification with sodium methoxide in dry methanol under controlled conditions to yield the C-12 mono-acyl derivative of phorbol (**1**). This product will acetylate readily in acetic anhydride/pyridine at 85°C to produce the 12-*O*-acyl-13,20-diacetate of phorbol (**1**) (Figure 7). The C-20 primary acetate moiety of such products is susceptible to transesterification under acid conditions which do not affect the C-13 acetate of the C-12 acyl functions, Normally $HClO_4$ in methanol will produce a 12,13-diester of the "A" series from such an intermediate in about 80% yield.[84,85] In the case of the "B" series of esters, the first step in their synthesis is the production of phorbol-12,13,20-triacetate (**2**) (Figure 1) from phorbol (**1**) with acetic anhydride/pyridine at room temperature.[64] Transesterification as before yields phorbol-12-acetate (**47**) which may be converted with acyl chloride in pyridine to the 12-*O*-acetyl-13,20-acyl analogue of phorbol (**1**). Acid-catalyzed transesterification of this compound produced esters of the "B" series in the normal manner[84,85] (Figure 7).

C. Naturally Occurring Esters of Phorbol(1)

Esters of phorbol (**1**) have been isolated from eight individual species of four genera of the plant family Euphorbiaceae (Table 1). These include two species of the genus *Croton, C. tiglium*[4] and *C. sparciflorus,*[86] two species of *Sapium, S. indicum*[87] and *S. japonicum,*[88] three *Euphorbia* species, *E. tirucalli,*[89] *E. frankiana,*[90] and *E. coerulescens,*[90] together with one species of *Ostodes, O. paniculata.*[91] Considering the importance of esters of this tigliane

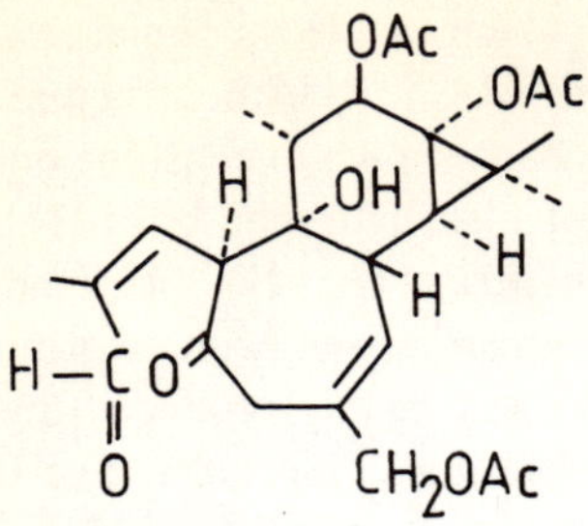

(36) 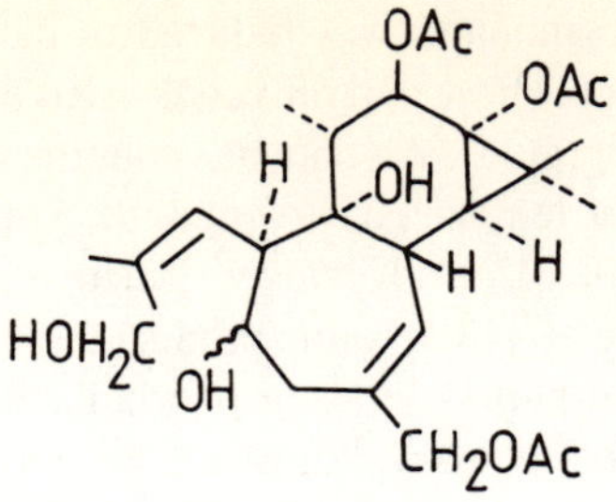 (37)

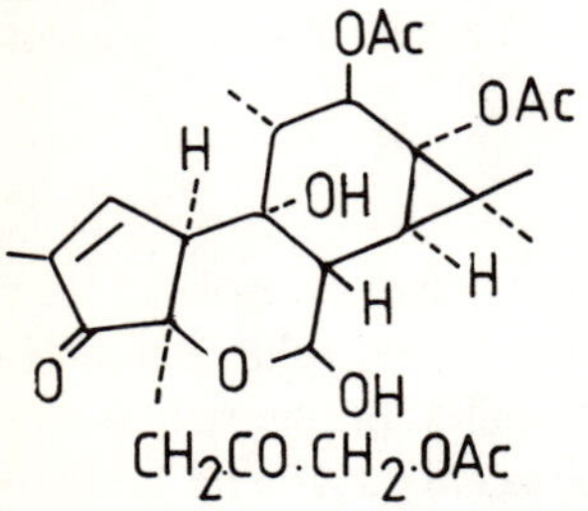

(38) 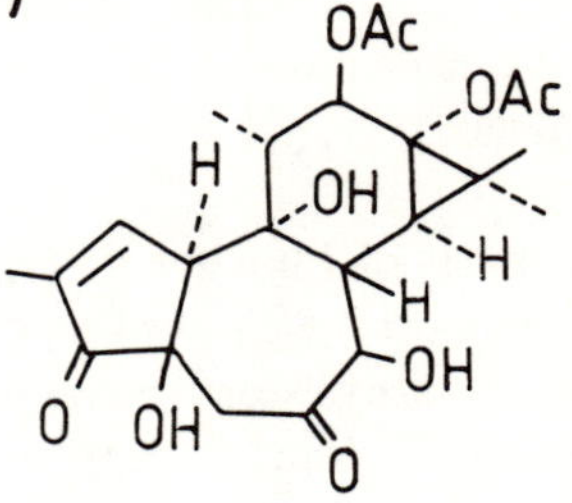 (39)

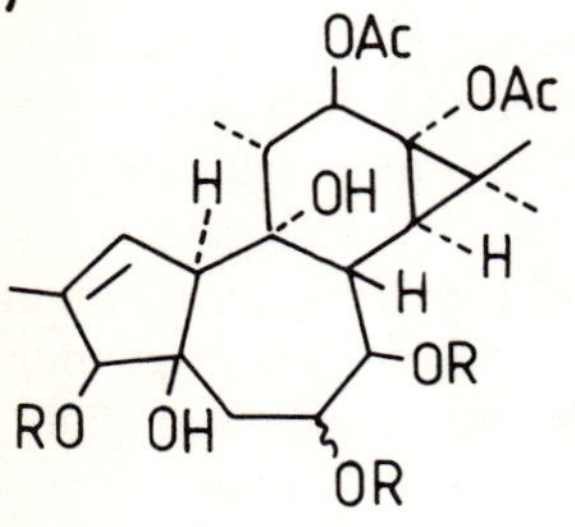

(40) 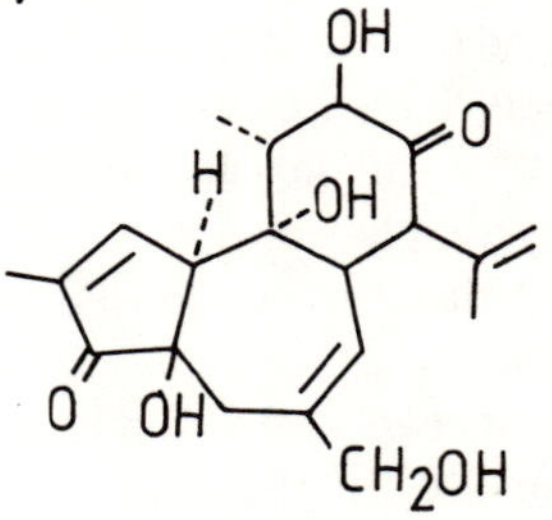 (41)

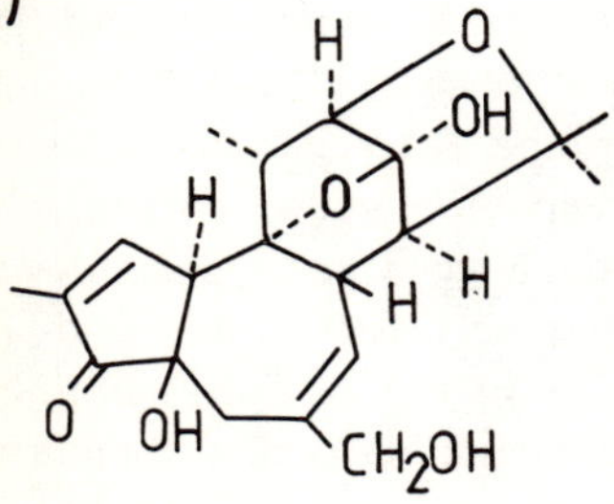

(42) 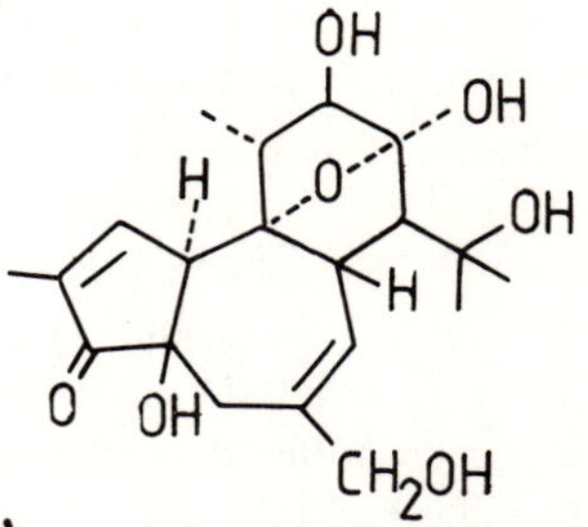 (43)

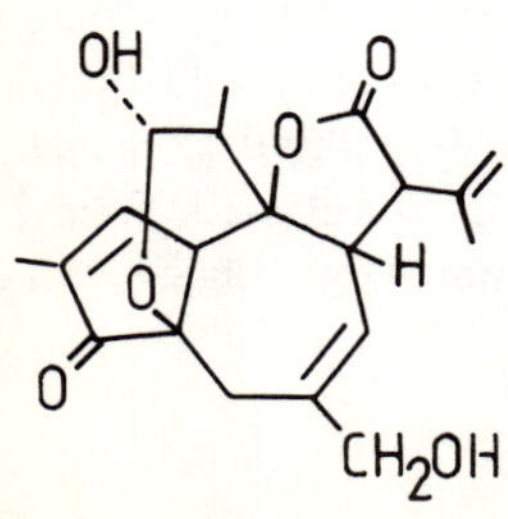

(44) 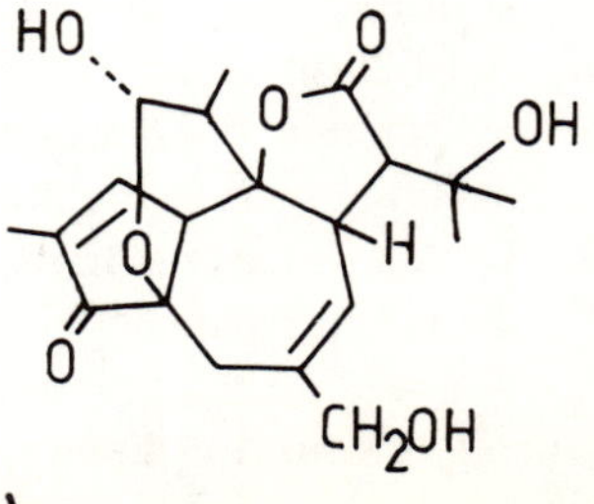 (45)

FIGURE 6. Products of ring opening reactions of phorbol (**1**).

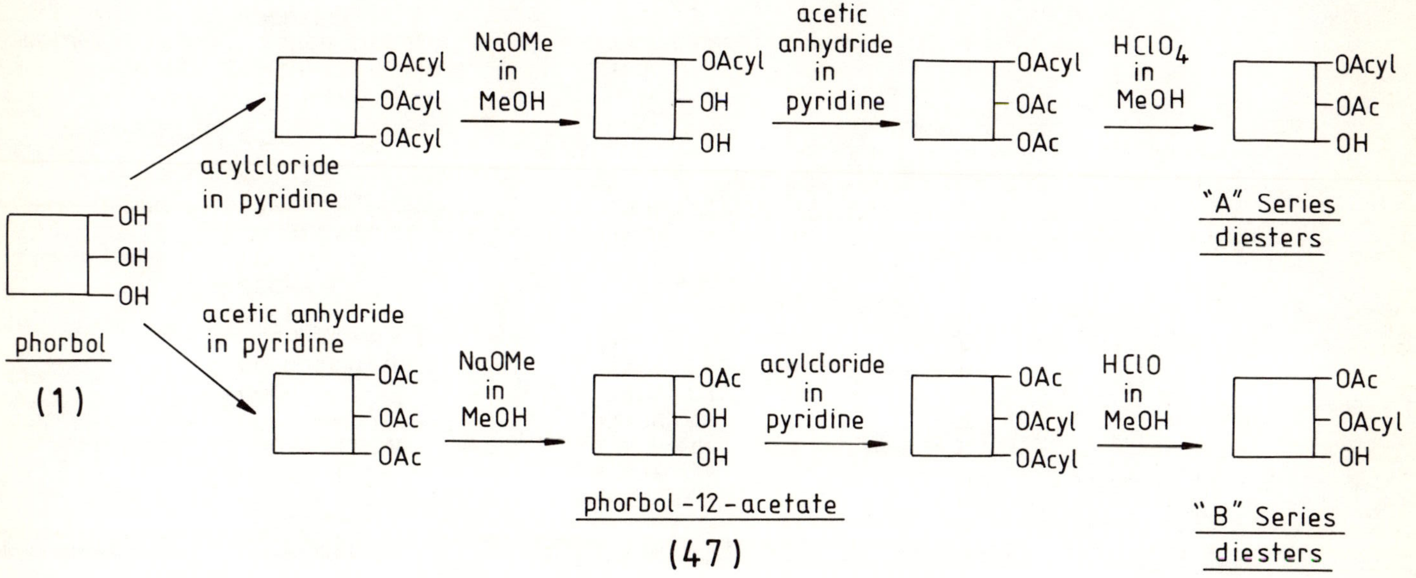

FIGURE 7. Partial synthesis of 12,13-diesters of phorbol (**1**).

Table 1
THE NATURALLY OCCURRING ESTERS OF PHORBOL

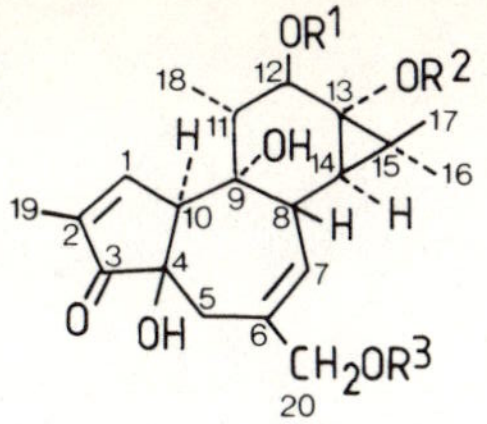

Plant	R^1	R^2	R^3	Ref.
Croton tiglium L.	$CO(CH_2)_8CH_3$	$COCH_3$	H	
	$CO(CH_2)_{10}{\cdot}CH_3$	$COCH_3$	H	
	$CO(CH_2)_{12}{\cdot}CH_3$	$COCH_3$	H	
	$CO(CH_2)_{14}{\cdot}CH_3$	$COCH_3$	H	
	$COC(CH_3){\cdot}CH{\cdot}CH_3$	$CO(CH_2)_6{\cdot}CH_3$	H	
	$COC(CH_3){\cdot}CH_2{\cdot}CH_3$	$CO(CH_2)_6{\cdot}CH_3$	H	
	$COCH_3$	$CO(CH_2)_8CH_3$	H	
	$COC(CH_3)CH{\cdot}CH_3$	$CO(CH_2)_8CH_3$	H	
	$COCH(CH_3){\cdot}CH_2{\cdot}CH_3$	$CO(CH_2)_8{\cdot}CH_3$	H	
	$COCH_3$	$CO(CH_2)_{10}{\cdot}CH_3$	H	
	$COCH(CH_3){\cdot}CH_2{\cdot}CH_3$	$CO(CH_2)_{10}{\cdot}CH_3$	H	4,73,93
C. sparciflorus Morong	$CO(CH_2)_{10}{\cdot}CH_3$	$COCH_3$	H	
	$CO(CH_2)_{10}{\cdot}CH_3$	$CO\ CH_3$	$CO(CH_2)_6{\cdot}(CH_2CH=CH)_3CH_2CH_3$	86
Sapium indicum Willd.	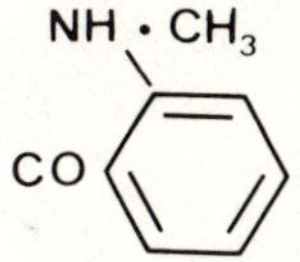 $NH \cdot CH_3$ CO	$COCH_3$	H	87
S. japonicum Pax et Hoffm.	$CO(CH=CH)_3{\cdot}(CH_2)_2{\cdot}CH_3$	$COCH_3$	H	88
Euphorbia frankiana Berger	$COCH(CH_3)CH_2CH_3$	$COC(CH_3)CHCH_3$	H	
	$COCH(CH_3)_2$	$COCH_3$	$COC(CH_3)CH{\cdot}CH_3$	90,95

E. coerulescens Haw.	$COCH(CH_3)CH_2CH_3$	$COC(CH_3)CHCH_3$	H	90,95
	$COCH(CH_3)_2$	$COCH_3$	$COC(CH_3)CHCH_3$	
E. tirucalli L.	$CO(CH=CH)_n\ (CH_2)_mCH_3$	$COCH_3$	H	
	m = 2, n = 2, 3, 4, 5			
	m = 4, n = 1, 2, 3, 4			
	$COCH_3$	$CO(CH=CH)_n(CH_2)_mCH_3$	H	89
Ostodes paniculata Blume	$CO(CH=CH)C_8H_{15}$	$COCH_3$		91

polyol to biochemical research their distribution in the plant kingdom is still restricted at the present time and confined to only one plant family. In fact, phorbol (**1**) itself is one of the rarer natural products of this family of diterpenes, which more commonly occur in either a deoxy or hydroxyl form of the tigliane nucleus.

Phorbol (**1**) has been obtained from plant sources in three distinct esterified forms. The first group of compounds are the 12,13-diesters which exhibit a long chain or higher molecular weight acyl function at C-12 of the nucleus the "A" series, the second group are the 12,13-positional isomers with a longer chain acyl function at C-13, the "B" series. The third group, known as the "cryptic" compounds, is characterized by the presence of an acyl group at the C-20 primary alcoholic residue of phorbol (**1**). The "cryptic" compounds are so called because they lack the well-known pro-inflammatory and tumor-promoting activities of the 12,13-diesters in vivo.[4] These triesters may be converted to the biologically active derivatives by acid-catalyzed or mild base-catalyzed transesterification reactions.

This reaction most probably explains the observation of Buchheim[92] that treatment of the hydrophobic fraction of croton oil with alkali restored its irritant properties. Phorbol esters were originally isolated from the oil and resin of the seeds of *Croton tiglium*.[4,73,93] Eleven tumor-promoting and pro-inflammatory 12-13-diesters were obtained from the plant, the most potent of which was compound A^1, later known as PMA or more commonly TPA. TPA remains the most potent tumor-promoting agent known. At a single repeated dose of 0.02 μM this compound produces an 82% tumor yield, when tested on mouse skin in traditional Berenblum type experiments, and has an irritant dose 50% of 0.016 μg[94] in the mouse-ear assay. A number of triesters were also detected from the same plant, but their structures were not completely elucidated because of separation difficulties from the oil.[4] Of the 12,13-diesters of phorbol (**1**) isolated, four belonged to the "A" series, while the others were of the "B" series[73,93] (Table 1). At a later date two esters of phorbol (**1**) were isolated from *C. sparciflorus,* one of which was a 12,13-diester of the "A" series, while the other was a cryptic compound. The structure of this triester was elucidated[86] as exhibiting an unsaturated C-12 fatty acid at position 20 of phorbol (**1**). The structures of similar cryptic phorbol (**1**) esters were obtained after isolation from *Euphorbia frankiana*[90] and *E. coerulescens,*[95] as well as from the African succulent species *E. tirucalli*.[89] An interesting 12,13-diester of the "A" series of phorbol (**1**) esters was recently isolated from *Sapium indicum*.[87] This compound was characterized by the presence of a 2-methylaminobenzoyl moiety at C-12 of the nucleus, and as such as the first nitrogen-containing derivative of phorbol (**1**). Because of its C-12 ester function this derivative exhibits a pronounced blue fluorescence in UV light and is accordingly of considerable importance for the study of the interaction of phorbol esters with membrane proteins in vitro. The naturally occurring phorbol esters are unstable compounds and are susceptible to oxidation, hydrolysis, transesterification, and epimerization reactions during their isolation. Even under controlled conditions of separation at low temperature, with redistilled solvents and vigorously excluding oxygen from the system, it is possible for artifacts to be produced (see Chapter 1). Recently Handa and coworkers[91] isolated a compound known as ostodin (**48**) (Figure 8) from *Ostodes paniculata,* which was structure-elucidated as the $\Delta^{5,6}$-7β-hydroperoxide of a phorbol diester. Ostodin (**48**) may be produced from the corresponding phorbol ester by means of autooxidation and was probably an artifact of the investigation.[91] The similarity in structure of the phorbol-esters within each group, either di- or triesters, makes their separation into individual compounds complicated. Partition techniques, in the form of countercurrent[4] methods or partition-thin layer chromatographic (TLC)[96] methods have been used with success for this purpose, as have high pressure liquid chromatographic (HPLC) techniques.[97] Phorbol esters are produced as friable resins which may be powdered and stored under nitrogen in ampoules. Some derivatives have been produced in the crystalline form,[4] but crystallization may induce loss of material due to oxidation and/or epimerization of the nucleus. (See Chapter 1.)

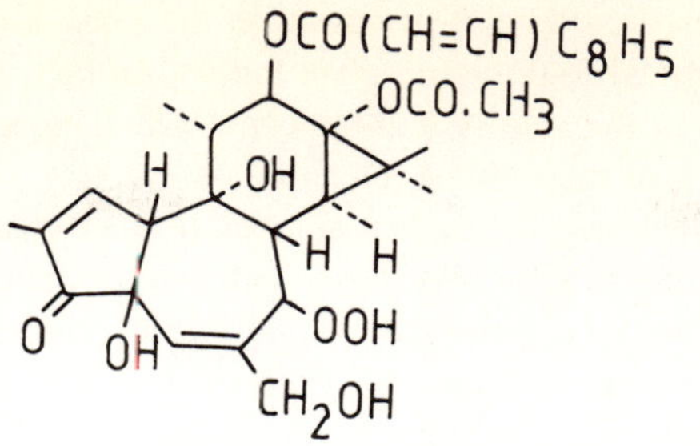

FIGURE 8. Ostodin (**48**) from *Ostodes paniculata* Blume.[91]

D. Identification of Phorbol Esters

The structure of phorbol (**1**), the parent polyol of the naturally occurring esters, was ascertained by traditional chemical methods[4,51] (Section I) and its absolute configuration deduced by X-ray techniques.[49] Esters of phorbol (**1**) as isolated from natural sources may be identified by partial synthesis and transesterification reactions (Section II) together with their spectroscopic data. Phorbol-12,13,20-triacetate (**2**), which was produced from phorbol (**1**) by acetylation,[64] provides a readily available material for comparison of the spectral characteristics of natural di- and triesters of this parent diterpene.

In the UV spectrum phorbol-triacetate (**2**) in methanol exhibits characteristic inflections at 230 and 330 nm due to the absorbance of the αβ-unsaturated carbonyl group of ring A.[90] The latter peak is of particular value for the detection of phorbol esters after HPLC using a micro-flow UV detector.[97] This spectrum is reminiscent of phorbol (**1**), which in ethanol as solvent exhibits maxima at 196 (ϵ = 10,600), 234 (ϵ = 5060), and 332 (ϵ = 72) nm.[65] In the circular dichroism spectra compound (**2**) exhibits a pronounced positive extreme between 225 to 250 nm and a less intense negative extreme between 330 to 340 nm. These Cotton effects are due, respectively, to $\pi - \pi^*$ and $\eta - \pi^*$ transitions of the αβ-unsaturated carbonyl group of ring A. The ring AB *cis* isomer of (**2**), 4α-phorbol-triacetate (**26**) may be distinguished from phorbol-triacetate (**2**) on the basis that this substance in methanol solution exhibits a less intense Cotton effect between 225 to 250 nm (Figure 9).[98] The IR spectra of esters of phorbol (**1**) including its triacetate (**2**) are very similar and not particularly characteristic.[4,90] These spectra, normally obtained from KBr discs or as solid films on NaCl discs, exhibit bands in the hydroxyl region at 3500 cm^{-1} carbonyl absorbance due to the ester moieties at between 1720 to 1740 cm^{-1}, and a ketone band at 1695 to 1715 cm^{-1}.

The mass spectra of the phorbol esters were of particular value for the determination of their empirical formula[59] because these compounds solvate strongly and are not suitable for elemental analysis. By means of electron-impact mass-spectrometry (EIMS), small molecular ions ($M^{+\cdot}$) are exhibited of 4% or less relative abundance. Nevertheless, such ions are suitable for accurate mass determinations using a high resolution instrument at temperatures of about 200°C and 70 eV. Phorbol-triacetate (**2**) is typical of this group of esters in that by means of EIMS it exhibits a small $M^{+\cdot}$ ion, together with fragment ions due to the sequential loss of acetate residues as CH_3COOH to produce the fragment at m/z 310 which corresponds to the deacylated nucleus. Due to the hydrocarbon nature of this fragment ion a large number of fragment ions are characteristically produced in this spectrum. The 12,13-diesters of phorbol (**1**) may be recognized due to the sequential loss of acyl moieties and the generation of ions at m/z 328 (20% relative abundance), together with an ion at m/z 310 due to the loss of water from this fragment. Accordingly the spectra of these compounds are of assistance in identifying the molecular weight of the relevant acyl substituents. Con-

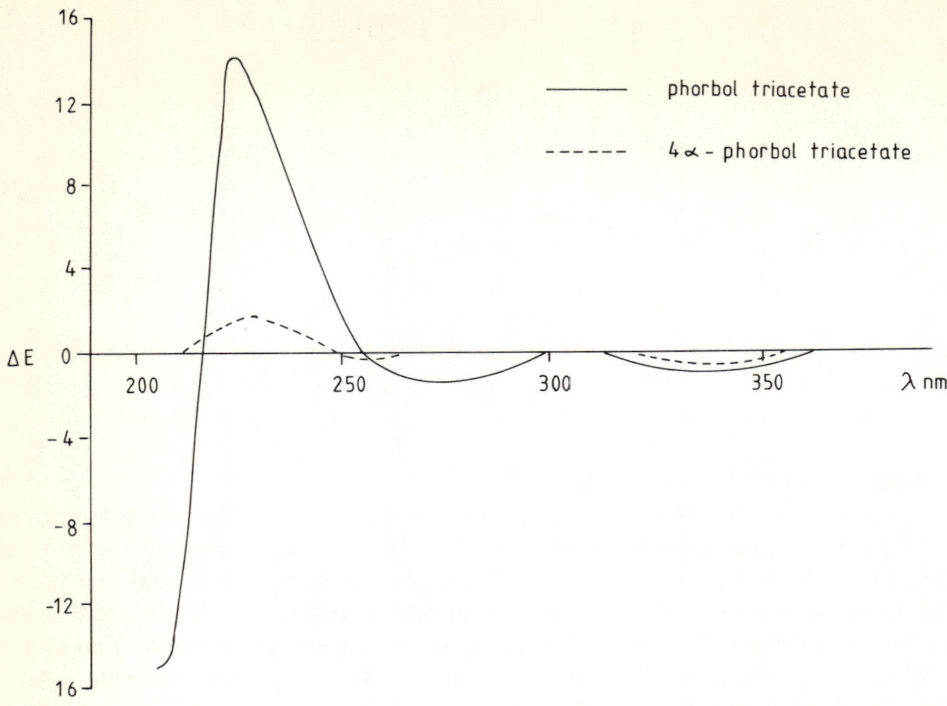

FIGURE 9. CD curves of 4α-phorbol-triacetate (**26**) and phorbol-triacetate (**2**), solvent methanol.

sidering the two series of phorbol-12,13-esters, some evidence concerning the positions of acylation may be obtained from their mass spectra. The spectra of these isomers are said to have distinct fragmentation patterns in that in the case of the "A" series of isomers the C-12 acyl moiety is characteristically eliminated from the molecular ion as the acyloxyradical (Figure 10), while in the case of the "B" series of esters the higher molecular weight acyl function is eliminated from the $M^{+\cdot}$ ion in the protonated RCOOH form. However, experience has demonstrated that the spectra of phorbol esters may be variable according to the conditions used and the distinction between the two groups of isomers can only with assurance be obtained by means of selective transesterification if the quantities of materials isolated permit. Softer ionization techniques have been used for the determination of the $M^{+\cdot}$ ion of phorbol esters. Field desorption (FD) MS is of particular use in this respect in that the $M^{+\cdot}$ and quasi $M^{+\cdot}$ ions are characteristically exhibited as the 100% relative abundance ions. Fragment ions in these spectra are uncommon and normally correspond to the loss of acyl functions from the quasi molecular ions.[99] Chemical ionization methods[87] (CIMS) are possibly the methods of choice for obtaining spectra of phorbol esters because they yield information pertaining to molecular weight and structure in one determination. In these spectra both $M^{+\cdot}$ and a quasi-$M^{+\cdot}$ ion are exhibited in relative abundance of greater than 20%, together with significant fragment ions originating from the $M^{+\cdot} - 1$ ion which are analogous to the fragment ions exhibited by EIMS. CIMS of phorbol-triacetate (**2**) and an ester of phorbol are shown in Figure 10. Fast atom bombardment[99] (FAB) methods of MS have also been attempted, but at present show no advantages over CIMS techniques for this group of compounds.

The nuclear magnetic resonance (NMR) spectra of phorbol esters have routinely been recorded and comparisons of such data are necessary for the identification of naturally occurring phorbol derivatives. the ^{13}C-NMR spectrum of phorbol-12,13,20-triacetate (**2**) was recorded by Neeman and Simmonds.[100] Signals in this spectrum were assigned by analogy to a series of model compounds in the normal manner, but the signals for the C-4 at 80

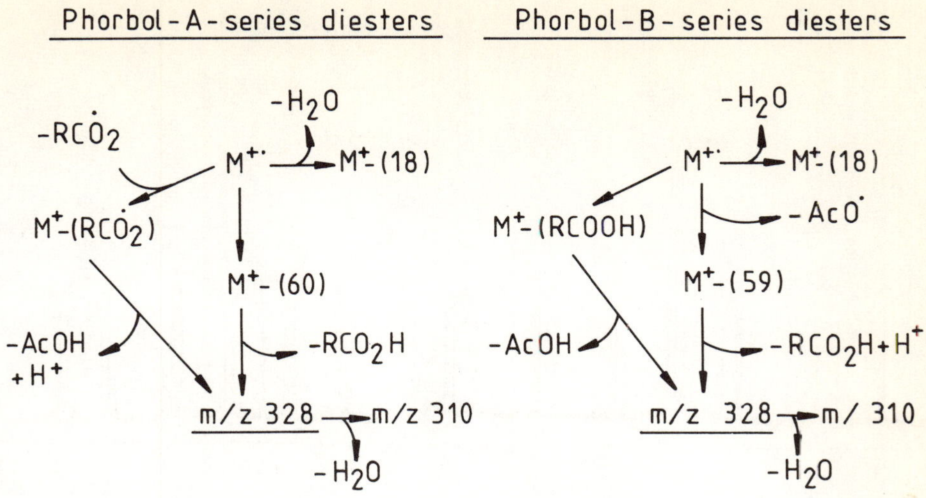

FIGURE 10. Fragmentation of A and B series esters by MS.

ppm and the C-9 at 75.3 ppm were only tentatively assigned. These were later reversed by Taylor et al.[101] following an investigation of several deoxyphorbol derivatives (Figure 11). The ^{1}H-NMR spectra of phorbol-esters are characteristic for this group of compounds. In the spectrum of phorbol 12,13,20-triacetate (**2**), the olifinic protons at C-1 and C-7 are exhibited at 7.54 ppm and 5.7 ppm, respectively, the former signal appearing as a distinct broad singlet and the latter as a doublet. The proton adjacent to the secondary hydroxyl at C-12 is exhibited downfield as a distinct doublet. Hydrolysis of the susceptible C-13 acyl function with 0.5 *M* KOH in methanol or NaOMe in methanol produces the corresponding C-13 hydroxyl derivative. In the ^{1}H-NMR of such C-12 esters of phorbol (**1**) there was a characteristic shift in the signal of the 1H-12 from about 5.63 ppm in the 12,13-diester to about 5.21 ppm in the monoester. This observation was confirmed by synthesis, and provides a relatively straightforward manner in which to assign esters of phorbol (**1**) to either the A or B series. The signal of the 2H allylic protons at C-20 are exhibited in the spectrum of phorbol-triacetate (**2**) at about 4.43 ppm and the chemical shift of this signal is characteristic of phorbol (**1**) triesters which exhibit a primary acyl function at the C-20 position. This acyl group is susceptible to hydrolysis with perchloric acid in dry methanol, producing the 12,13-diester in high yield. In the ^{1}H-NMR spectrum of the diester, the 2H signal for the allylic C-20 group is exhibited upfield at about 4 ppm when compared to the triesters. The ^{1}H-NMR spectrum of phorbol-12,13,20-triacetate (**2**) is shown in Figure 11, and the characteristic chemical shifts of 12,13-diesters and cryptic compounds from plant sources are summarized in Table 2. Phorbol (**1**) esters may be totally hydrolyzed by the action of alcoholic barium hydroxide[41] to the polyol (**1**). During the course of this reaction small quantities of the AB ring *cis* isomer 4α-phorbol (**20**) may be produced.[4] Although the mass spectrum, the IR, and UV spectra of (**20**) are similar to compounds (**1**), the ^{1}H-NMR spectra are quite distinct and provide a convenient means of distinguishing between the two isomers.[65,75] The chemical shifts of these two compounds are summarized in Table 3.

III. 12-DEOXYPHORBOL ESTERS

12-Deoxyphorbol (**53**) is one of the more common tigliane diterpenes to have been isolated from a relatively large number of plant species to date (Table 4). Compound (**53**) was characterized by the absence of the C-12 secondary hydroxyl group of phorbol (**1**) itself. Esters of 12-deoxyphorbol are widely distributed in species of the genus *Euphorbia*. Such

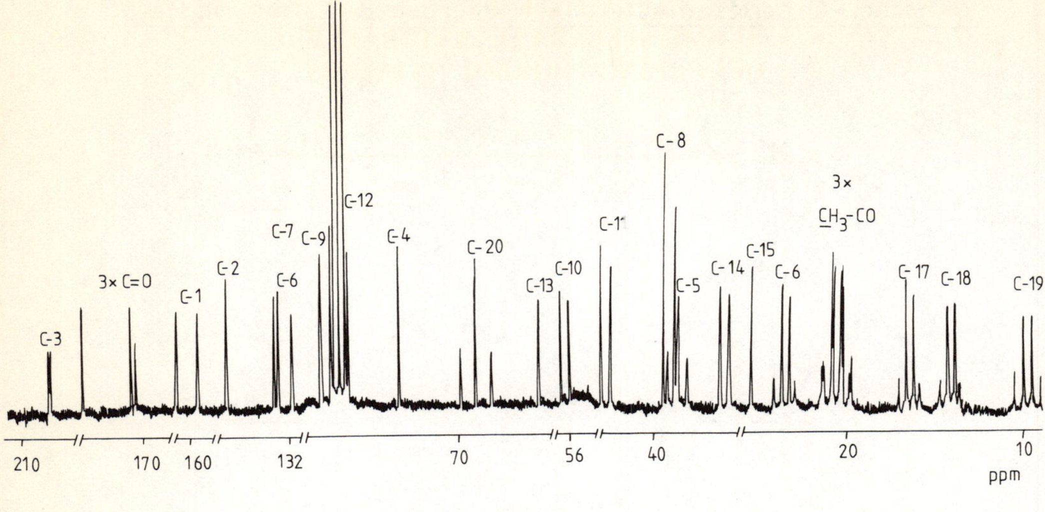

A

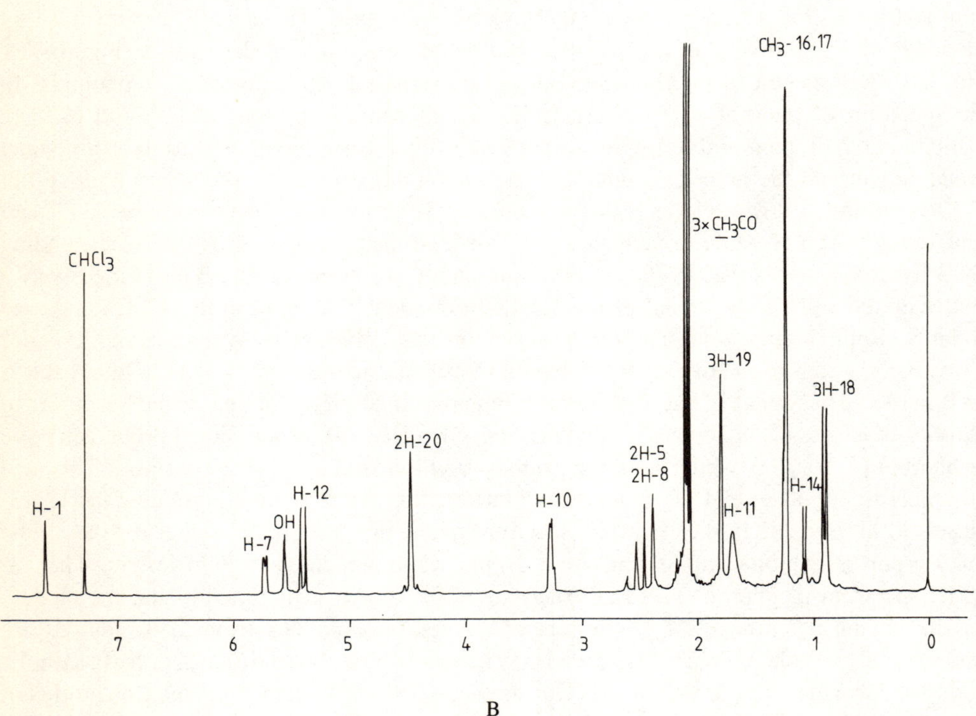

B

FIGURE 11. (A) ^{13}C-NMR spectrum of phorbol-12,13,20-triacetate (**2**). Solvent $CDCl_3$, 100.6 MHz. (B) ^{1}H-NMR spectrum of phorbol-12,13,20-triacetate (**2**). Solvent $CDCl_3$, 250 MHz.

compounds were initially isolated from a succulent species, *E. triangularis*[102-104] and later from eight other species of this genus.[105-112] Seven of these species were succulent African euphorbias,[106-112] but *E. helioscopia,* known as the ''sun spurge'', is an herbaceous variety common as a weed in European gardens.[105] As the result of a microchemical screen of about 60 species of the genus *Euphorbia,*[113] 12-deoxyphorbol (**53**) was detected as its synthetic diacetate in several species of the sections Euphorbium Bentham and Tithymalus Boiss.

Table 2
¹H-NMR SPECTRA OF NATURALLY OCCURRING PHORBOL (1) ESTERS

Protons of C no.	δ ppm for compounds (49—52) (49)	(50)	(51)	(52)
1	7.62	7.60	7.61	7.60
7	5.71	5.70	5.71	5.70
12	5.66	NR	NR	5.42
20	4.03	4.43	4.47	4.0
8	3.30	3.23	3.10	3.25
10	3.30	3.23	3.30	3.25
5	2.57	NR	2.57	2.52
19	1.77	1.78	1.80	1.78
16	1.25	1.28	1.23	1.28
17	1.25	1.28	1.23	1.28
18	0.93	NR	0.97	NR

(49) 12-*O*-[2-Methylaminobenzoyl]-phorbol-13-acetate, $CDCl_3$, 250 MHz.[87]

(50) 12-*O*-Dodecanoyl-13-*O*-acetyl-phorbol-20-linolenate, $CDCl_3$, 60 MHz.[86]

(51) 12-*O*-Isobutyl-13-*O*-acetyl-phorbol-20-angelate, $CDCl_3$, 60 MHz.[90]

(52) 12-*O*-Dodecanoyl-phorbol-13-acetate, $CDCl_3$, 60 MHz.[86]

Note: NR — not recorded. Signals for the acyl group protons are not included.

Table 3
¹H-NMR SPECTRA OF PHORBOL[65] (1) AND 4α-PHORBOL[75] (20), d_5-PYRIDINE/D_2O, 100 mHz

Protons of C no.	δ ppm of Phorbol (1)[65]	4α-Phorbol (20)[75]
1	7.88	7.42
7	6.17	5.85
12	5.03	4.92
20	4.29	4.38
5α	3.10 (5α, 5β)	3.95
5β		2.95
10	3.71	3.87
8	3.93	2.30
11	2.79	2.30
19	1.70 (19, 18, 16, 17)	1.82
18		1.58
16, 17		1.45
14	1.34	1.11

Table 4
NATURALLY OCCURRING ESTERS OF 12-DEOXYPHORBOL (53)

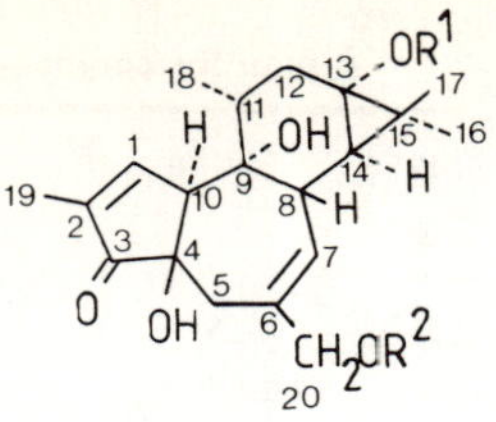

Species	R^1	R^2	Ref.
Euphorbia triangularis Desf.	$COCH(CH_3)CH_2CH_3$	H	102—104
	$COCH\cdot(CH_3)\cdot CH_3$	H	
	$COC(CH_3)\cdot CH\cdot CH_3$	H	
	$COCH(CH_3)CH_2\cdot CH_3$	$COCH_3$	
	$COCH(CH_3)\cdot CH_3$	$COCH_3$	
	$CO\cdot CH(CH_3)CH_2CH_3$	$COCH_3$	
E. resinifera Berg.	$COC(CH_3)C(CH_3)_2$	$COCH_3$	106
	$COCH(CH_3)\cdot CH_3$	$COCH_3$	
	$COCH_2C_6H_5$	$COCH_3$	
E. fortissima Leach.	$CO\cdot C(CH_3)CH\cdot CH_3$	H	107
	$COCH(CH_3)CH_3$	$COCH_3$	
	$COC(CH_3)CH\cdot CH_3$	$COCH_3$	
	$COCH(CH_3)CH_2\cdot CH_3$	$COCH_3$	
	$CO(CH_2)_{10}\cdot CH_3$	H	
	$COC_{11}H_{21}$	H	
	$COC_{11}H_{21}$	$COCH_3$	
E. coerulescens Haw.	$COCH(CH_3)CH_2CH_3$	H	108, 120
	$COC(CH_3)CH\cdot CH_3$	$COCH_3$	
	$COCH(CH_3)CH_2CH_3$	$COCH_3$	
	COC_6H_{13}	H	
	$CO(CH_2)_{10}\cdot CH_3$	H	
	$CH(CH_2)_{10}CH_3$	$COCH_3$	
E. polyacantha Bois.	COC_7H_{13}	$COCH_3$	109
	COC_9H_{15}	$COCH_3$	
E. poissonii Pax.	$COCH_2C_6H_4OH$	H	110, 121—123
	$COCH_2C_6H_5$	H	
	$COCH(CH_3)CH_3$	H	
	$COCH(CH_3)CH_2CH_3$	H	
	$COC(CH_3)CH\cdot CH_3$	H	
	$COCH_2C_6H_4OH$	$COCH_3$	
	$COCH_2C_6H_5$	$COCH_3$	
	$COCH(CH_3)CH_3$	$COCH_3$	
	$COC(CH_3)CH\cdot CH_3$	$COCH_3$	
	$COCH(CH_3)CH_2\cdot CH_3$	$COCH_3$	
E. helioscopia L.	$COC(CH_3)CH\cdot CH_3$	H	105
	$COCH_2C_6H_5$	$COCH_3$	
	$COC(CH_3)CHCH_3$	$COCH_3$	
	$COC_{11}H_{19}$	$COCH_3$	
E. unispina N.E.Br.	$COCH_2C_6H_4OH$	$COCH_3$	111
	$COCH_2C_6H_5$	$COCH_3$	
	$COC(CH_3)CH\cdot CH_3$	$COCH_3$	
	$COC(CH_3)CH_2CH_3$	$COCH_3$	
	$COCH_2C_6H_5$	H	
	$COC(CH_3)CHCH_3$	H	
	$COCH(CH_3)CH_2CH_3$	H	

Table 4 (continued)
NATURALLY OCCURRING ESTERS OF 12-DEOXYPHORBOL (53)

Species	R¹	R²	Ref.
E. balsamifera Aiton.	$COC_{11}H_{19}$	COC_7H_{11}	112
Baliospermum montanum	$CO(CH_2)_{14}CH_3$	H	115
Croton californicus	$CO(CH_2)_8CH_3$	$CO(CH_2)_8CH_3$	
	$CO(CH_2)_8CH_3$	$CO(CH_2)_{10}CH_3$	
	$CO(CH_2)_8CH_3$	$CO(CH_2)_{12}CH_3$	
	$CO(CH_2)_8CH_3$	$CO(CH_2)_{14}CH_3$	114
Pimelea prostrata (Thymelaeaceae)	$COCH_3$	H	116,124, 125

Esters of 12-deoxyphorbol (**53**) were later isolated from two other genera of the plant families Euphorbiaceae, *Croton,*[115] and *Baliosperum.*[115] Unlike phorbol (**1**) esters, 12-deoxyphorbol (**53**) was also isolated from *Pimelea prostrata,*[116] a member of the family Thymelaeaceae.

12-Deoxyphorbol (**53**) esters occur as two series of compounds in these plants, there being the C-13 tertiary monoesters and the 13-20-diesters. The primary acyl group at C-20 is present as acetate in most 13,20-diesters of this parent polyol. Longer chain, higher molecular weight residues at this position were initially detected from the latex of *E. balsamifera,*[112] but the structure of this ester was not completely elucidated due to paucity of material isolated. At a later date, Chavez and others[114] characterized a series of four 13,20-diesters of 12-deoxyphorbol (**53**) from *Croton californicus* which exhibited C10 to C16 carbon number acyl residues at C-20 of the polyol nucleus. More commonly, the 13,20-diesters of 12-deoxyphorbol (**53**) exhibit an aliphatic acyl residue at C-13, which varies in carbon number from C4 to C16. Aromatic acyl moieties including phenylacetate and *p*-hydroxyphenylacetate have also been identified as the esterifiying group of the C-13 position of 12-deoxyphorbol (**53**). 12-Deoxyphorbol phenylacetate, a C-13 monoester isolated from *Euphorbia resinifera, E. poissonii, E. helioscopia,* and *E. unispina,* has been of particular interest in that it was the most potent of this series of compounds when tested for proinflammatory activity[117] on mouse ears and the most potent compound used for the induction of human blood platelet aggregation.[118] This substance has been used for studies involving the mechanism of action of phorbol-ester-induced platelet aggregation[20] and has been shown to be responsible for the secretion of a biologically active substance from platelets which is not ADP, 5-HT, or prostaglandin related.[119]

The parent diterpene, 12-deoxyphorbol (**53**), is less stable during isolation[102] than is phorbol (**1**), but may be isolated as its diacetate. The esters of 12-deoxyphorbol within each group are difficult to separate by conventional techniques in that migration by adsorption chromatographic methods is dependent upon the presence or absence of a primary hydroxyl function at C-20. Partition methods similar to those used for the separation of esters of phorbol (**1**) have been used for the isolation of individual compounds within the two groups. These methods include countercurrent distribution,[102] partition-TLC,[96] and HPLC.[114] 12-Deoxyphorbol esters are obtained as resinous derivatives from such separations, which should be stored under nitrogen in an analogous manner to the phorbol (**1**)-esters. (See Chapter 1.)

A number of derivatives of 12-deoxyphorbol (**53**) involving hydroxylation or substitution of hydroxy groups in the tigliane nucleus have been identified from plants of the family Euphorbiaceae. The first of these new tigliane diterpenes was 12-deoxy-16-hydroxyphorbol (**54**) (Table 5). This compound in its esterified form was originally isolated from the latex

Table 5
ESTERS OF 12-DEOXY-16-HYDROXYPHORBOL (54)

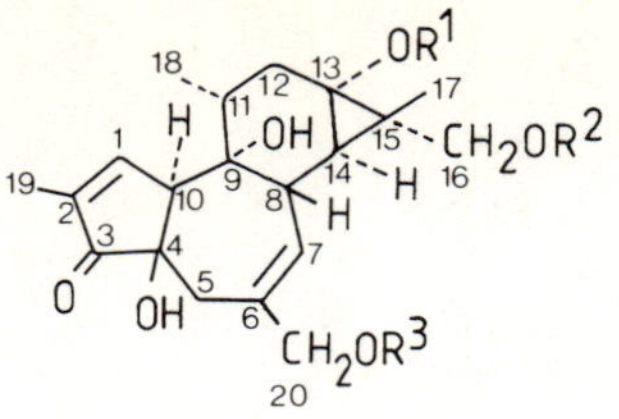

Species	R^1	R^2	R^3	Ref.
Euphorbia cooperi N.E.Br.	$COC(CH_3)CHCH_3$	$COCH(CH_3)CH_3$	H	103, 126
	$COC(CH_3)CHCH_3$	$COCH(CH_3)CH_3$	$COCH_3$	127
E. poissonii Pax.	$COCH_2C_6H_5$	$COCH(CH_3)CH_2CH_3$	H	
	$COCH_2C_6H_5$	$COCH(CH_3)CH_2CH_3$	$COCH_3$	128
E. triangularis Desf.	$COC(CH_3)CH \cdot CH_3$	$COCH(CH_3)CH_3$	$COCH_3$	104
E. unispina N.E.Br.	$COCH_2C_6H_5$	$COC(CH_3)CH \cdot CH_3$	H	
	$COCH_2C_6H_5$	H	H	129
Baliospermum montanum	$CO(CH_2)_{14} \cdot CH_3$	H	H	115

of *Euphoriba cooperii.*[103,126,127] These esters differ from those of (**53**) in that they are characterized by the presence of an extra primary hydroxyl group at C-16 of the nucleus. Consequently, esters of 12-deoxy-16-hydroxyphorbol (**54**) are unstable to strong alkali and give rise to the formation of crotophorbolone (**31**) (Figure 5) as an artifact of total hydrolysis.[126] The C-13 tertiary ester function of the toxic tigliane derivatives of *E. cooperii* was identified as angelate, the primary C-16 acyl function as isobutyrate, while the C-20 position was either acetylated or free hydroxyl. Similar di- and triesters of 12-deoxy-16-hydroxyphorbol (**54**) were at a later date obtained from the latices of *E. triangularis,*[104] *E. poissonii,*[128] and *E. unispina.*[129] From the genus *Baliospermum,* also of the family Euphorbiaceae, a monoester of this parent polyol was separated as well.[115] This compound (Table 5) was acylated with a 16-carbon fatty acid at the C-13 position of 12-deoxy-16-hydroxyphorbol (**54**). *Jatropha* is a genus of the family Euphorbiaceae common to parts of India, Africa, and South America. From this genus antitumor macrocyclic diterpenes (see Chapter 6) have been isolated. Recently, Adolf and others[133] isolated a series of C-13 monoesters from this plant. The nature of the acyl groups of two distinct groups of inseparable C-13 monoesters was not determined beyond establishing their character as high mol wt, highly unsaturated fatty acids. These two groups of mixtures were isolated from *J. curcas* and *J. gossypifolia,* both of which are well known from traditional medicine to be irritant and toxic plants. However, an early investigation of this genus from species collected in India had previously failed to produce compounds of the tigliane type.[131] The nature of the parent diterpene of the *Jatropha* irritant factors was established by the fact that on reduction with lithium aluminum hydride in diethyl ether followed by acetylation with acetic anhydride in pyridine 3-deoxo-12-deoxy-3ξ,16-dihydroxyphorbol-3,13,16,20-tetraacetate (**55**) was produced (Figure 12). It had previously been established that transesterification of 12-deoxyphorbol-16-hydroxylphorbol (**54**) C-13 monoesters did not yield the parent diterpene, and further that strong basic hydrolysis produced crotophorbolone (**31**). However, using mild conditions of acid catalyzed transesterification, Schmidt and others[128,129] were able to consign the positions of acylation of di- and triesters of this polyol (**54**) in combination with partial synthetic methods. Further 12-deoxyphorbol (**53**) analogues have been obtained from plants. These compounds were esters

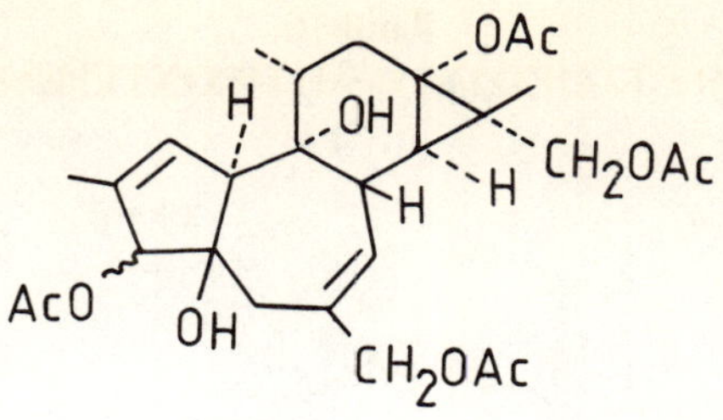

FIGURE 12. 12-Deoxy-3ξ,16-dihydroxyphorbol-3,13,16,20-tetraacetate (**55**).

Table 6
ESTERS OF 12,20-DIDEOXYPHORBOL (56)

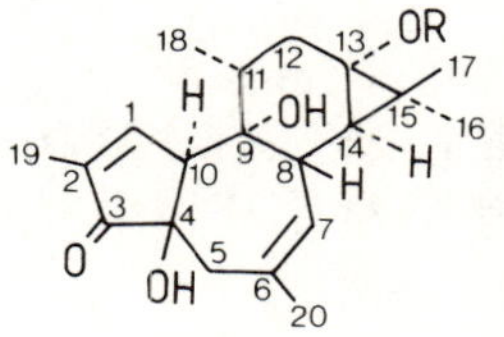

Species	R	Ref.
Euphorbia resinifera Berg.	$COCH(CH_3) \cdot CH_3$ $COC(CH_3) \cdot CH{\cdot}CH_3$	132

of novel tigliane diterpenes, 12,20-dideoxyphorbol (**56**) (Table 6), 12-deoxy-5-hydroxyphorbol (**57**) (Table 7), and 12-deoxy-5-hydroxy-6,7-epoxyphorbol (**58**) (Table 8). Esters of compound (**56**) were originally isolated from *Euphorbia resinifera,*[132] but at a later date esters of (**57**) were obtained from *Baliosperum montanum*[115] and *Stillingia sylvatica.*[133] Derivatives of compound (**58**) were isolated from *Hippomane mancinella,*[134] *B. montanum,*[115] and *Pimelea prostrata.*[125] From *H. mancinella*[134] an ester known as mancinellin which exhibited a 6,7-epoxy group was separated. A 16-carbon unsaturated aliphatic acyl group was shown to be present at the C-13 position of mancinellin. A further ester of (**58**) was also isolated from *B. montanum,*[115] but in this instance the C-13 aliphatic acyl moiety was a saturated fatty acid.

IV. 4-DEOXYPHORBOL ESTERS

4-Deoxyphorbol (**59**) is the least stable of the tigliane diterpenes. This compound readily epimerizes to its AB cis isomer 4-deoxy-4α-phorbol (**60**) when exposed to acid or alkaline conditions. Compound (**60**) was identified as a hydrolysis product of croton oil and can readily be synthesized from phorbol (**1**) by reaction with zinc in acetic acid.[75] The inherent instability of (**59**) even to the acidic conditions of the oil of *Croton* led to the speculation that esters of this polyol were converted in the oil to derivatives of 4-deoxy-4α-phorbol (**60**), the more stable epimer.[4] More recently, esters of both epimers have been isolated from natural sources and it would appear that both tiglianes are present in the plant as natural products. A group of highly irritant 12,13-diesters of (**59**) were originally isolated from the latex of *Euphorbia tirucalli.*[125-137] Like the 12,13-diesters of phorbol (**1**) these compounds occurred as isomers of the 12,13-glycol groups, where in one series the longer chain acyl group was present at C-12 and on acetate at C-13, and in the second series the positions

Table 7
ESTERS OF 12-DEOXY-5-HYDROXYPHORBOL (57)

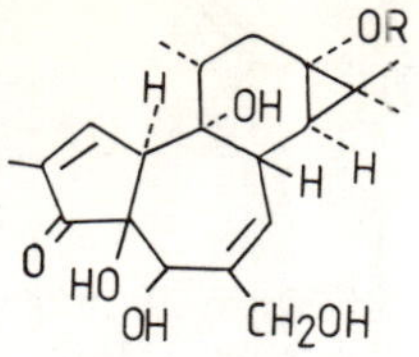

Species	R	Ref.
Baliosperum montanum	$CO(CH_2)_{12} \cdot CH_3$	115
Stillingia sylvatica L.	$CO \cdot (CH_2)_{14} \cdot CH_3$	133

Table 8
ESTERS OF 12-DEOXY-5-HYDROXY-6,7-EPOXYPHORBOL (58)

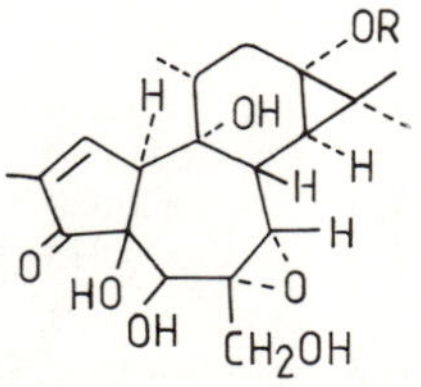

Species	R	Ref.
Hippomane mancinella	$CO \cdot (CH{=}CH)_3(CH_2)_8 \cdot CH_3$	134
Baliospermum montanum	$CO \cdot (CH_2)_{10} \cdot CH_3$	115
Pimelea prostrata	$CO \cdot (CH_2)_{12} \cdot CH_3$	125

were reversed. The higher molecular weight acyl function ranged in carbon number from C_8 to C_{14}. Also in an analogous manner to the phorbol (**1**) esters, triesters of the 12,13,20-hydroxyl groups were later obtained from *E. biglandulosa.*[138] The extra C-20 primary acyl derivative was identified as either acetate, propionate, or α-methylbutyrate. Recently, a 12,13-diester of (**59**) was isolated from the ripe fruits of *Sapium indicum.* This compound has previously been shown to occur in *E. tirucalli* latex.[136] Also from *S. indicum,* an unusual blue fluorescent diester of 4-deoxyphorbol (**59**) was separated.[140] This compound, sapintoxin-A was characterized by the presence of a 2-methylaminobenzoate moiety at C-12 of the tigliane nucleus (Table 9).

4-Deoxy-4α-phorbol (**60**) esters were detected in the oil of *Croton tiglium* at an early stage of its complete chemical evaluation.[4] Two further esters of (**60**) were later isolated from the latex of *E. tirucalli,*[135,136] and an aromatic ester known as sapinine was obtained from *S. indicum* by Miana et al.[141] and from *S. insigne* by Taylor and others[150] (Table 10). Further tigliane derivatives based upon 4-deoxyphorbol (**59**) have been isolated from species of the plant family Euphorbiaceae. These esters are derivatives of three novel tiglianes, 4-deoxy-16-hydroxyphorbol (**61**), 4-deoxy-5-hydroxyphorbol (**62**) and 4,20-dideoxy-5-hydroxyphorbol (**63**). Esters of 4-deoxy-16-hydroxyphorbol (**61**) were isolated from *C. flavens,*[142,143] and *S. insigne*[50] in the form of 12,13-diesters; a 12,13,20-triester was also

Table 9
ESTERS OF 4-DEOXYPHORBOL (59)

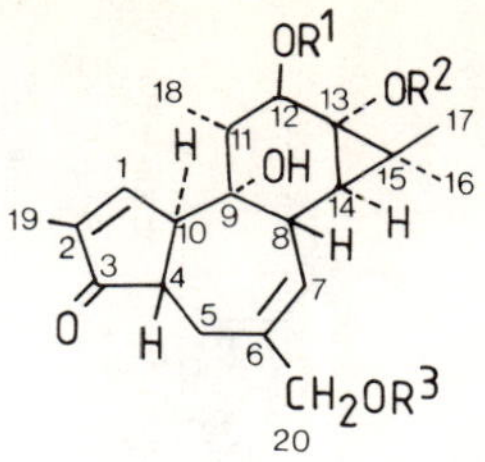

Species	R^1	R^2	R^3	Ref.
Euphorbia tirucalli	$COCH_3$	$CO(CH{=}CH)_n(CH_2)_2CH_3$ n = 2, 3, 4, 5	H	
	$COCH_3$	$CO(CH{=}CH)_n(CH_2)_2CH_3$ n = 1, 2, 3, 4	H	
	$CO(CH{=}CH)_n(CH_2)_2CH_3$ n = 2, 3, 4, 5	$COCH_3$	H	
	$CO(CH{=}CH)_n \cdot (CH_2)_2 \cdot CH_3$ n = 1, 2, 3, 4	$COCH_3$	H	135,136
Sapium indicum	$CO(CH{=}CH)_3(CH_2) \cdot CH_3$	$COCH_3$	H	139
	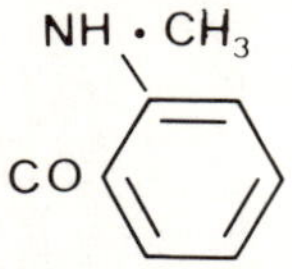 CO–C₆H₄–NH · CH₃ (N-methylanthraniloyl; structure)	$COCH_3$	H	140
E. biglandulosa	$CO(CH{=}CH)_2(CH_2)_2CH_3$	$COCH_3$	$COCH_3$	
	$CO(CH{=}CH)_2(CH_2)_2CH_3$	$COCH_2CH_3$	$COCH_3$	
	$CO(CH{=}CH)_2(CH_2)_2CH_3$	$COCH(CH_3)CH_3$	$COCH_3$	138

Table 10
ESTERS OF 4-DEOXY-4α-PHORBOL (60)

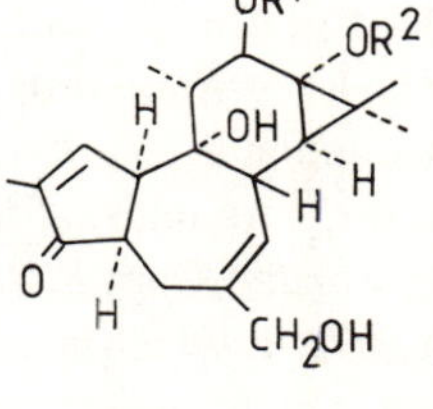

Species	R^1	R^2	Ref.
Euphorbia tirucalli	$CO(CH{=}CH)_3(CH_2)_2CH_3$	$COCH_3$	
	$COCH_3$	$CO(CH{=}CH)_4(CH_2)_4CH_3$	135, 136
Sapium indicum	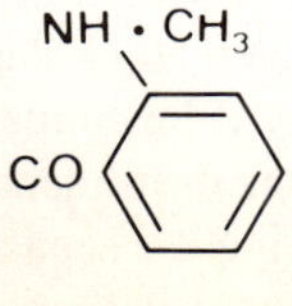 CO–C₆H₄–NH · CH₃ (N-methylanthraniloyl; structure)	$COCH_3$	141
S. insigne	$CO(CH_2)_4 \cdot CH_3$	$COCH_3$	150

Table 11
FURTHER DERIVATIVES OF 4-DEOXYPHORBOL (59)

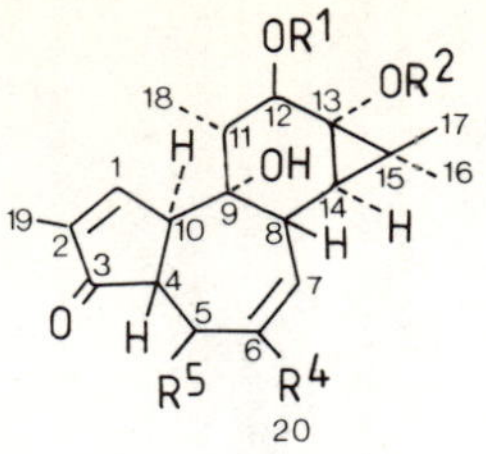

Species	R¹	R²	R³	R⁴	R⁵	Ref.
Croton flavens	$CO(CH_2)_{14}CH_3$	$COCH_3$	OH	CH_2OH	H	
	$CO(CH_2)_{14}CH_3$	$COCH_3$	OH	$CH_2OCO(CH_2)_8CH_3$	H	142,143
Sapium indicum	$CO(CH{=}CH)_3(CH_2)_2CH_3$	$COCH_3$	H	CH_2OH	OH	139
	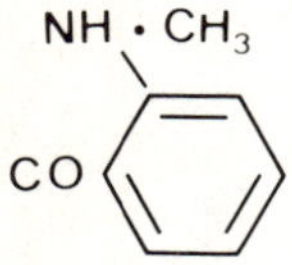 CO, $NH \cdot CH_3$ (ring structure)	$COCH_3$	H	CH_2OH	OH	144
	$CO(CH{=}CH)_3(CH_2)_3 \cdot CH_3$	$COCH_3$	H	CH_3	OH	139
	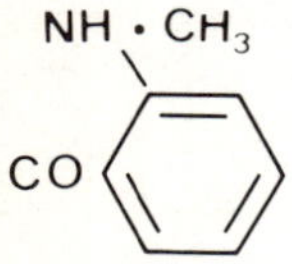 CO, $NH \cdot CH_3$ (ring structure)	$COCH_3$	H	CH_3	OH	144
S. insigne	$CO(CH{=}CH)_2(CH_2)_4CH_3$	$COCH_3$	OH	CH_2OH	H	150

obtained from the *Croton* species (Table 11). The extra primary hydroxyl group at C-16 of the nucleus was free of esterification in both cases. Two esters of 4-deoxy-5-hydroxyphorbol (**62**) were recently obtained from the fruits of *S. indicum.* the first of these compounds exhibited an aliphatic acyl moiety at C-12 of (**62**) while the second compound exhibited 2-methylaminobenzoate at this position. Similar esters of 4,20-dideoxy-5-hydroxyphorbol (**63**) were obtained from the same species.[139,144] An unusual tigliane ester, based on 4-deoxy-4α-16-hydroxyphorbol (**67**) was recently isolated from *S. insigne.*[150] This compound occurred in the form of two 12,13-diesters of (**67**) in which the saturated longer chain acyl groups were exhibited at C-12 and a common acetate moiety was present at C-13.

Phorbol (**1**) and 4-deoxyphorbol (**59**) are readily oxidized at the C-20 primary hydroxyl position to furnish their C-20 aldehydes, respectively (Table 12). It is possible that autoxidation of naturally occurring esters at this position would result in the presence of C-20 aldehydes in plant extracts. However, using carefully controlled extraction and separation methods at low temperature and vigorously excluding oxygen at all stages, a 12,13-diester of 4-deoxyphorbaldehyde (**64**) and its corresponding 4-deoxy-4α-epimer (**65**) were isolated from the fruits of *S. indicum.*[145] Chromatographic monitoring of the separation at each stage indicated that these C-20 aldehydes were not produced during the separation procedure and that they were possibly natural products. Alternatively, these derivatives may have been formed during postharvest autoxidation in the fruits themselves. The structures of these

Table 12
ESTERS OF 4-DEOXY-4α-PHORBALDEHYDE (65) AND 4-DEOXYPHORBALDEHYDE (64)

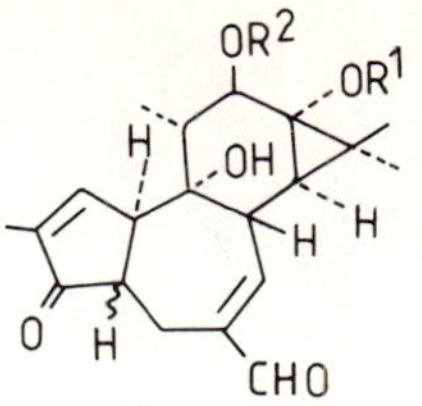

Species	R^1	R^2	Ref.
Sapium indicum	$COCH_3$	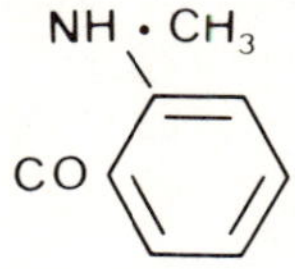	145

epimeric esters were confirmed by reduction with sodium borohydride to 4-deoxyphorbol (**59**) and 4-deoxy-4α-phorbol (**60**), respectively.[145]

V. 16-HYDROXYPHORBOL ESTERS

Okdua and others[146,147] in 1974 detected a highly oxygenated tigliane derivative from *Aleurites fordii* of the plant family Euphorbiaceae. This compound was identified as 16-hydroxyphorbol (**66**) from its spectral characteristics in the acetylated form as 13-acetyl-16-hydroxyphorbol (**67**). Two esters of 16-hydroxyphorbol (**66**) were obtained from *Aleurites fordii.* The first was a 12,13-diester of (**66**) which exhibited a long-chain acyl group at C-12 and an acetate at C-13, while the second compound was a monoacetate of the C-13 position (**67**). Two similar 12,13-diesters of this tigliane polyol were later isolated from *Croton flavens*[142,143] together with two new 12,13,20-triesters of 16-hydroxyphorbol (**66**). The C-20 acyl moiety in both cases was a saturated 10-carbon aliphatic acid (Table 13). The fact that esters of 16-hydroxyphorbol (**66**) might be expected to occur more widely in genera of the family Euphorbiaceae was emphasized by the recent isolation of esters of this diterpene from two *Jatropha* species.[130] The seed oil of *J. multifida* was fractionated into two factors, M_1 and M_2, by multistage Craig distribution and TLC methods. Both factors were shown to be mixtures of inseparable esters containing highly unsaturated (6,7 double bond equivalents) acids. From the oil of the seeds of *J. podagrica,*[130] a similar factor P_1 was also obtained. The nature of the esterifying acids in these mixtures was not established.

VI. IDENTIFICATION OF DEOXYPHORBOL AND HYDROXYPHORBOL ESTERS

Esters of deoxyphorbol and hydroxyphorbol may be identified after isolation from plant extracts by means of a combination of selective hydrolysis and spectroscopic methods. Total hydrolysis to the respective tigliane polyols was achieved in most cases using 1% sodium methoxide in dry methanol over a period of up to 24 hr. These conditions are required to remove the stable C-12 secondary acyl group. Because most of these tigliane diterpenes are unstable, the nucleus has been identified after the synthesis of their respective acetates following reaction in acetic anhydride and pyridine (2/1). Esters of 12-deoxyphorbol (**53**)

Table 13
ESTERS OF 16-HYDROXYPHORBOL (66)

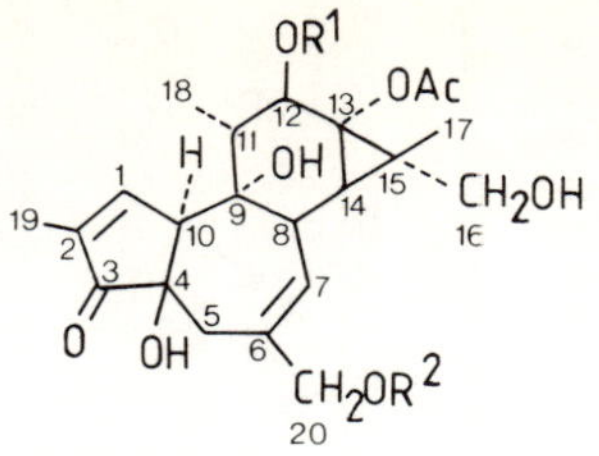

Species	R^1	R^2	Ref.
Aleurites fordii Hemsl.	$CO(CH_2)_{14} \cdot CH_3$	H	146,147
	H	H	
Croton flavens L.	$CO(CH_2)_{14}CH_3$	H	
	$CO(CH_2)_{14}CH_3$	$CO(CH_2)_8CH_3$	
	$CO(CH_2)_{12}CH_3$	H	
	$CO(CH_2)_{12}CH_3$	$CO(CH_2)_8CH_3$	142,143

are notoriously unstable during hydrolysis and the 13,20-diacetate of this compound is normally obtained in less than 40% yield.[102-104] during strong alkaline reaction it has been noted that 12-deoxy-16-hydroxyphorbol (**54**) esters undergo ring D opening to produce crotophorbolone (**31**).[120] 4-Deoxyphorbol (**59**) esters, however, readily undergo alkaline hydrolysis reactions to produce 4-deoxy-4α-phorbol (**60**) acetate after acetylation as a single product.[135] Accordingly, total hydrolysis of these AB ring *trans* tiglianes results in the epimerization of the 4β-orientated proton to form the 4α-epimer. In the case of esters of 16-hydroxyphorbol (**66**), 0.1 *M* sodium methoxide treatment followed by acetylation produced the tetraacetate of (**66**) within 1 hr.[130] As a result of these reactions it is not only possible to identify the parent tigliane by isolation from the reaction mixture, but also to obtain the acyl derivatives, normally following methylation, in pure form for identification by a combination of gas-liquid chromatography (GLC) and MS.[148] The spectral characteristics of some typical tigliane acetates are shown in Figure 13.

Naturally occurring esters of these tigliane diterpenes which exhibit a primary acyl moiety at C-20 of the nucleus are susceptible to acid-catalyzed transesterification by means of perchloric acid in methanol to produce the C-20 hydroxyl analogues in a manner similar to the triesters of phorbol (**1**).[84,85] The tertiary acyl group at C-13 of these compounds is susceptible to akaline-catalyzed transesterification in 0.1*M* KOH in dry methanol,[149] although in the case of 4-deoxy derivatives epimerization to the 4-deoxy-4α-analogues is expected both in acid and mild alkaline conditions.[149] Using a combination of acid and alkaline transesterifications it was possible to assign the positions of acylation of deoxy- and hydroxyphorbols by isolation of their reaction products.

The mass spectra of deoxy- and hydroxyphorbol esters have been of particular assistance for the identification of the nature of the tigliane nucleus and the mol wt of the acyl substituents. In the case of the 12-deoxyphorbol (**53**) esters, ions above m/z 294 are characteristic and assist in the identification of the acyl substituents; below the ion m/z 294 the spectra of 12-deoxyphorbol (**53**) esters are similar and not particularly characteristic, with the exception of the ion at m/z 212. The genesis of this fragment ion as determined by Chavez and co-workers[114] was verified by high-resolution mass measurements and the scheme shown in Figure 14 was suggested as a possible mechanism. The esters of 12-deoxyphorbol (**53**) from the mono- and diesterified series of compounds may be characterized from their

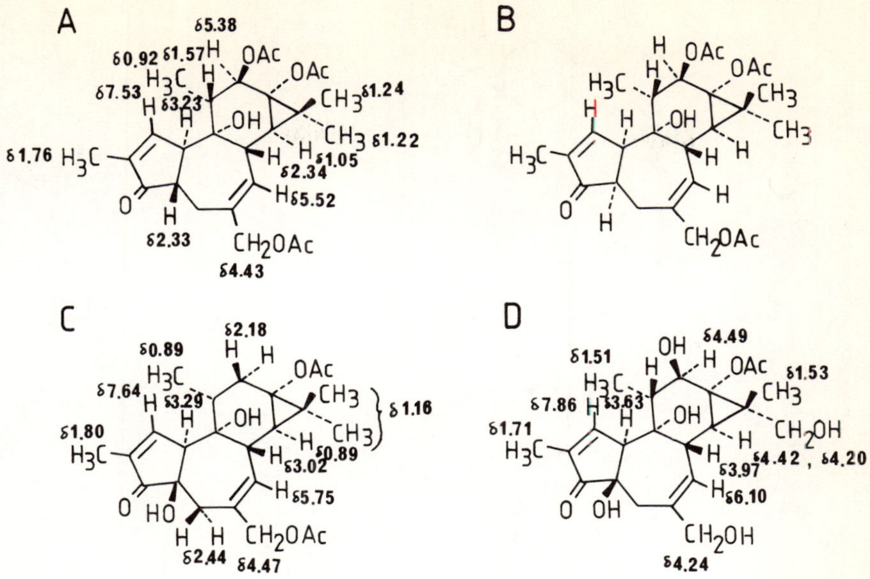

FIGURE 13. Spectral data of some tigliane hydrolysis and acetylation products.

A. 4-Deoxyphorbol (**59**) triacetate $C_{28}H_{34}O_6$;
MS $M^{+\cdot}$ m/z 474
UV λ_{max}^{MeOH} (ϵ) nm 198 (11,600), 230 (6050), 310 (140)
IR ν_{max}^{KBr} cm^{-1}, 3145, 1745, 1730, 1710, 1635
CD λ_{max}^{EtOH} nm ($\Delta\epsilon$), 202 (−1943), 241, (+3.10), 318, (−2.02)
^{1}H-NMR ($CDCl_3$) according to Reference 135

B. 4-Deoxy-4α-Phorbol (**60**)-triacetate $C_{28}H_{34}O_6$, MS, $M^{+\cdot}$ m/z 474

C. 12-Deoxyphorbol (**53**)-diacetate $C_{24}H_{32}O_7$; MP 138°C.
MS $M^{+\cdot}$ m/z 432.
UV γ_{max}^{MeOH} nm (ϵ), 196 (12,300), 235 (5200), 334.
IR υ_{max}^{KBr} cm^{-1} 3400, 1715, 1700, 1628.
CD $\gamma_{max}^{dioxane}$ nm ($\Delta\epsilon$), 276 (−0.194), 334 (−0.583), 343 (−0.61).
^{1}H-NMR ($CDCl_3$ 60 MHz) assigned according to Reference 121.

D. 13-Acetyl-16-hydroxyphorbol (**67**), m.p. 278—282°C.
MS m/z 422 ($M^{+\cdot}$, $C_{22}H_{30}O_8$).
UV γ_{max}^{MeOH}, nm (log ϵ), 232 (3.70).
IR υ_{max}^{KBr} cm^{-1}, 3550, 3500, 3400, 3250, 1705, 1695, 1630.
^{1}H-NMR (C_5D_5N) assigned according to Reference 147.

fragment patterns.[96] In the case of the monoesters of this parent polyol, the fragment ion at m/z 230 is derived from the molecular ion by the loss of the C-13 acyl substituent in the carboxyl form. A typical companion ion at m/z 312 is additionally exhibited in these spectra due to the loss of the elements of water from the fragment of m/z 330. The diesters of 12-deoxyphorbol (**53**) exhibit fragment ions in their mass spectra corresponding to the loss of both the C-13 and C-20 acyl moieties in the carboxyl form RCOOH. The ion at m/z 312 in these spectra accordingly corresponds to the deacylated tigliane nucleus, $C_{20}H_{24}O_3$, as determined from accurate mass measurements.[96] The fragmentation of both the mono- and diesters of 12-deoxyphorbol (**53**) in the region m/z 294 and above are summarized in Figure 15. In the mass spectra of 12-deoxy-16-hydroxyphorbol (**54**) there is exhibited a sequential loss of acyl residues to form a fragment ion at m/z 328 corresponding to the deacylated tigliane nucleus in an analogous manner to phorbol (**1**) diesters. However, in the spectra of esters of (**54**) the two acyl moieties are eliminated in the protonated form[129] (Figure 16).

Commonly in the EIMS of 4-deoxyphorbol (**59**) and 4-deoxy-4α-phorbol (**60**) esters the

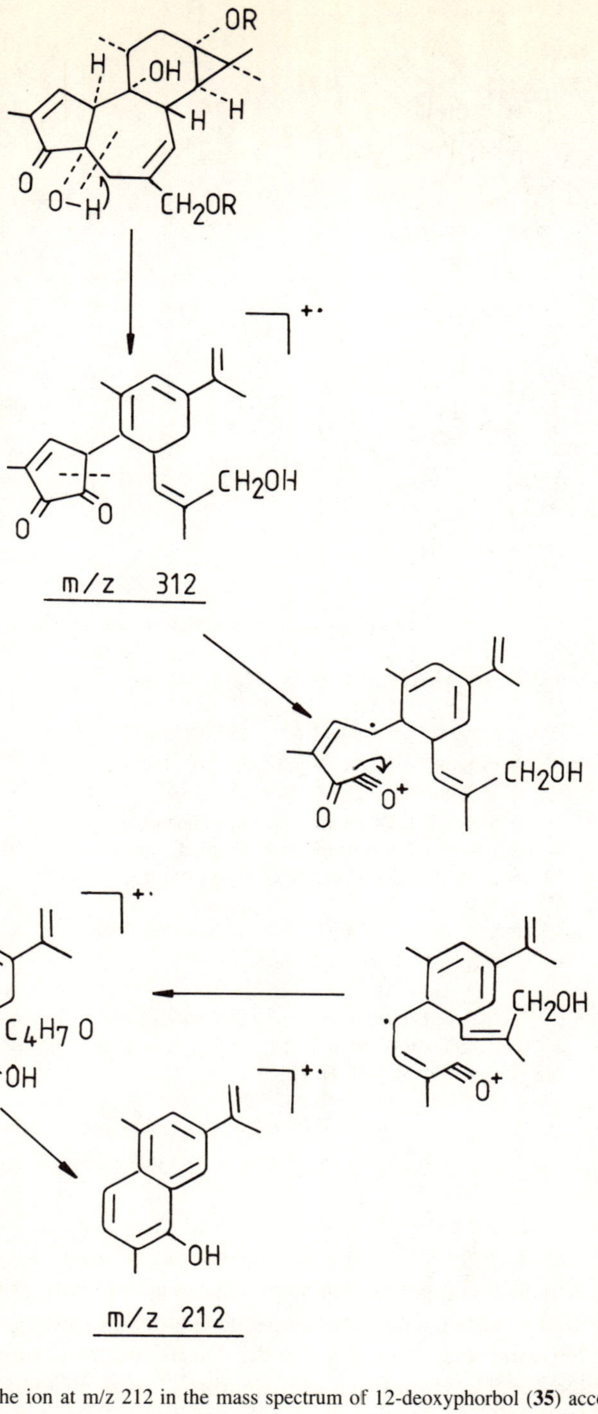

FIGURE 14. Genesis of the ion at m/z 212 in the mass spectrum of 12-deoxyphorbol (**35**) according to Chavez and others.[114]

highest mass fragment ion is the $M^{+\cdot}$-18. When $M^{+\cdot}$ ions are observed in these spectra, they are commonly of less than 1% relative abundance but may be observed up to 4% relative abundance, depending upon the conditions used.[131] CIMS is of particular assistance in the determination of the mol wt of 4-deoxyphorbol (**59**) esters in that both $M^{+\cdot}$ and quasi-$M^{+\cdot}+1$ ions are exhibited in high relative abundance.[131] By means of CIMS the esters of

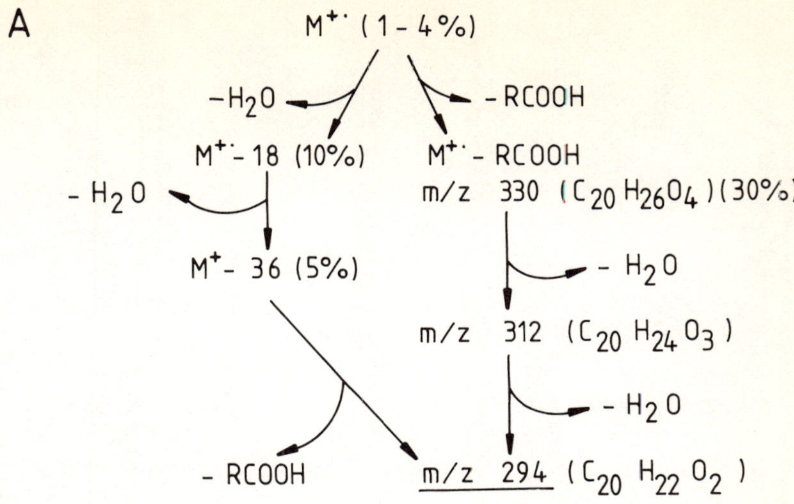

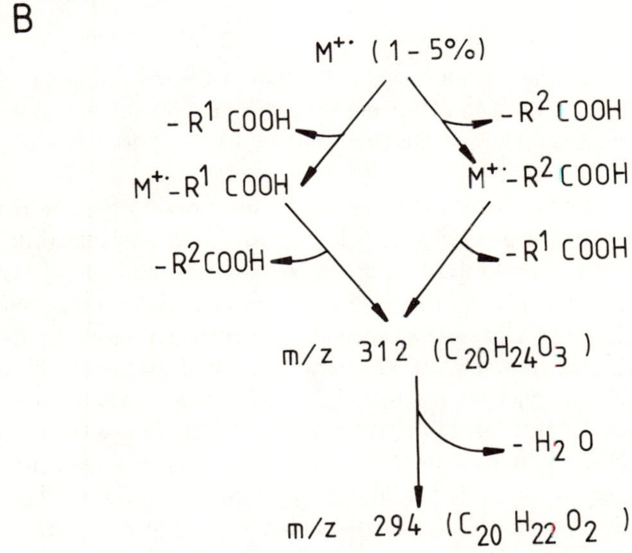

FIGURE 15. Fragmentation of (A) monoesters of 12-deoxyphorbol (**53**) and (B) diesters of 12-dexoyphorbol (**54**) according to Evans and others.[96]

both 4-deoxyphorbol (**59**) and 4-deoxy-4α-phorbol esters (**60**) are similar, exhibiting the sequential loss of acyl functions of the 12 and 13 positions from the $M^{+\cdot}+1$ quasi molecular ions. Like the esters of 12-deoxyphorbol (**53**) and unlike the phorbol (**1**) diesters, acyl functions are eliminated in the carboxyl form to generate typical fragments ions at m/z 312 and 313 together with the dehydroxylated fragment ion at m/z 294 and 295 (Figure 16). Hydroxylated 4-deoxyphorbol (**59**) derivatives such as 4-deoxy-16-hydroxyphorbol (**61**) and 4-deoxy-5-hydroxyphorbol (**62**) exhibit mass spectra which are similar to those exhibited by 12-deoxy-16-hydroxyphorbol (**54**) and phorbol (**1**) esters by means of CIMS, in that the diesters of these tigliane diterpenes all generate large fragment ions at m/z 310 and 311 due to the loss of acyl moieties plus the elements of water to produce a fragment corresponding to the deacylated tigliane nucleus.[129,131] The relative abundance of ions in the spectra of these tigliane diterpenes are seen to vary greatly with both the inlet temperature and ionization

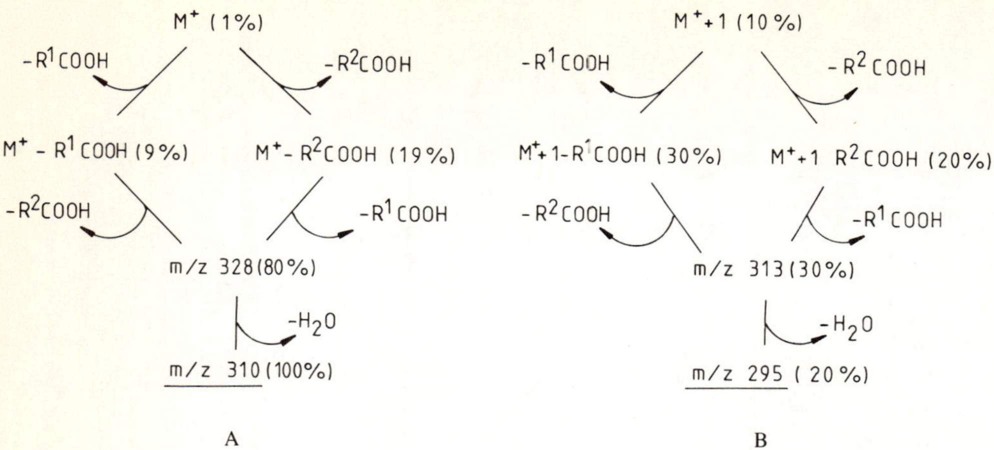

FIGURE 16. (A) Fragmentation of diesters of 12-deoxy-16-hydroxyphorbol (**54**) according to Schmidt,[129] by means of CIMS. (B) Fragmentation diesters of 4-dexoyphorbol (**59**) according to Taylor,[131] by means of CIMS.

potential used for EIMS and with temperature and the nature of the ionization gas used for CIMS. This variation has been shown to effect not only the $M^{+\cdot}$ or $M^{+\cdot}+1$ ions, which may be difficult to attain, but also many of the characteristic fragment ions produced from them.

The ^{1}H-NMR spectra of esters of phorbol (**1**) have been discussed in Section II.D. The structure of the deoxy- and hydroxyphorbol (1) analogues have been elucidated in part by comparison of their ^{1}H-NMR spectra to those of the previously isolated phorbol (**1**) esters. The major difference in ^{1}H-NMR spectra of these two groups of tigliane compounds is the disappearance of the 1H-12 proton adjacent to a secondary hydroxyl group in the spectra of esters of (**53**) and the appearance of a new 2H-12 signal[102] as a multiplet at about 2.0 ppm. Esters of the 13-*O*-acyl-12-deoxyphorbol and the 13,20-diacyl-12-deoxyphorbol may also be distinguished on the basis of their ^{1}H-NMR spectra.[102,103,109] the monoesters characteristically exhibit the allylic 2H-20 protons in their spectra at about 4.0 ppm, while the diesters in which this position is acylated normally exhibit this signal at about 4.5 ppm (Figure 17). The rest of the signals in the spectra of 12-deoxyphorbol (**53**) esters can be assigned from the spectrum of phorbol-triacetate (**2**). The ^{1}H-NMR spectra of esters of 12-deoxy-16-hydroxyphorbol (**54**) were characterized by the disappearance of the 3H-16 signal at about 1.2 ppm and the appearance of the new 2H AB quartet at about 4.15 ppm[128] due to the allylic protons of the extra primary hydroxyl at C-16 of the tigliane nucleus, which overlaps with the 2H-20 signal in these spectra (Figure 17). In a similar manner the β-oriented secondary hydroxy of esters of 12-deoxy-5-hydroxyphorbol (**57**) could be assigned from their ^{1}H-NMR spectra due to the absence of the characteristic 2H-5 signal at about 2.4 ppm and the exhibition of a new 1H singlet at about 4.3 ppm.[115,133] Esters of 12,30-dideoxyphorbol (**56**) were characterized by the appearance of an extra 3H signal at about 1.78 ppm in their ^{1}H-NMR and absence of the allylic 2H-20 signal at 4.0 ppm.[132] Mancinellin[134] and baliospermin[115] were both 5-hydroxy-6,7-epoxy analogues of 12-deoxyphorbol (**53**). Both of these derivatives could be identified because of the early work concerning the reactivity of phorbol (**1**). Reaction of phorbol (**1**) with perbenzoic acid led to the production of the 6β,7β-expoxide (**18**) together with smaller amount sof the 6α,7α-expoxide.[71] Accordingly, from the ^{1}H-NMR spectra,[115,134] mancillin and baliospermin could be assigned as the 6α,7α-epoxides of 12-deoxy-5-hydroxyphorbol (**57**) due to the absence of the 1H-7 olifinic signal at about 5.8 ppm in their spectra and the presence of a new doublet at 3.2 ppm. The chemical

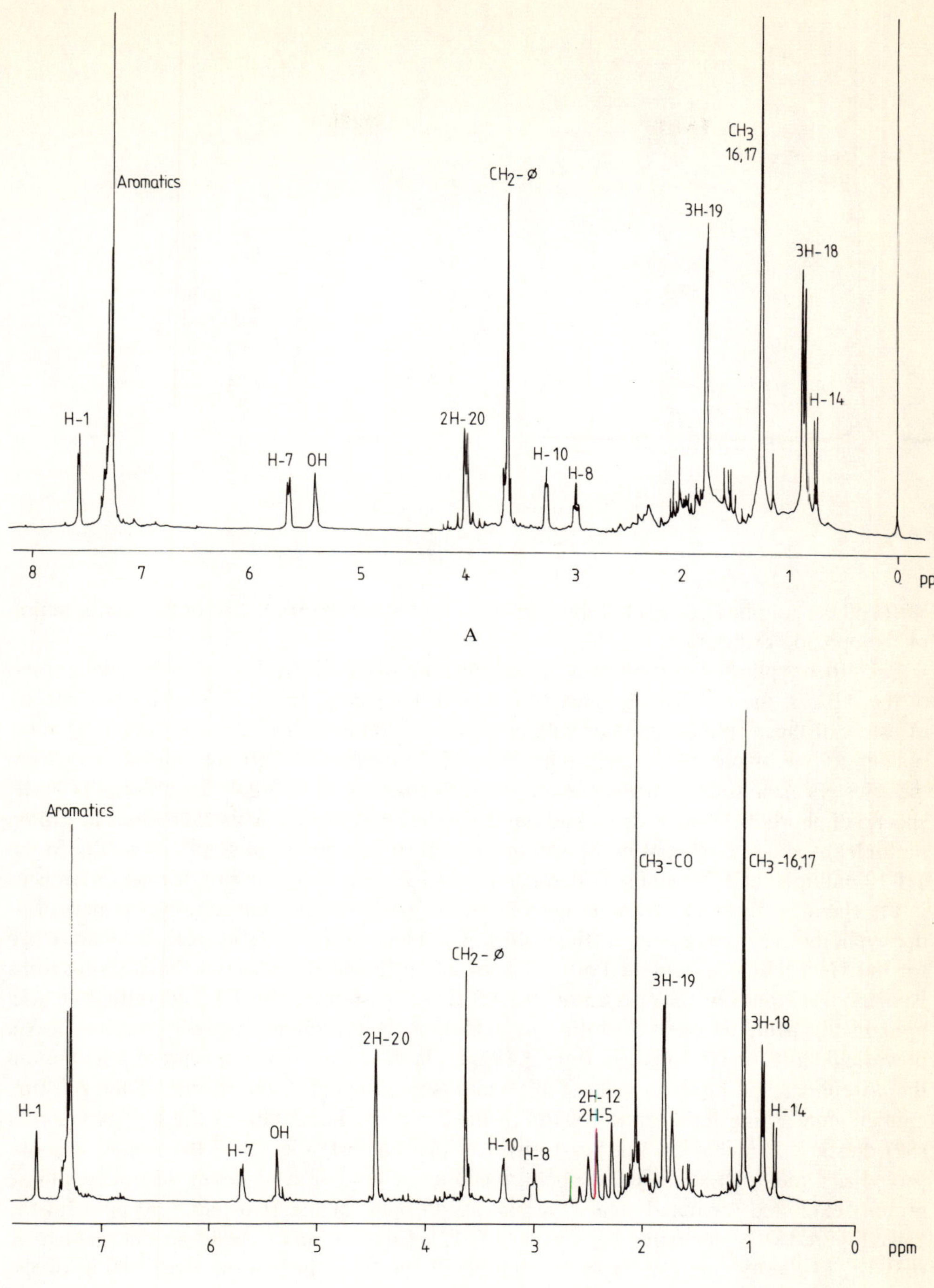

FIGURE 17. (A) 250 MHz ^{1}H-NMR spectrum of 12-deoxyphorbol-13-phenylacetate.[101] (B) 250 MHz ^{1}H-NMR spectrum of 12-deoxyphorbol-13-phenylacetate-20-acetate.[101](C) 100 MHz ^{1}H-NMR spectrum of 12-deoxy-16-hydroxyphorbol-13-phenylacetate-16-(2-methylbutanoate).[96]

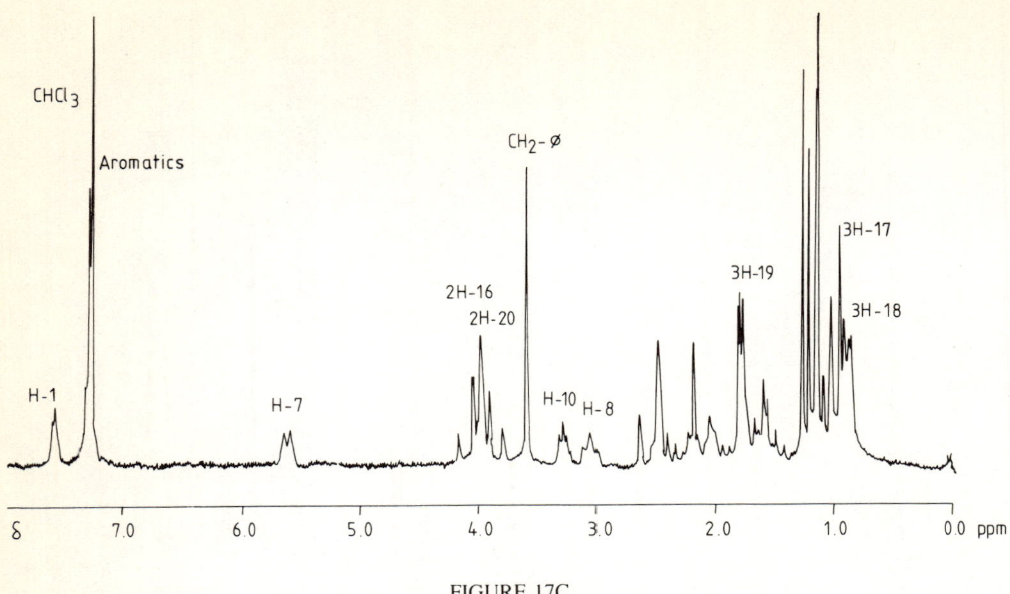

FIGURE 17C

shift and the coupling constant of this signal assisted in the determination of the configuration of the epoxide bridge.

The 4-deoxyphorbol (**59**) esters occur as either the AB *trans,* the biologically active forms, or the AB *cis,* the biologically inactive forms of this parent tigliane. The ^{1}H-NMR spectra of esters of these epimers are particularly interesting from the viewpoint of assigning compounds to one or other of these groups. The 12,13-diesters of (**59**) are quite distinct from the triesters in a similar manner to the di- and triesters of phorbol (**1**).[138] The ^{1}H-NMR spectra of phorbol (**1**) and 4-deoxyphorbol (**59**) esters may be distinguished by the one proton multiplet at about 2.85 ppm in the spectra of (**59**). Irradiation of this multiplet induced the 1H-10 multiplet at 3.28 ppm to collapse to a broad singlet.[149] Because of distinct differences in the chemical shifts of certain protons in the ^{1}H-NMR spectra, the AB ring epimers of 4-deoxyphorbol (**59**) have been distinguished. The most significant chemical shifts are those for the 1H-1 which is exhibited at 7.57 ppm in the ^{1}H-NMR spectra of the AB ring *trans* isomers, 7.10 ppm in the spectra of the AB ring *cis* isomers, the 1H-7 identified at 5.59 ppm in the *trans* derivatives shifted to 5.15 ppm in the *cis* analogues. There is also a downfield shift for 1H-10 signal from 3.28 ppm in the *trans*-epimer spectra to 3.54 ppm in the *cis*-epimer.[149] Finally, a distinct difference was observed in the spectra of the AB ring epimers concerning the chemical shifts of the 2H-5. In the spectra of the 4-deoxyphorbol (**59**) diesters,[149] including 4-deoxy-16-hydroxyphorbol (**61**) diester,[150] this signal was observed as a multiplet at about 2.50 ppm, while in the case of the 4-deoxy-4α-epimers these protons exhibited separated signals at about 3.45 ppm for the 1Hα and 2.50 ppm for the 1Hβ (Figure 18). 4-Deoxyphorbaldehyde (**64**)-12,13-diesters and 4-deoxy-4α-phorbaldehyde (**65**) 12, 13-diesters are similar in their 1H-NMR spectra to their respective C-20 alcohols, with the exception of the chemical shift of the aldehydic proton at this position which is exhibited downfield at 9.45 and 9.33 ppm, respectively, for these two tigliane derivatives.[145] Further derivatives of 4-deoxyphorbol (**59**) involving hydroxylation at the 5 or 16 positions or reduction of the C-20 primary alcohol position to methylene were evident from their ^{1}H-NMR spectra. Esters of tigliane derivatives hydroxylated at the secondary hydroxyl at C-5 exhibited a distinct 1H singlet in their ^{1}H-NMR spectra at about 5.00 ppm downfield,[144] while the presence of an extra primary hydroxyl at C-16 was ascertained from the presence of an AB quartet at about 4.0 ppm[150] in their ^{1}H-NMR spectra (Figure 18). The ^{1}H-NMR spectra of esters of 16-hydroxyphorbol (**66**) differ from those of phorbol (**1**) in that an extra

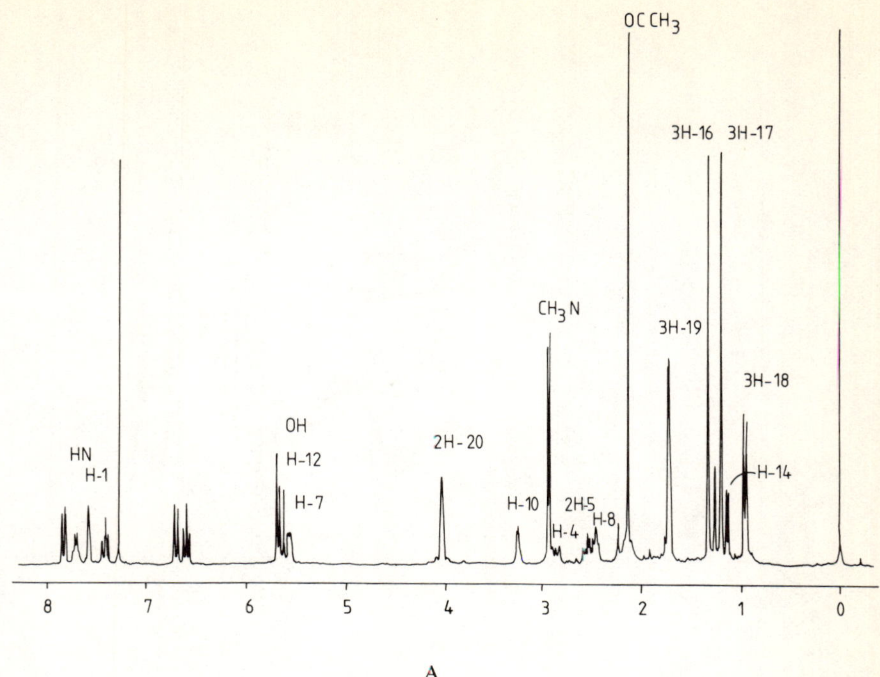

A

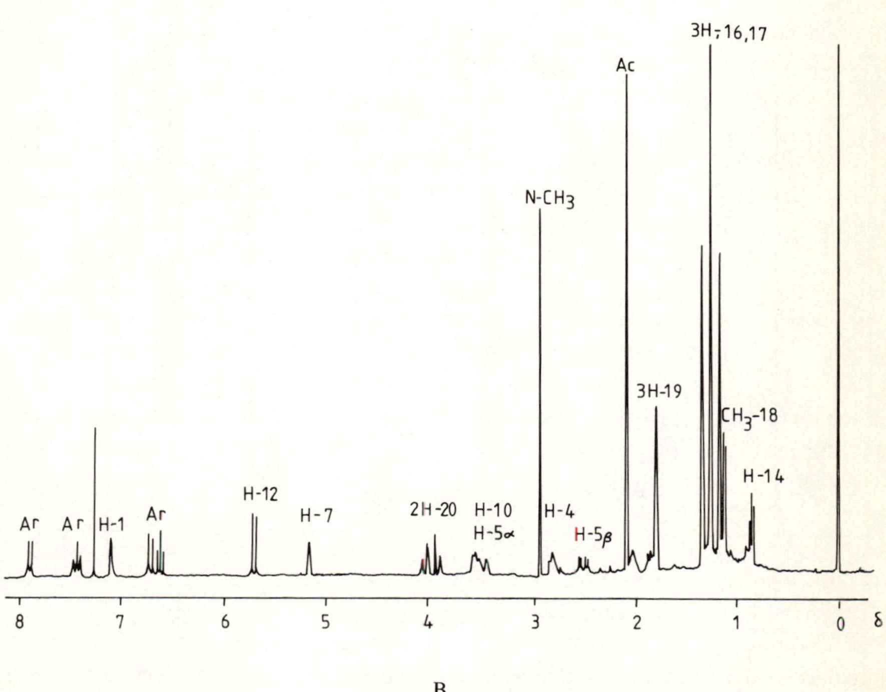

B

FIGURE 18. ¹H-NMR spectra of 4-deoxyphorbol (**59**) esters. (A) 12-*O*-[12-methylaminobenzoyl]-4-deoxyphorbol-13-acetate.[131] (B) 12-*O*-[2-methylaminobenzoyl]-4-deoxy-4α-phorbol-13-acetate.[131] (C) 12 *O*-[2Z, 4E)-*N*-deca-2,4,6-trienoyl]-4,20-dideoxy-5-hydroxyphorbol-13-acetate.[131] (D) 12-*O*-[*n*-deca-2E,4E-deinoyl]-4-deoxy-16-hydroxy-hporbol-13-acetate.

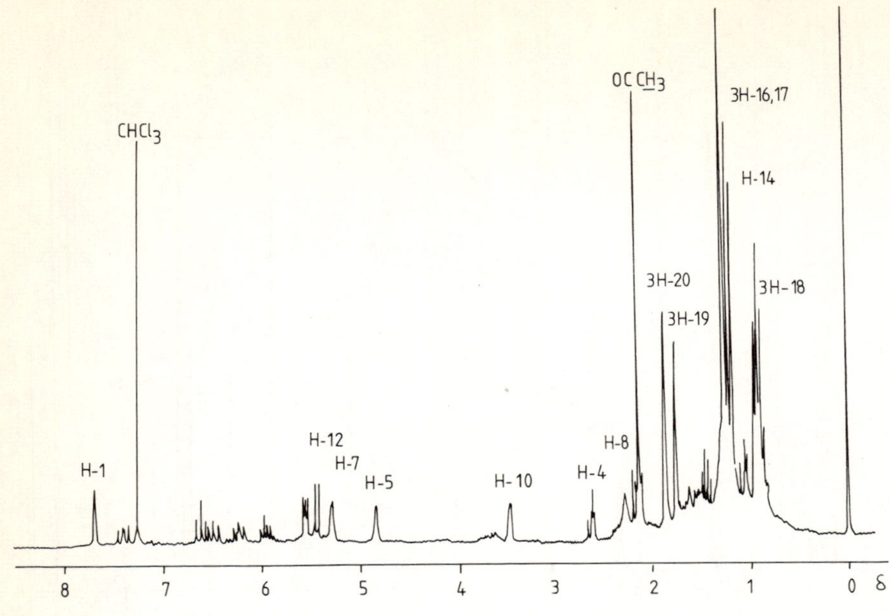

FIGURE 18C

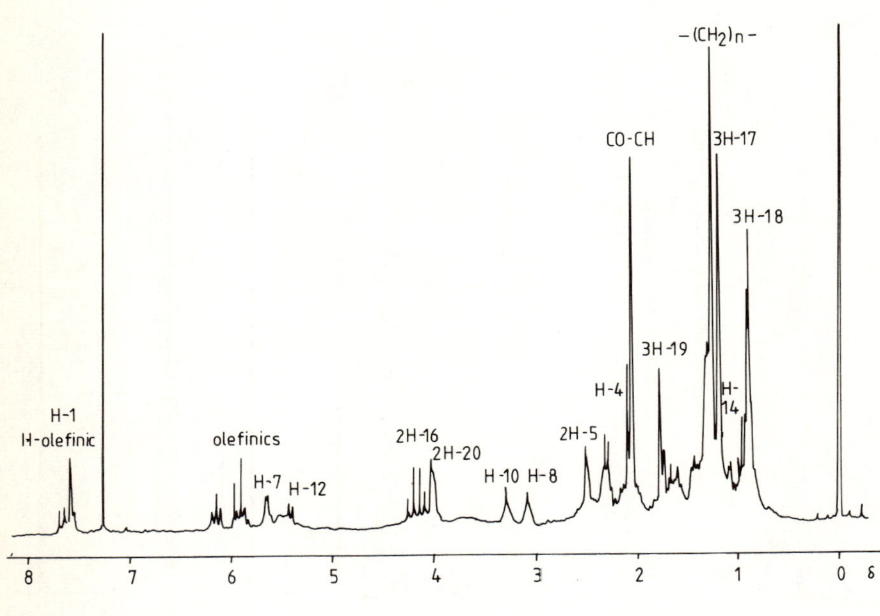

FIGURE 18D

2H signal is evident in their spectra at about 3.80 ppm[147,130] reminiscent of the signal for the allylic 2H-16 signal for esters of 12-deoxy-16-hydroxyphorbol (**54**) and 4-deoxy-16-hydroxyphorbol (**61**)-diesters. The chemical shifts for the nuclear protons of a series of deoxy- and hydroxyphorbol esters are recorded in Table 14.

The ^{13}C-NMR spectra of 4-deoxyphorbol (**59**) and 12-deoxyphorbol (**53**) esters were recorded by Taylor and others.[101] In the ^{13}C-NMR spectra of 4-deoxyphorbol (**59**) esters

Table 14
^{1}H-NMR SPECTRA; PROTON CHEMICAL SHIFTS δ PPM, RELATIVE TO TMS IN $CDCl_3$

Protons on	δ ppm of compounds (no)														
C no.	1	2	3	4	5	6	7	8	9	10	11	12	13	14	15
1	7.57	7.10	7.09	7.57	7.59	7.05	7.51	7.07	7.69	7.70	7.57	7.63	7.51	7.68	7.55
7	5.56	5.24	5.15	5.58	5.65	5.09	6.59	6.09	5.63	5.32	5.64	5.72	5.62	3.25	5.65
20	4.04	3.99	3.96	4.47	4.02	3.98	9.45	9.33	4.23	1.87	4.00	4.49	4.38	3.80	3.98
12	5.64	5.21	5.71	5.64	5.42	5.46	5.67	5.72	5.62	5.65	1.78	N/R	N/R	N/R	5.38
10	3.28	3.66	3.54	3.28	3.29	3.51	3.08	3.59	3.55	3.53	3.28	3.26	3.20	3.95	3.10
4	2.81	2.83	2.82	2.60	2.31	2.81	2.53	2.89	2.63	2.63	N/A	N/A	N/A	N/A	N/A
5	2.44	3.55α 2.55β	3.46α 2.51β	2.60	2.51	3.40α 2.5β	2.76	3.29α 3.14β	5.14	4.87	2.48	2.48	2.38	4.26	N/R
8	2.44	2.03	2.03	2.60	3.08	2.34	2.69	1.93	2.36	2.34	3.00	3.02	3.10	2.83	3.34
11	2.18	1.88	1.87	N/R	1.73	1.70	1.80	1.93	1.74	1.66	2.48	1.79	1.15	N/R	N/R
14	1.22	0.78	0.88	1.09	1.15	0.82	1.27	0.94	1.10	1.08	1.04	0.87	0.87	N/R	N/R
16	1.32	1.28	1.34	1.35	4.17	3.95	1.34	1.42	1.29	1.29	1.04	1.04	4.00	1.09	3.81
17	1.25	1.18	1.17	1.20	1.18	1.25	1.24	1.24	1.20	1.21	1.04	1.04	1.15	1.21	N/R
18	0.96	1.15	1.22	0.96	0.89	1.00	0.98	1.46	0.96	0.96	0.86	0.87	0.87	0.92	0.88
19	1.74	1.82	1.81	1.74	1.77	1.79	1.74	1.78	1.75	1.73	1.78	1.79	1.77	1.78	1.76

(1) A 4-deoxy-12,13-diester 250 MHz,[140] (2) 4-deoxy-4α-phorbol-12-monoester 60 MHz,[149] (3) 4-deoxy-4α-phorbol-12,13-diester 250 MHz,[149] (4) 4-deoxyphorbol-12,13,20-triester 250 MHz,[149] (5) 4-deoxy-16-hydroxyphorbol-12,13-diester 250 MHz,[150] (7) 4-deoxyphorbaldehyde-12,13-diester 250 MHz,[145] (8) 4-deoxy-4α-phorbaldehyde-12,13-diester 250 MHz,[145] (9) 4-deoxy-5-hydroxyphorbol-12,13-diester 90 MHz,[144] (10) 4,20-dideoxy-5-hydroxyphorbol-12,13-diester 250 MHz,[144] (11) 12-deoxyphorbol-13-monoester 60 MHz,[109] (12) 12-deoxyphorbol-13,20-diester 60 MHz,[109] (13) 12-deoxy-16-hydroxyphorbol-13,16,20-triester 100 MHz,[128] (14) 12-deoxy-5-hydroxy-6,7-epoxyphorbol-monoester 100 MHz,[134] (15) 16-hydroxyphorbol-12,13-diester 90 MHz.[147] N/A = not applicable; N/R = not recorded.

Table 15
NUCLEAR PROTON CHEMICAL SHIFTS FOR DEOXYPHORBOL ESTERS

	Chemical shifts δppm for compounds		
C no.	1	2	3
19	10.08q	10.35q	9.98q
18	15.00q	11.84q	15.26q
17	16.79q	16.55q	18.43q
16	23.63q	24.062	22.93q
15	25.68s	25.23s	22.99s
5	29.95t	26.39t	38.95t
14	35.54d	37.04d	36.28d
8	42.27d	41.00d	39.41d
4	42.57d	48.89d	73.59s
11	44.06d	43.11d	32.31d
10	54.04d	46.91d	55.73d
13	63.38s	65.41s	63.99s
20	68.76t	70.67t	69.57t
12	76.13d	74.99d	31.70t
9	77.77s	77.89s	75.88s
7	130.06d	128.66d	133.64d
6	136.37s	132.76s	132.83s
2	137.28s	143.32s	134.87s
1	159.35d	155.02d	161.04d
3	209.09s	210.76s	208.78s

Note: ^{13}C-NMR spectra of deoxyphorbol esters, nuclear carbon signals, $CDCl_3$, 100.6 mHz off resonance decoupled.[101] (1) A 4-deoxyphorbol (**59**)-triester; (2) a 4-deoxy-4α-phorbol (**60**)-triester; (3) a 12-deoxyphorbol- (**53**)-monoester.

there was an upfield shift in the signal for C-4 from a singlet at 73 ppm for phorbol triacetate (**2**) to a doublet at 43 ppm for the 4-deoxyphorbol derivative. Additionally, the signal for C-5 was observed upfield from 30 ppm for (**2**) to 39 ppm for (**59**) esters. The rest of the ^{13}C-NMR spectrum of esters of (**59**) are not significantly different from those of esters of phorbol (**1**). There are, however, differences in the ^{13}C-NMR spectra of esters of (**59**) and esters of 4-deoxy-4α-phorbol (**60**) which are attributed to the change in the molecular shape of the molecule. In the spectra of the 4-deoxy-4α-epimers compared with the 4-deoxyphorbol (**59**) esters there was a downfield shift (in ppm) of the signals for C-2 (143 from 137) and C-4 (467 from 42) and an upfield shift (in ppm) for the signals due to C-1 (155 from 159), C-5 (26 from 30), C-6 (133 from 136), C-10 (49 from 54), and C-18 (12 from 15). The major difference between the ^{13}C-NMR spectra of esters of 12-deoxyphorbol (**53**) and phorbol-triacetate (**2**) was the fact that the C-12 signal changed in both its chemical shift and its multiplicity to a triplet at 32 ppm from a doublet at 77 ppm. The signal for C-11 also appears further upfield at 32 ppm in the spectra of the 12-deoxyphorbol (**53**) esters. The nuclear proton chemical shifts for deoxyphorbol esters are recorded in Table 15.

The circular dichroism spectra (CD) of deoxyphorbol esters are not particulary characteristic and are similar to those of phorbol (**1**) esters (see Section II.D). These derivatives exhibit a pronounced positive Cotton effect between 230 and 240 nm and a negative Cotton effect between 300 and 330 nm due to the αβ-unsaturated keto group at C-3 of the tigliane nucleus. In the case of the 4-deoxy-4α-phorbol (**60**) analogues a large negative Cotton effect

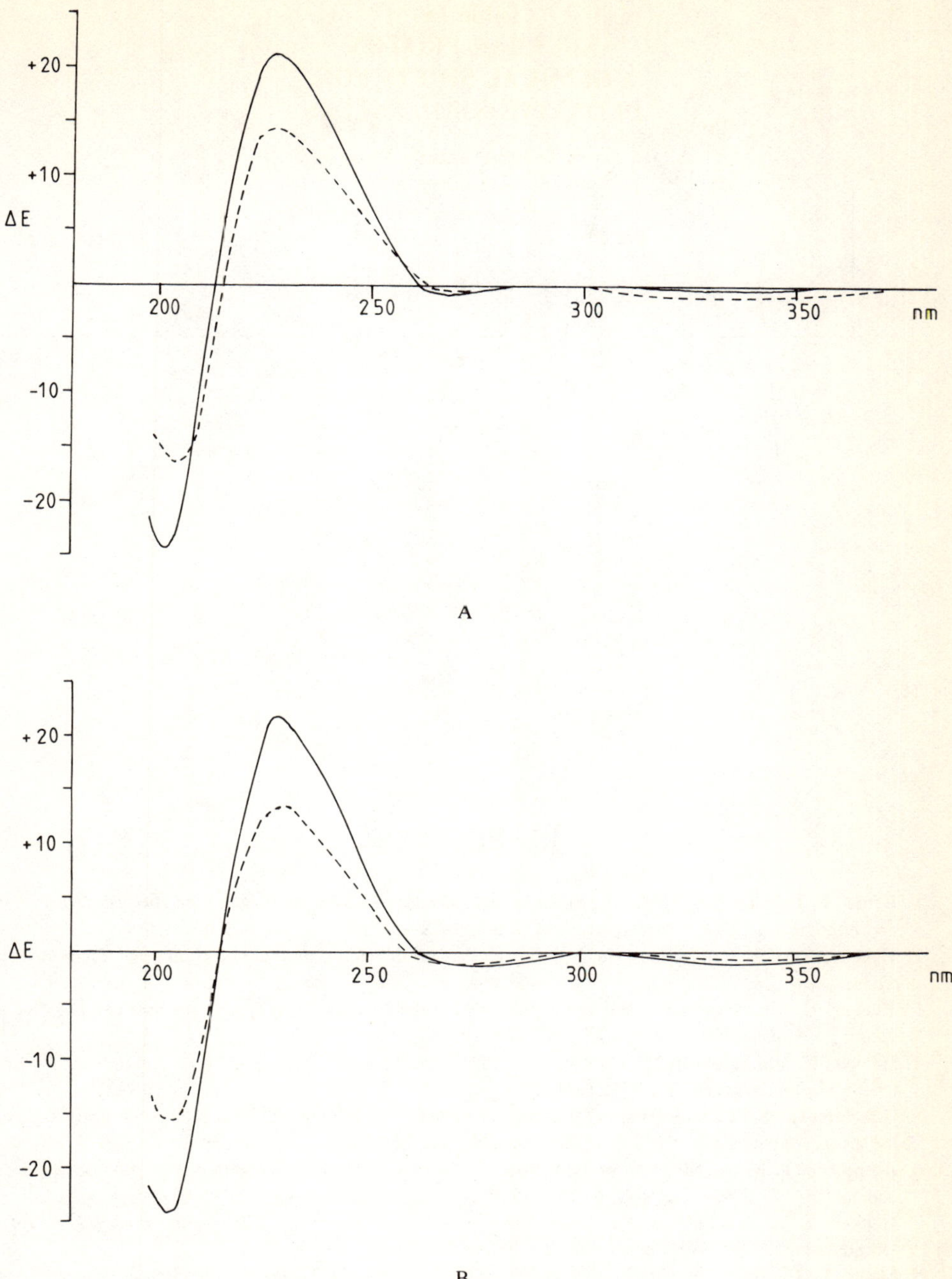

FIGURE 19. CD spectra of deoxyphorbol esters. (A) An ester of 12-deoxy-16-hydroxyphorbol (**54**). (B) An ester of 12-deoxyphorbol (**53**). (C) An ester of 4-deoxy (**59**) and 4-deoxy-4α-deoxyphorbol (**60**).

was recorded in the region of 240 and 250 nm which has been ascribed to the difference in the fusion of the AB ring systems. This particular effect was more noticeable in the spectra of 4-deoxy-4α-phorbol (**60**) esters than in the spectra of 4-deoxyphorbol (**59**) esters. Another point of distinction in the spectra of the esters of (**60**) was the occurrence of a positive extreme between 350 and 380 nm whereas in the spectra of esters of (**59**) a negative extreme was exhibited in this region. Typical CD curves of deoxyphorbol esters are shown in Figure 19.

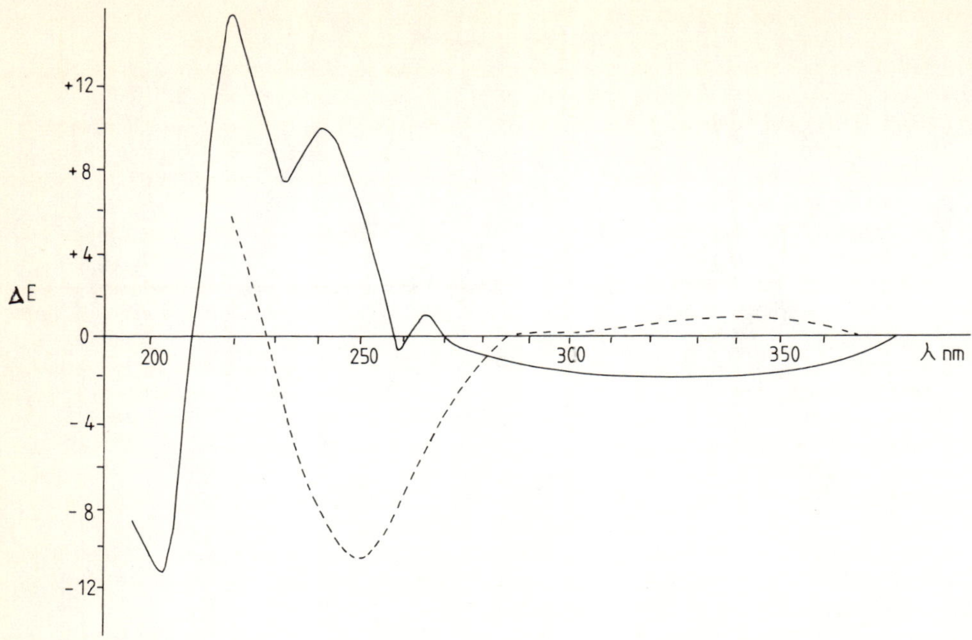

FIGURE 19C

REFERENCES

1. **Evans, F. J. and Soper, C. J.,** The tigliane, daphnane and ingenane diterpenes, their chemistry, distribution and biological activities, *J. Nat. Prod.*, 41, 193, 1978.
2. **Hecker, E.,** Co-carcinogenic principles of the seed oil of *Croton tiglium* and from other Euphorbiaceae, *Cancer Res.*, 28, 2338, 1968.
3. **Hecker, E.,** Co-carcinogenes and tumor-promoters of the diterpene-ester type as possible carcinogenic risk factors, *J. Cancer Res. Clin. Oncol.*, 99, 103, 1981.
4. **Hecker, E. and Schmidt, R.,** Phorbol esters, the irritants and co-carcinogens of *Croton tiglium, Fortschr. Chem. Organ. Naturst.*, 31, 377, 1971.
5. **Bohrman, J. S.,** Identification and assessment of tumor-promoting and co-carcinogenic agents, state of the art *in vitro* methods, *CRC Crit. Rev. Toxicol.*, 11, 121, 1983.
6. **Blumberg, P. M.,** *In vitro* studies on the mode of action of the phorbol esters, potent tumor-promoters. I and II. *CRC Crit. Rev. Toxicol.*, 8, 153, 1980.
7. **Hecker, E.,** Co-carcinogens and co-carcinogenesis, in *Handbuch der Allegemeine Pathologie*, Vol. 6, Grundmann, E., Ed., Springer-Verlag, Berlin, 1975, 651.
8. **Slaga, T. J., Ed.,** *Tumor Promotion and Skin Carcinogenesis*, CRC Press, Boca Raton, Fla., 1982, 1.
9. **Rohrschneider, L. R. and Boutwell, R. K.,** Phorbol-esters, fatty acids and tumor-promotion, *Nature (London) New Biol.*, 243, 212, 1973.
10. **Sivak, A. and Van, Duuren, B. L.,** Cellular interactions of phorbol-myristate acetate in tumor-promotion, *Chem. Biol. Interact.*, 3, 401, 1971.
11. **Kubinski, H., Stangestalien, M. A., Baird, W., and Boutwell, R. K.,** Interaction of phorbol-esters with cellular membranes *in vitro, Cancer Res.*, 33, 3103, 1973.
12. **Witz, G., Van-Duuren, B. L., and Banerjee, S.,** The interaction of phorbol myristate acetate, a tumor-promoter, with rat liver plasma membrane — a fluorescence study, *Proc. Am. Assoc. Cancer Res.*, 16, 30, 1975.
13. **Wenner, C. E., Hackney, E. J., Kimelberg, H. K., and Mayhew, E.,** Membrane effects of phorbol-esters, *Cancer Res.*, 34, 1731, 1974.
14. **Hecker, E.,** Structure-activity relationships in diterpene esters irritant and co-carcinogenic to mouse skin, in *Carcinogenesis*, Vol. 2, *Mechanisms of Tumor-Promotion and Co-Carcinogenesis*, Slaga, T. J., Sivak, A., and Boutwell, R. K., Eds., Raven Press, New York, 1978, 11.

15. **Wilson, S. R. and Huffman, J. C.**, The structural relationships of phorbol and cortisol, a possible mechanism for the tumor-promoting activity of phorbol, *Experientia*, 32, 1489, 1976.
16. **Lee, L. S. and Weinstein, I. B.**, Tumor-promoting phorbol-esters inhibit binding of epidermal growth factor to cellular receptors, *Science*, 202, 313, 1978.
17. **Slaga, T. G. and Scribner, J. D.**, Inhibition of tumor-initiation and promotion by anti-inflammatory agents, *J. Natl. Cancer Inst.*, 51, 1723, 1974.
18. **Troll, W., Klassen, A., and Janoff, A.**, Tumorgenesis in mouse skin, inhibition by synthetic inhibitors of proteases, *Science*, 169, 1211, 1970.
19. **Williamson, E. M. and Evans, F. J.**, Inhibition of erythema induced by pro-inflammatory esters of 12-deoxyphorbol, *Acta Pharmacol. Toxicol.*, 48, 47, 1981.
20. **Williamson, E. M., Westwick, J., Kakkar, V., and Evans, F. J.**, Studies on the mechanism of action of 12-deoxyphorbol phenyl-acetate, a potent platelet aggregating tigliane-ester, *Biochem. Pharmacol.*, 30, 2691, 1981.
21. **Edwards, M. C., Nouri, A. M. E., Gordon, D., and Evans, F. J.**, Tumor-promoting and non-promoting pro-inflammatory phorbol-esters act as human lymphocyte mitogens with different sensitivities to inhibition by cyclosporin-A., *Mol. Pharmacol.*, 23, 703, 1983.
22. **Driedger, P. E. and Blumberg, P. M.**, Specific binding of phorbol-ester tumor-promoters, *Proc. Natl. Acad. Sci. U.S.A.*, 77, 567, 1980.
23. **Shoyab, M. and Todaro, G. J.**, Specific high affinity cell membrane receptors for biologically active phorbol and ingenol esters, *Nature (London)*, 288, 451, 1980.
24. **Horowitz, A. D., Greenebaum, E., and Weinstein, I. B.**, Identification of receptors for phorbol ester tumor promotors in intact mammalian cells and of an inhibitor of receptor binding in biologic fluids, *Proc. Natl. Acad. Sci. U.S.A.*, 78, 2315, 1981.
25. **Sando, J. J., Hilfiger, M. L., Salomon, D. S., and Farrar, J. J.**, Specific receptors for phorbol-esters in lymphoid cell populations, role in enhanced production of T-cell growth factor, *Proc. Natl. Acad. Sci. U.S.A.*, 78, 1189, 1981.
26. **Jaken, B., Tashjian, A. H., and Blumberg, P. M.**, Characterization of phrobol-ester receptors and their down-modulation in GH_4C_1 rat pituitary cells, *Cancer Res.*, 41, 2175, 1981.
27. **Dunphy, W. G., Kochenburger, R. J., Castagna, M., and Blumberg, P. M.**, Kinetics and subcellular localization of specific [^{3}H]-phorbol-12,13-dibutyrate binding by mouse brain, *Cancer Res.*, 41, 2640, 1981.
28. **Goodwin, B. J. and Weinberg, J. B.**, Receptor-mediated modulation of human monocyte, neutrophil, lymphocyte and platelet function by phorbol-diesters, *J. Clin. Invest.*, 70, 699, 1982.
29. **Salomon, D. S.**, Inhibition of epidermal growth factor binding to mouse embryonal carcinoma cells by phorbol esters mediated by specific phorbol-ester receptors, *J. Biol. Chem.*, 256, 7958, 1981.
30. **Mori, T., Takai, Y., Minakuchi, R., Yu, B., and Nishizuka, Y.**, Inhibitory action of chlorpromazine, dibucaine and other phospholipid-interacting drugs on calcium-cativated phospholipid-dependent protein kinase, *J. Biol. Chem.*, 255, 8378, 1980.
31. **Castagna, M., Takai, Y., Kaibuchi, K., Sano, K., Kikkawa, U., and Nishizuka, Y.**, Direct activation of calcium-activated, phospholipid dependent protein kinase by tumor-promoting phorbol-esters, *J. Biol. Chem.*, 257, 7848, 1982.
32. **Yamanishi, J., Takai, Y., Kaibuchi, K., Sano, K., Castagna, M., and Nishizuka, Y.**, Synergistic functions of phorbol-ester and calcium serotonin release from human platelets, *Biochem. Biophys. Res. Commun.*, 112, 778, 1983.
33. **Niedel, J. E., Kuhn, L. J., and Vandenbark, G. R.**, Phorbol-diester receptor co-purifies with protein kinase-C, *Proc. Natl. Acad. Sci. U.S.A.*, 80, 36, 1983.
34. **Kikawa, U., Takai, Y., Tanaka, Y., Miyake, R., and Nishizuka, Y.**, Protein kinase C as a possible receptor protein of tumor-promoting phorbol-esters, *J. Biol. Chem.*, 258, 11442, 1983.
35. **Gilmore, T. and Martin, G. S.**, Phorbol-ester and diacylglycerol induce protein phosphorylation at tyrosine, *Nature (London)*, 306, 487, 1983.
36. **Naka, M., Nishikawa, M., Adelstein, R. S., and Hidaka, H.**, Phrobol-ester induced activation of human platelets is associated with protein kinase C phosphorylation of myrosin light chains, *Nature (London)*, 306, 490, 1983.
37. **Umezawa, K., Weinstein, I. B., Horowitz, A., Fujika, H., Matsuchima, T., and Sugimura, T.**, Similarity of teleocidin-B and phorbol-ester tumor-promoters in effects on membrane receptors, *Nature (London)*, 290, 411, 1981.
38. **Fujiki, H., Mori, M., Nakayasu, M., Terada, M., and Sugimura, T.**, A possible naturally occurring tumor-promoter, teleocidin B from *Streptomyces*, *Biochem. Biophys. Res. Commun.*, 90, 976, 1979.
39. **Hecker, E., Bartsch, H., Bresch, H., Gschwendt, M., Härle, E., Kreibich, G., Kubinyi, H., Schairer, H. U., Szczepaniski, V-Ch., and Thielmann, H. W.**, Structure and stereochemistry of the tetracyclic diterpene phorbol from *Croton Tiglium*, *Tetrahedron Lett.*, 3165, 1967.

40. **Evans, F. J. and Taylor, S. E.,** Pro-inflammatory, tumor-promoting and anti-tumor diperpenes of the plant families Euphorbiaceae and Thymelaeaceae, *Fortschr. Chem. Organ. Naturst.*, 44, 1, 1983.
41. **Bohm, R., Flaschentrager, B., and Lendle, L.,** Über die Wirksamkeit von Substanzen aus dem Kroton-Öl, *Arch. Exp. Pathol. Pharmakol.*, 177, 212, 1935.
42. **Hecker, E., Szczepanski, V-Ch., Kubinyi, H., Bresch, H., Härle, E., Schairer, H. U., and Bartsch, H.,** Über die Wirkstoffe des Crotonöls. VII., Phorbol, *Z. Naturforsch.*, 216, 1204, 1966.
43. **Arroyo, E. R. and Holcomb, J.,** Isolation and structure elucidation of a highly active principle from *Croton* oil, *Chem. Ind.*, 350, 1965.
44. **Arroyo, E. R. and Holcomb, J.,** Structural studies of an active principle from *Croton tiglium* L., *J. Med. Chem.*, 8, 672, 1965.
45. **Hecker, E., Kubinyi, E., Szczepanski, V-Ch., Härle, E., and Bresch, H.,** Phorbol — ein neues tetracyclisches diterpen aus *Croton öl, Tetrahedron Lett.*, 1837, 1965.
46. **Pettensen, R. C., Ferguson, G., Crombie, L., Games, M. L., and Pointer, D. J.,** The structure and stereochemistry of phorbol, diterpene parent of co-carcinogens of *Croton* oil, *Chem. Commun.*, 716, 1967.
47. **Hoppe, W., Brandl, F., Strell, I., Röhrl, M., Gassmann, I., Hecker, E., Bartsch, H., Kreibach, G., and Szczepanski, V-Ch.,** X-ray structure analysis of neophorbol, *Agnew. Chem. Int. Ed. Engl.*, 6, 807, 1967.
48. **Pettersen, R. C., Birnbaum, G. I., Gerguson, G., Islam, K. M. S., and Sime, J. G.,** X-ray investigation of several phorbol derivatives, the crystal and molecular structure of phorbol-bromofuroate chloroform solvate, *J. Chem. Soc.*, 980, 1968.
49. **Hoppe, W., Zechmeister, K., Röhrl, M., Brandl, F., Hecker, E., Kreibach, G., and Bartsch, H.,** The structure determination of a solvate of phorbol, the diterpene parent of the tumor-promoters from *Croton* oil, *Tetrahedron Lett.*, 667, 1969.
50. **Pack, G. R.,** The molecular structure and charge distribution of phorbol, parent compound of the tumour-promoting phorbol-diesters, *Cancer Biochem. Biophys.*, 5, 183, 1981.
51. **Crombie, L. M., Games, M. L., and Pointer, D. J.,** Chemistry and structure of phorbol, the diterpene parent of the cocarcinogens of *Croton* oil, *J. Chem. Soc.*, 1347, 1968.
52. **Kauffmann, T. and Neumann, H.,** Zer Konstitution des phorbols. I. Über die reduzierende Gruppe des Phorbols, *Chem. Ber.*, 92, 1715, 1959.
53. **Shairer, H. U., Thielmann, H. W., Gschwendt, M., Kreibach, G., Schmidt, R., and Hecker, E.,** Zur Chemie des Phorbols. IX. Polybenzoate und Acetate des Phorbols und Phorbol-3-ols und funktionelle Derivate der Allyl ruppierung des Phorbols, *Z. Naturforsch.*, 23b, 1430, 1968.
54. **Bartsch, H., Bresch, H., Gschwendt, M., Härle, E., Kreibich, G., Kubinyi, H., Schairer, H. U., Szczepanski, V-Ch., Thielmann, H. W., and Hecker, E.,** Kombination Wirksamer Trennverfahren mit modernen analytischen Methoden in der naturstoff Chemic, Isolierung und Strurkeraufklärung der biologisch aktiven Substanzen aus *Crotonöl, Z. Anal. Chem.*, 221, 424, 1966.
55. **Hecker, E.,** Die cocarcinogene wirkung der Phrobol-ester, in *Molekulane Biologie des malignen Wachstumn*, Holzer, H. A. and Holldorf, A. W., Eds., Springer-Verlag, Berlin, 1966, 105.
56. **Kreibich, G. and Hecker, E.,** Über die Verätherung des Phorbols, *Angew, Chem. Int. Ed. Engl.*, 6, 973, 1967.
57. **Kreibich, G. and Hecker, E.,** Zur Chemie des Phorbols. V. Über einige Äther des Phorbols, *Z. Naturforsch.*, 23b, 1444, 1968.
58. **Schaafsma, S. E., Steinberg, H., and De Boer, Th.,** Isomerisation and oxidative dimerisation of 1-substituted cyclopropanols, *Recl. Trav. Chim. Pays-Bas.*, 85, 73, 1966.
59. **Hecker, E. and Kubinyi, H.,** Über die Wirkstoffe des Crotonöls. IV. Reindarstellung und Charakterisierung des entzündlilchen und cocarcinogenen Wirkstoffe B_1 und B_2, *Z. Krebsforsch.*, 67, 176, 1965.
60. **Kreibich, G. and Hecker, E.,** On the active principles of *Croton* oil. X. Preparation of tritium labelled *Croton* oil factor A_1 and other tritium labelled phorbol derivatives, *Z. Krebsforsch.*, 74, 448, 1970.
61. **Schmidt, R. and Hecker, E.,** Untersuchungen über die Beziehungen zwischen Struktur und Wirkung von Phorbol-Estern, in *Aktuelle Probleme aus dem Gebiet der Cancerologie*, Vol. 3, Lettré, H. and Wagner, G., Eds., Springer-Verlag, Berlin, 1971, 98.
62. **Theilmann, H. W. and Hecker, E.,** Zur Chemie des Phorbols. XIV. Die Flaschenträger-reaction, *Liebigs Ann. Chem.*, 728, 158, 1969.
63. **Bartsch, H. and Hecker, E.,** Zur Chemie des Phorbols. XII. Phorbobutanon, *Z. Naturforsch.*, 24b, 99, 1969.
64. **Boehm, R., Flaschenträger, B. and Lendle, L.,** Über die Wirksamkeit von Substanzen aus dem *Krotonöl, Arch. Exp. Pathol. Pharmakol.*, 177, 212, 1935.
65. **Hecker, E., Szczepanski, V-Ch., Kubinyi, H., Bresch, H., Härle, E., Schairer, H. U. and Bartsch, H.,** Über die Wirkstoffe des Crotonöls. VII. Phorbol, *Z. Naturforsch.*, 21b, 1204, 1966.
66. **Szczepanski, V-Ch., Schairer, H. U., Gschwendt, M., and Hecker, E.,** Zur Chemie des Phorbols. III. Mono- und diacetates des Phrobols, *Liebigs Ann. Chem.*, 705, 199, 1967.

67. **Hoppe, W., Brandl, F., Strell, I., Röhrl, M., Gassmann, J., Hecker, E., Bartsch, H., Kreibish, G., and Szczepanski, V-Ch.**, Röntgenstruckturanalyse des Neophorbols, *Angew. Chem. Int. Ed. Engl.*, 6, 809, 1967.
68. **Hecker, E., Bartsch, H., Kreibich, G. and Szczepanski, V-Ch.**, Zur Chemie des Phorbols. X. Schweratom-derivate zur Röntgen-Strukturanalyse des Phorbols, *Tetrahedron Lett.*, 3165, 1967.
69. **Theilmann, W., Jacobi, P., and Hecker, E.**, Zur Chemie des Phorbols. XVIII. Bromierung des Phorbol-penta-acetate, *Liebigs Ann. Chem.*, 765, 171, 1972.
70. **Flaschenträger, B.**, Über den Giftstoff im Krotonöl, in *Festschrift Heinrich Zangger*, Vol. 2, Rascher, Zurich, 1935, 857.
71. **Bartsch, H. and Hecker, E.**, Zur Chemie des Phorbols. XIII. Über eine Acyloin-umlagerung des 12-Deoxy-12-oxo-phorbol-13,20-triacetate, *Liebigs Ann. Chem.*, 725, 142, 1969.
72. **Bartsch, H. and Hecker, E.**, Zur Chemie des Phorbols. XI. Crotophorbolon-Enolacetat und Acetoxycrotophorbolon, *Z. Naturforsch.*, 24b, 91, 1969.
73. **Hecker, E.**, Isolation and characterisation of the co-carcinogenic principles from croton oil, in *Methods in Cancer Research*, Busch, H., Ed., Academic Press, New York, 1971, 439.
74. **Hecker, E., Härle, E., Schairer, H. U., Jacobi, P., Hoppe, W., Gassmann, J., Röhrl, M., and Abel, H.**, Lumi-phorboltriacetat, ein käfigartiges Derivat des Diterpens 4α-Phorbol, *Angew. Chem. Int. Ed. Engl.*, 7, 890, 1968.
75. **Jacobi, P., Härle, E., Schairer, H. U., and Hecker, E.**, Zur Chemie des Phorbols. XVI. 4α-Phorbol, *Liebigs Ann. Chem.*, 741, 13, 1970.
76. **Schmidt, R. and Hecker, E.**, Ein neues Diterpen und Ferrsäureester aus Crotonöls, *Fette. Seifen Anstrichmittel*, 73, 676, 1971.
77. **Schmidt, R. and Hecker, E.**, Neue Phorbolester aus Crotonöl, *Fette, Seifen, Anstrichm.*, 70, 851, 1968.
78. **Härle, E. and Hecker, E.**, Zurchemie des Phorbols. XVII. Lumiphorbol und einige seiner Derivate, *Liebigs Ann. Chem.*, 748, 134, 1971.
79. **Flaschenträger B. and Wigner, G.**, Über den Giftstoff des Krotonöls. V. Die Gewinnung von Krotonharz, dünnem öl und Phorbol aus dem Crotonöl durch Alkoholyse, *Helv. Chim. Acta*, 25, 569, 1942.
80. **Gschwendt, M., Härle, E., and Hecker, E.**, Zur Chemie des Phorbols. VII. Die α,β-ungesaltigte tertiäre 1,2-Ketol-gruppierung im Phorbol, *Z. Naturforsch.*, 23b, 1579, 1968.
81. **Thielmann, H. W. and Hecker, E.**, Zur Chemie des Phorbols. XV. Oxydation der Δ^6-Doppelbindung des Phorbols mit Osmium-tetroxid und spaltung einiger 6β,7β-Diole, *Liebigs Ann. Chem.*, 735, 113, 1970.
82. **Gschwendt, M. and Hecker, E.**, Zur Chemie des Phorbols. VIII. Die Oxydation von Phorbol mit Bleitetra-acetat, *Z. Naturforsch.*, 23b, 1584, 1968.
83. **Gschwendt, M. and Hecker, E.**,. Zur Chemie des Phorbols. IX. Oxydation von Phorbol mit Natrium-perjodat, *Z. Naturforsch.*, 24b, 80, 1969.
84. **Bresch, H., Kreibich, G., Kubinyi, H., Schairer, H. U., Thielmann, H. W. and Hecker, E.**, Über die Wirkstoffe des Crotonöls. IX. Partialsynthese von Wirkstoffen des Crotonöls, *Z. Naturforsch.*, 23b, 538, 1968.
85. **Hecker, E., Kubinyi, H., Schairer, H. U., Szczepanski, V-Ch., and Bresch, H.**, Partialsynthese einiger co-carcinogener Wirkstoffe aus Crotonöls, *Angew. Chem. Int. Ed. Engl.*, 4, 1072, 1965.
86. **Upadhyay, R. R. and Hecker, E.**, A new cryptic irritant and co-carcinogen from the seeds of *Croton sparciflorus, Phytochemistry*, 15, 1070, 1976.
87. **Taylor, S. E., Evans, F. J., Gafur, M. A., and Choudhury, A. K.**, Sapintoxin-D, a new phorbol-ester from *Sapium indicum, J. Natl. Prod.*, 44, 729, 1981.
88. **Ohigashi, H. and Mitsui, T.**, Studies on the biologically active substances of *Sapium japonicum, Bull. Inst. Chem. Res. Kyoto Univ.*, 50, 239, 1972.
89. **Fürstenberger, G. and Hecker, E.**, New highly irritant *Euphorbia* factors from the latex of *Euphorbia tirucalli* L., *Experientia*, 33, 986, 1977.
90. **Evans, F. J.**, A new phorbol triester from the latices of *Euphorbia franckiana* and *E. Coerulescens, Phytochemistry*, 16, 395, 1977.
91. **Handa, S. S., Kingborn, A. D., Cordell, G. A., and Farnsworth, N. R.**, Plant anticancer agents. XXII. Isolation of a phorbol diester and its $\Delta^{5,6}$-7β-hydroperoxide derivative from *Ostodes paniculata, J. Natl. Prod.*, 46, 123, 1983.
92. **Buchheim, R.**, Über die pharmakologische Gruppe des Crotonöls, *Virchow's Arch. Path. Anat. Physiol.*, 12, 1, 1857.
93. **Hecker, E.**, Co-carcinogenic principles from the seed oil of *Croton tiglium* and from other Euphorbiaceae, *Cancer Res.*, 28, 2338, 1968.
94. **Hecker, E., Immich, H., Bresch, H., and Shairer, H. U.**, Über die Wirkstoffe des Crotonöls. VI. Entzandungsteste am Mäuseohr, *Z. Krebsforsch.*, 68, 366, 1966.
95. **Evans, F. J.**, The irritant toxins of the blue *Euphorbia, Euphorbia coerulescens, Toxicon*, 16, 51, 1978.
96. **Evans, F. J., Schmidt, R. J., and Kinghorn, A. D.**, A microtechnique for the identification of diterpene ester inflammatory toxins, *Biomed. Mass. Spectrom.*, 2, 126, 2975.

97. **Berry, D. L.,** Separation of a series of phorbol-ester tumor-promoters by L.C., *Chromatogr. Rev.*, 3, 5, 1977.
98. **Evans, F. J. and Kinghorn, A. D.,** A screening method for co-carcinogens, *J. Pharm. Pharmacol.*, 25, 145P, 1973; see also **Kinghorn, A. D.,** Some biologically Active Constituents of the Genus *Euphorbia*, Ph.D. thesis, University of London, London, 1975.
99. **Evans, F. J.,** unpublished results, 1975.
100. **Neeman, M. and Simmonds, O. D.,** Carbon-13-nuclear magentic resonance spectroscopy of phorbol, *Can. J. Chem.*, 57, 2071, 1979.
101. **Taylor, S. E., Scutt, A., and Evans, F. J.,** ^{13}C-NMR and ^{1}H-NMR spectroscopy of phorbol and deoxy-phorbol-esters, *J. Pharm. Pharmacol.*, 34, 109P, 1982; see also **Taylor, S. E.,** Toxic Diterpenes from Indian Euphorbiaecea, Ph.D, thesis, University of London. London, 1982.
102. **Gshwendt, M. and Hecker, E.,** Tumor-promoting compounds from *Euphorbia triangularis*, Mono and diesters of 12-deoxyphorbol, *Tetrahedron Lett.*, 3509, 1969.
103. **Gschwendt, M. and Hecker, E.,** Tumor promovierende Diterpenfettsäureester aus *Euphorbia triangularis* und *E. cooperi, Fette, Seifen, Anstrichm.*, 73, 221, 1971.
104. **Gschwendt, M. and Hecker, E.,** Hautreizende und co-carcinogene Fraktionen aus *E. triangularis, Z. Krebsforsch.*, 81, 193, 1974.
105. **Schmidt, R. and Evans, F. J.,** Skin irritants of the sun spurge *(Euphrobia helioscopia L.), Contact Derm.*, 6, 204, 1980.
106. **Hergenhahn, M., Kusomoto, M. S., and Hecker, E.,** Diterpene esters from Euphorbium and their irritant and co-carcinogenic activity, *Experientia*, 30, 1438, 1974.
107. **Kinghorn, A. D. and Evans, F. J.,** Skin irritants of *Euphorbia fortissima, J. Pharm. Pharmacol.*, 27, 329, 1975.
108. **Evans, F. J.,** The skin irritants of the blue *Euphorbia, (E. coerulescens), Toxicon*, 16, 51, 1978.
109. **Evans, F. J., Kinghorn, A. D., and Schmidt, R. J.,** Some naturally occurring skin irritants, *Acta Pharmacol. Toxicol.*, 37, 250, 1975.
110. **Evans, F. J. and Schmidt, R. J.,** Structure and potency of the aromatic diterpenes of *Euphorbia poissonii* Pax., *Acta Pharmacol. Toxicol.*, 45, 181, 1979.
111. **Schmidt, R. J. and Evans, F. J.,** Aliphatic diterpene esters from *Euphorbia poissonii* Pax and *E. unispina* N.E. Br., *Lloyida*, 40, 225, 1977.
112. **Kinghorn, A. D.,** Some biologically Active Constituents of the Genus *Euphorbia*, Ph.D. thesis, University of London, London, 1975.
113. **Evans, F. J. and Kinghorn, A. D.,** A comparative phytochemical study of the diterpenes of some species of the genera *Euphorbia* and *Elaeophorbia* (Euphorbiaceae), *Bot. J. Linn. Soc.*, 74, 23, 1977.
114. **Chavez, P. I., Jolad, S. D., Hoffmann, J. J., and Cole, J. R.,** Four new 12-deoxyphorbol diesters from *Croton californicus, J. Nat. Prod.*, 45, 745, 1982.
115. **Ogura, M., Koike, K., Cordell, G. A., and Farnsworth, N. R.,** Potential anticancer agents. VIII. Constituents of *Baliospermum montanum* (Euphorbiaceae), *Planta Med.*, 33, 128, 1978.
116. **Casmore, A. R., Seelye, R. N., Carin, B. F., Mack, H., Schmidt, R., and Hecker, E.,** The structure of prostatin a toxic tetracyclic diterpene ester from *Pimelea prostrata, Tetrahedron Lett.*, 1737, 1976.
117. **Schmidt, R. J. and Evans, F. J.,** The skin irritant effects of phorbol and related polyols, *Arch. Toxicol.*, 44, 279, 1980.
118. **Westwick, J., Williamson, E. M., and Evans, F. J.,** Structure-activity relationships of 12-deoxyphorbol esters on human platelets, *Thrombosis Res.*, 20, 683, 1980.
119. **Edwards, M., Westwick, J., and Evans, F. J.,** Release of an aggregating substance by human platelets in response to phorbol-ester stimulation, *J. Pharm. Pharmacol.*, 34, 40P, 1982.
120. **Evans, R. J. and Kinghorn, A. D.,** New diesters of 12-deoxyphorbol, *Phytochemistry*, 14, 1669, 1975.
121. **Schmidt, R. J. and Evans, F. J.,** A new aromatic ester diterpene from *Euphorbia poissonii, Phytochemistry*, 15, 1778, 1976.
122. **Schmidt, R. J. and Evans, F. J.,** Two minor diterpenes from *Euphorbia* latex, *Phytochemistry*, 17, 1436, 1978.
123. **Evans, F. J. and Schmidt, R. J.,** Two new toxins from the latex of *Euphorbia poissonii, Phytochemistry*, 15, 333, 1976.
124. **McCormick, I. R. N., Nixon, P. E., and Waters, T. N.,** On the structure of prostratin, an X-ray study, *Tetrahedron Lett.*, 1735, 1976.
125. **Zayed, S., Hafez, A., Adolf, A., and Hecker, E.,** New tigliane and daphnane derivatives from *Pimelea prostrata* and *P. simplex, Experientia*, 33, 1554, 1977.
126. **Gschwendt, M. amd Hecker, E.,** Tumor-promoting compounds from *Euphorbia cooperi*, di and triesters of 16-hydroxy-12-deoxy-phorbol, *Tetrahedron Lett.*, 567, 1970.
127. **Gschwendt, M. and Hecker, E.,** Hautreizende und co-carcinogene aus *Euphorbia cooperi, Z. Krebsforsch.*, 80, 335, 1973.

128. **Schmidt, R. J. amd Evans, F. J.,** Candeltoxins A and B, two new aromatic esters of 12-deoxy-16-hydroxy-phorbol from the latex of *Euphorbia poissonii, Experientia,* 33, 1197, 1977.
129. **Schmidt, R. J.,** Chemical and Biological Studies on the Tigliane and Daphnane Diterpenes of Three *Euphorbia* Species, Ph.D. thesis, University of London, London, 1978.
130. **Adolf, W., Opferkuch, H. J., and Hecker, E.,** Irritant phorbol derivatives from four *Jatropha* species, *Phytochemistry,* 23, 129, 1984.
131. **Taylor, S. E.,** Toxic Diterpenes from Indian Euphorbiaceae, Ph.D. thesis, University of London, London, 1982.
132. **Hergenhahn, M., Adolf, W., and Hecker, E.,** Resiniferatoxin and other novel polyfunctional diterpenes from *Euphorbia resinifera* and *E. unispina, Tetrahedron Lett.,* 1595, 1975.
133. **Adolf, W. and Hecker, E.,** New irritant diterpene esters from the roots of *Stillingia sylvatica* L. (Euphorbiaceae), *Tetrahedron Lett.,* 21, 2887, 1980.
134. **Adolf, W. and Hecker, E.,** On the irritant and co-carcinogenic principles of *Hippomane manicinella, Tetrahedron Lett.,* 1587, 1975.
135. **Fürstenberger, G. and Hecker, E.,** The new diterpene 4-deoxyphorbol and its highly unsaturated irritant diesters, *Tetrahedron Lett.,* 925, 1977.
136. **Fürstenberger, G. and Hecker, E.,** Zum Wirkungsmechanismus co-carcinogener Pflanzeninhaltsstoffe, *Planta Med.,* 22, 241, 1972.
137. **Kinghorn, A. D.,** Characterization of an irritant 4-deoxyphorbol-diester from *Euphorbia tirucalli, J. Nat. Prod.,* 42, 112, 1979.
138. **Falsone, G. and Crea, A. E. G.,** Drei neue 4-Deoxyphorbol-triesters aus *Euphorbia biglandulosa, Liebigs Ann. Chem.,* 1116, 1979.
139. **Taylor, S. E., Gafur, M. A., Choudhury, A. K., and Evans, F. J.,** Sapatoxins, aliphatic ester tigliane diterpenes from *Sapium indicum, Phytochemistry,* 21, 405, 1982.
140. **Taylor, S. E., Gafur, M. A., Choudhury, A. K., and Evans, F. J.,** Sapintoxin-A, a new biologically active nitrogen containing phorbol-ester, *Experientia,* 37, 681, 1981.
141. **Miana, G. A. R., Schmidt, R., Hecker, E., Shamma, M., Moniot, J. L., and Kiammudin, M.,** 4α-Sapinine a novel diterpene ester from *Sapium indicum, Z. Krebsforsch.,* 326, 727, 1977.
142. **Weber, J. and Hecker, E.,** Co-carcinogens of the diterpene ester type from *Croton flavens* and oesophageal cancer in Curacao, *Experientia,* 34, 1595, 1978.
143. **Hecker, E. and Weber, J.,** Co-carcinogens from *Croton flavens* and the high incidence of oesophageal cancer in Curacao, in *Proc. 7th Int. Symp. Biol. Characterisation of Human Tumors,* Davis, W., Ed., *Excerpta Med.,* Budapest, 1977.
144. **Taylor, S. E., Gafur, M. A., Choudhury, A. K., and Evans, F. J.,** Nitrogen containing phorbol derivatives of *Sapium indicum, Phytochemistry,* 20, 2749, 1981.
145. **Taylor, S. E., Gafur, M. A., Choudhury, A. K., and Evans, F. J.,** 4-Deoxyphorbol and 4α-deoxyphorbol aldehydes, new diterpenes of their esters, *Tetrahedron Lett.,* 22, 3321, 1981.
146. **Okuda, T., Yoshida, T., Koike, S., and Toh, N.,** The toxic constituents of the fruits of *Aleurites fordii, Chem. Pharm. Bull.,* 22, 971, 1974.
147. **Okuda, T., Yoshida, T., Koike, S., and Toh, N.,** New diterpene esters from *Aleurites fordii, Phytochemistry,* 14, 509, 1975.
148. **Abo, K. A.,** Cytotoxic and Pro-Inflammatory Diterpene-Esters of Nigerian *Euphorbia kamerunica,* Ph.D. thesis, University of London, London, 1982.
149. **Edwards, M. C., Taylor, S. E., Williamson, E. M., and Evans, F. J.,** New phorbol and deoxyphorbol esters, isolation and relative potencies in inducing platelet aggregation and erythema of skin, *Acta Pharmacol. Toxicol.,* 53, 177, 1983.
150. **Taylor, S. E., Williamson, E. M., and Evans, F. J.,** New deoxyphorbol-esters from *Sapium insigne, Phytochemistry,* 22, 1231, 1983.

Chapter 8

THE DAPHNANE POLYOL ESTERS

Richard J. Schmidt

TABLE OF CONTENTS

I. INTRODUCTION

The term "daphnane" has been adopted for the hydrocarbon structure "13,15-seco tigliane". It has also been utilized in *Chemical Abstracts* to describe the hypothetical parent compound of certain alkaloids found in plants of the family Daphniphyllaceae, but does not appear to have been adopted by authors of research publications in that field. Nevertheless, there arises the possibility of confusion.[1]

As a group, the daphnane-type polyol esters are structurally more varied and interesting than either the tiglianes or the ingenanes. All but one of the currently known compounds have an orthoester structure, which appears to be the natural consequence of the presence of three α-orientated hydroxyl functions on the C-9, C-13, and C-14 positions of the daphnane nucleus. The one compound lacking the orthoester structure, namely proresiniferatoxin, appears to be the immediate biosynthetic precursor of resiniferatoxin which is an orthoester (see Section III. C).

Perhaps as unusual as the orthoester function is the more recently characterized 1α-alkyldaphnane type of structure. This appears to be a further elaboration of an orthoester-type precursor in which a specifically functionalized long-chain alkyl orthoester is a necessary prerequisite for macrocyclic cyclization (see Section III.E1).

In general, like the tiglianes and ingenanes, the daphnanes may elicit skin inflammation on contact with mammalian skin, may exhibit cocarginogenic activity on mouse skin, and may have antitumor activity (see Chapter 1). However, it is the antitumor activity that has stimulated the most interest. In particular, an experimental tumor system measuring activity against P-388 lymphocytic leukemia in mice has been commonly used to monitor fractionation of the plant extracts from which the compounds have been extracted.[2-10] The skin irritant properties of many of the daphnane polyol esters are remarkable, resiniferatoxin and related resiniferonol orthoesters[11,12] and Daphnopsis factors R_1 and R_5 and related 1α-alkyldaphnane-type compounds[13] are among the most potent pro-inflammatory compounds known to man. Cocarcinogenicity has been studied the least, available data showing that the daphnane polyol esters resemble the tigliane polyol esters in that pro-inflammatory potency is not directly related to cocarcinogenic potency.[12,13,15] Cocarcinogenicity may also co-occur with skin irritant activity and with antileukemic activity in the same compound, for example, in mezerein[2,16] and in simplexin.[8,15]

In common with the tiglianes and ingenanes, the daphnane polyol esters also exhibit piscicidal activity; in Japan, the killie fish *(Oryzias latipes)* has been used to monitor fractionation procedures,[17-21] while goldfish have been used in the U.S.[5,22] An alternative bioassay procedure measuring nematocidal activity against the rice white tip nematode *(Aphelenchoides besseyi)* has also been reported.[23]

II. STRUCTURE ELUCIDATION OF DAPHNETOXIN

The first example of a compound with a daphnane hydrocarbon skeleton to be described was daphnetoxin. Stout et al.[22] isolated daphnetoxin from the bark of *Daphne mezereum* L. (family Thymelaeaceae) and from commercially available "mezeron bark" (derived form *D. mezereum, D. laureola* L., and *D. gnidium* L.). The similarity between daphnetoxin and then-recently described phorbol esters from croton oil *(Croton tiglium* L., family Euphorbiaceae) was recognized principally from proton magnetic resonance (^{1}H-NMR) data, but the spectral details of daphnetoxin were not given. Ronlán and Wickberg[24] then reported the presence of daphnetoxin in the seeds of *D. mezereum* (in which it co-occurred with a larger amount of another orthoester named mezerein), and also published the results of a detailed ^{1}H-NMR study of daphnetoxin and mezerein. It was considered remarkable that the daphnane hydrocarbon skeleton of these compounds from members of the plant family

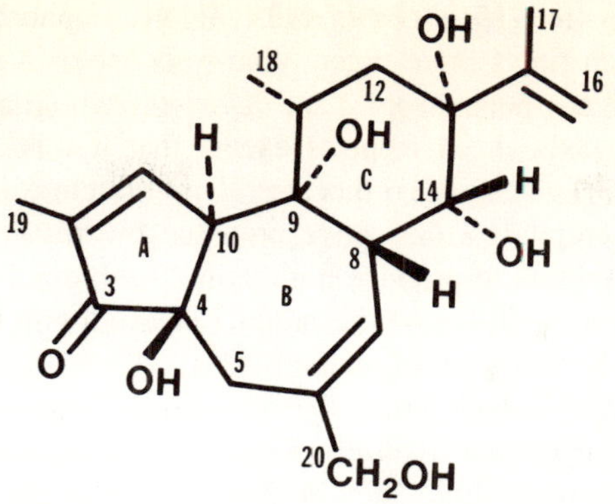

FIGURE 1. Absolute configuration and numbering of resiniferonol.[25]

Thymelaeaceae was so similar to the tigliane hydrocarbon skeleton of the phorbol esters derived from *C. tiglium* L., a member of the family Euphorbiaceae. The only difference between the two hydrocarbon skeletons is the opened cyclopropane ring (ring D) of the tigliane structure, being present as an isopropenyl side chain on C-13 of the daphnane-type esters. Biosynthetically (see Chapter 4), stereochemically, and in terms of their biological activities, the daphnane-type compounds are closely related to both the tigliane and ingenane types of compounds also discussed in this volume. Of the five classes of daphnane polyol esters currently recognized, the daphnetoxin- and 12-hydroxydaphnetoxin-types are found in both the Euphorbiaceae and the Thymelaeaceae, providing chemosystematic evidence for a close relationship between these two families (see Chapter 4). The 5-deoxydaphnetoxin and 1α-alkyldaphnane types are at present only known from the family Thymelaeaceae; resiniferonol esters have been reported only from members of the family Euphorbiaceae.

III. ESTERS OF DAPHNANE POLYOLS

A. 5β-Hydroxyresiniferonol-6α,7α-Oxide ("Daphnetoxin Type") Esters

The rather unwieldy nomenclature is the unfortunate result of a failure by the authors of the early literature on daphnetoxin and mezerein to provide a trivial name for the parent polyol of the compounds they isolated and characterized; the term resiniferonol was not coined until 1975 when Hergenhahn et al.[25] reported the isolation of resiniferonol orthoesters from *Euphorbia resinifera* Berg (see Section III.C). The term "daphnetoxin-type" has been used to described this class of compounds and is perhaps sufficiently acceptable to warrant its retention. Figure 1 gives the stereochemistry and numbering of the carbon atoms of resiniferonol.

The first member of this group to be characterized was daphnetoxin. Its structure was unequivocally determined by X-ray crystallography of daphnetoxin bisbromoacetate.[22] In its ^{1}H-NMR spectrum,[24] the signals for the isopropylene side chain are evident at δ1.80 (3H, s) and δ5.00 (2H, ABq), replacing the 6H signal associated with the geminal dimethyl group of ring D in the tigliane type of compound. In addition, the signal at about δ5.8 associated with the olefinic proton on C-7 of phorbol-12,13,20-triacetate is no longer evident in the ^{1}H-NMR spectrum of daphnetoxin, being replaced by a signal further upfield at δ3.4 to 3.5. This is consistent with the replacement of the $\Delta^{6,7}$ function with an epoxy group. The 1H doublet signal at δ4.57 is characteristic of the orthoester function, especially when associated with the presence of signals for an isopropylene group and the absence of the high field, but often obscured, signal for the proton on C-14 in the tigliane type of compound.

Shortly after the isolation of daphnetoxin, a further daphnetoxin-like compound was isolated from the sandbox tree *(Hura crepitans* L., family Euphorbiaceae). The compound which was noted to be ten times more potent than rotenone as a piscicide, was given the name huratoxin.[17] Chemical, spectroscopic, and X-ray crystallographic studies showed it to be identical with daphnetoxin in all respects except that it was a 9,13,14-tetradeca-2,4-dienoate rather than a 9,13,14-benzoate orthoester. X-ray structure determination was carried out on *p*-bromobenzoyl crepital, a derivative prepared by treating huratoxin with *p*-bromobenzoyl chloride in pyridine to produce huratoxin-20-*p*-bromobenzoate which was then ozonized in ethyl acetate at $-70°C$ and the product cleaved with triphenylphosphine. The structure was solved by the heavy atom method.[18,19]

Huratoxin has also been isolated from the roots of *Stellera chamaejasme* L. (family Thymelaeaceae),[21] an ether extract of *Wikstroemia monticola* Skottsb. (family Thymelaeaceae),[26] and an acetone extract of the latex of the manchineel tree *(Hippomane mancinella* L., family Euphorbiaceae).[27] In the latter species, it co-occurs with the closely related mono-orthoester Hippomane factor M_2 (from which is could not be separated) and a complex mixture of their C-20 alkyl acylates given the name Hippomane factor M_x'. These C-20 acylates also could not be chromatographically separated from one another, and were characterized by a quantitative gas-liquid chromatographic (GLC) analysis of their methyl esters.[27] Adolf and Hecker[27] also isolated a tigliane polyol monoester from *H. mancinella*. This compound was named Hippomane factor M_3 or mancinellin and was thought to be a biosynthetic precursor of Hippomane factor M_2.

Excoecariatoxin, a further daphnetoxin-type orthoester, has been reported from the latex of *Excoecaria agallocha* L. (family Euphorbiaceae)[20] and has been found also in both *Gnidia kraussiana* Meissner and *Wikstroemia monticola* Skottsb. of the family Thymelaeaceae.[10,26] *W. monticola* also afforded further such orthoesters which were given the names wikstrotoxins A, B, and D.[26] One of these, wikstrotoxin D, appears to be identical with simplexin, a daphnetoxin-type orthoester originally found in the desert rice flower *(Pimelea simplex* F. Muell., family Thymelaeaceae),[28,29] but now known to occur in a number of other species, all belonging to the family Thymelaeaceae (see Table 1). Montanin, a homologue of simplexin, has been found in the roots of *Baliospermum montanum* Muell. Arg.[6] and *Cunuria spruceana* Baillon,[30] both of which belong to the family Euphorbiaceae; this poses an interesting question regarding fatty acid biosynthesis. Both simplexin and montanin produce 5,20-acetonides when treated with *p*-toluenesulfonic acid in acetone, this being a characteristic feature of all known tigliane- and daphnane-type esters with a 5β-hydroxy-6α,7α-oxide grouping and a primary hydroxyl function on C-20. Such compounds lacking the 6,7-oxide grouping, having instead a $\Delta^{6,7}$, produce 4,5-acetonides.[13]

The structures and sources of naturally occurring daphnetoxin-type orthoesters are given in Table 1; comparative ^{13}C-NMR and ^{1}H-NMR data are given in Tables 6 and 7.

B. 5β,12β-Dihydroxyresiniferonol-6α,7α-Oxide ("12-Hydroxydaphnetoxin Type") Esters

To avoid the somewhat cumbersome chemical description, this group of compounds may be conveniently described as the "12-hydroxydaphnetoxin type", by analogy with the "daphnetoxin type" described above.

The first compound of this type to be characterized was mezerein which occurs in *Daphne mezereum* L. (family Thymelaeaceae) together with small amounts of daphnetoxin.[2,16,24] The structure was confirmed X-ray crystallographically by Nyborg and la Cour[32] using direct methods, showing that a secondary aromatic ester was located on C-12 rather than on C-5 as had previously[33] been suggested.

As well as differences ascribable to the cinnamylideneacetic and benzoic ester function of mezerein, the principal difference between its ^{1}H-NMR spectrum and that of daphnetoxin is the signal associated with the proton(s) on C-12. In the case of mezerein, this signal is a 1H singlet at δ5.12; in the case of daphnetoxin, this signal is a 2H multiplet at δ2.36.

Table 1
DAPHNETOXIN TYPE ESTERS

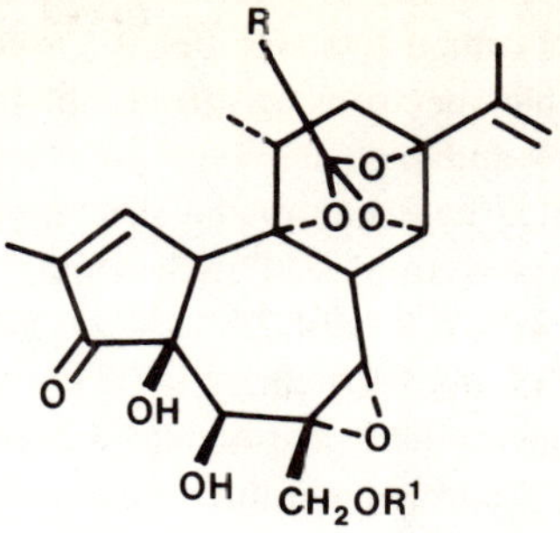

Name	R	R[1]	Source
Daphnetoxin	$-C_6H_5$	H	*Daphne mezereum* L.[22] *D. laureola* L.[22] *D. gnidium* L.[22]
Excoecariatoxin	$-(CH=CH)_2.(CH_2)_4.CH_3$	H	*Excoecaria agallocha* L.[20] *Gnidia kraussiana* Meissner[10] *Wikstroemia monticola* Skottsb.[26]
Simplexin	$-(CH_2)_8.CH_3$	H	*Pimelea simplex* F. Muell.[15,28,31]
Pimelea factor P_1, Pimelea factor S_8, Daphnopsis factor R_3, Wikstrotoxin D			P. prostrata Willd.[8,14,31] *P. trichostachya* Lindley[28] *Daphnopsis racemosa* Griseb.[13] *Stellera chamaejasme* L.[21] *W. monticola* Skottsb.[26]
Wikstrotoxin B	$-(CH=CH)_2.(CH_2)_6.CH_3$	H	*W. monticola* Skottsb.[26]
Montanin	$-(CH_2)_{10}.CH_3$	H	*Baliospermum montanum* Muell. Arg.[6] *Cunuria spruceana* Baillon[30]
Huratoxin	$-(CH=CH)_2.(CH_2)_8.CH_3$	H	*Hura crepitans* L.[17—19]
Hippomane factor M_1			*Hippomane mancinella* L.[27] *S. chamaejasme* L.[21] *W. monticola* Skottsb.[26]
Pimelea factor P_4	$-C_{13}H_{27}$	H	*P. prostrata* Willd.[14,31]
Wikstrotoxin A	$-(CH=CH)_2.(CH_2)_9CH_3$	H	*W. monticola* Skottsb.,[26]
Hippomane factor M_2	$-(CH=CH)_3.(CH_2)_8.CH_3$	H	*H. mancinella* L.[27]
Hippomane factors M'_x	$-(CH=CH)_n.(CH_2)_8.CH_3$ n = 2, 3	$CO.(CH_2)_n.CH_3$ n = 14, 16, 18, 20, 22, 24	*H. mancinella* L.[27]

The observation that this signal is a singlet rather than a doublet in the case of mezerein is worthy of note, evidently being brought about by the angle between the two protons on C-11 and C-12, and supporting the conclusion that the ester of C-12 is β- rather than α-oriented. The downfield shift of the signal for the proton on C-8 in the ^{1}H-NMR spectrum of mezerein when compared with that of daphnetoxin is most probably caused by the *endo* configuration of the C-12 substituent.[24]

Although the fragmentation of mezerein by electron-impact mass spectrometry (EIMS) has been reported by Ronlán and Wickberg,[24] it would appear to be of little value in providing evidence for the nature of the esterifying acids. This may well have been the result of recording the spectrum at an inappropriate source temperature or accelerating voltage. Certainly in the case of resiniferonol orthoesters (see below), source temperature was found to be critical if an easily interpretable spectrum was to be obtained.[34]

As in the case of the tigliane- and ingenane-type polyol esters, IR and UV spectra of mezerein provide useful supportive evidence of the structure, but are of only minor value in the structure elucidation of new examples of such compounds. Mezerein exhibits[24] UV absorption maxima at 227 (logϵ 4.24), 234 (4.29), 241 (4.27), and 314 (4.60) nm, and IR absorbances at 3520, 1714, 1698, and 1626/cm.

After mezerein, the next compound to be isolated and characterized was 12-hydroxydaphnetoxin itself following mild hydrolysis of three toxic principles from a benzene extract of the leaves of *Lasiosiphon burchellii* Meissner (family Thymelaeaceae). Again, X-ray crystallography of the compound as its tri-bromoacetate derivative confirmed the structure unequivocally.[35] The identities of the three esters from the plant material were not determined, but one was noted to be a benzoate ester while another was a cinnamate ester.

Several *Gnidia* species (family Thymelaeaceae) then yielded a series of orthoesters of the 12-hydroxydaphnetoxin type but varying in the nature and number of ester functions.[3,4] These compounds were isolated by bioassay-guided fractionation of crude extracts in search of antileukemic activity, and were found to significantly increase survival time of mice with P-388 lymphocytic leukemia. More recently, gnidilatidin, one of the orthoesters isolated from *G. latifolia* hort. ex Meissner,[4] has been found in a methanol extract of the roots of *Stillingia sylvatica* L. (family Euphorbiaceae).[1] This and a number of other skin irritant orthoesters, designated Stillingia factors S_1 to S_6, and being of the 12-hydroxydaphnetoxin type, have been described by Adolf and Hecker.[1]

Also from the Thymelaeaceae, Kogiso et al.[23] reported the isolation of a nematocidal compound named odoracin from *Daphne odora* Thunb.; an identical compound (or possibly an isomer) was subsequently reported to occur in the roots of *D. genkwa* Siebold & Zucc. and was given the name yuanhuacine.[36] Subsequent investigation of the flowers of *D. genkwa* resulted in the isolation and characterization of further 12-hydroxydaphnetoxin-type orthoesters named genkwadaphnin and yuanhuafin.[7,37,38] Genkwadaphnin was found to have significant inhibitory activity against P-388 leukemia in mice,[7] while yuanhuafin is reportedly abortifacient in monkeys.[38] More recently, Rizk et al.[39] have reported the occurrence of mezerein-like orthoesters in *Thymelaea hirsuta* Endl., one of which was identical with gnidicin[3] previously isolated from *Gnidia lamprantha* Gilg.

Jolad et al.,[26] during an investigation of an ether extract of *Wikstroemia monticola* Skottsb. stems and bark for antitumor activity, isolated and characterized several daphnane polyolorthoesters. One, wikstrotoxin C, was of the 12-hydroxydaphnetoxin type but was unusual in having a 4,5-hemiorthoacetate group. Evidence for this was obtained from ^{1}H- and ^{13}C-NMR spectra. In the ^{1}H-NMR spectrum, a 3H singlet at δ1.83 appeared to be too far upfield for a normal acetate ester function. Since the signals for the C-20 protons were apparently unaffected, and the location of a secondary acetate on C-12 was clear, a hemiorthoacetate group on C-4 and C-5 seemed plausible. The ^{13}C-NMR spectrum exhibited a signal at δ111.4 consistent with the presence of a quaternary carbon of a hemiorthoacetate function.

Table 2 summarized the structures and botanical sources of the naturally occurring 12-hydroxydaphnetoxin type orthoesters; comparative ^{13}C-NMR and ^{1}H-NMR data are given in Tables 6 and 7.

C. Resiniferonol Esters

This group of daphnane derivatives most closely resembles the classical phorbol-type

Table 2
12-HYDROXYDAPHNETOXIN TYPE ESTERS

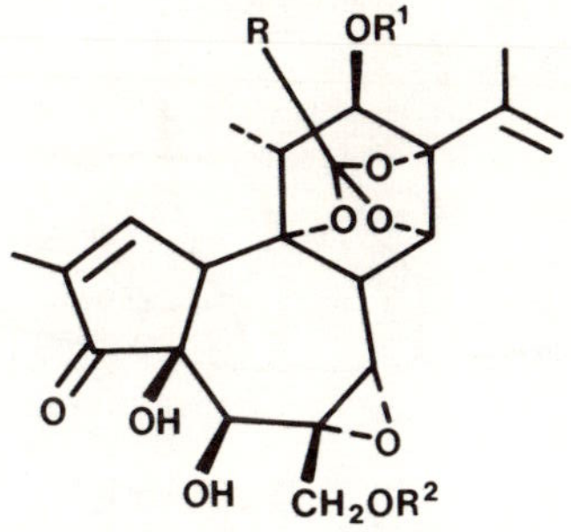

Name	R	R^1	R^2	Source
Yuanhuafin	$-C_6H_5$	$CO.CH_3$	H	*Daphne genkwa* Siebold & Zucc.[37,38]
Genkwadaphnin	$-C_6H_5$	$CO.C_6H_5$	H	*D. genkwa* Siebold & Zucc.[37,38]
Gnidicin	$-C_6H_5$	$CO.CH{=}CH.C_6H_5$	H	*Gnidia lamprantha* Gilb.[3] *Thymelaea hirsuta* Endl.[39]
Gniditrin	$-C_6H_5$	$CO.(CH{=}CH)_3.(CH_2)_2.CH_3$	H	*G. lamprantha* Gilg.[3] *D. tangutica* Maxim.[40]
Gnididin	$-C_6H_5$	$CO.(CH{=}CH)_2.(CH_2)_4CH_3$	H	*G. lamprantha* Gilg.[3]
Mezerein	$-C_6H_5$	$CO.(CH{=}CH)_2.C_6H_5$	H	*D. mezereum* L.[2,16,24,32,33]
Unnamed compound	$-C_6H_5$	$CO.CH{=}CH.(CH_2)_{13}.CH_3$	H	*T. hirsuta* Endl.[39]
Stillingia factor S_1	$-(CH{=}CH)_2.(CH_2)_2.CH_3$	$CO.(CH{=}CH)_3.(CH_2)_3.CH_2OH$	H	*Stillingia sylvatica* L.[1]
Stillingia factor S_2	$-(CH{=}CH)_2.(CH_2)_2.CH_3$	$CO.(CH{=}CH)_3.(CH_2)_3.CH_2O.CO.(CH_2)_{12}.CH_3$	H	*S. sylvatica* L.[1]
Stillingia factor S_3	$-(CH{=}CH)_2.(CH_2)_2.CH_3$	$CO.(CH{=}CH)_3.(CH_2)_3.CH_2O.CO.(CH_2)_{11}.CH_3$	H	*S. sylvatica* L.[1]
Odoracin, Yuanhuacine	$-(CH{=}CH)_2.(CH_2)_4.CH_3$	$CO.C_6H_5$	H	*D. odora* Thunb.[23] *D. genkwa* Siebold & Zucc.[36]

Table 2 (continued)
12-HYDROXYDAPHNETOXIN TYPE ESTERS

Name	R	R[1]	R[2]	Source
Gnidiglaucin	$-(CH_2)_8.CH_3$	$CO.CH_3$	H	*G. glauca* Steudel[4]
Gnidilatidin, Stillingia factor S_6	$-(CH{=}CH)_2.(CH_2)_4.CH_3$	$CO.C_6H_5$	H	*G. latifolia* hort. ex Meissner[4] *G. kraussiana* Meissner[10] *S. sylvatica* L.[1]
Gnidilatin	$-(CH_{28}).CH_3$	$CO.C_6H_5$	H	*G. latifolia* hort. ex Meissner[4] *G. kraussiana* Meissner[10]
Stillingia factor S_4	$-(CH{=}CH)_2.(CH_2)_4.CH_3$	$CO.(CH{=}CH)_3.(CH_2)_2.CH_2O.CO.C_{13}H_{27}$	H	*S. sylvatica* L.[1]
Stillingia factor S_5	$-(CH{=}CH)_2.(CH_2)_4.CH_3$	$CO.(CH{=}CH)_3.(CH_2)_2.CH_2O.CO.C_{14}H_{27}$	H	
Subtoxin A	$-(CH{=}CH)_2.(CH_2)_8.CH_3$	$CO.CH_3$	H	*P. simplex* F. Muell.[28] *Stellera chamaejasme* L.[21]
Wikstrotoxin C[a]	$-(CH{=}CH)_2.(CH_2)_8.CH_3$	$CO.CH_3$	H	*Wikstroemia monticola* Skottsb.[26]
Gnidilatidin-20-palmitate	$-(CH{=}CH)_2.(CH_2)_4.CH_3$	$CO.C_6H_5$	$CO.(CH_2)_{14}.CH_3$	*G. latifolia* hort. ex Meissner[4]
Gnidilatin-20-palmitate	$-(CH_2)_8.CH_3$	$CO.C_6H_5$	$CO.(CH_2)_{14}.CII_3$	

[a] Wikstrotoxin C is subtoxin A -4,5-hemiorthoacetate:

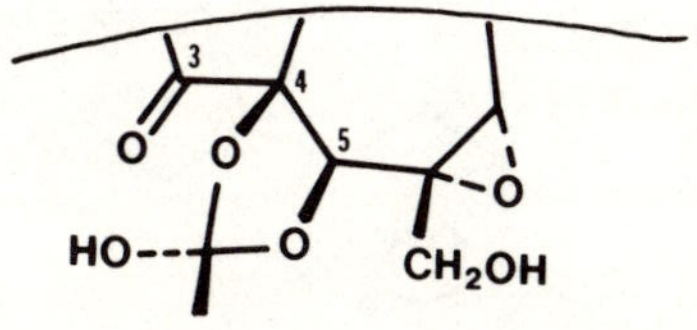

tigliane derivatives, being characterized by the presence of the $\Delta^{1,2}$ and $\Delta^{6,7}$ bonds. Otherwise, the presence of an isopropylene side chain on C-13 and a 9,13,14-orthoester renders these compounds typical of the daphnane types known to date. The first of this group to be characterized was resiniferatoxin, which was isolated from *Euphorbia resinifera* Berg,[25] but which is now known to occur in two other species, *E. unispina* N.E. Br.[25,41] and *E. poissonii* Pax.[42] The structure of resiniferatoxin was deduced principally from ^{1}H-NMR data and simple chemical modification reactions. It is interesting because the C-20 ester is phenolic. Hergenhahn et al.[25] concluded that this phenolic acyl group was (3-hydroxy-5-methoxy)phenylacetic acid. This was accepted by Evans and Schmidt[42] on the basis of ^{1}H-NMR evidence in which the resonances of the three isolated protons on the aromatic ring of the phenolic ester were thought to be responsible for the broad 3H singlet at about δ6.8 Adolf et al.[12] subsequently revised the structure of this acyl group to (4-hydroxy-3-methoxy)phenylacetic acid after partially synthesizing resiniferatoxin from resiniferonol-9,13,14-orthophenylacetate and homovanillic acid. It is difficult to reconcile this conclusion with the ^{1}H-NMR spectrum of resiniferatoxin unless one refers to published[43] spectra of homovanillic acid and other 3,4-disubstituted phenylacetic acids.

Resiniferatoxin occurs in the latex of *E. resinifera* with its 20-desacyl derivative,[25] and in both *E. unispina* and *E. poissonii* together with tinyatoxin.[41,42,44] Tinyatoxin differs from resiniferatoxin only in the nature of the C-20 phenolic acylate (Table 3). Both compounds produce the 20-desacyl derivative, namely, resiniferonol-9,13,14-orthophenylacetate, on mild acid-catalyzed transesterification. Resiniferonol-9-13,14-orthophenylacetate-20-acetate is readily produced by acetylation of the 20-desacyl compound but has also been isolated as a natural product from the latex of *E. poissonii*.[45,46]

The ^{1}H-NMR spectra of resiniferatoxin and tinyatoxin are comparable to those of both 12-deoxyphorbol esters and other daphnane polyol orthoesters. The proton on C-7, which is olefinic, resonates at about δ5.8 to 5.9 ppm downfield from tetramethylsilane (TMS, the internal standard). This is in contrast to the upfield position at δ3.4 to 3.5 in the case of compounds having a 6(7)-epoxy groups. The geminal protons on C-5 are evident at about δ2.1 to 2.3, appearing to be nonequivalent to one another. In contrast, compounds having a C-5 hydroxy group exhibit a 1H singlet at about δ4.2 to 4.5; this signal is clearly absent from the spectra of both resiniferatoxin and tinyatoxin.

Electron-impact mass spectra of resiniferatoxin and other resiniferonol orthoesters exhibit fragments associated with the sequential loss of ester functions and hydroxyl groups as free acids and water, respectively.[11,42,44] Resiniferatoxin (70 eV; 210°C) exhibits a base peak molecular ion at m/z 628 ($C_{37}H_{40}O_9$) and significant fragment ions at m/z 492 (9%, $M^{+\cdot}$ − 136), 446 (4%, $M^{+\cdot}$ − 182), 310 (73%, $M^{+\cdot}$ − 136 − 182), each of which being associated with a smaller fragment ion corresponding with the loss of 18 mass units (H_2O). Tinyatoxin (70 eV; 200°C) exhibits a molecular ion at m/z 598 (5%, $C_{36}H_{38}O_8$), with significant fragment ions at m/z 462 (3%, $M^{+\cdot}$ − 136), 446 (5%, $M^{+\cdot}$ − 152), and 310 (100%, $M^{+\cdot}$ − 152 — 136) associated with smaller fragment ions corresponding to loss of 18 mass units (H_2O).

IR spectra of resiniferatoxin and tinyatoxin provide only supportive evidence for structure; resiniferatoxin exhibits absorbances at 3460, 3090, 3060, 3040, 2970, 2950, 2880, 1735, 1715, 1610, 1520, 1500, 1460, 1440, 1380, 1335, 1275, 1240, 1150, 1080, and 1030/cm, while tinyatoxin exhibits absorbances at 3440, 2980, 2950, 2890, 1740, 1715, 1630, 1605, 1520, 1500, 1455, 1380, 1340, 1270, 1240, 1135, 1100, 1080, 1050, and 1030/cm.[42,44]

The circular dichroism (CD) spectra of resiniferatoxin and tinyatoxin, because of their similarity to the CD spectrum of phorbol-12,13,20-triacetate in having a negative maximum Cotton effect at about 200 nm and a positive maximum at about 225 nm, provide evidence for an αβ-unsaturated cyclopentenone (ring A) being trans-linked to the B ring. CD spectra of resiniferatoxin and tinyatoxin are reproduced in Figure 2.

Table 3
RESINIFERONOL ESTERS

Resiniferonol-9,13,14-orthoesters

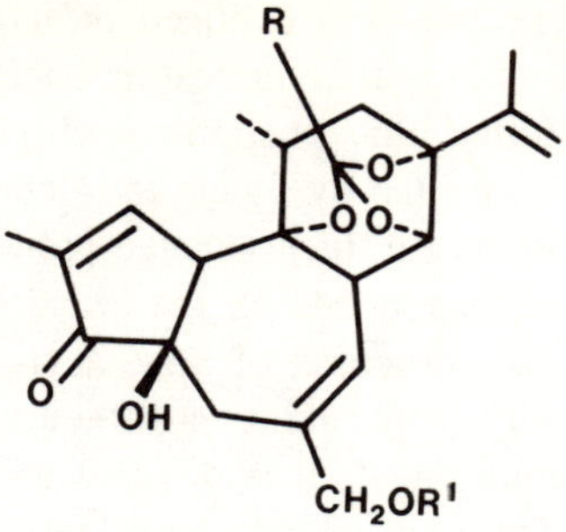

Name	R	R[1]	Source
Resiniferonol-9,13,14-ortho-phenylacetate	$-CH_2.C_6H_5$	H	*Euphorbia resinifera* Berg[25]
Resiniferonol-9,13,14-ortho-phenyl-acetate-20-acetate	$-CH_2.C_6H_5$	$CO.CH_3$	*E. poissonii* Pax[45,46]
Tinyatoxin	$-CH_2.C_6H_5$	Ester 1	*E. poissonii* Pax[41,42] *E. unispina* N.E. Br.[41,44]
Resiniferatoxin	$-CH_2.C_6H_5$	Ester 2	*E. resinifera* Berg[25] *E. poissonii* Pax[42] *E. unispina* N.E. Br.[25,41]

Resiniferonol-14,20-diester

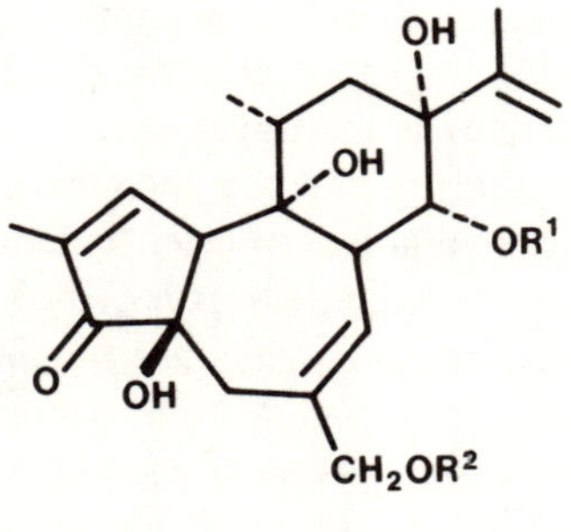

Name	R[1]	R[2]	Source
Proresiniferatoxin	$CO.CH_2.C_6H_5$	Ester 2	*E. resinifera* Berg[25] *E. poissonii* Pax[47]

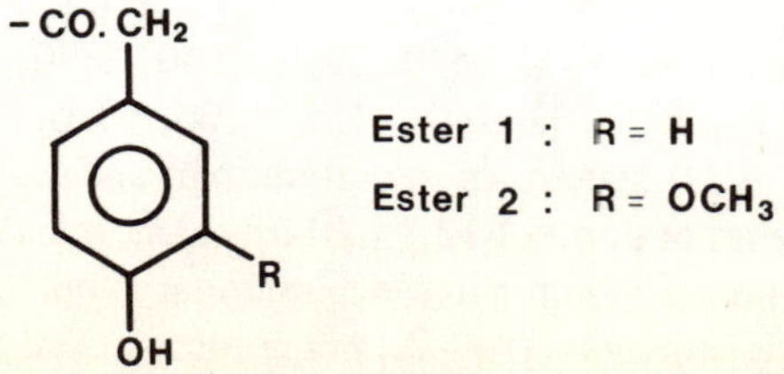

Ester 1 : R = H
Ester 2 : R = OCH_3

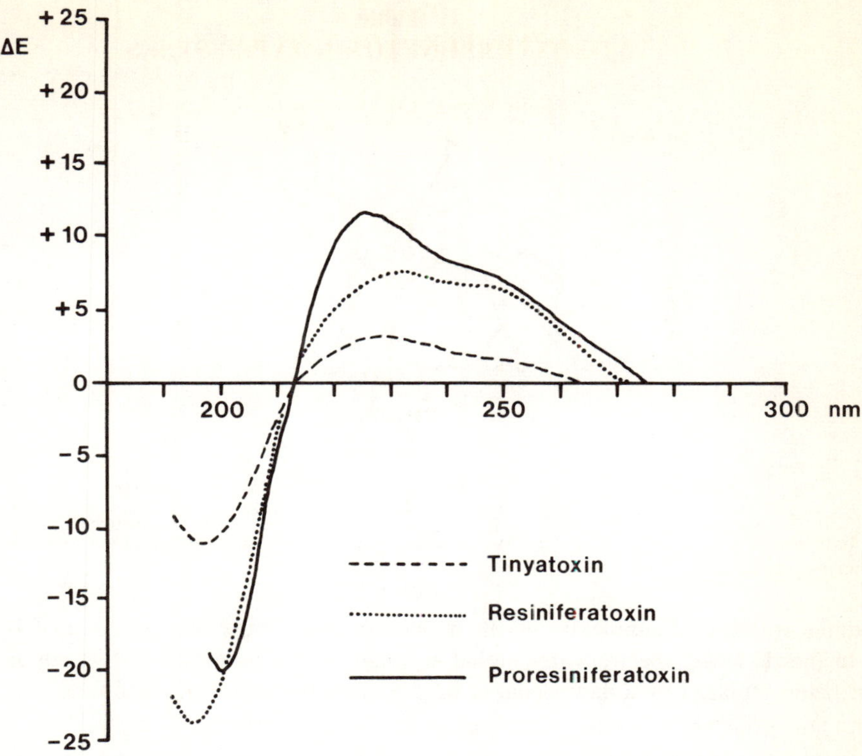

FIGURE 2. CD spectra of tinyatoxin, resiniferatoxin, and proresiniferatoxin.[44]

An unusual daphnane derivative, not having an orthoester function, has been found to co-occur with resiniferatoxin in the latices of *Euphorbia resinifera* Berg[25] and *E. poissonii* Pax.[47] It has been named proresiniferatoxin in recognition of its apparent biosynthetic relationship with resiniferatoxin on the basis of in vitro observations that it is readily converted into resiniferatoxin by mild acid catalyzed intramolecular esterification. Its structure was deduced from its ^{1}H-NMR spectrum by comparison with the spectra of resiniferatoxin and of 12-deoxyphorbol esters. The EIMS was identical with that of resiniferatoxin with an additional weak molecular ion 18 mass units higher than the molecular ion of resiniferatoxin ($M^{+\cdot}$ 646, 1%, $C_{37}H_{42}O_{10}$). The CD spectrum was essentially similar to those of resiniferatoxin and tinyatoxin (see Figure 2), confirming the stereochemistry in the region of the chromophore. The principal difference in the ^{1}H-NMR spectrum was the slight downfield shift of the signal associated with the methylene protons of the phenylacetate ester from δ3.2 to δ3.5-3.6, indicating a "normal" ester rather than an orthoester function. Interestingly, Schmidt[44] could not detect proresiniferatoxin in the latex of *E. unispina;* nor was it reported to co-occur with resiniferatoxin in *E. unispina* in the investigation carried out by Hergenhahn et al.[25] Similarly, a "protinyatoxin" could not be detected in the latices of either *E. poissonii* or *E. unispina*.[44] This poses some interesting questions concerning the significance of the occurrence of proresiniferatoxin in *E. resinifera* and *E. poissonii*.

Table 3 gives structures and botanical sources of resiniferonol esters. Comparative ^{1}H-NMR data are given in Table 7.

D. Resiniferonol-6α,7α-Oxide ("5-Deoxydaphnetoxin Type") Esters

To date, only one compound of this type, 5-deoxysimplexin, has been characterized. It was given the name Daphnopsis factor R_4 after its isolation from the roots of *Daphnopsis racemosa* Griseb. (family Thymelaeaceae).[13] Daphnopsis factor R_4 exhibits spectroscopic

Table 4
5-DEOXYDAPHNETOXIN TYPE ESTERS

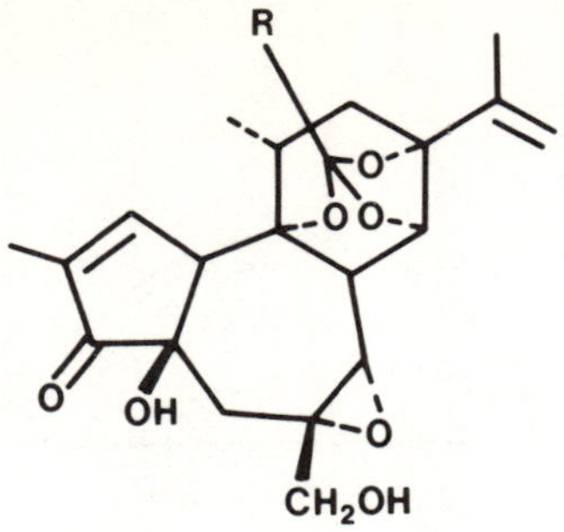

Name	R	R[1]	Source
5-Deoxysimplexin Daphnopsis factor R_4	$-(CH_2)_8 \cdot CH_3$	H	*Daphnopsis racemosa* Griseb.[13]

data similar to those of simplexin, but its molecular weight differs by a deficit of 16 mass units. In the ^{1}H-NMR spectrum, the signal at δ4.24 for the proton on C-5 in simplexin is absent, being replaced by a new signal at δ2.1 attributable to a methylene function. There is also a small diamagnetic shift of the signal for the C-20 protons.[13]

Table 4 gives the structure of Daphnopsis factor R_4: ^{1}H-NMR data are given in Table 7.

E. 1α-Alkyldaphnane Type Esters

Members of this group of compounds have, to date been found to occur naturally only in the plant family Thymelaeaceae. A few resemble the daphnetoxin type compounds in all but two features. First, the long-chain aliphatic orthoester is attached at its sub-terminal carbon atom to C-1 of the daphnane nucleus, producing a macrocyclic fourth ring. Secondly, and probably as an inevitable result of macrocyclic cyclization, the $\Delta^{1,2}$ bond is saturated. The remaining majority of 1α-alkyldaphnane derivatives have as their parent alcohols a series of polyols varying in their degree of hydroxylation in an analogous manner to the ingenane polyols. To facilitate classification, the name macrinol is proposed here for the parent polyol of the first such compound to have been described, namely gnidimacrin.[5] Polyols related to macrinol are named in the same manner as are the various ingenane polyols (see Chapter 9).

1. Daphnetoxin Type Esters

The four esters of this type to have been described to date are the enantiomeric Pimelea factors S_6 and S_7 from *Pimelea simplex* F. Muell.,[15,48] Pimelea factor P_6 from *Pimelea prostrata* Willd.,[14,48] and Synaptolepis factor K_1 from an unidentified species of *Synaptolepis*.[48] ^{1}H-NMR spectra of these compounds (see Table 7) show a similarity to the spectrum of daphnetoxin, the most significant difference being the lack of a signal at about δ7.6 for the olefinic C-1 proton of daphnetoxin. It is interesting to note that the signal for the geminal protons on C-20 of Synaptolepis factor K_1 and of Pimelea factor S_7 appears as an AB quartet at about δ3.8, but as a singlet in the case of Pimelea factor P_6.

The structures and sources of the known daphnetoxin type 1α-alkyldaphnane esters are summarized in Table 5 below. ^{1}H-NMR data are given in Table 7.

2. Macrinol Esters

The first macrinol esters to be characterized were gnidimacrin and its 20-palmitate from

Table 5
1α-ALKYLDAPHNANE TYPE ESTERS

Daphnetoxin type

Name	R	Source
Pimelea factor S_6	(C-1)–*$CH(CH_3).(CH_2)_7$–	*Pimelea simplex* F. Muell.[15]
Pimelea factor S_7	(C-1)–*$CH(CH_3).(CH_2)_7$–	*P. simplex* F. Muell.[15,48]
Synaptolepis factor K_1	(C-1)–$C_{13}H_{26}.CH{=}CH$–	*Synaptolepis* sp.[48]
Pimelea factor P_6	(C-1)–$CH(CH_3).(CH_2)_6.CH$– | $C_6H_5.CO.O$	*P. prostrata* Willd.[14,48]

Macrinol esters

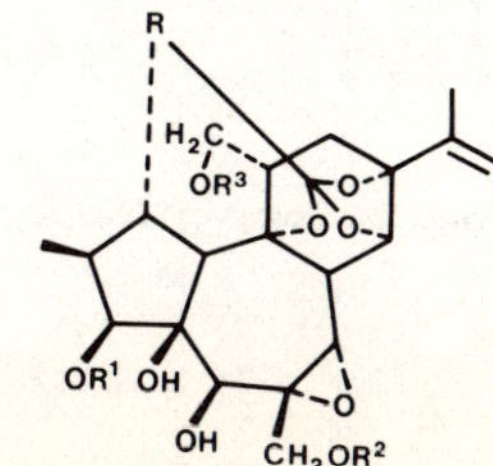

Table 5 (continued)
1α-ALKYLDAPHNANE TYPE ESTERS

Name	R	R[1]	R[2]	R[3]	Source
Gnidimacrin	(C-1)–$CH(CH_3).(CH_2)_6.CH(OH)$–	$O.CO.C_6H_5$	H	$O.CO.C_6H_5$	*G. subcordata* Meissner[5]
Gnidimacrin-20-palmitate	(C-1)–$CH(CH_3).(CH_2)_6.CH(OH)$–	$O.CO.C_6H_5$	$CO.(CH_2)_{14}.CH_3$	$O.CO.C_6H_5$	*G. subcordata* Meissner[5]

12-Hydroxy-18-deoxymacrinol esters

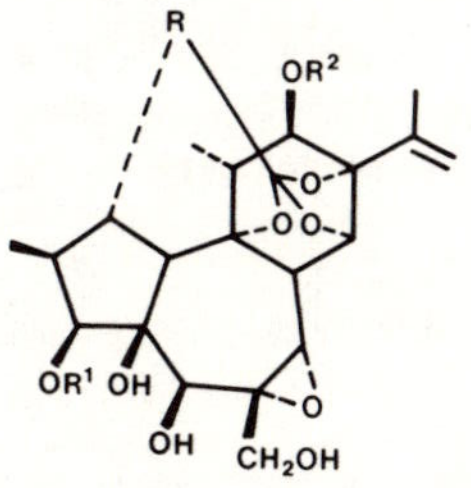

Name	R	R[1]	R[2]	Source
Linifolin A	(C-1)–$CH(CH_3).(CH_2)_7$–	$O.CO.C_6H_5$	$O.COCH_3$	*P. linifolia* Smith[49]

18-Deoxymacrinol esters

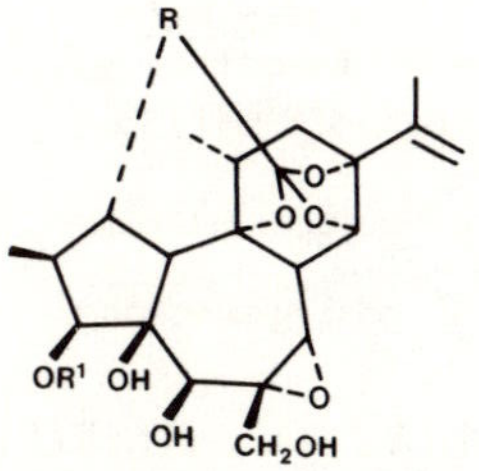

Name	R	R^1	Source
Pimelea factor P_2, Daphnopsis factor R_1, Linifolin B	(C-1)–$CH(CH_3).(CH_2)_7$–	$O.CO.C_6H_5$	*P. prostrata* Willd.[8,14,48] *P. ligustrina* Labill.[49] *P. linifolia* Smith[48,49] *Daphnopsis racemosa* Griseb.[13,48] *Dirca occidentalis* A. Gray[9] *G. kraussiana* Meissner[10] *Stellera chamaejasme* L.[21]
Pimelea factor P_3, Daphnopsis factor R_5,	(C-1)–$CH(CH_3).(CH_2)_7$–	$O.CO.C_6H_5$	*P. prostrata* Willd.[14] *D. racemosa* Griseb.[13]
Kraussianin	(C-1)–$CH(CH_3).(CH_2)_6.CH(OH)$–	$O.CO.C_6H_5$	*G. kraussiana* Meissner[10]
Dircin	(C-1)–$CH(CH_3).(CH_2)_6.CH(O.CO.CH_3)$–	$O.CO.C_6H_5$	*D. occidentalis* A. Gray[9]

5,18-Dideoxymacrinol esters

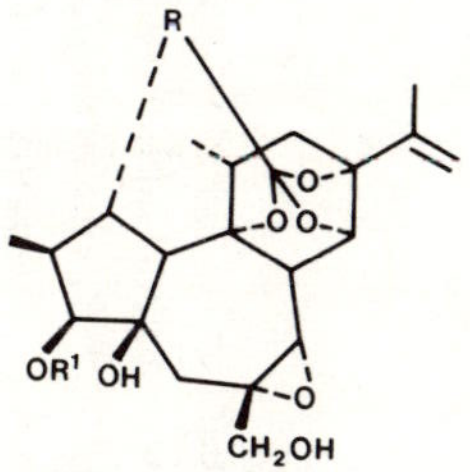

Name	R	R^1	Source
Daphnopsis factor R_6	(C-1)–$CH(CH_3).(CH_2)_7$–	$O.CO.C_6H_5$	*Daphnopsis racemosa* Griseb.[13]
Daphnopsis factor R_7	(C-1)–$CH(CH_3).(CH_2)_7$–	$O.CO.C_6H_5$	*D. racemosa* Griseb.[13]

Gnidia subcordata Meissner.[5] The structure of gnidimacrin was unequivocally determined by X-ray crystallography of its 20-*p*-iodobenzoate derivative at liquid nitrogen temperature. It was solved by the heavy atom method, and showed gnidimacrin to have a trans-linked A/B ring junction, 4β- and 5β-hydroxyl groups, and an α-orientated 6(7)-epoxy function.

This signals in the ^{1}H-NMR spectrum of gnidimacrin are similar to those of daphnetoxin in many respects. Characteristic signals at δ3.29 and δ4.29 are associated with the protons on C-7 and C-14, respectively. The signals for the isopropylene side chain on C-13 are again evident as a 3H singlet and a 2H AB quartet, suggesting hindered rotation of this group. The geminal protons of the new ester function on C-18 also occur as a 2H AB quartet, but are further perturbed by the proton on C-11.

The structures of gnidimacrin and its 20-palmitate are given in Table 5. ^{1}H-NMR data are given in Table 7.

3. *12-Hydroxy-18-Deoxymacrinol Ester*

Only one example of an ester of this type is known, namely, linifolin A from *Pimelea linifolia* Smith.[49] Although it resembles the 12-hydroxydaphnetoxin type compounds described above in having an ester function on C-12, it differs form such compounds in also having a hydroxyl rather than a keto function on C-3. As with other 1α-alkyldaphnanes, linifolin A lacks a signal in its ^{1}H-NMR spectrum at about δ7.6 that is ascribable to an olefinic C-1 protein. Linifolin A was recognized to be a 12β-acetoxy derivative of linifolin B (otherwise known as Pimelea factor P_2) when its ^{1}H-NMR spectrum was compared with those of 12-hydroxydaphnetoxin, daphnetoxin, mezerein, and linifolin B.

Linifolin A has piscicidal properties but does not show antileukemic activity.[49]

The structure of linifolin A is given in Table 5; ^{1}H-NMR data are given in Table 7.

4. *18-Deoxymacrinol Esters*

This group of esters appears to be the most widely distributed type of 1α-alkyldaphnane. Pimelea factor P_2, otherwise named Daphnopsis factor R_1 and linifolin B, has been found in *Pimelea prostrata* Willd,[8,14,48] *P. ligustrina* Labill.,[49] *P. linifolia* Smith,[48,49] *Daphnopsis racemosa* Griseb.,[13,48] *Dirca occidentalis* A. Gray,[9] *Gnidia kraussiana* Meissner,[10] and *Stellera chamaejasme* L.[21] An identical compound has also been given the name gnidilatimacrin.[48] The enantiomeric Pimelea factor P_3, otherwise named Daphnopsis factor R_5, has been reported from *Pimelea prostrata*[4] and *Daphnopsis racemosa*.[13] Dircin (which may also be described as 29β-acetoxy Pimelea factor P_2) and kraussianin (which is the 29β-hydroxy compound) have also been reported from *Dirca occidentalis*[9] and *Gnidia kraussiana*,[10] respectively.

Although structurally related to gnidimacrin (see Section III.E.2), linifolin B did not possess antileukemic activity but was piscicidal.[49]

The ^{1}H-NMR spectra of these compounds lack a signal for an olefinic C-1 proton, suggesting the 1α-alkyldaphnane structure. The chemical shift of the proton on C-10, together with its multiplicity and coupling constant, provides evidence for a 6α-7α-oxide group, an α-oriented proton on C-10, a β-oriented proton on C-1, and a β-oriented C-19 methyl group.[48] Pimelea factor P_2 afforded a 5,20-acetonide when treated with acetone and *p*-toluenesulfonic acid; after alkaline transesterification of Pimelea factor P_2, the product was found to produce a triester on acetylation with acetic anhydride/pyridine, and a 3,4,5,20-diacetonide when treated with acetone/*p*-toluenesulfonic acid, confirming that the secondary hydroxyl groups of the parent polyol were located on C-3 and C-5 and were both β-oriented.[48]

The structures and sources of the known 18-deoxymacrinol esters are summarized in Table 5 below. Comparative ^{1}H-NMR data are given in Table 7.

5. *5,18-Dideoxymacrinol Esters*

The enantiomers Daphnopsis factors R_6 and R_7 from *Daphnopsis racemosa* Griseb.[13] are

Table 6
^{13}C-NMR DATA OF THREE DAPHNANE POLYOL ESTERS

Carbon no.	Huratoxin[26] (22.63 MHz)[a] δ	Wikstrotoxin C[26] (22.63 MHz) δ	Genkwadaphnin[7] (25.20 MHz) δ
1	161.2 d	160.4	160.1
2	136.7 s	136.9	136.9
3	209.8 s	209.4	209.2
4	84.4 s	83.7	78.5
5	82.0 d	80.4	71.3
6	60.5 s	60.4	61.0
7	48.2 d	47.5	64.0
8		36.7	35.7
	36.7 d		
9	72.3 s	72.2	72.4
10	64.2 d	67.3	47.4
11	34.9 d	35.4	44.1
12	36.5 t	44.0	78.9
13	79.6 s	78.2	84.3
14	71.9 d	72.1	80.7
15	146.2 s	143.1	142.9
16	111.2 t	113.3	113.8
17	20.4 q	18.7	9.9
18	9.9 q	9.9	18.4
19	18.9 q	18.2	18.8
20	65.0 t	65.1	64.8

the only two esters of this type to have been described. Their molecular weights differed from those of the co-occurring Daphnopsis factors R_1 and R_5 (see Section III.E.4). by a deficit of 16 mass units. Signals at about δ4.1 for the C-5 protons in the ^{1}HNMR spectra of R_1 and R_5 were missing from the spectra of R_6 and R_7, suggesting that R_6 and R_7 were 5-deoxy derivatives of R_1 and R_5.[13]

The structure of Daphnopsis factors R_6 and R_7 are given in Table 5 below. H-NMR data are given in Table 7.

IV. SUMMARY AND CONCLUDING REMARKS

Daphnane polyol esters have been found in members of both the families Euphorbiaceae and Thymelaeaceae, but appear to be more widely distributed in the latter family. This provides evidence for a close relationship between these two families.

As with the tiglianes and ingenanes, their isolation requires the use of a multistage, bioassay-guided fractionation procedure. Most commonly utilized has been a bioassay procedure involving the assessment of antileukemic activity against P-388 lymphocytic leukemia in mice. Less commonly, skin irritant activity on mouse ears or piscicidal activity has been monitored during fractionation of plant extracts. One group has advocated the assessment of nematocidal activity for this purpose. The cocarcinogenic activity of a few of the compounds has been investigated, but this particular biological activity does not appear to have been used to monitor fractionation procedures.

A number of daphnane polyol esters have had their structures elucidated unequivocally by X-ray crystallographic methods. These include daphnetoxin, mezerein, 12-hydroxydaphnetoxin, and gnidimacrin. This has in turn enabled the signals on their ^{1}H-NMR spectra to be assigned with certainty, making ^{1}H-NMR the most generally useful technique in the investigation of new compounds. Only a few of ^{13}C-NMR spectra have been recorded to date. Tables 6 and 7 give most of the available ^{13}C- and ^{1}H-NMR data of the naturally

Table 7
^{1}H-NMR DATA OF DAPHNANE POLYOL ESTERS[a]

Protons on

C-3	C-7	C-1	C-5	C-8	C-20	C-10	C-19	C-16	C-17	C-18	C-11	C-12	C-14
Daphnetoxin[24]													
		7.62 1H, m	4.26 1H, s	3.02 1H, d	3.83 2H, s	3.91 1H, m						2.36 1H, m	4.57 1H, d
Simplexin[8] ($CDCl_3$, 400 MHz, TMS)													
	3.414 1H, s	7.592 1H, d J = 0.5 Hz	4.233 1H, s	2.880 1H, d J = 2.4 Hz	3.86, 3.75 2H, ABq J_{AB} = 12.5 Hz	3.717 1H, m	1.781 3H, s	4.88, 5.00 1H, s 1H, s	1.754 3H, s	1.139 3H, d J = 7.1 Hz	2.424 1H, m	(β)2.191 1H, dd J = 14.4, 8.7 Hz (α)1.622 1H, d J = 14.4 Hz	4.345 1H, d J = 2.4 Hz
Simplexin[21] ($CDCl_3$, 100 MHz, TMS)													
	3.44 1H, s	7.62 1H, q J = 1 Hz	4.23 1H s	2.90 1H, d J = 3 Hz	3.82 2H, s	3.75 1H, m	1.80 3H, s	4.88, 5.00 1H, s 1H, s	1.80 3H, s	1.15 3H, d J = 7 Hz			4.35 1H, d J = 3 Hz
Montanin[6] $CDCl_3$, 60 MHz, TMS)													
	3.42 1H, s	7.58 1H, m	4.23 1H, s	2.88 1H, d J = 2.6 Hz	3.82 2H, bs		1.77 3H, bs	4.88, 5.00 1H, s 1H, s	1.77 3H, bs				4.37 1H, d J = 2.5 Hz
Huratoxin[6,19,21] ($CDCl_3$, 60 MHz, TMS)													
	3.46 1H, s	7.61 1H, m	4.23 1H, s	2.94 1H, d J = 2.5 Hz	3.84 2H, bs	3.84 1H, m	1.80 3H, bs	4.92, 5.04 1H, t 1H, s J = 1 Hz	1.80 3H, bs	1.18 3H, d J = 6 Hz			4.45 1H, d J = 2.5 Hz

Pimelea factor P_4[14,31] ($CDCl_3$, TMS)

3.44 1H, s	7.66 1H, bs	4.22 1H, s	2.90 1H, d J = 3 Hz	3.81 ± 0.03 2H, ABq J_{AB} = 12 Hz	3.75 1H, m	1.80 3H, s	4.92, 5.04 1H, s 1H, s	1.80 3H, s	1.14 3H, d J = 7 Hz			4.40 1H, d J = 3 Hz

Wikstrotoxin A[26] ($CDCl_3$, 250 MHz, TMS)

3.45 1H, s	7.63 1H, bs	4.25 1H, s	2.94 1H, d J = 2.3 Hz	3.86, 3.79 2H, ABq J_{AB} = 12.5 Hz	3.82 1H, d J = 2.8 Hz	1.79 3H, s	5.02, 4.90 1H, s 1H, s	1.79 3H, s	1.18 3H, d J = 7 Hz	2.49 1H, m J = 7 Hz	(β)2.23 1H, dd J = 14.2, 8.6 Hz (α)1.68 1H, d J = 14.2 Hz	4.43 1H, d J = 2.3 Hz

Genkwadaphnin[7] ($CDCl_3$, TMS)

3.62 1H, s	7.50 1H, s	4.16 1H, s	3.80 1H	3.80 2H	3.80 1H	1.72 3H, s	5.05 2H	1.90 3H, s	1.43 3H, d J = 7 Hz	2.63 1H, q J = 7 Hz	5.28 1H, s	5.05 1H

Gnididin[3] ($CDCl_3$, 100 MHz, TMS)

3.63 1H		4.27 1H, s	3.63 1H	3.92 2H	3.92 1H	1.79 3H, bs	5.02 2H, s	1.88 3H, s	1.38 3H, d J = 7 Hz	2.50 1H, q J = 7 Hz	5.11 1H, s	

Mezerein[24]

	7.50 1H, m	4.20 1H, s	3.64 1H, d	3.91, 3.75 1H, s 1H, s	3.88 1H, m						5.12 1H, s	4.49 1H, d

Stillingia factor S_1[1] ($CDCl_3$, 90 MHz, TMS)

3.53 1H, s	7.6 1H, m	4.26 1H, s	3.76 1H, d J = 2 Hz	3.86 2H	3.8—3.9 1H	1.80 3H, dd	5.10 2H, bs	1.88 3H, s	1.16 3H, dd	2.5 1H, q	3.90 1H, s	4.76 1H, d J = 2 Hz

Table 7 (continued)
^{1}H-NMR DATA OF DAPHNANE POLYOL ESTERS[a]

Protons on													
C-3	C-7	C-1	C-5	C-8	C-20	C-10	C-19	C-16	C-17	C-18	C-11	C-12	C-14
Gnidiglaucin[4] ($CDCl_3$, 100 MHz, TMS)													
	3.47 1H, s	7.47 1H, bs	4.17 1H, bs	3.39 1H, d J = 2 Hz	3.70, 3.84 2H, ABq J_{AB} = 12 Hz	3.70 1H, bs	1.77 3H, bs	4.92 2H, bs	1.82 3H, bs	1.34 3H, d J = 7 Hz	2.50 1H, q J = 7 Hz	5.12 1H, s	4.80 1H, d J = 2 Hz
Gnidilatidin[4] ($CDCl_3$, 100 MHz, TMS)													
	3.56 1H, s	7.50 1H, bs	4.13 1H, s	3.56 1H, d J = 2 Hz	3.80 2H	3.80 1H	1.72 3H, bs	4.92 2H, bs	1.82 3H, s	1.34 3H, d J = 7 Hz	2.50 1H, q J = 7 Hz	5.12 1H, s	4.80 1H, d J = 2 Hz
Gnidilatidin-20-palmitate[4] ($CDCl_3$, 100 MHz, TMS)													
	3.64 1H, s	7.59 1H, bs	4.24 1H, s	3.71 1H, d J = Hz	3.87, 4.72 2H, ABq J_{AB} = 12 Hz	3.93 1H, bs	1.77 3H, bs	5.01 2H, s	1.88 3H, s	1.32 3H, d J − 7 Hz	2.50 1H, q J = 7 Hz	5.23 1H, s	4.90 1H, d H = 2 Hz
Gnidilatin[4] ($CDCl_3$, 100 MHz, TMS)													
	3.53 1H, s	7.46 1H, bs	4.11 1H, s	3.53 1H, d J = 2 Hz	3.55, 3.73 2H, ABq J_{AB} = 11 Hz	3.78 1H, bs	1.72 3H, bs	4.90 2H, bs	1.80 3H, s	1.33 3H, d J = 7 Hz	2.48 1H, q J = 7 Hz	5.10 1H, s	4.74 1H, d J = 2 Hz
Gnidilatin[10] ($CDCl_3$, 60 MHz, TMS)													
	3.60 1H, s	7.50 1H, bs	4.21 1H, s	3.60 1H, d J = 2.5 Hz	3.86 2H, bs	3.86 1H, s	1.78 3H, bs	4.98 2H, bs	1.84 3H, s	1.35 3H, d J = 7.2 Hz	2.52 1H, q J = 7.2 Hz	5.18 1H, s	4.83 1H, d J = 2.5 Hz

Gnidilatin-20-palmitate[4] ($CDCl_3$, 100 MHz, TMS)

3.40 1H, s	7.46 1H, bs	4.12 1H, s	3.51 1H, d J = 2 Hz	3.77, 4.73 2H, ABq J_{AB} = 12 Hz	3.70 1H, bs	1.70 3H, bs	4.90 2H, s	1.78 3H, s	1.31 3H, d J = 7 Hz	2.47 1H, q J = 7 Hz	5.10 1H, s	4.71 1H d J = 2 Hz

Subtoxin A[21] ($CDCl_3$, 100 MHz, TMS)

3.51 1H, s	7.56 1H, q J = 1 Hz	4.24 1H, s	3.52 1H, d J = 2Hz	3.81 2H	3.81 1H	1.80 3H, bs	4.99, 4.96 2H	1.80 3H, bs	1.30 3H, d J = 7 Hz	2.38 1H, q J = 7 Hz	4.97 1H, s	4.76 1H, d J = 3 Hz

Wikstrotoxin C[26] ($CDCl_3$, 250 MHz, TMS)

3.55 1H, s	7.57 1H, bs	4.25 1H, s	3.51 1H, d J = 2.4 Hz	3.92, 3.79 2H, ABq J_{AB} = 12.5 Hz	3.83 1H, s	1.79 3H, s	5.01, 4.95 1H, s 1H, s	1.79 3H, s	1.29 3H, d J = 7.3 Hz	2.37 1H, q J = 7.3 Hz	4.97 1H, s	4.76 1H, d J = 2.4 Hz

Tinyatoxin[42,44] ($CDCl_3$, 90 MHz, TMS)

5.84 1H, bs	7.42 1H, bs	2.12, 2.24 2H, ABq J_{AB} = 6.5 Hz	3.06 1H, bs	4.55 2H, s	3.06 1H, bs	1.82 3H, bs	4.72 2H, s	1.51 3H, s	0.95 3H, d J = 7 Hz	2.69 1H, m	2.5 2H	4.18 2H, d J = 2.5 Hz

Resiniferatoxin[25] ($CDCl_3$, 100 MHz, TMS)

5.84 1H, m	7.42 1H, m	2.12, 2.23 2H, AB	3.05 1H, m	4.53 2H, AB	3.05 1H, m	1.81 3H, m	4.70 2H, AB	1.52 3H, bs	0.96 3H, d J = 7.5 Hz	2.60 1H, m	2.0 2H, m	4.18 1H, d J = 7 Hz

Resiniferatoxin[42,44] ($CDCl_3$, 60 MHz, TMS)

5.92 1H, bs	7.49 1H, bs	2.42, 2.06 2H, ABq J_{AB} = 7.3 Hz	3.10 1H, m	4.60 2H, s	3.10 1H, m	1.86 3H, m	4.77 2H, s	1.54 3H, s	0.98 3H, d J = 6.7 Hz	(2.8 —— 1H	—— 2.0) 2H	4.24 1H, d J = 2.7 Hz

Proresiniferatoxin[44,47] ($CDCl_3$, 60 MHz, TMS)

5.49 1H, m	7.52 1H, bs	2.35 1H, s	3.42 1H, m	4.46 2H, s	3.02 1H, bs	1.78 3H, m	5.14 2H, s	1.70 3H, s	0.94 3H, d J = 6 Hz	(2.4 —— 1H	—— 2.0) 2H	5.08 1H, m

Table 7 (continued)
^{1}H-NMR DATA OF DAPHNANE POLYOL ESTERS[a]

Protons on													
C-3	C-7	C-1	C-5	C-8	C-20	C-10	C-19	C-16	C-17	C-18	C-11	C-12	C-14
Daphnopsis factor R_4[13] ($CDCl_3$, 90 MHz, TMS)													
	3.51 1H, s	7.56 1H, m	2.10 2H, d	2.89 1H, d J = 2.5 Hz	3.65 2H, s	3.87 1H, m	1.80 3H, s	5.02, 4.90 1H, s 1H, bs	1.80 3H, s	1.14 3H, d J = 7 Hz			4.40 1H, d J = 2.5 Hz
Pimelea factor S_6[15] ($CDCl_3$, 90 MHz, TMS)													
	3.41 1H, s		4.05 1H, s	2.86 1H, d J = 2 Hz	3.80 ± 0.03 2H, ABq J_{AB} = 12 Hz	3.28 1H, d J = 10 Hz	0.95 3H, d J = 7 Hz	5.00, 4.94 2H	1.77 3H, s	1.13 2H, d J = 6 Hz			4.32 1H, d J = 2 Hz
Pimelea factor S_7[48] ($CDCl_3$)													
	3.34 1H, s		4.05 1H, s	2.93 1H, d J = 3 Hz	3.80 ± 0.03 2H, ABq J_{AB} = 12 Hz	3.10 1H, d J = 12 Hz	1.13 3H, d J = 6 Hz	5.05, 4.94 2H	1.71 3H, s	0.95 3H, d J = 7 Hz			4.24 1H, d J = 3 Hz
Synaptolepis factor K_1[48] ($CDCl_3$)													
	3.42 1H, s		4.05 1H, s	2.94 1H, d J = 3 Hz	3.80 ± 0.02 2H, ABq J_{AB} = 12 Hz	3.00 1H, d J = 12 Hz	1.14 3H, d J = 7 Hz	5.05, 4.92 2H	1.80 3H, s	0.90 3H, d J = 6 Hz			4.35 1H, d J = 3 Hz
Pimelea factor P_6[48] ($CDCl_3$)													
	3.42 1H, s		4.12 2H, s	2.90 1H, d J = 3 Hz	3.82 2H, s	3.35 1H, d J = 10 Hz	1.08 3H, d J = 7 Hz	5.00, 4.90 2H	1.76 3H, s	0.92 3H, d J = 6 Hz			4.32 1H, d J = 3 Hz

						Gnidimacrin[5] ($CDCl_3$)							
4.93 1H, d J = 6 Hz	3.29 1H, s		3.99 1H, s	2.97 1H, d J = 2.5 Hz	3.76 2H, s			5.09, 4.85 1H, s; 1H, s	1.76 3H, s	4.30 1H, dd J = 10.5, 10.5 Hz 4.88 1H, bd J = 10.5 Hz			4.29 1H, d J = 2.5 Hz
						Linifolin A[49] ($CDCl_3$)							
	3.43 1H, s		4.10 1H, s	3.48 1H, d J = 2 Hz	3.84 2H, d J = 6 Hz	3.11 1H, d J = 12 Hz	1.03 3H, d J = 6 Hz	4.96, 4.88 2H	1.78 3H, s	0.84 3H, d J = 6 Hz		4.93 1H, s	4.60 1H, d J = 2 Hz
						Pimelea factor P_2[8] ($CDCl_3$, 400 MHz, TMS)							
5.041 1H, d J = 5 Hz	3.337 1H, s	2.356 1H, dd J = 11, 11 Hz	4.089 1H, s	2.869 1H, d J = 2.2 Hz	3.86, 3.77 2H, ABq J = 12 Hz	3.082 1H, d J = 11 Hz	1.038 3H, d J = 6.7 Hz	4.95, 4.83 1H, s; 1H, s	1.716 3H, s	1.424 3H, d J = 6.6 Hz	2.633 1H, m	(β)2.116 1H, dd J = 14.8 Hz (α)1.607 1H, d J = 14 Hz	4.261 1H, d J = 2.2 Hz
						Pimelea factor P_2[48,49] ($CDCl_3$)							
	3.32 1H, s		4.10 1H, s	2.88 1H, d J = 3 Hz	3.76 ± 0.12 2H, ABq J_{AB} = 12 Hz	3.10 1H, d J = 13 Hz	1.04 3H, d J = 7 Hz	4.94, 4.85 2H	1.75 3H, s	0.82 3H, d J = 7 Hz			4.24 1H, d J = 3 Hz
						Kraussianin[10] ($CDCl_3$, 60 MHz, TMS)							
4.97 1H, d J = 4.4 Hz	3.32 1H, s		4.03 1H, s	3.01 1H, d J = 1.8 Hz	3.80 2H, bs	2.88 1H, d J = 11.2 Hz	1.19 3H, d J = 8 Hz	5.09, 4.86 1H, s 1H, s	1.76 3H, s	0.97 3H, d J = 6.4 Hz			4.31 1H, d J = 1.8 Hz
						Dircin[9] ($CDCl_3$, 400 MHz, TMS)							
4.88 1H, d J = 6 Hz	3.32 1H, s		4.05 1H, s	2.96 1H, d J = 3 Hz	3.81 2H, ABq J_{AB} = 12 Hz	2.95 1H, d J = 12 Hz	1.08 3H, d J = 7 Hz	5.07, 4.88 1H, s 1H, s	1.77 3H, bs	0.98 3H, d J = 7.5 Hz			4.19 1H, d J = 3 Hz

Table 7 (continued)
^{1}H-NMR DATA OF DAPHNANE POLYOL ESTERS[a]

Protons on

C-3	C-7	C-1	C-5	C-8	C-20	C-10	C-19	C-16	C-17	C-18	C-11	C-12	C-14
Daphnopsis factor R_6[13] ($CDCl_3$, 90 MHz, TMS)													
4.85 1H, d	3.43 1H, s			2.90 1H, d J = 2.5 Hz	3.58 2H, s	3.10 1H, d J = 10 Hz	1.04 3H, d J = 6 Hz	4.97, 4.86 1H, s 1H, s	1.75 3H, s	0.86 3H, d J = 7 Hz			4.31 1H, d J = 2.5 Hz
Daphnopsis factor R_7[13] ($CDCl_3$, 90 MHz, TMS)													
4.75 1H, d J = 5 Hz	3.42 1H, s			2.99 1H, d J = 2.5 Hz	3.59 2H, bs	2.90 1H, d J = 12 Hz	1.07 3H, d J = 6 Hz	4.96, 4.85 1H, s 1H, bs	1.76 3H, s	0.97 3H, d J = 6 Hz			4.24 1H, d J = 2.5 Hz

[a] δ ppm downfield from internal standard; s = singlet; d = doublet; q = quartet; m = multiplet; b = broad.

occurring compounds described above. Detailed EIMS fragmentation studies of daphnane polyol esters do not appear to have been published in the literature. Nevertheless, the technique is invariably utilized in the investigation of a newly isolated compound, providing evidence for the molecular weight of the compound and the number and gross nature of the esterifying acids.

CD measurements have only rarely been reported except in the case of the resiniferonol esters, details of stereochemistry being clarified by techniques such as acetonide formation and the detailed consideration of ^{1}H-NMR shifts, multiplicities, and coupling constants. As with the tiglianes and the ingenanes generally, the daphnanes fail to crystallize because of their resinous nature.

Because of their unique structure/activity profiles, the daphnane polyol esters are likely to become increasingly sought after for use as biochemical research tools.

REFERENCES

1. **Adolf, W. and Hecker, E.,** New irritant diterpene-esters from the roots of *Stillingia sylvatica* L. (Euphorbiaceae), *Tetrahedron Lett.*, (21), 2887, 1980.
2. **Kupchan, S. M. and Baxter, R. L.,** Mezerein: antileukemic principle isolated from *Daphne mezereum* L., *Science,* 187, 652, 1974.
3. **Kupchan, S. M., Sweeny, J. G., Baxter, R. L., Murae, T., Zimmerly, V. A., and Sickles, B. R.,** Gnididin, gniditrin, and gnidicin, novel potent antileukemic diterpenoid esters from *Gnidia lamprantha, J. Am. Chem. Soc.*, 97, 672, 1975.
4. **Kupchan, S. M., Shizuri, Y., Sumner, W. C., Haynes, H. R., Leighton, A. P., and Sickles, B. R.,** Isolation and structural elucidation of a new potent antileukemic diterpenoid ester from *Gnidia* species, *J. Org. Chem.*, 41, 3850, 1976.
5. **Kupchan, S. M., Shizuri, Y., Murae, T., Sweeny, J. G., Haynes, H. R., Shen, M.-S., Barrick J. C., Bryan, R. F., van der Helm, D., and Wu, K. K.,** Gnidimacrin and gnidimacrin 20-palmitate, novel macrocyclic antileukemic diterpenoid esters from *Gnidia subcordata, J. Am. Chem. Soc.*, 98, 5719, 1976.
6. **Ogura, M., Koike, K., Cordell, G. A., and Farnsworth, N. R.,** Potential anticancer agents. VIII. Constituents of *Baliospermum montanum* (Euphorbiaceae), *Planta Med.*, 33, 128, 1978.
7. **Kasai, R., Lee, K.-H., and Huang, H.-C.,** Genkwadaphnin, a potent antileukemic diterpene from *Daphne genkwa, Phytochemistry,* 20, 2592, 1981.
8. **Pettit, G. R., Zou, J.-C., Goswami, A., Cragg, G. M., and Schmidt, J. M.,** Antineoplastic agents. LXXXVIII. *Pimelea prostrata, J. Nat. Prod.*, 46, 563, 1983.
9. **Badawi, M. M., Handa, S. S., Kinghorn, A. D., Cordell, G. A., and Farnsworth, N. R.,** Plant anticancer agents. XXVII. Antileukemic and cytotoxic constituents of *Dirca occidentalis* (Thymelaeaceae), *J. Pharm. Sci.*, 72, 1285, 1983.
10. **Borris, R. P. and Cordell, G. A.,** Antineoplastic principles of *Gnidia kraussiana, J. Natl. Prod.*, 47, 270, 1984.
11. **Schmidt, R. J. and Evans, R. J.,** Investigations into the skin-irritant properties of resiniferonol orthoesters, *Inflammation,* 3, 273, 1979.
12. **Adolf, W., Sorg, B., Hergenhahn, M., and Hecker, E.,** Structure-activity relations of polyfunctional diterpenes of the daphnane type. I. Revised structure for resiniferatoxin and structure-activity relations of resiniferonol and some of its esters, *J. Nat. Prod.*, 45, 347, 1982.
13. **Adolf, W. and Hecker, E.,** On the active principles of the Thymelaeaceae. II. Skin irritant and cocarcinogenic diterpenoid factors from *Daphnopsis racemosa, J. Med. Plant Res.*, 45, 177, 1982.
14. **Zayed, S., Adolf, W., and Hecker, E.,** On the active principles of the Thymelaeaceae. I. The irritants and cocarcinogens of *Pimelea prostrata, J. Med. Plant Res.*, 45, 67, 1982.
15. **Hafez, A., Adolf, W., and Hecker, E.,** Active principles of the Thymelaeaceae. III. Skin irritant and cocarcinogenic factors from *Pimelea simplex, Planta Med.*, 49, 3, 1983.
16. **Schildknecht, H., Edelman, G., and Maurer, R.,** Zur Chemie des Mezereins, des entzündlichen und co-carcinogenen Giftes aus dem Seidelbast *Daphne mezereum, Chem. Z.*, 94, 347, 1970.
17. **Sakata, K., Kawazu, K., and Mitsui, T.,** Studies on a piscicidal constituent of *Hura crepitans.* I. Isolation and characterization of huratoxin and its piscicidal activity, *Agric. Biol. Chem.*, 35, 1084, 1971.

18. **Sakata, K., Kawazu, K., and Mitsui, T.,** Studies on a piscicidal constituent of *Hura crepitans*. II. Chemical structure of huratoxin, *Agric. Biol. Chem.*, 35, 2113, 1971.
19. **Sakata, K., Kawazu, K., Mitsui, T., and Masaki, N.,** Structure and stereochemistry of huratoxin, a piscicidal constituents of *Hura crepitans, Tetrahedron Lett.*, 16, 1141, 1971.
20. **Ohigashi, H., Katsumata, H., Kawazu, K., Koshimizu, K., and Mitsui, T.,** A piscicidal constituent of *Excoecaria agallocha, Agric. Biol. Chem.*, 38, 1093, 1974.
21. **Niwa, M., Takamizawa, H., Tatematsu, H., and Hirata, Y.,** Piscicidal constituents of *Stellera chamaejasme* L., *Chem. Pharm. Bull.*, 30, 4518, 1982.
22. **Stout, G. H., Balkenhol, W. G., Poling, M., and Hickernell, G. L.,** The isolation and structure of daphnetoxin, the poisonous principle of *Daphne* species, *J. Am. Chem. Soc.*, 92, 1070, 1970.
23. **Kogiso, S., Wada, K., and Munakata, K.,** Odoracin, a nematicidal constituent from *Daphne odora, Agric. Biol. Chem.*, 40, 2119, 1976.
24. **Ronlán, A. and Wickberg, B.,** The structure of mezerein, a major toxic principle of *Daphne mezereum* L., *Tetrahedron Lett.*, (49), 4261, 1970.
25. **Hergenhahn, M., Adolf, W., and Hecker, E.,** Resiniferatoxin and other esters of novel polyfunctional diterpenes from *Euphorbia resinifera* and *unispina, Tetrahedron Lett.*, (19), 1595, 1975.
26. **Jolad, S. D., Hoffmann, J. J., Timmermann, B. N., Schram, K. H., Cole, J. R., Bates, R. B., Klenck, R. E., and Tempesta, M. S.,** Daphnane diterpenes from *Wikstroemia monitola:* wikstrotoxins A — D, huratoxin, and exoecariatoxin, *J. Nat. Prod.*, 46, 675, 1983.
27. **Adolf, W. and Hecker, E.,** On the irritant and cocarcinogenic principles of *Hippomane mancinella, Tetrahedron Lett.*, (19), 1587, 1975.
28. **Freeman, P. W., Ritchie, E., and Taylor, W. C.,** The constituents of Australian *Pimelea* spp. I. The isolation and structure of the toxin of *Pimelea simplex* and *P. trichostachya* form B responsible for St. George's disease of cattle, *Aust. J. Chem.*, 32, 2495, 1979
29. **Roberts, H. B., McClure, T. H., Ritchie, E., Taylor, W. C., and Freeman, P. W.,** The isolation and structure of the toxin of *Pimelea simplex* responsible for St George's disease of cattle, *Aust. Vet. J.*, 51, 325, 1975.
30. **Gunasekera, S. P., Cordell, G. A., and Farnsworth, N. R.,** Potential anticancer agents. XIV. Isolation of spruceanol and montanin from *Cunuria spruceana* (Euphorbiaceae), *J. Nat. Prod.*, 42, 658, 1979.
31. **Zayed, S., Hafez, A., Adolf, W., and Hecker, E.,** New tigliane and daphnane derivatives from *Pimelea prostrata* and *Pimelea simplex, Experientia,* 33, 1554, 1977.
32. **Nyborg, J. and la Cour, T.,** X-ray diffraction study of molecular structure and conformation of mezerein, *Nature (London),* 257, 824, 1975.
33. **Schildknecht, H. and Maurer, R.,** Die Struktur des Mezereins aus der Frucht des Seidelbastes *(Daphne mezereum), Chem. Z.*, 94, 849, 1970.
34. **Schmidt, R. J. and Evans, F. J.,** unpublished observation 1977.
35. **Coetzer, J. and Pieterse, M. J.,** The isolation of 12-hydroxy-daphnetoxin, a degradation product of a constituent of *Lasiosiphon burchellii, J. Suid-Afrik. Chem. Inst.*, 24, 241, 1974.
36. **Ying, B.-P., Wang, C.-S., Chou, P.-N., Pan, P.-C., and Liu, J.-S.,** Studies on the active principles of the root of Yuan-Hua *(Daphne genkwa).* I. Isolation and structure of yuanhuacine, *Hua Hsueh Hsueh Pao,* 35, 103, 1977, *Chem. Abstr.*, 89:39369.
37. **Wang, C., Huang, H., Xu, R., Dou, Y., Wu, X., and Li, Y.,** Isolation and structure of a new diterpene orthoester, yuanhuafin, *Yaoxue Tongbao,* 17, 174, 1982; *Chem. Abstr.*, 97:107012.
38. **Wang, C., Huang, H., Xu, R., Dou, Y., Wu, X., Li, Y., and Ouyang, S.,** Studies on the active principles of Yuan-Hua roots. III. Isolation and structure of yuanhuafin, *Huaxue Xuebao,* 40, 835, 1982; *Chem. Abstr.*, 98:14335.
39. **Rizk, A. M., Hammouda, F. M., Ismail, S. E., El-Missiry, M. M., and Evans, F. J.,** Irritant resiniferonol derivatives from *Thymelea hirsuta* L. *Experientia,* 40, 808, 1984.
40. **Zhuang, L.-G., Seligmann, O., Jurcic, K., and Wagner, H.,** Inhaltsstoffe von *Daphne tangutica, J. Med. Plant Res.*, 45, 172, 1982.
41. **Schmidt, R. J. and Evans, F. J.,** Structure and potency of some "Tinya" toxins, *J. Pharm. Pharmacol.*, 27, 50P, 1975.
42. **Evans, F. J. and Schmidt, R. J.,** Two new toxins from the latex of *Euphorbia poisonii, Phytochemistry,* 15, 333, 1976.
43. **Pouchert, C. J. and Campbell, J. R.,** *The Aldrich Library of NMR Spectra,* Vol. 6, Aldrich Chemical Co., 1974.
44. **Schmidt, R. J.,** Chemical and biological Studies on the Tigliane and Daphnane Diterpenes of Three *Euphorbia* Species, Ph.D. thesis, University of London, London, 1977.
45. **Schmidt, R. J. and Evans, F. J.,** Two minor diterpenes from *Euphorbia* latex, *Phytochemistry,* 17, 1436, 1978.
46. **Evans, F. J. and Schmidt, R. J.,** The succulent euphorbias of Nigeria. III. Structure and potency of the aromatic ester diterpenes of *Euphorbia poissonii* Pax, *Acta Pharmacol. Toxicol.*, 45, 181, 1979.

47. **Schmidt, R. J. and Evans, F. J.,** A new aromatic ester diterpene from *Euphorbia poisonii*, *Phytochemistry*, 15, 1778, 1976.
48. **Zayed, S., Adolf, W., Hafez, A., and Hecker, E.,** New highly irritant 1-alkylcaphnane derivatives from several species of Thymelaeaceae, *Tetrahedron Lettt.*, (39), 3481, 1977.
49. **Tyler, M. I. and Howden, M. E. H.,** Piscicidal constituents of *Pimelea* species, *Tetrahedron Lett.*, (22), 689, 1981.

Chapter 9

THE INGENANE POLYOL ESTERS

Richard J. Schmidt

TABLE OF CONTENTS

I. INTRODUCTION

Esters of ingenol and other ingenane polyols probably account for the greatest proportion of the skin irritant compounds found in the plant family Euphorbiaceae. As a class of compounds, they are comparatively more homogeneous than the tiglianes and the daphnanes, their parent polyols differing from one another only in the number and position of hydroxyl groups. However, since the ingenanes have proved to be technically the most difficult to separate from one another and to characterize, there have been somewhat fewer reports of their successful isolation and characterization than might otherwise have been expected. A greater variety of structural types may yet be found.

Notwithstanding the apparent structural dissimilarity of the ingenanes to the tiglianes and daphnanes, their biological activities are remarkably similar, reflecting their presumed biosynthetic relationship (see Chapter 4). Being chemically inconspicuous, their biological reactivities have had to be exploited in monitoring fractionation procedures utilized in their isolation from plant material. In general, the compounds elicit an inflammatory reaction when applied to mammalian skin. This provides a convenient means of monitoring the fractionation of a plant extract; to this end a mouse ear irritancy assay has been developed.[1,2] Cocarcinogenic activity on mouse skin also appears to be a feature of at least some of the ingenanes,[3-5] but although this activity (as measured by the Berenblum experiment),[6] was monitored in the isolation of the phorbol esters from croton oil (*Croton tiglium* L.),[3,7] the time-consuming nature of the Berenblum experiment renders it unsuitable for routine use in the monitoring of fractionation procedures. In fact, cocarcinogenic activity has been little studied with individual ingenane polyol esters, but as with the tiglianes, its occurrence and magnitude is likely to be related to the nature of both the ester functions and the parent polyol (see Chapter 10).

A third biological activity was recognized more recently. Kupchan et al.[8] found that an aqueous alcoholic extract of *Euphorbia esula* L. had significant inhibitory activity against several experimental tumors in rodents. Bioassay-guided fractionation of the extract culminated in the isolation and characterization of ingenol-3,20-dibenzoate. This compound was to increase survival time of mice with P-388 lymphocytic leukemia significantly. Sayed et al.[9] have also found cytotoxic activity in an ether extract of *Euphorbia paralias* L. latex. The ingenol and 20-deoxyingenol esters isolated from this fraction were found to be as potent as podophyllin in inhibiting the uptake of ^{3}H-thymidine by TLX/5 mouse lymphoma cells.

Ingenane polyol esters, in common with tigliane and daphnane polyol esters, are also known to exhibit piscicidal activity, this property being exploited in some laboratories to monitor fractionation of extracts thought to contain such compounds. In Japan, the killie fish (*Oryzias latipes*) is commonly used, having recently been used by Hirota et al.,[10] for example, in the investigation of *Euphorbia cotinifolia* L.

II. STRUCTURE ELUCIDATION OF INGENOL

In contrast to the structure elucidation of phorbol, partial and proposed structures for which were a feature of the chemical literature for a number of years prior to the final surrender of phorbol to X-ray crystallography, the structure elucidation of ingenol, as its 3,5,20-triacetate, was quickly achieved. The ingenol-3,5,20-triacetate was derived from the naturally occurring ingenol-3-hexadecanoate present in both the latex of *Euphorbia ingens* E. Meyer and in *Euphorbia lathyris* L. seed oil following mild base-catalyzed transesterification, acetylation of the parent diterpene polyol, and crystallization from ether/petroleum ether. Using X-ray crystallography, the structure was determined by a direct method not requiring any structural chemical information. The procedure employed was based on a

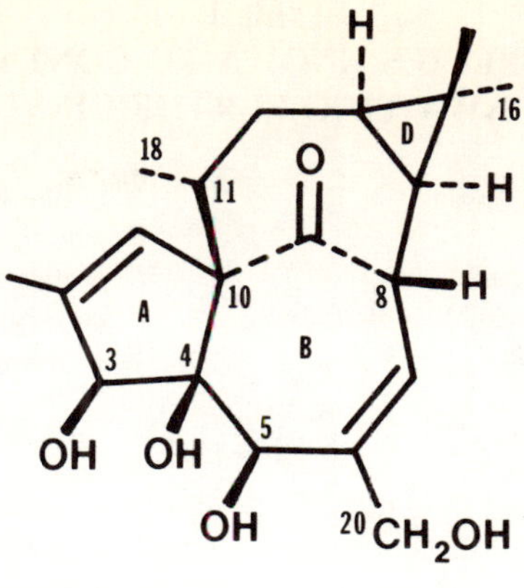

FIGURE 1. Absolute configuration and numbering of ingenol.[11]

cyclic application of the statistical triple product phase relations and the Sayre equation.[11] Figure 1 gives the stereochemistry and numbering of the carbon atoms of ingenol.

Ingenol-3,5,20-triacetate was the first example of a compound with an ingenane hydrocarbon skeleton. The similarity of ingenol to phorbol (a tigliane polyol) was evident, the principal difference being the 7-membered ring C with the C-11 of tigliane hydrocarbon skeleton of phorbol being linked to C-10 rather than to C-9. Other differences are evident in the functional groups. Ingenol lacks the C-3 keto group found in ring A of phorbol, having instead a 3β-OH group. Also, ingenol has a C-9 keto function in place of the α-OH group found in phorbol. The only other difference lies in the position and number of hydroxyl groups around the molecule. Otherwise the $\Delta^{1,2}$ and $\Delta^{6,7}$ bonds in rings A and B and the 4β-OH group and remaining stereochemical features are identical with those of phorbol (see Chapter 7).

Because of technical difficulties encountered in the separation of naturally occurring mixtures of ingenol esters, some of the early literature on the occurrence of ingenol in members of the genus *Euphorbia* refers to the preparation and subsequent detection of the semisynthetic ingenol-3,5,20-triacetate.[12-15] This procedure has been used by Evans and Kinghorn[16] and Upadhyay et al.[17] to screen *Euphorbia* species for diterpene polyol esters. It is significant that the presence of ingenol in plants other than those in the genus *Euphorbia* has yet to be reported (see Chapter 1). Table 1 lists those species of *Euphorbia* that have been reported to yield ingenol-3,5,20-triacetate, but which have not been investigated in any more detail. From the point of view of structure elucidation, the spectroscopic and spectrometric characteristics of ingenol-3,5,20-triacetate are generally typical of other ingenol esters.

In electron-impact mass spectrometry (EIMS), the molecular ion is weak (relative abundance about 1%). The esters are lost principally as the acids; ingenol-3,5,20-triacetate exhibits sequential losses of 60 mass units corresponding with acetic acid. The deacylated ingenane nucleus manifests itself as a fragment ion (30% relative abundance) at m/z 294 ($C_{20}H_{22}O_2$), this particular ion being a useful indicator of the degree of hydroxylation of the parent polyol since it appears at m/z 292 in hydroxyingenols and at m/z 296 in deoxyingenols. In the case of ingenol-3,5,20-triacetate, a diagnostic fragment ion at m/z 372 arising from a loss of 42 mass units is also observed (see Figure 2).[12,18] The base peak is seen at m/z 121 in the case of ingenol-3,5,20-triacetate, but not necessarily in the case of other ingenane polyol esters, and has the composition C_8H_9O. This differentiates ingenol esters from phorbol esters and may represent[19] the stable ion shown in Figure 2. Major fragments at m/z 81, 83, and 97 are predominantly hydrocarbon in composition.[12,18] Spectra giving this fragmentation picture

Table 1
***EUPHORBIA* SPECIES KNOWN TO CONTAIN INGENOL ESTERS BUT AWAITING FURTHER INVESTIGATION**

E. candelabrum Trémaut[16]	*E. nivulia* Buch.-Ham.[16]
E. corallioides L.[16]	*E. palustris* L.[17]
E. deightonii Croizat[16]	*E. pentagona* Haw.[16]
E. desmondi Keay & Milne-Redh.[12,16]	*E. polyacantha* Boiss.[16]
E. drupifera Thonn.[16]	*E. pseudo-grantii* Pax[16]
E. dulcis L.[17]	*E. robbiae* Turrill[16]
E. erythraea Hemsley[16]	*E. royleana* Boiss.[16]
E. grandifolia Haw.[16]	*E. segetalis* L.[17]
E. kotschyana Fenzl[16]	*E. seguieriana* Necker[17]
E. lactea Haw.[16]	*E. sibthorpii* Boiss.[16]
E. megalantha Boiss.[17]	*E. sikkimensis* Boiss.[16]
E. memoralis R. A. Dyer[16]	*E. striatella* Boiss.[17]
E. myrsinites L.[16,17]	*E. stricta* L.[16]
E. neriifolia L.[16]	*E. szovitsii* Fischer & Meyer[17]

are normally obtained by direct insertion with a source temperature of 170 to 180°C, and a 70 eV accelerating voltage.

The proton magnetic resonance (^{1}H-NMR) spectrum of ingenol-3,5,20-triacetate resembles that of phorbol-12,13,20-triacetate. The olefinic protons at C-1 and C-7 resonate at δ6.08 and δ6.27 ppm respectively, downfield from tetramethylsilane (TMS, internal standard). The geminal protons on C-20 appear as an AB quartet signal in the case of ingenol-3,5,20-triacetate and other ingenol-20-acylates, but as a singlet in ingenol esters having a free C-20 hydroxyl group. This generalization also hold for other ingenane polyols (see Table 8). As would be expected, other differences include the absence of a 1H signal for the proton on C-10, and the presence of 1H signals at δ4.96 and δ5.39 corresponding to the protons on C-3 and C-5 (on which the secondary acetates are located). It is difficult to assign with certainty the position of the acyl group in an ingenol monoester, and especially to differentiate between 3-*O*-acyl and 5-*O*-acylingenol esters. Hirota et al.[10] assign the signals for the protons on C-3 and C-5 on the basis of their appearance, pointing out that the signal is a broad singlet for the proton at C-5 but a sharp singlet for the proton at C-3. It was claimed that double resonance experiments supported this conclusion, but details of the experiments were not published. Ott and Hecker[20] have provided data from double resonance experiments on 13-hydroxyingenol esters in support of this generalization. A further useful observation is that the signal associated with the proton on C-5 in the case of 5-*O*-acyl ingenane polyols appears at about δ5.4 to δ5.6.

No ^{13}C-NMR spectra of ingenol esters have yet been recorded.

The IR spectrum of ingenol-3,5,20-triacetate, and indeed of all tigliane, ingenane, and daphnane polyol esters, is not particularly helpful in structure elucidation. While mass spectrometric (MS) and nuclear magnetic resonance (NMR) techniques are of value in detecting homologues and impurities, IR spectroscopy is much less sensitive in this respect. Nevertheless, absorptions characteristic of a tertiary hydroxyl group (3420 to 3430/cm), ester functions (1740 to 1750/cm), βγ-unsaturated ketone (1705 to 1715/cm), and olefinic bonds (1640/cm) are readily observable in the spectra of ingenol esters. Because of the resinous nature of the compounds. IR spectra can conveniently be recorded from solid films deposited from dichloromethane solutions onto sodium chloride discs.

UV absorption spectra of ingenol esters are to be found in the literature, and usually provide useful supportive evidence for the presence of the C-9 keto function from the low intensity absorption maximum at about 280 to 300 nm. A far more useful technique is circular dichroism (CD). CD spectra give information concerning stereochemistry in the

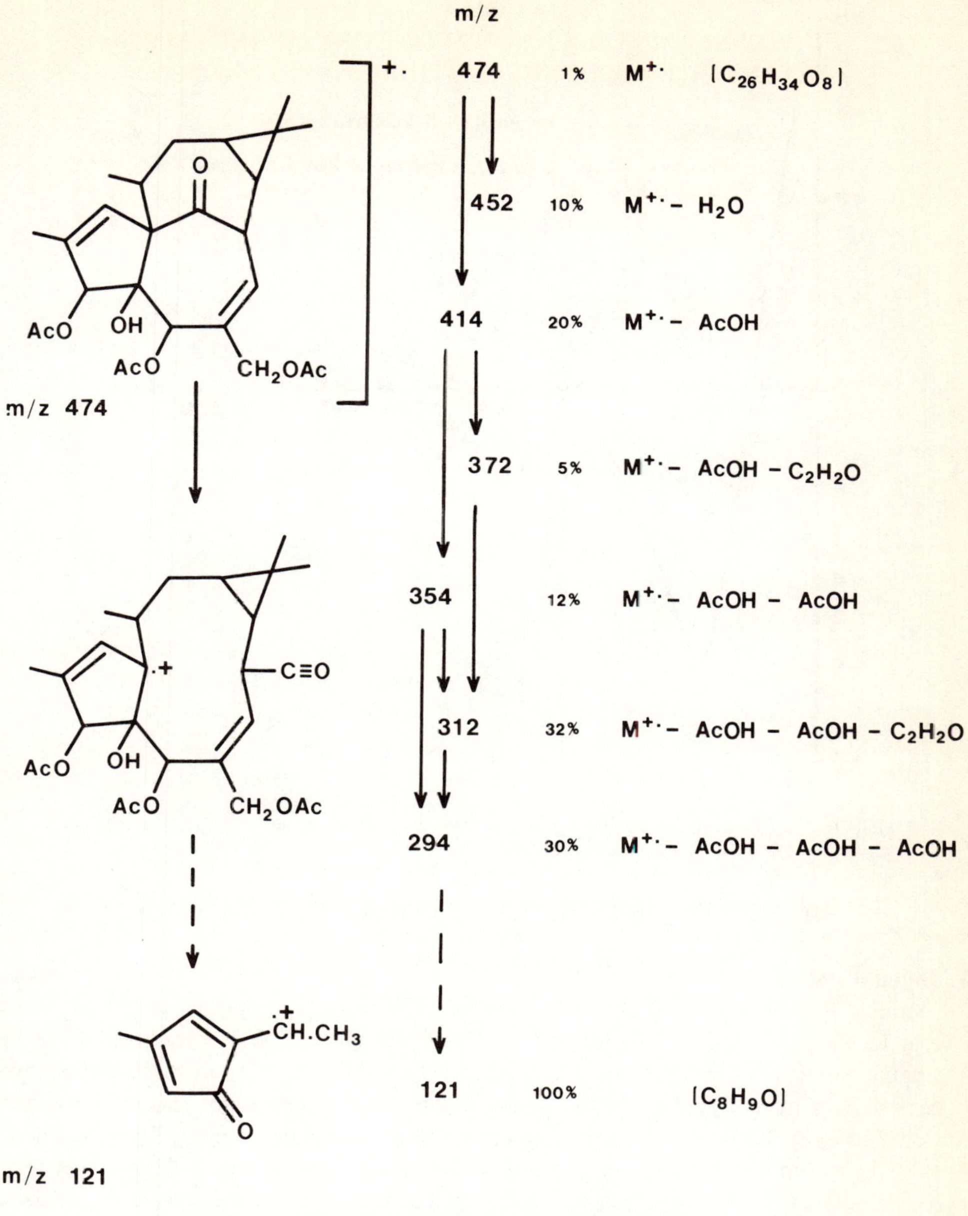

m/z

474 1% $M^{+\cdot}$ $[C_{26}H_{34}O_8]$

452 10% $M^{+\cdot} - H_2O$

414 20% $M^{+\cdot} - AcOH$

372 5% $M^{+\cdot} - AcOH - C_2H_2O$

354 12% $M^{+\cdot} - AcOH - AcOH$

312 32% $M^{+\cdot} - AcOH - AcOH - C_2H_2O$

294 30% $M^{+\cdot} - AcOH - AcOH - AcOH$

121 100% $[C_8H_9O]$

AcO OH AcO CH_2OAc m/z 474

AcO OH AcO CH_2OAc C≡O

$\dot{C}H.CH_3$ O m/z 121

FIGURE 2. EIMS fragmentation of ingenol-3,5,20-triacetate.[12,18,19]

region of the chromophores. Structurally, ingenol is immediately distinguishable from most of the other tigliane and daphnane polyols (with the exception of some of the 1α-alkyl-daphnanes) known to date in its lack of an αβ-unsaturated ketone function in the A ring. The CD spectra of ingenane polyol esters are therefore quite distinct form CD spectra of the other classes of esters. The negative maximum Cotton effect at 200 to 230 nm and the small positive maximum Cotton effect at about 300 nm appear to be characteristic features, being associated with n-π* transitions of the carbonyl functions of the esters and the C-9 keto group, respectively. CD spectra of acetates of ingenol and 5-deoxyingenol are reproduced in Figure 3. Specific rotations using the D lines of sodium are also routinely recorded in some laboratories; esters of ingenol and other ingenane polyols appear to be dextrorotatory.[10,21]

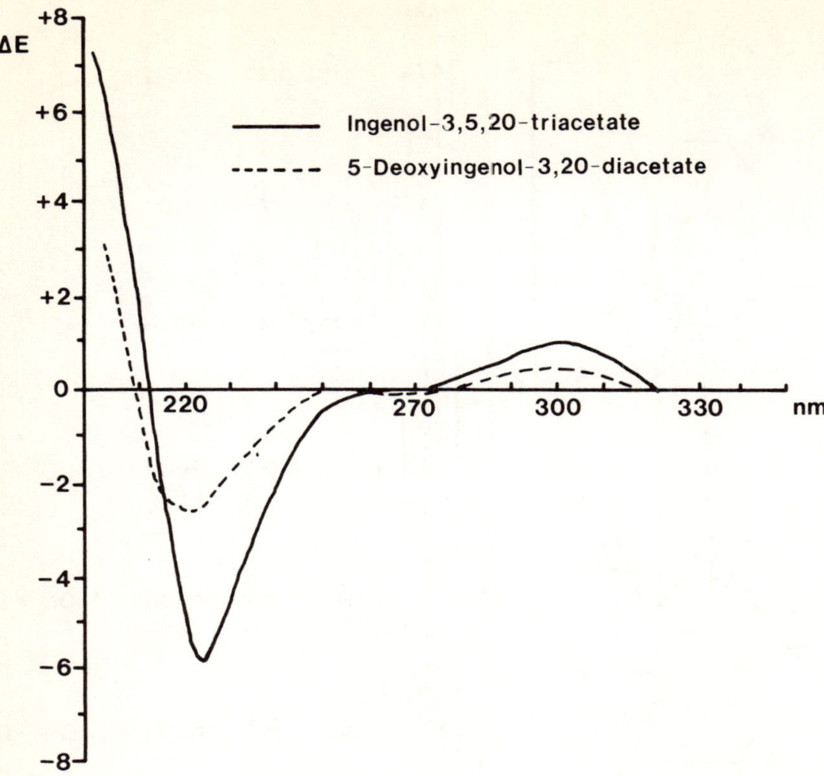

FIGURE 3. CD spectra of ingenol-3,5,20-triacetate and 5-deoxyingenol-3,20-diacetate. (Photo courtesy of Dr. Fred. J. Evans.)

III. ESTERS OF INGENOL AND RELATED POLYOLS

A. Ingenol Esters

Chemical screening carried out by Evans and Kinghorn[16] on species originating from several locations around the world, and by Upadhyay et al.[17] on species growing in the Azarbaijan province of Iran suggests that ingenol is the most common parent diterpene polyol to be found in irritant species of the genus *Euphorbia.* A number of species have been investigated more thoroughly for irritant diterpene esters, the majority being found to contain more or less complex mixtures of ingenol esters, often co-occurring with esters of other ingenane polyols. The structures and sources of the currently known, naturally occurring ingenol ester are summarized in Table 2.

Ingenol esters have been isolated as either monoesters (esterifed at C-3, C-5, or C-20) or diesters. Whether C-20 ingenol monoesters are actually artifacts of the isolation procedure is not clear. Certainly, silica gel used for chromatographic separation of ingenol esters is known to cause acyl shifts.[34,38,39] Acyl shifts may also occur on attempted base-catalyzed transesterification in KOH/CH_3OH, with facile shift of ester groups from C-3 to C-5 or C-20 rather than to the solvent hydroxyl group. The significance of this was discussed by Lin et al.[21] with regards to earlier assignments of the locations of acyl groups in ingenol and 16-hydroxyingenol esters made by Hirota et al.[10] and Lin and Kinghorn,[35] respectively.

While most of the ingenol esters differ from one another only in the length, degree of branching, and degree of unsaturation of their aliphatic acyl groups, a few are conspicuously different. The most unusual are the milliamines A to G from *Euphorbia milii* Des Moul. with their anthraniloyl tripeptide esters. All showed pro-inflammatory activity in the mouse

Table 2
NATURALLY OCCURRING INGENOL ESTERS

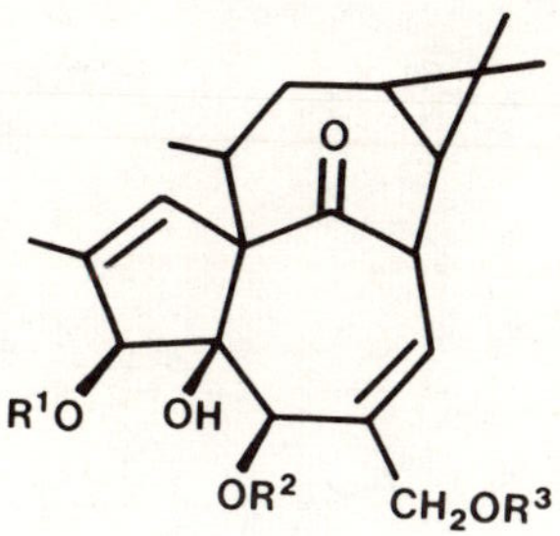

Name	R^1	R^2	R^3	Source
3-*O*-acyl				
	CH_3 · CO.C=CH · CH_3	H	H	*Euphorbia paralias* L.[9] *E. virgata* Waldst. & Kit.[22]
	CO.$(CH_2)_8$.CH_3	H	H	*E. kamerunica* Pax[23]
	CO.CH.$(CH_2)_6$.CH_3 · CH_3	H	H	Euphorbium[5]
	CO.CH.$(CH_2)_3$.CH.CH_2.CH_3 ·　· CH_3　CH_3	H	H	Euphorbium
Euphorbia factor E_1	CO.$(CH{=}CH)_2$.$(CH_2)_4$.CH_3	H	H	*E. esula* L.[24] *E. helioscopia* L.[24] *E. lathyris* L.[25,31] *E. tirucalli* L.[26]

Table 2 (continued)
NATURALLY OCCURRING INGENOL ESTERS

Name	R¹	R²	R³	Source
Euphorbia factor E_2 Euphorbia factor I_6	$CO.(CH{=}CH)_3.(CH_2)_2.CH_3$	H	H	*E. esula* L.[24] *E. ingens* E. Meyer[27] *E. lathyris* L.[25,31] *E. tirucalli* L.[26]
	$CO.CH.(CH_2)_7.CH_3$ · CH_3	H	H	Euphorbium[5] *E. virgata* Waldst. & Kit.[22] Euphorbium[5]
	$CO.CH.(CH_2)_3.CH.(CH_2)_2.CH_3$ · · CH_3 CH_3	H	H	
	$CO.(CH_2)_{10}.CH_3$	H	H	*E. esula* L.[24] *E. kamerunica* Pax[23]
	$CO.CH.(CH_2)_8.CH_3$ · CH_3	H	H	Euphorbium[5]
	$CO.CH.(CH_2)_3.CH.(CH_2)_3.CH_3$ · · CH_3 CH_3	H	H	Euphorbium
	$CO.(CH{=}CH)_2.(CH_2)_6.CH_3$	H	H	*E. lathyris* L.[25,31]
Euphorbia factor E_3	$CO.(CH{=}CH)_3.(CH_2)_4.CH_3$	H	H	*E. esula* L.[24] *E. lathyris* L.[25,31] *E. tirucalli* L.[26]
	$CO.(CH{=}CH)_4.(CH_2)_2.CH_3$	H	H	*E. tirucalli* L.[26]
	$CO.CH.(CH_2)_3.CH.(CH_2)_4.CH_3$ · · CH_3 CH_3	H	H	Euphorbium[5]
	$CO.(CH{=}CH)_4.(CH_2)_4.CH_3$	H	H	*E. tirucalli* L.[26]
Euphorbia factor L_6	$CO.(CH{=}CH)_5.(CH_2)_2.CH_3$	H	H	*E. esula* L.[28]

				E. jolkini Boiss.[29,30]
				E. lathyris L.[25,31]
				E. tirucalli L.[26]
Euphorbia factor I_1	$CO.(CH_2)_{14}.CH_3$	H	H	*E. ingens* E. Meyer[4,11,27]
				E. lathyris L.[11,25]
Milliamine C	Peptide 1	H	H	*E. milii* Des Moul.[29,30,33]
Milliamine G	Peptide 2	H	H	*E. milii* Des Moul.[33]
5-*O*-acyl				
	H	$CO.(CH{=}CH)_2.CH_3$	H	*E. kamerunica* Pax[23]
	H	$CO.CH{=}CH.(CH_2)_4.CH_3$	H	
20-*O*-acyl				
	H	H	$CO.CH(CH_3)_2$	*E. cotinifolia* L.[10]
	H	H	$CO.(CH_2)_4.CH_3$	*E. serrata* L.[34]
	H	H	$CO.C_6H_5$	*E. virgata* Waldst. & Kit.[22]
	H	H	$CO.(CH_2)_6.CH_3$	*E. peplus* L.[40]
	H	H	$CO.CH_2.C_6H_5$	*E. virgata* Waldst. & Kit.[22]
Euphorbia factor I_2	H	H	$CO.(CH_2)_{14}.CH_3$	*E. ingens* E. Meyer[4,27]
				E. lathyris L.[25]
				E. serrata L.[34]
Milliamine B	H	H	Peptide 1	*E. milii* Des Moul.[30,32,33]
3,20-di-*O*-acyl				
	CH_3 · CO.C=CH · CH_3	H	$CO.CH_3$	*E. canariensis* L.[21,35]
				E. hermentiana Lemaire[21]
				E. kamerunica Pax[23]
	$CO.CH{=}CH.(CH_2)_4.CH_3$	H	$CO.CH_3$	*E. kamerunica* Pax[23]
	$CO.(CH{=}CH)_2.(CH_2)_4.CH_3$	H	$CO.CH_3$	*E. biglandulosa* Desf.[36]
				E. kansui Liou[30,37]
	$CO.(CH{=}CH)_3.(CH_2)_2.CH_3$	H	$CO.CH_3$	*E. biglandulosa* Desf.[36]
Milliamine A	Peptide 1	H	$CO.CH_3$	*E. milii* Des Moul.[30,32,33]
	$CO.CH(CH_3).CH_2.CH_3$	H	$CO.CH_2.CH_3$	*E. cotinifolia* L.[10]
	$CO.CH_2.CH_3$	H	$CO.CH(CH_3)_2$	*E. cotinifolia* L.[10]
	$CO.CH(CH_3)_2$	H	$CO.CH(CH_3)_2$	*E. cotinifolia* L.[10]
	$CO.C_6H_5$	H	$CO.C_6H_5$	*E. esula* L.[8]

Table 2 (continued)
NATURALLY OCCURRING INGENOL ESTERS

Name	R^1	R^2	R^3	Source
5,20-di-*O*-acyl				
Milliamine D	H	Peptide 1	$CO.CH_3$	*E. milii* Des Moul.[33]
Milliamine F	H	Peptide 2	$CO.CH_3$	*E. milii* Des Moul.[33]
3,5-di-*O*-acyl				
Milliamine E	Peptide 1	$CO.CH_3$	H	*E. milii* Des Moul.[33]

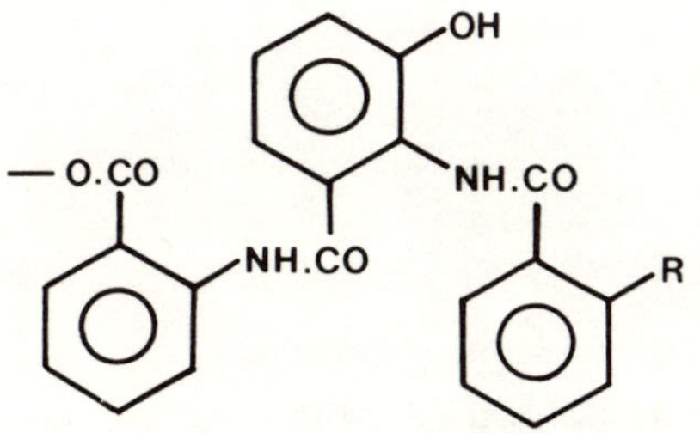

—O.CO, NH.CO, OH, NH.CO, R

Peptide 1 : R = $N(CH_3)_2$

Peptide 2 : R = H

ear assay, but neither milliamine A nor milliamine C was found to be a tumor promoter.[33] A little less unusual is the ingenol-3,20-dibenzoate found in *Euphorbia esula* L.[8] This compound was found to have significant inhibitory activity against P-388 lymphocytic leukemia in mice[8] and is also a skin irritant.[24] Interestingly, when *E. esula* was investigated by Seip and Hecker,[24] it was found not to contain ingenol-3,20-dibenzoate, thus providing evidence for the existence of chemical races in this species.

B. 5-Deoxyingenol Esters

Evans and Kinghorn[18] first reported the occurrence of esters of 5-deoxyingenol in the latices of *Euphorbia biglandulosa* Desf. and *Euphorbia myrsinites* L., where they co-occurred with esters of ingenol. The presence of the ingenane polyols was demonstrated following hydrolysis and subsequent acetylation of the naturally occurring esters; the identity of the esters was not determined.

The EIMS of 5-deoxyingenol-3,20-diacetate closely resembled that of ingenol-3,5,20-triacetate.[18] Both exhibited a high intensity fragment ion at m/z 121. A low intensity molecular ion having an accurately determined mass corresponding with a composition of $C_{24}H_{32}O_6$ was observed together with other fragments consistent with the presence of two acetate esters. A fragment ion at m/z 296 suggested the presence of a deoxyingenol.

The ^{1}H-NMR spectrum was compared with that of ingenol-3,5,20-triacetate. Differences attributable to the lack of a C-5 acetoxy group were evident. Specifically, the presence of only two acetate methyl singlets at δ2.24 and δ2.11, the upfield shift of the signal for the olefinic proton on C-7 to δ5.82 from δ6.25 with only a minimal change in the position of the signal for the olefinic proton on C-1, and the disappearance of the signal for a proton on C-5 adjacent to an acetoxy function were evident. The position for the new signal associated with the geminal protons on C-5 was not clearly recognizable because it was obscured by signals further upfield.

The IR spectrum of 5-deoxyingenol-3,20-diacetate resembled that of ingenol-3,5,20-triacetate in its major absorptions (3430, 1740, 1705, and 1640/cm). The CD spectrum was also qualitatively similar to that of ingenol-3,5,20-triacetate (see Figure 3) confirming the stereochemical similarity of the two compounds in the region of the chromophoric carbonyl group on C-9.

To date, there are no reports in the literature on the characterization of individual naturally occurring 5-deoxyingenol esters. However, in addition to the two species mentioned above *(E. biglandulosa* and *E. myrsinites),* the occurrence of esters of 5-deoxyingenol in the latices of *E. peplus* L. and *E. sibthorpii* Boiss. has also been reported.[16]

C. 20-Deoxyingenol Esters

The first report of the occurrence of 20-deoxyingenol was when the two isomers, 20-deoxyingenol-3-benzoate and 20-deoxyingenol-5-benzoate were isolated from an ethanol extract of the dried roots of *E. kansui* Liou. The structure of the parent polyol was determined by ^{1}H-NMR in deuteropyridine when the signal for the hydroxymethylene protons on C-20 of ingenol was not visible, being replaced by a 3H broad singlet at δ1.98. Otherwise, the spectrum was similar to that of ingenol. Final proof of structure was obtained following the hydrogenation of 20-deoxyingenol-3,5-diacetate with Pd/C in ethyl acetate. The product was identical with one of the two isomeric products produced from ingenol-3,5,20-triacetate by hydrogenolysis of the C-20 ester function and hydrogenation of the $\Delta^{6,7}$ bond under the same conditions.[37] Interestingly, the $\Delta^{1,2}$ bond was not hydrogenated under the conditions used.

Esters of 20-deoxyingenol have, to date, been found in three other species of *Euphorbia* (see Table 3). 20-Deoxyingenol-3-angelate was reported to be the major skin irritant constituent of *E. paralias* L. in which is co-occurred with smaller amounts of 20-deoxyingenol-

Table 3
NATURALLY OCCURRING 20-DEOXYINGENOL ESTERS

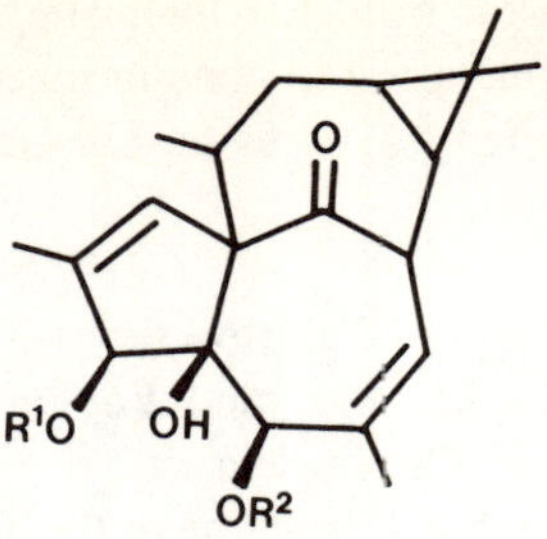

Name	R^1	R^2	Source
3-*O*-acyl			
	CH_3 · CO.C=CH · CH_3	H	*Euphorbia paralias* L.[9] *E. peplus* L.[40]
	$CO.(CH_2)_4.CH_3$	H	*E. paralias* L.[9]
	$CO.C_6H_5$	H	*E. kansui* Liou[37]
	$CO.(CH{=}CH)_3.(CH_2)_2.CH_3$	H	*E. biglandulosa* Desf.[36]
5-*O*-acyl			
	H	$CO.C_6H_5$	*E. kansui* Liou[37]
	H	$CO.(CH{=}CH)_3.(CH_2)_2.CH_3$	*E. biglandulosa* Desf.[36]

3-hexanoate and ingenol-3-angelate.[9] Falsone et al.[36] reported the occurrence of 3- and 5-*O*-tridecenoate esters of 20-deoxyingenol in *E. biglandulosa* Desf. Rizk et al.[40] found 20-deoxyingenol-3-angelate as a minor constituent of *E. peplus* L. herb growing in Egypt, in which it co-occurs with ingenol-20-octanoate.

The CD spectrum of 20-deoxyingenol-3-angelate was recorded by Sayed et al.[9] It was found to differ from that of ingenol-3,5,20-triacetate, having a *positive* maximum Cotton effect at 220 nm ($[\theta] = +20493$) and a smaller positive maximum at 309 nm ($[\theta] = +2185$). EIMS[9] produced an unusually intense (60% relative abundance) molecular ion at m/z 414, a $M^{+\cdot} - 100$ fragment ion at m/z 296 (40% relative abundance), and a base peak at m/z 121 consistent with the loss of an unsaturated 5-carbon acyl group from a deoxyingenol. The chemical-ionization mass spectrum (CIMS) (CH_4, 150°) exhibited a quasimolecular $M^{+\cdot} + 1$ ion at m/z 415 together with the expected fragment ions at 414, 315, 297, and 296.

Table 3 summarizes the presently known naturally occurring 20-deoxyingenol esters and their botanical sources.

D. 16-Hydroxy-20-Deoxyingenol Esters

The latex of *E. hermentiana* Lemaire has been shown to contain minor amounts of the 3-angelate-16-acetate diester of the previously unknown polyol 16-hydroxy-20-deoxyingenol.[21] The parent polyol was obtained by complete hydrolysis of the diester; subsequent acetylation provided the 3,5,16-triacetate. The EIMS exhibited a weak molecular ion at m/z 472. A fragment ion at m/z 413 provided evidence for the presence of an acetate ester; the fragment ion at m/z 372 suggested that a second ester with a mol wt of 100 was also present. Fragment ions at m/z 330, 312, 294, and 121 were suggestive of an ingenane polyol with the same degree of hydroxylation as ingenol. The identity of the parent polyol became

Table 4
NATURALLY OCCURRING 16-HYDOXY-20-DEOXYINGENOL ESTERS

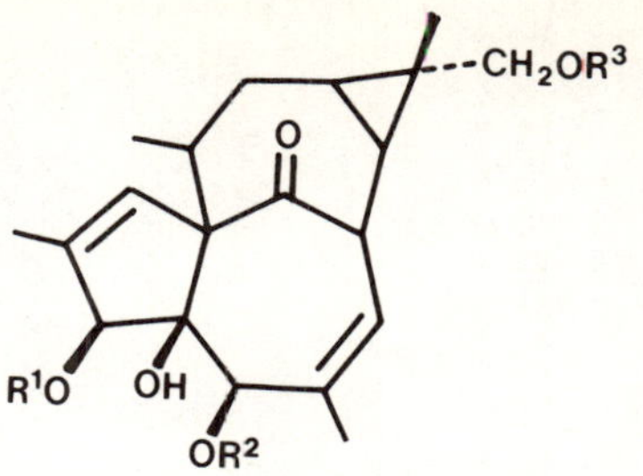

Name	R¹	R²	R³	Source
3,16-di-*O*-acyl				
	CH_3 · CO.C=CH · CH_3	H	$CO.CH_3$	*Euphorbia hermentiana* Lemaire[21]

evident from the ^{1}H-NMR spectrum. The upfield shift of the 1H doublet for the olefinic C-7 proton from its usual position at δ6.0—δ6.3 to δ5.74 had previously been observed in the spectra of 20-deoxyingenol esters. The continuing presence of a 2H AB quartet at about δ4.2 was suggestive of the methylene-oxy group of a primary ester, its location on C-16 being rationalized by the observation that the signal for the proton on C-14 had become shifted downfield to δ0.91 from its position at δ0.73 in the spectrum of ingenol-3,5,20-triacetate.

Attempts to selectively remove the primary esters group from the angelate/acetate diester by mild base-catalyzed transesterification produced a somewhat unexpected acyl shift of the remaining angelate monoester. This was evident from the ^{1}H-NMR spectrum of the product, where the sharp signal for the methine proton on C-3 was diamagnetically shifted from δ5.47 to δ3.47 while the broad signal for the corresponding C-5 proton suffered a paramagnetic shift from δ3.69 to δ5.24. This provided the necessary information to assign the positions of the esters in the original diester.

Measurement of the specific rotation of 16-hydroxy-20-deoxyingenol-3-angelate-20-acetate showed it to be dextrorotatory; absorbance maxima in the IR spectrum were also typical of other ingenane polyol esters. Interestingly, no absorbance at about 300 nm in the UV spectrum was reported.

Table 4 gives the structure and botanical source of the only 16-hydroxy-20-deoxyingenol ester yet described.

E. 13-Hydroxyingenol Esters

As well as ingenol and 20-deoxyingenol esters, the roots of *E. kansui* Liou were found to contain a 13-hydroxyingenol diester, namely 13-hydroxyingenol-13-dodecanoate-20-hexanoate.[30,41] In an attempt to obtain the parent alcohol by reductive cleavage of 13-hydroxyingenol-13-dodecanoate using $LiAlH_4$ in ether, a novel tetracyclic structure was produced as a result of an intramolecular aldol condensation. This suggested the location of the ester function on C-13. The position of the hydroxyl function on C-13 was confirmed by methylation of the dodecanoate ester with Ag_2O and CH_3I in dimethylformamide to produce a tetramethyl ether, followed by reductive cleavage of the ester to produce cyclopropanol

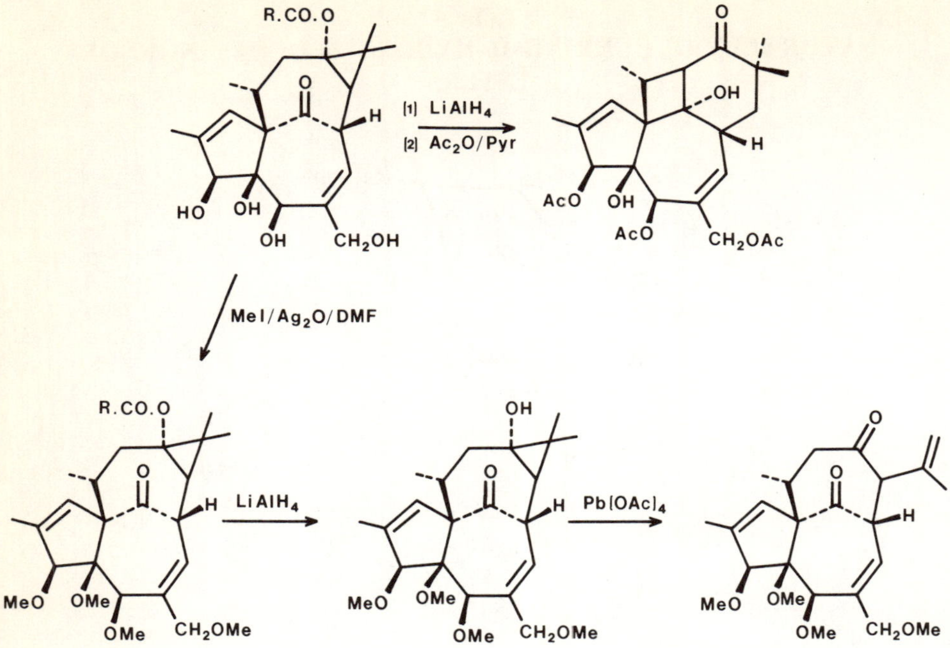

FIGURE 4. Products of reductive cleavage and other reactions in a 13-*O*-acyl ingenol monoester.[30,41]

function which cleaved with lead tetraacetate in benzene, leaving an isopropenyl and a keto function (see Figure 4).[41]

Detailed EIMS and ^{1}H-NMR data were later published following isolation of 13-hydroxyingenol esters from the roots of *E. cyparissias* L.[20] The ^{1}H-NMR spectrum of Euphorbia factor Cy_{13}, a 5,13-diacyl 13-hydroxyingenol diester, resembled those of 5-acyl ingenol esters generally; that of Euphorbia factor Cy_{14}, a 3,13-diacyl 13-hydroxyingenol diester, resembled those of 3-acyl ingenol esters. The ^{1}H-NMR spectrum of Euphorbia factor Cy_{12}, a 13,20-diacyl 13-hydroxyingenol diester, was found to differ slightly for unknown reasons from that of an isomeric compound[41] isolated from *E. kansui* Liou. Generally, the 2H signal from the protons on C-20 appears as an AB quartet in the 20-acylated compounds, and the signal for the proton on C-5 appears at about δ5.4 in the 5-acylated compound. The EIMS of Euphorbia factor Cy_{14},[20] as in the case of 16-hydroxyingenol-3,5,16,20-tetraacetate, exhibited a base peak at m/z 310 rather than at m/z 121.

Uemura et al.[41] recorded a CD spectrum for 13-hydroxyingenol-13-dodecanoate, quoting just one Cotton effect at 290 nm ($\Delta E = -1.5$). Ingenol-3,5,20-triacetate produces a *positive* maximum at 300 nm, suggesting that the compound isolated by Uemura et al.[41] could have been stereochemically distinct from ingenol-3,5,20-triacetate (although the presence of the new ester at C-13 could also be the cause of this anomaly.) Uemura et al.[41] did consider the stereochemistry of their 13-hydroxyingenol ester and concluded from their chemical and ^{1}H-NMR data that the hydroxy functions on C-3, C-4, and C-5 have the same relative orientations as in ingenol; also that the relationship between C-4 and C-5 is *cis,* and that the C-5 hydroxyl group has a β-equatorial orientation. Finally, Dreiding models suggested that the C-8, C-11, and C-14 protons were respectively, β- β-, and α-oriented. Ott and Hecker[20] did not record CD data for their compounds isolated from *E. cyparissias* L. Presently known naturally occurring 13-hydroxyingenol esters and their botanical sources are given in Table 5.

F. 16-Hydroxyingenol Esters

Esters of 16-hydroxyingenol have been found in four *Euphorbia* species. Opferkuch and

Table 5
NATURALLY OCCURRING 13-HYDROXYINGENOL ESTERS

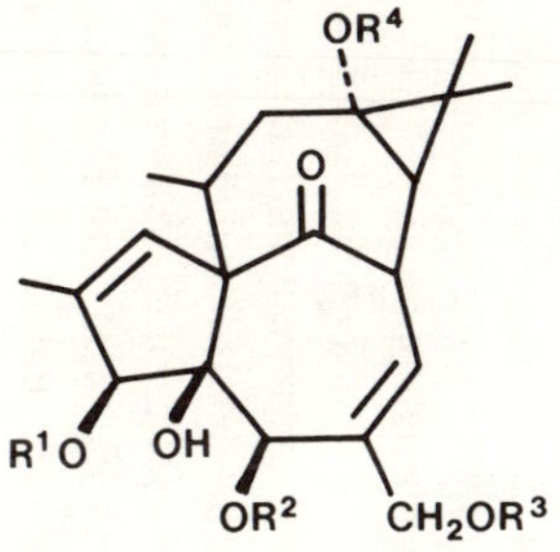

Name	R^1	R^2	R^3	R^4	Source
3,13-di-*O*-acyl					
	$CO.CH.CH(CH_3)_2$ · CH_3	H	H	$CO.C_9H_{19}$	*Euphorbia cyparissias* L.[20]
Euphorbia factor Cy_{14}	$CO.CH.CH(CH_3)_2$ · CH_3	H	H	$CO.C_{11}H_{23}$	*E. cyparissias L.*[20]
5,13-di-*O*-acyl					
Euphorbia factor Cy_{13}	H	$CO.CH.CH(CH_3)_2$ · CH_3	H	$CO.C_{11}H_{23}$	*E. cyparissias* L.[20]
13,20-di-*O*-acyl					
	H	H	$CO.(CH_2)_4CH_3$	$CO.(CH_2)_{10}.CH_3$	*E. kansui* Liou[30,41]
Euphorbia factor Cy_{12}	H	H	$CO.CH.CH(CH_3)_2$ · CH_3	$CO.C_{11}H_{23}$	*E. cyparissias* L.[20]

Hecker[27] first reported the occurrence of 16-hydroxyingenol-3-decatrienoate-16-angelate in the latex of *E. ingens* E. Meyer. The parent polyol was obtained by base catalyzed trans-esterification; subsequent acetylation afforded a tetraacetate.

The structure of the parent polyol was proposed on the basis of comparative ^{1}H-NMR studies of its tetraacetate with the 3,5,20-triacetate of ingenol. It was observed the signal associated with protons on C-13 and C-14 appeared downfield at δ0.9 to 1.4, as compared with δ0.73 in the case of C-13 and C-14 protons of ingenol-3,5,20-triacetate. Furthermore, the positions of the signals associated with C-8, C-11, and C-12 protons of 16-hydroxyingenol-3,5,26,20-tetraacetate were unaffected by the new acetate group, providing evidence for its location on C-16 rather than on C-17. As would be expected, a new 2H signal at δ4.22, ascribable to the protons on C-16, was also evident.

EIMS of 16-hydroxyingenol-3,5,16-20-tetraacetate is typical of other ingenane polyol esters. Fragment ions at m/z 310 and 292 suggest a hydroxyingenol. The base peak was observed at m/z 310 rather than at m/z 121 as in ingenol-3,5-20- and 20-deoxyingenol-3,5-diacetate. The four acetates do not all cleave as acetic acid, one acetate group leaving preferentially as a C_2H_2O fragment[42] from m/z 412 to m/z 370. This pair of fragment ions was also observed in the EIMS of the naturally occurring 16-hydroxyingenol-3-angelate-5,16,20-triacetate, but the fragment at m/z 310 was of only 20% relative abundance when compared with the base peak at m/z 83. Other mass spectra of 16-hydroxyingenol diesters[21] also exhibited base peaks at m/z 83. This difference in base peak is most probably the result of the use of different operating conditions in the instruments rather than a consequence of the slight difference in structure of the two compounds.

IR and UV spectra were unremarkable, but served to confirm the similarity with ingenol-3,5,20-triacetate with regards to functional groups and chromophores.

The irritant fraction of the latex of *E. lactea* Haw. has also been found to yield 16-hydroxyingenol-3,5,16,20-tetraacetate on base-catalyzed methanolysis followed by acetylation.[42] The nature of the individual esters could not be determined. Three esters of 16-hydroxyingenol have, however, been isolated from the latex of *E. hermentiana* Lemaire using droplet countercurrent chromatography.[21] One was identical with Euphorbia factor I_5 previously isolated from *E. ingens* E. Meyer by Opferkuch and Hecker;[27] another proved to be the first naturally occurring 16-hydroxyingenol tetraester to be described. *E. canariensis* L. has also been reported to contain two 16-hydroxyingenol esters.[35]

Table 6 lists currently known naturally occurring 16-hydroxyingenol esters and their botanical sources.

G. 13,19-Dihydroxyingenol Esters

Esters of 13,19-dihydroxyingenol were reported to co-occur with 13-hydroxyingenol esters in *E. cyparissias* L.[20] No other natural source of these compounds has yet been reported.

The compounds in *E. cyparissias* were identified as (2,3-dimethylbutyrate) triesters of 13,19-dihydroxyingenol on the basis of chemical and spectroscopic data. The EIMS of Euphorbia factor Cy_6, a 3,13,19-triacyl 13,19-dihydroxyingenol triester, exhibited the expected losses of ester functions as free acids; the base peak was observed at m/z 326. IR and UV spectra were unremarkable, resembling those of ingenol-3,5,20-triacetate. In the ^{1}H-NMR spectrum, observable differences, as compared with the spectrum of a 13-hydroxyingenol ester, were the downfield shift of the signal associated with the proton on C-1, a new signal at δ4.63 ascribable to the geminal methylene-oxy protons adjacent to the new ester function on C-19, together with the loss of the signal for the C-19 methyl group and the appearance of the C-14 proton signal as a doublet at δ1.26. The ^{1}H-NMR spectrum of the 13,19,20-triacyl 13,19-dihydroxyingenol triester, Euphorbia factor Cy_2, when compared with that of Euphorbia factor Cy_6, exhibited a complex of signals at about δ4.7 attributable to a diamagnetically shifted signal for the C-3 proton and a paramagnetically shifted signal

Table 6
NATURALLY OCCURRING 16-HYDROXYINGENOL ESTERS

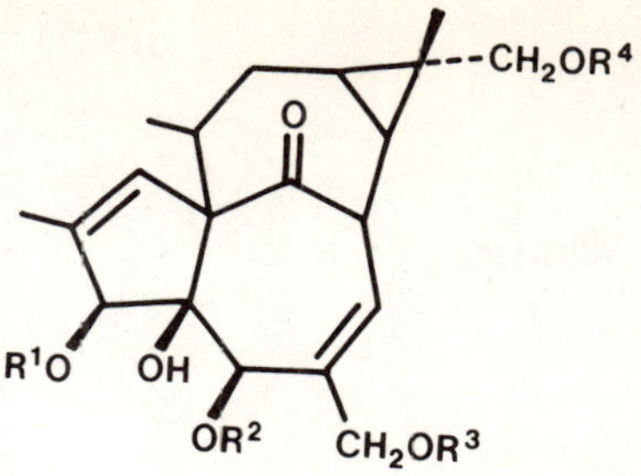

Name	R^1	R^2	R^3	R^4	Source
3,16-di-*O*-acyl					
	CH_3 $CO.C{=}CH$ CH_3	H	H	$CO.C_6H_5$	*Euphorbia canariensis* L.[35]
	$CO.(CH{=}CH)_3.(CH_2)_2.CH_3$	H	H	CH_3 $CO.C{=}CH$ CH_3	*E. hermentiana* Lemaire[21] *E. ingens* E. Meyer[27]
3,16,20-tri-*O*-acyl					
	CH_3 $CO.C{=}CH$ CH_3	H	$CO.CH_3$	$CO.CH_3$	*E. hermentiana* Lemaire[21]
	CH_3 $CO.C{=}CH$ CH_3	H	$CO.CH_3$	$CO.C_6H_5$	*E. canariensis* L.[35]
3,15,16,20-tetra-*O*-acyl					
	CH_3 $CO.C{=}CH$ CH_3	$CO.CH_3$	$CO.CH_3$	$CO.CH_3$	*E. hermentiana* Lemaire[21]

for the protons on C-20. Euphorbia factor Cy_4, a 5,13,19-triacyl 13,19-dihydroxyingenol triester exhibited signals in its ^{1}H-NMR spectrum resembling those of other 5-acylated ingenane polyol esters, especially the 1H singlet at about δ5.4. These finding are consistent with the conclusion that 13,19-dihydroxyingenol is the parent polyol of the Euphorbia factor Cy_2, Cy_4 and Cy_6. CD measurements were not recorded.

Table 7 summarizes the structures of known 13,19-dihydroxyingenol esters of natural origin.

IV. SUMMARY AND CONCLUDING REMARKS

Esters of ingenol and other ingenane polyols have been found, to date, only in the genus *Euphorbia* (family Euphorbiaceae). This would suggest that their occurrence is of chemosystematic significance (see Chapter 4). Their isolation from plant material requires the use

Table 7
NATURALLY OCCURRING 13, 19-DIHYDROXYINGENOL ESTERS

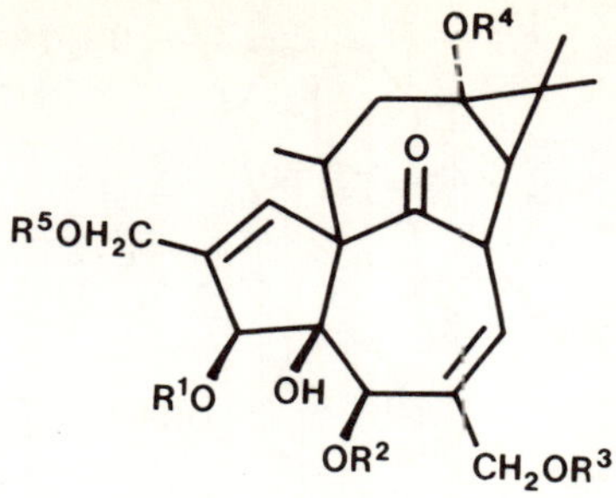

Name	R^1	R^2	R^3	R^4	R^5	Source
3,13,19-tri-*O*-acyl						
Euphorbia factor Cy_6	Ester	H	H	Ester	Ester	*Euphorbia cyparissias* L.[20]
5,13,19-tri-*O*-acyl						
Euphorbia factor Cy_4	H	Ester	H	Ester	Ester	*E. cyparissias* L.[20]
13,19,20-tri-*O*-acyl						
Euphorbia factor Cy_2	H	H	Ester	Ester	Ester	*E. cyparissias* L.[20]

Ester = $CO.CH(CH_3).CH(CH_3)_2$

of a multistage, bioassay-guided fractionation procedure. Both a mouse ear irritancy assay and an antitumor test system measuring an increasing survival time of mice against P-388 lymphocytic leukemia appear to be in common use, Japanese researchers in particular favoring an assay procedure measuring piscicidal activity. Chromatographic fractionation typically involves the use of both partition and sorption procedures.

While melting points (if appropriate), IR and UV spectra, specific rotations, and the newer MS techniques such as field desorption and chemical ionization all have supportive and/or confirmatory value, the most useful instrumental methods for characterizing newly isolated compounds are EIMS and ^{1}H-NMR spectroscopy. Typically, the EIMS can give an indication of the molecular weight of the esterifying acids and the degree of hydroxylation of the parent polyol and, to a limited extent, the nature and position of the ester groups. Final proof of structure of the acyl groups usually requires their prior removal from the parent polyol either hydrolytically or reductively. Table 8 gives details of proton magnetic resonances of a selection of ingenane polyols and their esters.

Table 8
^{1}H-NMR DATA OF INGENOL AND SOME OF ITS DERIVATIVES[a]

Protons on

C-7	C-1	C-5	C-8	C-20	C-3	C-19	C-16	C-17	C-18	C-14
Ingenol[11] ($CDCl_3$, TMS)										
6.00 1H, d	5.84 1H	4.32 1H	4.20 1H	4.09 2H	3.79 1H	1.85 3H	(1.16 & 3H	1.10) 3H	0.98 3H	
Ingenol-3,5,20-triacetate[11,27] ($CDCl_3$, 60 MHz, TMS)										
6.24 1H, d	6.08 1H	5.38 1H, s	4.25 1H, d J = 4.6 Hz	4.58, 4.18 2H, ABq J_{AB} = 12.6 Hz	4.97 1H, s	1.76 3H	(1.12 & 3H	1.09) 3H	1.00 3H	
Ingenol-3-(2′,4′,6′-decatrienoate)[27] ($CDCl_3$, 60 MHz, TMS)										
		4.04 1H	4.30 1H	4.13 2H	5.63 1H	1.80 3H	(108 & 3H, s	1.05) 3H, s		
Milliamine C[33] ($CDCl_3$, 90 MHz, TMS)										
6.05 1H, m	6.12 1H, m	(4.90 1H	— 1H	3.90) 2H	5.77 1H, s	1.80 3H, m	1.04 3H, s	1.04 3H, s	1.03 3H, d J = 7 Hz	
Milliamine G[33] ($CDCl_3$, 90 MHz, TMS)										
6.06 1H, m	6.13 1H, m	(4.70 1H	— 1H	3.70) 2H	5.81 1H, s	1.83 3H, m	1.04 3H, s	1.04 3H, s	1.05 3H, d J = 7 Hz	
Ingenol-20-isobutyrate[10] ($CDCl_3$, 90 MHz, TMS)										
6.06 1H, d J = 5 Hz	5.89 1H, q J = 2 Hz	3.65 1H, bs	4.08 1H, m	4.70, 4.49 2H, ABq J_{AB} = 12 Hz	4.39 1H, s	1.82 3H, d J = 2 Hz	(1.12 & 3H, s	1.06) 3H, s	0.96 3H, d J = 7 Hz	

Table 8 (continued)
^{1}H-NMR DATA OF INGENOL AND SOME OF ITS DERIVATIVES[a]

Protons on										
C-7	C-1	C-5	C-8	C-20	C-3	C-19	C-16	C-17	C-18	C-14
Milliamine B[33] ($CDCl_3$, 90 MHz, TMS)										
6.14 1H, m	5.90 1H, m	3.63 1H, d J = 12 Hz	4.05 1H, m	4.93, 4.70 2H, ABq J_{AB} = 13 Hz	4.31 1H, d J = 5 Hz	1.80 3H, m	1.08 —— 3H, s	1.04) 3H, s	0.94 3H, d J = 7 Hz	
Milliamine E[33] ($CDCl_3$, 90 MHz, TMS)										
6.21 1H, m	6.21 1H, m	5.48 1H, bs	4.23 1H, m	3.92 2H, s	5.23 1H, s	1.82 3H, m	(1.04 —— 3H, s	1.02) 3H, s	1.04 3H, d J = 7 Hz	
Ingenol-3-angelate-20-acetate[21] ($CDCl_3$, 60 MHz, TMS)										
6.10 1H, m	6.03 1H, d J = 1.5 Hz	3.88 1H, bs	4.19 1H, m	4.74, 4.49 2H, ABq J_{AB} = 13.4 Hz	5.58 1H, s	1.80 3H, d J = 1.3 Hz	4.35, 4.05 2H, ABq J_{AB} = 12.6 Hz	1.13 3H, s	0.96 3H, d J = 7 Hz	
Ingenol-3-propionate-20-isobutyrate[10] ($CDCl_3$, 90 MHz, TMS)										
6.09 1H, d J = 5 Hz	6.01 1H q J = 2 Hz	3.84 1H, bs	4.08 1H, m J = 10 Hz, 5 Hz	4.75, 4.45 2H, ABq J_{AB} = 13 Hz	5.44 1H, s	1.78 3H, d J = 2 Hz	(1.09 & 3H, s	1.06) 3H, s	0.99 3H, d J = 8 Hz	
Ingenol-3-octenoate-20-acetate[23] ($CDCl_3$, 250 MHz, TMS)										
6.13 1H, d J = 2.2 Hz	6.06 1H, d J = 1.5 Hz	3.89 1H, s	4.15 1H, m	4.63 2H, ABq J_{AB} = 12.5 Hz		1.81 3H, d J = 1.6 Hz	(0.97 & 3H, s	0.96) 3H, s	0.87 3H, m	

Milliamine A[33] ($CDCl_3$, 90 MHz, TMS)

6.14 1H, m	6.14 1H, m	3.90 1H, m	4.12 1H, m	4.82, 4.43 2H, ABq J_{AB} = 12 Hz	5.74 1H, s	1.80 3H, m	(1.05 & 3H, s	1.03) 3H, s	1.03 3H, d J = 7 Hz	

Milliamine D[33]($CDCl_3$, 90 MHz, TMS)

6.28 1H, m	5.95 1H, m	5.56 1H, bs	4.25 1H, m	4.55, 4.28 2H, ABq J_{AB} = 12 Hz	3.77 1H, d J = 6 Hz	1.77 3H, m	(1.15 & 3H, s	1.07) 3H, s	0.98 3H, d J = 7 Hz	

Milliamine F[33] ($CDCl_3$, 90 MHz, TMS)

6.30 1H, m	6.01 1H, m	5.63 1H, bs	4.29 1H, m	4.57, 4.36 2H, ABq J_{AB} = 12 Hz	3.83 1H, d J = 5 Hz	1.81 3H, m	(1.18 & 3H, s	1.09) 3H, s	1.04 3H, d J = 7 Hz	

5-Deoxyingenol-3,20-diacetate[18] ($CDCl_3$, 60 MHz, TMS)

5.82 1H, d	6.09 1H, s		4.30 1H, d	4.04, 3.95 2H, ABq J_{AB} = 2.75 Hz	4.93 1H, s	1.74 3H, d	(1.10 3H	— 3H	1.05) 3H	0.92 1H

20-Deoxyingenol[37] (C_5D_5D, 100 MHz)

5.90 1H, bd J = 4 Hz	6.22 1H, q J = 1Hz	3.86 1H, bs	4.60 1H, dd J = 12 Hz, 4 Hz	1.98 3H, bs	5.02 1H, s	1.90 3H, bs	(1.26 & 3H, s	1.10) 3H, s J = 8 Hz	1.16 3H, d	

20-Deoxyingenol-3-angelate[40] ($CDCl_3$, 60 MHz, TMS)

5.75 1H, d J = 7.5 Hz	6.05 1H, s	3.43 1H, s	4.03 1H, m	1.80 3H, s	5.46 1H, s	1.80 3H, s	(1.10 & 3H, s	1.05) 3H, s H = 7.8 Hz	0.89 3H, d	

16-Hydroxy-20-deoxyingenol-3-angelate-16-acetate[21] ($CDCl_3$, 360 MHz, TMS)

5.74 1H, d J = 3.5 Hz	6.05 1H, d J = 1.4 Hz	3.69 1H, bs	4.13 1H, m	1.78 3H, s	5.47 1H, s	1.80 3H, bs	4.28, 4.13 2H, ABq J_{AB} = 11.9 Hz	1.12 3H, s	0.98 3H, d J = 7.1 Hz	0.91 1H, m

Table 8 (continued)
¹H-NMR DATA OF INGENOL AND SOME OF ITS DERIVATIVES[a]

Protons on										
C-7	C-1	C-5	C-8	C-20	C-3	C-19	C-16	C-17	C-18	C-14
13-Hydroxyingenol-3-(2′,3′-dimethylbutyrate)-13-isododecanoate[20] ($CDCl_3$, 90 MHz, TMS)										
6.05 1H, bs	6.05 1H, bs	4.08 1H, s		4.18 2H, s	5.47 1H, s	1.80 3H, s				
13-Hydroxyingenol-5-(2′,3′-dimethylbutyrate)-13-isododecanoate[20] ($CDCl_3$, 90 MHz, TMS)										
6.18 1H, d J = 6 Hz	5.62 1H, bs	5.42 1H, s	4.37 1H, dd J = 6 Hz, 12 Hz	4.03 2H, bs	1.83 3H, s					
13-Hydroxyingenol-13-isododecanoate-20-(2′,3′-dimethylbutyrate)[20] ($CDCl_3$, 90 MHz, TMS)										
6.07 1H, d J = 4 Hz	5.88 1H, d J = 1.5 Hz	3.67 1H, d J = 10 Hz	4.01 1H, dd	4.64 ± 0.05 2H, ABq J_{AB} = 12 Hz	4.43 1H, bs	1.85 3H, d J = 1.5 Hz				1.30 1H, d
16-Hydroxyingenol-3,5,16,20-tetraacetate[27] ($CDCl_3$, 60 MHz, TMS)										
6.25 1H, m	6.01 1H	5.40 1H, bs	4.20 1H, m	4.4 ± 0.2 2H, ABq J_{AB} = 13 Hz	5.00 1H, s	1.77 3H, d	4.22 2H, s	1.15 3H, s	1.00 3H, s	
16-Hydroxyingenol-3-angelate-5,16,20-triacetate[21] ($CDCl_3$, 60 MHz, TMS)										
6.12 1H, m	6.06 1H, s	5.41 1H, bs		2H, ABq J_{AB} = 12.4 Hz	5.05 1H, s	1.76 3H, bs	4.28, 4.08 2H, ABq J_{AB} 12 Hz	1.13 3H, s	0.99 3H, d J = 7.1 Hz	

16-Hydroxyingenol-3-angelate-16,20-diacetate[21] ($CDCl_3$, 60 MHz, TMS)

6.10 1H, m	6.03 1H, d J = 1.5 Hz	3.88 1H, bs	4.19 1H, m	4.49, 4.74 2H, ABq J_{AB} = 13.4 Hz	5.58 1H, s	1.80 3H, d J = 1.3 Hz	4.35, 4.05 2H, ABq J_{AB} = 12.6 Hz	1.13 3H, s	0.97 3H, d J = 7 Hz	

16-Hydroxyingenol-3-(2′,4′,6′-decatrienoate)-16-angelate[21] ($CDCl_3$, 60 MHz, TMS)

(7.77 —— 1H	5.76 1H	4.07 1H, s		4.13 2H, s	5.63 1H, s	1.80 3H, bs	4.48, 4.13 2H, ABq J_{AB} − 12.1 Hz	1.15 3H, s	0.98 3H, d J = 6.6 Hz	

Euphorbia factor Cy_6[20] ($CDCl_3$, 90 MHz, TMS)

6.01 1H, d J = 4 Hz	6.32 1H, s	4.08 1H, bs		4.15 2H, bs	5.64 3H, s	4.63 2H, bs				1.26 1H, d

Euphorbia factor Cy_4[20] ($CDCl_3$, 90 MHz, TMS)

6.21 1H, d J = 6 Hz	6.03 1H, s	5.45 1H, s	4.42 1H, dd	4.05 2H, s	4.19 1H, s	4.71 2H, s				

Euphorbia factor Cy_2[20] ($CDCl_3$, 90 MHz, TMS)

6.08 1H, d J = 4 Hz	6.32 1H, s	3.64 1H, d	4.02 1H, dd	(4.95 2H	— 1H	4.45) 2H				

[a] δ ppm downfield from internal standard; s = singlet; d = doublet; q = quartet; m = multiplet; b = broad.

REFERENCES

1. **Hecker, E., Immrich, H., Bresch, H., and Schairer, H. U.**, Über die Wirkstoffe des Crotonols. VI. Entzundungsteste am Mauseohr, *Z. Krebsforsch.*, 68, 366, 1966.
2. **Evans, F. J. and Schmidt, R. J.**, An assay procedure for the comparative irritancy testing of esters in the tigliane and daphnane series, *Inflammation*, 3, 215, 1979.
3. **Adolf, W., Opferkuch, H., and Hecker, E.**, Über die dieterpenoiden Inhaltstoffe des Samenöls von *Euphorbia lathyris* und ihre tumorpromovierende Wirkung, *Fette, Seifen, Anstrichm.*, 70, 850, 1968.
4. **Hecker, E.**, New phorbol esters and related cocarcinogens, in *Proc. 10th Int. Cancer Congr.*, Vol. 5, Clark, R. L., Ed., Year Book Medical Publishers, Chicago, 1971, 213.
5. **Hergenhahn, M., Kusumoto, S., and Hecker, E.**, Diterpene esters from "Euphorbium" and their irritant and cocarcinogenic activity, *Experientia*, 30, 1438, 1974.
6. **Berenblum, I. and Shubik, P.**, The role of croton oil applications, associated with a single painting of a carcinogen in tumour induction of the mouse skin, *Br. J. Cancer*, 1, 379, 1947.
7. **Hecker, E.**, Cocarcinogene Wirkstoffe aus Euphorbiaceen, *Planta Med.*, 16, 24S, 1968.
8. **Kupchan, S. M., Uchida, I., Braufman, A. R., Dailey, R. G., and Fei, B. Y.**, Antileukemic principles isolated from Euphorbiaceae plants, *Science*, 191, 571, 1976.
9. **Sayed, M. D., Riszk, A., Hammounda, F. M., El-Missiry, M. M., Williamson, E. M., and Evans, F. J.**, Constituents of Egyptian Euphorbiaceae. IX. Irritant and cytotoxic ingenane esters from *Euphorbia paralias* L., *Experientia*, 36, 1206, 1980.
10. **Hirota, M., Ohigashi, H., Oki, Y., and Koshimizu, K.**, New ingenol-esters as piscicidal constituents of *Euphorbia cotinifolia* L., *Agric. Biol. Chem.*, 44, 1351, 1980.
11. **Zechmeister, K., Brandl, F., Hoppe, W., Hecker, E., Opferkuch, H. J., and Adolf, W.**, Structure determination of the new tetracyclic diterpene ingenol-triacetate with triple product methods, *Tetrahedron Lett.*, 47, 4075, 1970.
12. **Evans, F. J. and Kinghorn, A. D.**, Ingenol from *Euphorbia desmondii*, *Phytochemistry*, 13, 1011, 1974.
13. **Upadhyay, R. R., Zarintan, M. H., and Ansarin, M.**, Isolation of ingenol from the irritant and cocarcinogenic latex of *Euphorbia seguieriana*, *Planta Med.*, 30, 32, 1976.
14. **Upadhyay, R. R., Zarintan, M. H., and Ansarin, M.**, Irritant constituents of Iranian plants. Ingenol from *Euphorbia seguieriana*, *Planta Med.*, 30, 196, 1976.
15. **Upadhyay, R. R., Ansarin, M., and Zarintan, M. H.**, Isolation of ingenol from the irritant latex of *Euphorbia serrata* L., *Curr. Sci.*, 45, 500, 1976.
16. **Evans, F. J. and Kinghorn, A. D.**, A comparative phytochemical study of the diterpenes of some species of the genera *Euphorbia* and *Elaeophorbia* (Euphorbiaceae), *Bot. J. Linn. Soc.*, 74, 23, 1977.
17. **Upadhyay, R. R., Bakhtavar, F., Mohseni, H., Sater, A. M., Saleh, N., Tafazuli, A., Dizaji, F. N., and Mohaddes, G.**, Screening of *Euphorbia* from Azarbaijan for skin irritant activity and for diterpenes, *Planta Med.*, 38, 151, 1980.
18. **Evans, F. J. and Kinghorn, A. D.**, A new ingenol type diterpene from the irritant fractions of *Euphorbia myrsinites* and *Euphorbia biglandulosa*, *Phytochemistry*, 13, 2324, 1974.
19. **Kinghorn, A. D.**, Some Biologically Active Constituents of the Genus *Euphorbia*, Ph.D. thesis, University of London, London, 1975.
20. **Ott, H. H. and Hecker, E.**, Highly irritant ingenane type diterpene esters from *Euphrobia cyparissias* L., *Experientia*, 37, 88, 1981.
21. **Lin, L.-J., Marshall, G. T., and Kinghorn, A. D.**, The dermatitis-producing constituents of *Euphorbia hermentiana* latex, *J. Nat. Prod.*, 46, 723, 1983.
22. **Upadhyay, R. R., Samiyeh, R., and Tafazuli, A.**, Tumor promoting and skin irritant diterpene esters of *Euphorbia virgata* latex, *Neoplasma*, 28, 555, 1981.
23. **Abo, K. A. and Evans, F. J.**, Ingenol esters from the pro-inflammatory fraction of *Euphorbia kamerunica*, *Phytochemistry*, 21, 725, 1982.
24. **Seip, E. H. and Hecker, E.**, Skin irritant ingenol esters from *Euphorbia esula*, *Planta Med.*, 46, 215, 1982.
25. **Fürstenberger, G. and Hecker, E.**, Zum Wirkungsmechanismus cocarcinogener Pflanzeninhaltsstoffe, *Planta Med.*, 22, 241, 1972.
26. **Fürstenberger, G. and Hecker, E.**, New highly irritant Euphorbia factors from the latex of *Euphorbia tirucalli* L., *Experientia*, 33, 986, 1977.
27. **Opferkuch, H. J. and Hecker, E.**, New diterpenoid irritants from *Euphorbia ingens*, *Tetrahedron Lett.*, (3), 261, 1974.
28. **Upadhyay, R. R., Bakhtavar, F., Ghaisarzadeh, M., and Tilabi, J.**, Co-carcinogenic and irritant factors of *Euphorbia esula* L. latex, *Tumori*, 64, 99, 1978. ·
29. **Uemura, D. and Hirata, Y.**, Isolation and structures of irritant substances obtained from *Euphorbia* species (Euphorbiaceae), *Tetrahedron Lett.*, (11), 881, 1973.

30. **Hirata, Y.,** Toxic substances of Euphorbiaceae, *Pure Appl. Chem.*, 41, 175, 1975.
31. **Adolf, W. and Hecker, E.,** Further new diterpene esters from the irritant and cocarginogenic seed oil and latex of the caper spurge *(Euphorbia lathyris* L.), *Experientia*, 27, 1393, 1971.
32. **Uemura, D. and Hirata, Y.,** The isolation and structure of two new alkaloids, milliamines A and B, obtained from *Euphorbia millii, Tetrahedron Lett* , 39, 3673, 1971.
33. **Marston, A. and Hecker, E.,** On the active principles of the Euphorbiaceae. VI. Isolation and biological activities of seven milliamines from *Euphorbia milii, Planta Med.*, 47, 141, 1983.
34. **Upadhyay, R. R., Ansarin, M., Zarintan, M. H., and Shakui, P.,** Tumor promoting constituent of *Euphorbia serrata* latex, *Experientia*, 32, 1196, 1976.
35. **Lin, L. J. and Kinghorn, A. D.,** Three new ingenane derivatives from the latex of *Euphorbia canariensis* L., *J. Agric. Fd Chem.*, 31, 396, 1983.
36. **Falsone, G., Crea, A. E. G., and Noack, E. A.,** Über Inhaltsstoffe von Euphorbiaceae. VII. 20-Desoxyingenolmonoester und Ingenoldiester aus *Euphorbia biglandulosa* Desf., *Arch. Pharm.*, 315, 1026, 1982.
37. **Uemura, D., Ohwaki, H., Hirata, Y., Chen, Y.-P., and Hsu, H.-Y.,** Isolation and structures of 20-deoxyingenol new diterpene, derivatives and ingenol derivative obtained from "kansui", *Tetrahedron Lett.*, (29), 2527, 1974.
38. **Hecker, E.,** Cocarcinogens from Euphorbiaceae and Thymelaeaceae, in *Pharmacognosy and Phytochemistry*, Wagner, H. and Hörhammer, L., Eds., Springer-Verlag, Heidelberg, 1971, 147.
39. **Sorg, B. and Hecker E.,** Zur Chemie des Ingenols. II. Ester des Ingenol und des $\Delta^{7,8}$-Isoingenols, *Z. Naturforsch.*, 37B, 748, 1982.
40. **Rizk, A. M., Hammouda, F. M., El-Missiry, M. M., Radwan, H. M., and Evans, F. J.,** Constituents of Egyptian Euphorbiaceae. XIII. Biologically active diterpene esters from *Euphorbia peplus, Phytochemistry*, 24, 1605, 1984.
41. **Uemura, D., Hirata, Y., Chen, Y.-P., and Hsu. H.-Y.,** New diterpene, 13-oxyingenol, derivative isolated from *Euphorbia kansui* Liou, *Tetrahedron Lett.*, 29, 2529, 1974.
42. **Upadhyay, R. R. and Hecker, E.,** Diterpene esters of the irritant and cocarcinogenic latex of *Euphorbia lactea, Phytochemistry*, 14, 2514, 1975.

Chapter 10

THE BIOCHEMICAL MECHANISM OF ACTION OF PHORBOL ESTERS

Alastair Aitken

TABLE OF CONTENTS

I. INTRODUCTION

The aim of this chapter is to describe the current state of knowledge of the mechanism of action of phorbol esters. These compounds have been shown to be responsible for eliciting a wide range of responses in different tissues. The biological effects of administration of phorbol esters include tumor promotion,[1,2] cell proliferation,[3] activation of blood platelets,[4,5] lymphocyte mitogenesis,[6] inflammation[7,8] (erythema of the skin), prostaglandin production,[9] and stimulation of degranulation in neutrophils.[10]

Much of what is known about how the effects of phorbol esters are transmitted inside cells comes from the study of tumor-promoting and platelet-aggregating phorbol esters. These have been shown to interact with and activate a recently discovered protein kinase, first described by Nishizuka and colleagues,[11] and termed C kinase or protein kinase C. This is a Ca^{2+}- and phospholipid-dependent protein kinase of apparently ubiquitous tissue distribution.[12] The evidence[13] that phorbol esters causing tumor promotion and other biological effects substitute for the natural activator of this protein kinase, diacylglycerol, will be discussed in detail. Diacylglycerol (diolein) is a product of phosphatidylinositol hydrolysis (Figure 1), which has been shown to be an important mechanism by which the effects of many neurotransmitters and hormones are transmitted (Table 1). This activator of protein kinase C, diacylglycerol, may be considered a ''second messenger'' acting in an analogous manner to cyclic AMP (cAMP). Phosphatidylinositol breakdown does not only give rise to diacylglycerol, but the other product, inositol trisphosphate, is implicated as a second messenger for the release of Ca^{2+} from internal stores.[14,15] The precise details of the events occurring when an agonist causes stimulation of phosphatidylinositol metabolism are still being elucidated,[14,15] but for a recent detailed treatment of phosphatidylinositol turnover the reader is referred to the review of Berridge.[16]

Since the present review concentrates on the interaction of phorbol esters with protein kinase C, readers of this article should not feel that undue emphasis has been given to the importance of diacylglycerol at the expense of the effects of the other second messenger, inositol trisphosphate, which raises intracellular Ca^{2+} levels. Evidence has accumulated that there are some effects of phorbol esters on Ca^{2+} levels inside the cells, but probably still transmitted via protein kinase C.[16]*

The importance of reversible phosphorylation as a major intracellular regulatory mechanism has been fully realized only within the last decade.[17,18] Not only is this a mechanism for mediating neural and hormonal regulation of enzyme activity in a wide variety of metabolic processes, but the recent discovery of tyrosine-specific protein kinases from growth factor receptors and RNA tumor virus gene products[19,20] has further emphasized the biological importance of this covalent modification of proteins.

A brief overview of the mechanism of phosphatidylinositol turnover as a means of generating second messengers for the transmission of intracellular signals follows in the next section. This article will then concentrate on description of the properties of protein kinase C and the activation of this enzyme by phorbol esters. The subsequent biochemical events that lead to tumor promotion and platelet aggregation, in particular, will then be described (see also Chapter 2). The current literature on substrates of protein kinase C involved in other effects will also be reviewed.

II. PHOSPHATIDYLINOSITOL TURNOVER

Before considering the mode of action of phorbol esters in the activation of protein kinase

* TPA has recently been shown to inhibit muscarinic receptor-mediated phosphatidylinositol hydrolysis and Ca^{2+} mobilization[117] and prevent increase in intracellular Ca^{2+} in neutrophils in response to chemotactic factors.[118]

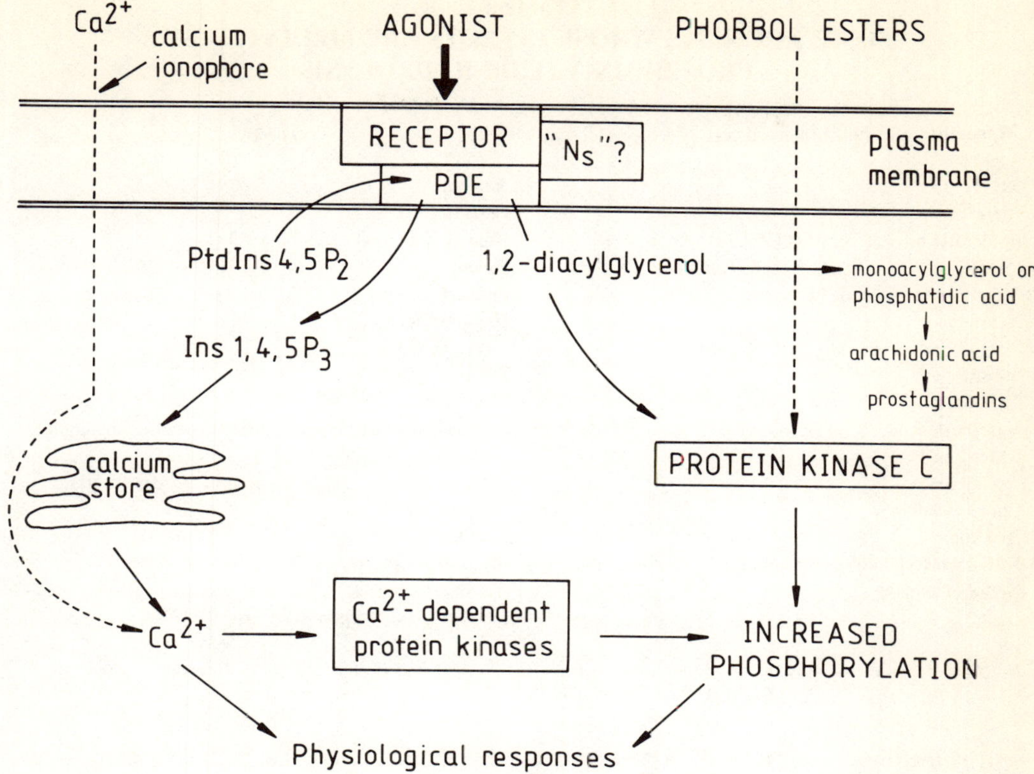

FIGURE 1. Phosphoinositide hydrolysis. On binding to its receptor, the agonist stimulates the activity of a phosphodiesterase (PDE) or phospholipase C. There is recent evidence that a guanine nucleotide-binding regulatory protein may be involved and is designated N_s to indicate possible analogy with the regulatory protein in the adenylate cyclase receptor.[112] The products from phosphatidylinositol-4,5-bisphosphate (PtdIns $4,5P_2$) are inositol-1,4,5-trisphosphate (Ins $1,4,5P_3$), which releases calcium from (probably) the endoplasmic reticulum, and 1,2-diacylglycerol, which activates protein kinase C. The direct stimulation of protein kinase C by phorbol esters is illustrated. Calcium ionophores may increase intracellular calcium levels directly, leading to synergistic effects of the two artificial stimuli. Other details are in the text.

C, the role of this enzyme should be briefly mentioned in the overall mechanism of action of the large number of neurotransmitters and hormones whose effects are mediated through hydrolysis of polyphosphoinositides (Table 1). As seen from Table 1, the type of receptor implicated in phosphatidylinositol hydrolysis includes the α_1-adrenergic, H_1-histaminergic, muscarinic cholinergic, and the V_1-vasopressin receptors. The breakdown of the phosphatidylinositide, PtdIns $4,5P_2$ to produce inositol 1,4,5-trisphosphate (Ins $1,4,5P_3$) responsible for the mobilization of Ca^{2+} from internal stores, and 1,2-diacylglycerol, which activates protein kinase C is illustrated in Figure 1.

The hypothesis that is currently the generally accepted version of events is as follows. On binding to the receptor on the cell surface, the agonist causes hydrolysis of PtdIns $4,5P_2$, catalyzed by a phospholipase C or phosphodiesterase to produce the two "second messengers" inositol — 1,4,5-trisphosphate (Ins $1,4,5P_3$) and diacylglycerol. The former is responsible for triggering release of Ca^{2+} ions from internal stores.[15,16] The increased intracellular level of Ca^{2+} then causes, among other events, activation of Ca^{2+}-dependent protein kinases.[21]

In the other branch of the response to the agonist, the 1,2-diacylglycerol activates the Ca^{2+}- and phospholipid-dependent protein kinase C.[22] The phosphorylation of specific intracellular proteins in a number of tissues has been demonstrated in response to the activation of this protein kinase, and will be discussed in subsequent sections.

Table 1
SOME AGONISTS WHOSE EFFECTS ARE MEDIATED BY PHOSPHOINOSITIDE HYDROLYSIS

Hormones and neurotransmitters	Tissue
ACTH	Adrenal
Angiotensin	Hepatocytes
Epidermal growth factor (EGF)	Human epidermal carcinoma A431 cells
Caerulein	Pancreas
Thyrotropin-releasing hormone	Pituitary
Neurotensin	Hypothalamus
Platelet-derived growth factor (PDGF)	Swiss 3T3 cells
Substance P	Parotid, hypothalamus
Thrombin	Platelets
Vasopressin	Liver/hepatocytes, hippocampus, sympathetic ganglia
f-Met-Leu-Phe	Neutrophils, leukocytes
Acetylcholine	Adrenal medulla, pancreas, parotid, synaptosomes
Cholescystokin	Cerebral cortex
Histamine	Cerebral cortex
5-Hydroxytryptamine (serotonin)	Fly salivary gland
Noradrenaline	Iris smooth muscle
Light	*Limulus* photoreceptors[114,115]

Note: References to individual agonists may be found in the text and in the reviews of Berridge,[16] Cockroft,[10] Fisher et al.,[32] and Michell.[15,44]

It is the direct activation of protein kinase C by phorbol esters that is the currently held view of the mechanism by which these tumor-promoting and platelet-activating phorbol esters exert their effects on tissues. It should be stressed that the stimulation of kinase C by this mechanism does not in many instances appear to affect the other ''branch'' of the normal cascade system whereby agonists would also activate Ca^{2+}-dependent intracellular events by the production of Ins $1,4,5P_3$. Roach and Goldman[23] have shown that tumor-promoting phorbol esters inactivate (by phosphorylation through kinase C) glycogen synthase in rat hepatocytes. This inactivation is uncoupled from alterations in phosphorylase kinase activity (a Ca^{2+}- and calmodulin-dependent protein kinase that also phosphorylates glycogen synthase[24]).

In addition to phorbol esters, other types of tumor promoters, e.g., the indole alkaloid teleocidin and the polyacetate aplysiatoxin, have been shown to have cocarcinogenic effects.[25,26] The tumor-promoting compound mezerein, structurally related to phorbol esters, has been recently shown to stimulate the activity of protein kinase C.[27] All three classes of molecule have hydrophilic and hydrophobic domains and these other types of tumor promoter may bind to and activate protein kinase C in a similar manner to the tumor-promoting phorbol esters.[25] A monosaccharide lipid A precursor from *Escherichia coli*[28] has recently been shown to stimulate turnover of PtdIns $4,5P_2$. This lipopolysaccharide precursor activated protein kinase C but the effect appeared to be related to its ability to substitute for phospholipid (phosphatidylserine). As well as functioning as a second messenger in the stimulation of protein kinase C activity, diacylglycerol may also be a precursor for release of arachidonic acid, which is required not only for synthesis of prostaglandins but also leukotrienes and thromboxane.[29] There is evidence for preferential breakdown of those phosphoinositides that carry arachidonic acid in the 2 position[30] the most common fatty acid at this position. After hydrolysis of these phosphoinositides, arachidonic acid is subsequently released from diacylglycerol by a lipase[31] or, it has been suggested, from phosphatidic acid by a specific phospholipase A_2 after diacyglycerol has been converted to phosphatidic acid.[29,32] In addition, the turnover of phosphoinositides may increase the level of cGMP by stimulation of guanylate

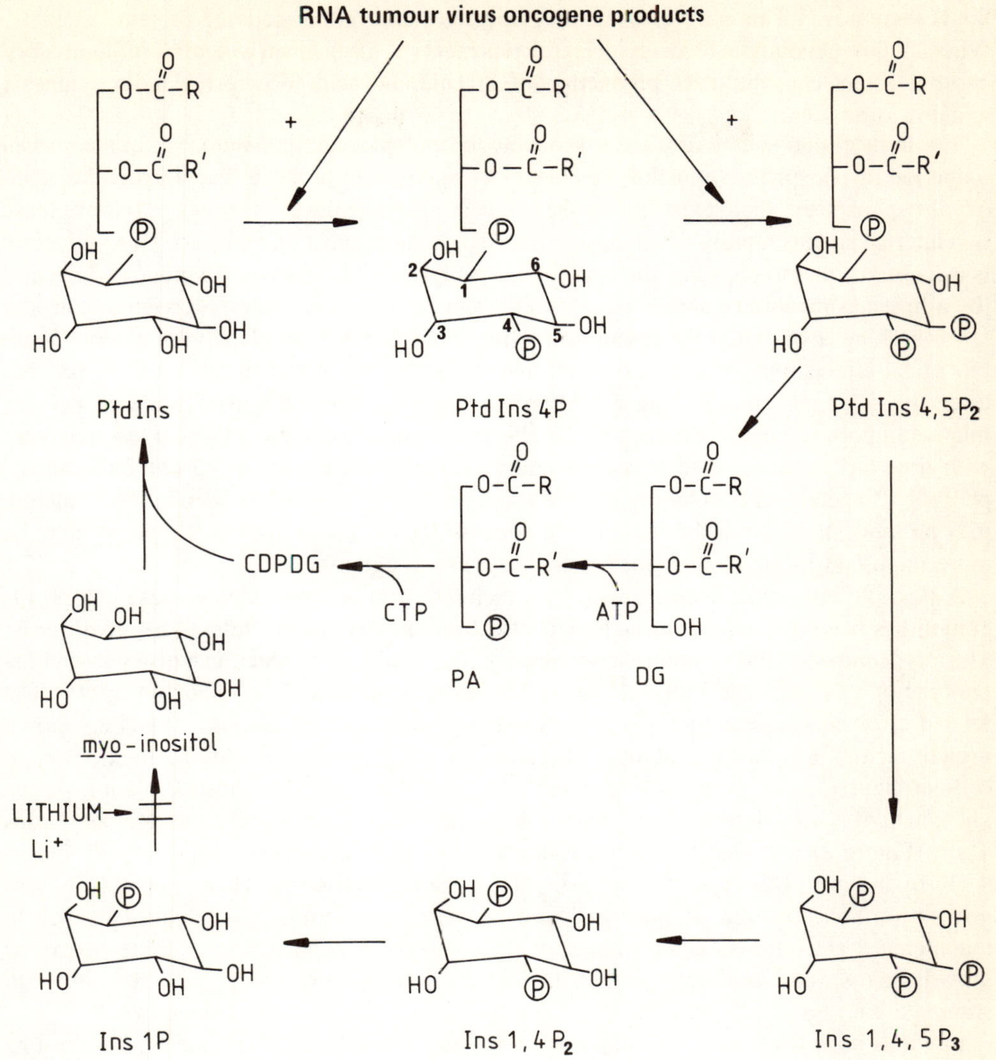

FIGURE 2. Recycling of products of phosphoinositide hydrolysis. Inositoltrisphosphate (Ins 1,4,5P_3) is recycled via inositol to phosphatidylinositol (PtdIns). Lithium ions inhibit inositol-1-phosphatase, which reduces the pool of PtdIns. *De novo* synthesis of inositol via glucose-6-phosphate to inositol-1-phosphate would also be reduced. Diacylglycerol (DG) is recycled via phosphatidic acid (PA) and cytidine diphosphate diacylglycerol (CDPDG). The stimulation of the pool of phosphatidylinositol-4,5-bisphosphate (PtdIns 4,5P_2) via phosphatidylinositol-4-phosphate (PtdIns4P) by "tyrosine" specific protein kinases is also shown.

cyclase.[16,22] Arachidonic acid and related compounds have been implicated in this stimulation. Nishizuka[33] has discussed the role of cyclic nucleotide-dependent protein kinases as feedback regulators to modulate the phosphatidylinositol turnover response.

Certain phorbol esters have been reported to stimulate prostaglandin E_2 production in human rheumatoid synovial cells.[34] Either these phorbol esters are activating mechanisms for elevation of diacylglycerol or phosphatidic acid levels which would provide arachidonic acid (Figure 2), or these phorbol esters, perhaps mediated by their effects on protein kinase C, are stimulating release of arachidonic acid from other phospholipid precursors such as phosphatidylcholine, phosphatidylethanolamine, and phosphatidylinositol via a Ca^{2+}-dependent phospholipase A_2.

In some cases when cells are exposed to phorbol esters, turnover of membrane phospholipids increases.[35] This may generate diacylglycerol and further activate protein kinase C. Whether this generation of diacylglycerol represents a mechanism whereby inflammatory phorbol esters can stimulate production of arachidonic acid for prostaglandin synthesis remains to be seen.[34]

The intracellular source of Ca^{2+} is probably the endoplasmic reticulum.[36,37] In the normal induction of phosphatidylinositol breakdown by agonists to produce diacylglycerol and inositoltrisphosphate, there exists inside the cell degradative pathways that would remove these two internal signals rapidly when the external signal, the agonist, is removed. Diacylglycerol is converted either to phosphatidic acid (by a diacylglycerol kinase) or to a monoacylglycerol (by a lipase as mentioned above; see Figures 1 and 2). This is an important point to consider when seeking a basis for the tumor-promoting and other effects of phorbol esters. These biological effects may be due to the fact that the activation of protein kinase C by phorbol esters is of longer duration than would be activation by diacylglycerol. Evidence that the relative importance of each branch of the phosphatidylinositol hydrolysis pathway may vary with time has been obtained in a comparative study of the effects of 12-*O*-tetradecanoylphorbol-13-acetate on platelet myosin phosphorylation[38] (discussed in Section IV). Calcium may be more involved in the initiation of some of the effects, while diacylglycerol may be more important in the maintenance of the response to agonists.

A possible biochemical mechanism by which RNA tumor viruses may cause cell proliferation has been discovered through their effect on metabolites of PtdIns 4,5P_2 hydrolysis. The "tyrosine-specific" protein kinase activity, the oncogene product, can also cause phosphorylation of PtdIns and PtdIns 4P as well as phosphorylation of 1,2-diacylglycerol.[14] The related tyrosine kinase activity in growth hormone receptors for insulin[39,40] and epidermal growth factor[41] also has this ability. These enzymes act by either modifying the activity of cellular lipid kinases or by acting directly as lipid kinases. The phosphorylation of the phosphatidylinositol intermediates would have the effect of increasing the pool of PtdIns 4,5P_2 (Figure 2) and lead to a greater intensity of effect of agonists that stimulate PtdIns 4,5P_2 hydrolysis. This type of synergistic effect has been observed where a combination of growth hormone (insulin or epidermal growth factor) and PtdIns 4,5P_2 agonist has led to much greater activation of cell proliferation than one type of agonist alone.[14] Platelet-derived growth factor[42] and epidermal growth factor[41] have been shown to possess the ability to stimulate both tyrosine kinase activity[43] and to enhance PtdIns 4,5P_2 breakdown.[41,42]

Li^+ ions have been shown to inhibit myoinositol 1-phosphatase[44] (Figure 2). This results in an accumulation of inositol 1-phosphate (Ins-P) and a reduction in the resynthesis of phosphatidylinositol.[45] The end result will be a lowering of the pool of PtdIns 4,5P_2. The biochemical consequences of Li^+ administration may therefore be a lowering of the sensitivity of the receptors for agonists whose effects are mediated by hydrolysis of phosphoinositides. Small amounts of Li^+ administered to patients have been shown to alleviate the symptoms of manic depression.[44] The use of lithium ions is also proving to be an extremely useful biological tool for the study of the metabolism of polyphosphoinositides.[44,45]

In blood platelets it has been possible to mimic the normal secretory effects by using a Ca^{2+} ionophore and phorbol ester (or 1-oleoyl-2-acetylglycerol) together.[46] The latter compound is a derivative of diacylglycerol that activates protein kinase C. Using Ca^{2+} ionophores or kinase C activators alone produced very little secretion. Similar synergistic effects have been observed in a number of other tissues.[16]

In most systems so far studied the stimulation of phosphatidylinositol metabolism is independent of an increase in the intracellular level of calcium. However, in a few cases e.g., in neutrophils,[10] stimulation of phosphatidylinositol turnover does appear to be dependent on calcium. There is evidence that besides formation of Ins 1,4,5-trisphosphate (releasing Ca^{2+} from internal stores), phosphatidylinositol metabolism may also be responsible for increasing the permeability of the plasma membrane to calcium.[15,16]

This overview of events occurring when the agonists listed in Table 1 increase the metabolic turnover of PtdIns 4,5P_2, has outlined the involvement of this mechanism in intracellular regulation of a very wide range of biological events. On binding to their receptors these agonists can: (1) cause increased phosphorylation of specific proteins (through activation of protein kinase C) (2) cause increased intracellular Ca^{2+} levels (which may also affect phosphorylation of proteins by activating distinct Ca^{2+}-dependent protein kinases and a phosphatase,[18,21] (3) induce sodium influx, (4) affect cAMP and cGMP levels, and (5) stimulate prostaglandin, thromboxane, and leukotriene production.

The events occurring in one branch of this activation pathway, namely, the effects of diacylglycerol and phorbol esters on the activity of protein kinase C, will now be considered in more detail.

III. PHYSICOCHEMICAL PROPERTIES OF PROTEIN KINASE C

Since Castagna et al.[13] first showed that phorbol esters could activate this enzyme, recent studies on the co-purification of the phorbol ester "receptor" and protein kinase C from a variety of tissues,[47-50] particularly the purification to homogeneity of bovine brain protein kinase C[50], have produced firm evidence that phorbol esters exert their effects by substituting for diacylglycerol. This evidence, which has been accumulated mainly in the study of phorbolesters involved in tumor promotion and platelet aggregation, will be considered in detail in this section. The properties of protein kinase C from different sources will be discussed along with phorbol-ester binding studies. The substrate specificity of protein kinase C, which has been shown to phosphorylate proteins with a wide range of function in different tissues will be reviewed in Section IV.

A. Purification and Properties of Protein Kinase C

This enzyme was first purified from rat brain by Kikkawa et al.[51] in the group of Nishizuka and from bovine heart by Wise et al.[52] Protein kinase C has recently been purified to homogeneity from bovine brain by Parker et al.,[50] who have shown it to be a monomeric protein of *M* 79,000, in fairly close agreement with the value obtained by most other groups. The purified enzyme is dependent on Ca^{2+} (with a K_a in the micromolar range) and is dependent on phospholipid (particularly phosphatidylserine). Five other phospholipids (phosphatidyl-inositol, -ethanolamine, -choline, phosphatidic acid, and sphingomyelin) were shown to be ineffective in stimulating protein kinase C phosphorylation of proteins;[53] indeed phosphatidyl choline and sphingomyelin inhibited the enzyme.[54] Although it is not a calmodulin-dependent enzyme, protein kinase C is inhibited by the phenothiazine and other type of inhibitors of Ca^{2+}-calmodulin-regulated enzymes.[55,56] The purified enzyme from bovine brain[50] exhibits two apparent K_a values for phosphatidylserine of approximately 0.6 to 2 and 35 to 80 μg/mℓ. Single K_a values of 6 to 20 μg/mℓ were reported for the enzyme from other tissues.[57,58] In the presence of 1,2-diacylglycerol (diolein) the K_a for Ca^{2+} is lowered from 10 to 1 μ*M*.[54]

The enzyme from bovine[50] and rat[59] brain has been shown to bind close to 1 mol tritiated-phorbol dibutyrate, furnishing proof that protein kinase C is the phorbol ester receptor. The bovine[50] and rat[59] brain preparations were shown to bind this phorbol ester with K_a values of 15 and 8 n*M*, respectively. Diacylglycerol has also been shown to competitively inhibit this phorbol ester binding.[60]

In the presence of diacylglycerol or phorbol dibutyrate the K_a values for phosphatidylserine were lowered two- to threefold in bovine brain protein kinase C.[50]

The kinetic properties of the enzyme isolated from these other sources appear to be similar, although large differences in specific activity are apparent. Protein kinase C from EL4 thymoma cells[61] was reported to be a protein of M_r 70,000. The enzyme has been partially

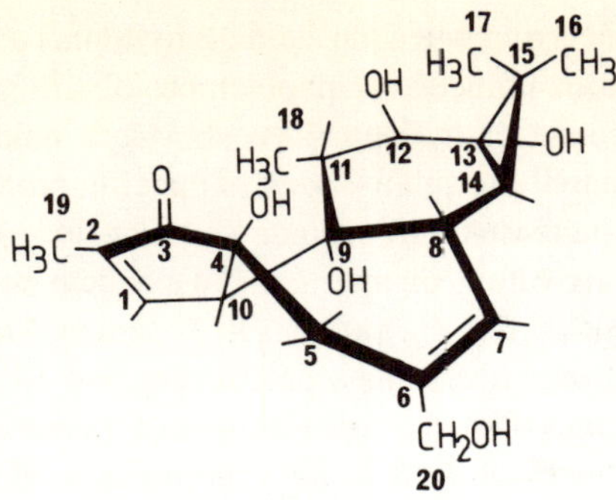

FIGURE 3. Phorbol. Illustrating the "parent" compound of the derivatives described in the text. 4α-Phorbol derivatives are inactive (the 5- and 7-membered rings are *trans*-fused in phorbol and *cis*-fused in 4α-phorbol, leading to quite a different shape of molecule).

purified from chicken[53] and pig brain[62] and to homogeneity from spleen[57] with an M_r of 67,000, although the spleen enzyme may be partially proteolyzed and is a form which is not activated by diacylglycerol.

Protein kinase C appears to be present in all tissues, from mammals to insects,[12] and is present in comparatively large amounts in brain, where it is partially localized in synaptosomes. Nishisuka[33] has evidence for the existence in protein kinase C of two functionally distinct domains. He suggests the presence of a hydrophilic domain that carries the active site and a hydrophobic domain that may bind to membranes. The two domains may be cleaved by a Ca^{2+}-dependent thiol proteinase (active at 10^{-6} to 10^{-5} M Ca^{2+}). This proteolysis produces an enzyme fragment of approximate M_r of 51,000, which is fully active in the absence of Ca^{2+}, phospholipid, and diacylglycerol.[63] A possible physiological significance of this proteolysis is not yet established.

There is evidence that, on activation by either diacylglycerol or phorbol esters, protein kinase C, which is mainly associated with the cytoplasm, becomes attached to the inside of the plasma membrane.[61,64,65] Although this active form of the enzyme is now membrane associated, it is still capable of phosphorylating specific intracellular proteins.

Protein kinase C is normally assayed by following the incorporation of ^{32}P-radioactivity from ^{32}P-ATP/Mg^{2+} into histone type III-S or histone H1 by incubation with the enzyme preparation in the additional presence of Ca^{2+}, phosphatidylserine, and diacylglycerol (or phorbol esters).[13,48,50] The incubation is carried out at 30°C at pH 7.5 (tris or HEPES buffer).

B. Structural Correlations of Phorbol Esters with Biological Activity

Studies on a wide range of phorbol esters with platelet-aggregating, tumor-promoting, lymphocyte mitogenic, and inflammatory properties (including stimulation of production of prostaglandins) has not only shown certain structural correlations between the ability to provoke these different effects, but has also indicated some possible differences in the biochemical mechanisms of these phorbol esters in eliciting the above responses. The biological effects of the most potent phorbol ester derivatives are seen at very low doses (nanogram levels) comparable to growth factors and hormones (Figure 3).

The structural correlations of a large number of tumor-promoting and nontumor-promoting phorbol esters have been studied with respect to their activity as human lymphocyte mitogens.[66] This mitogenic activity of the phorbol esters was inhibited by the anti-inflammatory glucocorticosteroid dexamethasone. These steroids have been shown to induce the release and synthesis of a phospholipase A_2-inhibiting polypeptide from leukocytes.[67] In the study, two other phospholipase inhibitors, chloroquine and *p*-bromophenacyl bromide are also found to be inhibitors of phorbol ester-induced lymphocyte proliferation.[66]

Tetradecanoyl phorbol acetate (TPA) has been shown to enhance interleukin-2 (IL-2) release by lectin-treated human T cells.[6] This mimics the effect of the lymphocyte-activating factor IL-1 (interleukin-1), thus replacing the requirement for monocytes/macrophages in the process. TPA has also been shown recently to induce the expression and phosphorylation of the IL-2 receptor in lymphocytes.[69] Phorbol-12,13 diesters showed both inflammatory and tumor-promoting responses, while the 12-deoxyphorbol-13,20-diesters were inflammatory but not tumor promoting.[66] The C-13 monoester equivalents were potent inflammatory agents.[70] This has led to suggestions that distinct biochemical targets (receptors) exist for inflammatory and tumor-promoting effects.[71,72]

Structural correlations of these phorbol esters, including derivatives based on phorbol, 4-deoxyphorbol, and 12-deoxyphorbol, showed that these structures correlated with their ability to stimulate human lymphocyte mitogenesis but not with the in vivo inflammatory activity. Human and primate lymphocytes are stimulated to divide by phorbol esters alone, although comitogenic effects have been observed in these and other species. The conclusion has been reached from the study of the activity of 12-deoxyphorbol and other phorbol esters on human lymphocytes[71] that two trigger sites many exist on lymphocytes, only one of which correlates with in vivo tumor promotion.[72]

These results contradict conclusions drawn from previous studies that lymphocyte mitogenic activity serves as a good model for cocarcinogenic properties. The activation of lymphocyte mitogenesis by the nontumor-promoting pro-inflammatory phorbol esters differs from that of the tumor promoters in their ability to induce competence to respond to IL-2.[66] Dexamethasone inhibits IL-2 production. Dexamethasone also inhibits the tumor-promoting and pro-inflammatory activities of phorbol esters.

The noncorrelating phorbol esters (i.e., nontumor-promoting pro-inflammatory 13,20-diesters) were found in that study to be exceptional in that their effects were sensitive to inhibition by cyclosporin A.[66]

The ability of phorbol esters to cause aggregation of blood platelets has been studied with a range of structurally distinct compounds.[73] 12-Deoxyphorbolphenylacetate was the most potent aggregating agent of platelets among a range of 12-deoxyphorbol derivatives studied. An ester function at C-13 and a free (primary) hydroxy group at C-20 were essential for activity, while esterification at C-12 was not necessary.[73] It was noted that 12-deoxyphorbol esters have very little if any tumor-promoting activity, in contrast to TPA. The platelet aggregating and inflammatory responses of a large number of 4-deoxyphorbol and related derivatives have been correlated.[73] The tertiary C-4 hydroxy groups was shown not to be essential for activity although the 4-deoxy analogues were shown to be less potent. 5-Hydroxy derivatives (which can undergo intramolecular hydrogen bonding) had much lower inflammatory and platelet aggregating abilities. The presence of a free primary hydroxy group at C-20 was also essential for these activities.[73]

IV. PROTEIN KINASE C SUBSTRATE SPECIFICITY

A. Interaction with the Intracellular Effects of Growth Hormones and RNA Tumor Viruses

Protein kinase C is an enzyme that specifically phosphorylates serine and threonine residues in proteins. In common with many other classes of protein kinase, for example, tyrosine specific,[19,43] and cAMP[76] and cGMP[77] dependent enzymes, protein kinase C has been shown to have the ability to autophosphorylate.[78] In this case there is no known physiological significance of the reaction.

There would seem to be a clear distinction between protein kinases that phosphorylate

Table 2
AMINO ACID SEQUENCES SURROUNDING THE SITES OF PHOSPHORYLATION OF SOME SUBSTRATES OF PROTEIN KINASE C

Substrate	Sequence	Ref.
Rat mammary gland, ATP citrate lyase	Lys/Arg Thr-Ala-Ser(P)-Phe-Ser-Glu-Ser-Arg	120
Epidermal growth factor	Arg-Arg-Arg-His-Ile-Val-Arg-Lys-Arg-Thr(P)-Leu-Arg-Arg	82
Calf thymus histone H1	38 Ala-Lys-Arg-Lys-Ala-Ser(P)-Gly-Pro-Pro-Val-Ser	11
Calf thymus histone H2B	32 36 * Gly-Lys-Lys-Arg-Lys-Arg-Ser(P)-Arg-Lys-Glu-Ser(P)-Tyr	11
Bovine myelin basic protein	115 Arg-Phe-Ser(P)-Trp	116

* Slow rate of phosphorylation.[36]

serine and threonine and those that are specific for tyrosine residues. The latter class(es) of protein kinase has been shown to be an intrinsic part of the growth hormone receptors.[43,79,80] RNA tumor virus gene products have also been shown to be tyrosine-specific protein kinases that are involved in cellular transformation.[19,20,81] If protein kinase C could be shown to affect the state of phosphorylation of substrates for these tyrosine-specific protein kinases this could provide a molecular mechanism for at least some of the effects of phorbol esters; this has, in fact, been seen in a number of systems. A combination of "tyrosine" kinase and PtdIns4,5P_2 hydrolysis stimulus leads to synergistic effects in cell proliferation.[14,16]

Protein kinase C has been shown to phosphorylate growth hormone receptors.[82-85] The best studied example is epidermal growth factor (EGF) receptor which is phosphorylated by protein kinase C on a threonine residue nine amino acids from the cytoplasmic end of a proposed transmembrane domain. The threonine phosphorylation is in a basic region and the sequence surrounding the site of phosphorylation was deduced:[82] Arg-Arg-Arg-His-Ile-Val-Arg-Lys-Arg-Thr(P)-Leu-Arg-Arg.

The in vivo phosphorylation of EGF receptor mediated by protein kinase C has been shown in A431 cells to lead to a reduction in the ability of EGF to bind to its receptor and to reduce the tyrosine protein kinase activity. Both TPA and the indole alkaloid tumor promoter, teleocidin, stimulated this phosphorylation. EGF has been shown to increase the hydrolysis of PtdIns4,5P_2 (see Section II). The role of this EGF receptor phosphorylation by kinase C may therefore be a feedback mechanism to modulate the response of the EGF receptor.

Gilmore and Martin[86] have shown that an M_r 42,000 polypeptide that is phosphorylated at a tyrosine residue in avian sarcoma virus transformed cells and in cells stimulated with growth hormones (EGF, PDGF, and multiplication-stimulating activity), also becomes phosphorylated at tyrosine residues in cells treated with TPA. Exogenously added 1-oleoyl-2-acetyl-glycerol also stimulates the phosphorylation of the M_r 42,000 protein on tyrosine. The most probable explanation was that this tyrosine phosphorylation was not mediated directly by stimulation of protein kinase C activity, but that C-kinase (with a specificity for serine and threonine phosphorylation) was stimulating tyrosine-specific protein kinase(s).

The large number of basic amino acids surrounding the site of phosphorylation is immediately apparent in the EGF receptor.[82] Other sites of phosphorylation in the basic proteins histone H1, H2B, and myelin basic protein (all phosphoserine residues) are preceded by basic amino acid residues (see Table 2). The primary structure requirement of protein kinase C would therefore appear to be similar to those of some other protein kinases, particularly cAMP[76]- and cGMP[77]-dependent protein kinases. There may in fact be some overlap in structural specificity of cAMP-dependent and C-kinase. (A further example is ribosomal protein S6 — see below.) The same serine residue in ATP-citrate lyase in the sequence Lys/Arg, Thr-Ala-Ser(P)-Phe-Ser-Glu-Arg is phosphorylated by both kinases.[120] The state of phosphorylation of this site is increased by insulin.

Protein kinase C has been shown to phosphorylate acetyl CoA carboxylase but on a different site than that phosphorylated by cAMP-dependent protein kinase;[120] the phosphorylation does not result in activation of the enzyme while tyrosine hydroxylase phosphorylation by protein kinase C does result in activation of this enzyme. This phosphorylation is on the same site labeled with cAMP-dependent protein kinase.[121]

Administration of insulin has been shown to lead to the stimulation of the phosphorylation of an M_r 47,000 protein in rat hippocampus.[87] This substrate, protein F1, is probably identical to the central nervous system B-50 protein (M_r 48,000), which is a substrate for protein kinase C.[88] This protein, and the following ribosomal protein, are examples of substrates whose state of phosphorylation is increased in response to phorbol esters, although it is not yet clear whether the phosphorylation is mediated directly by protein kinase C or by indirect activation of an additional protein kinase.

TPA can stimulate phosphorylation of ribosomal protein S6 in hepatoma cells on serine residues.[89] This H35 rat hepatoma cell line has been shown to increase ornithine decarboxylase activity and polyamine synthesis in response to tumor-promoting phorbol esters.[90] The growth hormones insulin and insulin-like growth factor have also been shown to enhance phosphorylation in protein S6. A function for ribosomal protein S6 phosphorylation has been suggested in a role of transition into the G1 phase of the cell cycle and in altering the affinity of the ribosome for certain classes of mRNA.[90]

Wettenhall and colleagues[119] have shown that within the palindromic sequence Arg-Arg-Leu-Ser-Ser-Leu-Arg the serine residues are phosphorylated by cyclic-AMP-dependent protein kinase, a protease-activated kinase and protein kinase C. Cyclic-AMP-dependent kinase preferentially phosphorylates the first serine, while the latter kinases appear to prefer the other site.

Phosphorylation of vinculin (a cytoskeletal protein) in fibroblasts and Swiss 3T3 cell cultures is increased in response to TPA and phorbol-dibutyrate (PDB).[91,92] The latter, which is a less potent phorbol ester, resulted in a smaller increase in phosphorylation (inactive phorbol esters showed no increase in phosphorylation). Significant increases in phosphorylation were seen in three peptides, on serine and threonine residues. This protein is also a substrate for the transforming protein of Rous sarcoma virus when it is phosphorylated on a tyrosine residue.[19] Vinculin has been postulated as having a role in the morphological changes seen in Rous sarcoma virus transformation. This work establishes a link between the morphological changes induced by phorbol esters and those seen following viral transformation.

TPA administration induces terminal differentiation into macrophages in normal bone marrow promyelocytes and in myeloid leukemic cells.[93,94] A large increase in the phosphorylation of two cytosolic proteins of M_r 17,000 and 27,000 in the promyelocytic cell line HL-60 has been noted, in response to biologically active phorbol esters.[93,94] Increase in the intracellular level of Ca^{2+} and activation of an Na^+/H^+ exchange carrier are the main ionic events occurring that are responsible for the onset of cell proliferation.[16,95] Protein kinase C appears to phosphorylate and activate this neutral Na^+/H^+ exchanger.[96] Induction of a murine pre-lymphocyte cell line differentiation is accomplished by activation of this Na^+ uptake system. In 3T3 cells, administration of phorbol esters has been shown to increase internal pH.[97]

B. Phorbol Esters, Protein Kinase C, and Platelet Aggregation

The mechanism of aggregation of blood platelets by structurally distinct phorbol esters has been studied in detail.[71,73,74] Although protein kinase C activity is not affected by calmodulin the enzyme is sensitive to calmodulin inhibitors such as chlorpromazines (mentioned in Section III). These compounds have also been shown to inhibit platelet aggregation and other effects of phorbol esters. Membrane stabilizing agents, such as imipramine, inhibited platelet aggregation by phorbol esters.[74] Cyclooxygenase inhibitors (e.g., indomethecin) and free radical scavengers had no inhibitory effects, while platelet aggregation was susceptible to inhibition by prostaglandins such as PGI_2 and PGE_1. Phospholipase A_2 inhibitors (such as propranolol and mepacrine) also affect phorbol ester-induced platelet aggregation as well as induction of erythema.[73,74] In common with other aggregating agents, the effects of the phorbol esters were dependent on levels of cAMP, divalent cations, and an intact microtubule system.[73]

In blood platelets, in response to stimulation by 12-*O*-tetradecanoyl phorbol-13-acetate (TPA), there is a marked increase in phosphorylation of two proteins[98] of M_r 20,000 and 40,000. The protein of M_r 20,000 has been shown to be identical to myosin light chain.[38] The major phosphorylation site has been shown by peptide mapping to be different from that phosphorylated by the Ca^{2+}-calmodulin-dependent myosin light chain kinase. The latter is the site phosphorylated when platelets are stimulated with thrombin and when purified human platelet myosin is phosphorylated by myosin light chain kinase.[38]

The M_r 40,000 protein appears to be a specific substrate for protein kinase C and the state of phosphorylation of this protein is also increased by thrombin[98] on an identical peptide.

The phosphorylation of both the M_r 20,000 (platelet myosin light chain) and M_r 40,000 proteins therefore appears to be important, although their exact role(s) in the regulation of platelet function is (are) not yet known. A role for the M_r 40,000 protein phosphorylation in serotonin release has been suggested by the group of Nishisuka.[78]

It is interesting to note that phosphorylation of the myosin light chain was much faster in response to thrombin than TPA.[38] This suggests that the effects of thrombin on the stimulation of PtdIns $4,5P_2$ hydrolysis to inositol trisphosphate (which leads to activation of Ca^{2+}-calmodulin-dependent myosin light chain kinase) is faster than the protein kinase C effects.[38] Human platelet myosin light chain in vivo, in response to TPA, showed an additional minor site of phosphorylation that appeared to be identical to that phosphorylated in response to the thrombin activation of platelets (the Ca^{2+}-calmodulin activated-myosin light chain kinase site).[38] This indicates a possible effect of protein kinase C on PtdIns $4,5P_2$ hydrolysis which would subsequently raise Ca^{2+} levels to activate the former Ca^{2+}-calmodulin-dependent protein kinase.

The effects of the two second messenger signals that result from phosphatidylinositol hydrolysis can be mimicked in platelets by the addition of Ca^{2+} ionophores activating synergistically with activators of protein kinase C.[15] Addition of the calcium ionophore A23187 together with 12-*O*-tetradecanoyl phorbol-13-acetate also produces a greater response in lymphocytes,[99] in insulin-secreting pancreatic islet cells,[100] in the cell line BALB/c3T3,[101] in adrenal glomulerosa,[102] in neutrophils,[103] and in the release of acetylcholine from guinea pig ileum.[104] This synergism between effectors of protein kinase C and Ca^{2+} levels has been best studied in platelets, however, where addition of the calcium ionophore together with 1-oleoyl-2-acetylglycerol results in a full secretory response.[5,15,16,46,78]

C. Involvement with Other Hormone Receptor Systems

Protein kinase C phosphorylates both rat liver[105] and skeletal muscle[106] glycogen synthase in vitro. Phosphate is incorporated into at least two sites. This phosphorylation is accompanied by a decrease in the activity of the enzyme, according to the work of Roach and colleagues,[105,106] in keeping with the well-studied effects of phosphorylation on glycogen synthase by a number of other protein kinases, including the β-adrenergic, second messenger-stimulated cAMP-dependent protein kinase. Ca^{2+}-dependent protein kinases also inactivate glycogen synthase but phosphorylation on a separate site by another protein kinase (GSK5) has no direct effect on the activity.[18,107,108] In a similar study by Imazu et al.,[109] no decrease in activity of glycogen synthase was observed in either liver or muscle enzymes in response to protein kinase C.

In rat hepatocytes in response to phorbol esters, Roach and Goldman[23] have shown that the state of phosphorylation in sites on glycogen synthase associated with protein kinase C is increased while those associated with other Ca^{2-}-dependent protein kinases (including phosphorylase kinase) were unaffected. This would indicate that one part of the phosphatidylinositol hydrolysis pathway, namely, activation of protein kinase C, is involved, while the other branch, stimulation of release of Ca^{2+} from internal stores, is unaffected. Cuidad et al.,[105] studied glycogen synthase phosphorylation in rat hepatocytes in response to adrenaline, vasopressin, and glucagon. They reported changes in the phosphorylation located on two different peptides and concluded that regulation of liver glycogen synthase by multisite phosphorylation involved other protein kinases in addition to cyclic nucleotide and Ca^{2+}-dependent enzymes, and that regulation of protein phosphatase activity may also be involved. The precise role of protein kinase C therefore remains to be elucidated.

Vasopressin, angiotensin II, and α-adrenergic agonists cause increased phosphorylation of ten proteins in hepatocytes,[110] three of which were phosphorylated in response to TPA alone, while the phosphorylation state of the others was increased by the ionophore A23187 which would exert its effect through increased intracellular calcium levels.[110]

In turkey erythrocytes the β-adrenergic receptor has been shown to be phosphorylated in response to TPA.[111] This results in a time-dependent desensitization of isoprenaline (β-adrenergic agonist) stimulated adenylate cyclase activity. The effects appear to be directed to the uncoupling of receptor interactions with N_s (the stimulatory guanine nucleotide regulatory protein of adenylate cyclase). The authors did not shown that the protein kinase C directly phosphorylated the β-adrenergic receptors. Compared to many other cell types, the agonist-induced desensitization of adenylate cyclase in turkey erythrocytes has some differences and it is not yet known it this is a general mechanism.[111] It is interesting, however, to note the effects of phorbol esters (through protein kinase C?) on the adenylate cyclase N_s protein since it has been reported that the effects of the agonists that stimulate PtdIns $4,5P_2$ hydrolysis may also be mediated by a guanine nucleotide-binding regulatory protein.[112]

Protein kinase C phosphorylates a number of cardiac sarcolemma proteins from chicken heart, including phospholamban.[53] This protein, which is also phosphorylated by a Ca^{2+}-calmodulin-dependent protein kinase,[113] regulates the sarcoplasmic reticulum Ca^{2+}-ATPase, and provides evidence for the involvement of protein kinase C in the regulation of cardiac contraction.

V. CONCLUSION

This review has considered in detail the activation of protein kinase C by phorbol esters as the biochemical basis of their mode of action. There is strong evidence that protein kinase C is the phorbol ester "receptor", certainly for the involvement of this protein kinase in the effects of phorbol esters on platelet aggregation and tumor promotion. The possibility exists, however, since there is not complete structure/function correlation between all the biologically active phorbol esters, that some effects of phorbol esters (notably inflammation and lymphocyte mitogenesis) may be mediated by "receptors" for particular phorbol ester derivatives other than or additional to protein kinase C.

REFERENCES

1. **Blumberg, P. M.,** *In vitro* studies on the mode of action of the phorbol esters. Potent tumour promoters. I., *CRC Crit. Rev. Toxicol.*, 8, 153, 1980.
2. **Blumberg, P. M.,** *In vitro* studies on the mode of action of the phorbol esters. Potent tumour promoters. II. *CRC. Crit. Rev. Toxciol.*, 8, 199, 1981.
3. **Dicker, P. and Rozengurt, E.,** Phorbol esters and vasopressin stimulate DNA synthesis by a common mechanism, *Nature (London)*, 287, 607, 1980.
4. **Billah, M. M. and Lapetina, E. G.,** Rapid decrease of phosphatidyl inositol 4,5-bisphosphate in thrombin-stimulated platelets, *J. Biol. Chem.*, 257, 12705, 1982.
5. **De Chaffoy de Courcelles, D., Roevens, P., and van Belle, H.,** 12-O-tetradecanoylphorbol 13-acetate stimulates inositol lipid phosphorylation in intact human platelets, *FEBS Lett.*, 173, 389, 1984.
6. **Touraine, J. L., Hadden, J. W., Touraine, F., Estensen, R. D., and Good, R. A.,** Phorbol myristate acetate: a mitogen selective for a T-lymphocyte subpopulation, *J. Exp. Med.*, 145, 460, 1977.
7. **Evans, F. J. and Taylor, S. E.,** Pro-inflammatory tumour promoting and antitumour diterpenes of the plant families Euphorbiaceae and Thymellaeaceae, in *Progress in the Chemistry of Natural Products*, Herz, W., Grisebach, H., and Kirby, G. W., Eds., Springer-Verlag, New York, 1983.
8. **Hecker, E. and Schmidt, R.,** Phorbol esters, the irritants and co-carcinogens of *Croton tiglium* L., *Prog. Chem. Org. Nat. Prod.*, 31, 377, 1974.
9. **Billah, M. M., Lapetina, E. G., and Cuatrecasas, P.,** Phospholipase A_2 activity specific for phosphatidic acid. A possible mechanism for the production of arachidonic acid in platelets, *J. Biol. Chem.*, 256, 5399, 1981.
10. **Cockcroft, S.,** Does phosphatidylinositol breakdown control the Ca^{2+}-gating mechanism?, *Trends Pharmacol. Sci.*, 2, 340, 1981.
11. **Takai, Y., Kishimoto, A., Inoue, T., and Nishizuka, Y.,** Studies on a cyclic-nucleotide-independent protein kinase and its proenzyme in mammalian tissues. I. Purification and characterization of an active enzyme from bovine cerebellum, *J. Biol. Chem.*, 252, 7610, 1977.
12. **Kuo, J. F., Anderson, R. G. G., Wise B. C., Mackerlova, L., Salmonsson, I., Brackett, N. L., Katoh, N., Shoji, M., and Wrenn, R. W.,** Calcium dependent protein kinase. Widespread occurrence in various tissues and phyla of the animal kingdom and comparison of effects of phospholipid, calmodulin and trifluoperazine, *Proc. Natl. Acad. Sci. U.S.A.*, 77, 7039, 1980.
13. **Castagna, M., Takai, Y., Kaibuchi, K., Sano, K., Kikkawa, U., and Nishizuka, Y.,** Direct activation of calcium-activated, phospholipid-dependent protein kinase by tumour promoting phorbol esters, *J. Biol. Chem.*, 257, 7847, 1982.
14. **Michell, R. H.,** Oncogenes and inositol lipids, *Nature (London)*, 308, 770, 1984.
15. **Michell, R. H.,** Ca^{2+} and protein kinase C: two synergistic cellular signals, *Trends Biochem. Sci.*, 8, 263, 1983.
16. **Berridge, M. J.,** Inositol trisphosphate and diacylglycerol as second messengers, *Biochem. J.*, 220, 345, 1984.
17. **Cohen, P.,** The role of protein phosphorylation in neural and hormonal control of cellular activity, *Nature (London)*, 296, 613, 1982.
18. **Cohen, P., Aitken, A., Damuni, Z., Hemmings, B. A., Ingebritsen, T. S., Parker, P. J., Picton, C., Resink, T. J., Stewart, A. A., Tonks, N. K., and Woodgett, J.,** Protein phosphorylation and the neural and hormonal regulation of enzyme activity, in *Posttranslational Covalent Modifications of Proteins*, Connor Johnson, B., Ed., Academic Press, New York, 1983, 19.
19. **Hunter, T. and Sefton, B. M.,** Protein kinases and viral transformation, in *Molecular Action of Toxins and Viruses*, Cohen, P. and Van Heyningen, S., Eds., Elsevier, Amsterdam, 1982, 337.
20. **Bishop, J. M.,** Cellular oncogenes and retroviruses, *Annu. Rev. Biochem.*, 52, 301, 1983.
21. **Klee, C. B. and Newton, D. L.,** Calmodulin modulated protein phosphorylation, in *Posttranslational Covalent Modifications of Proteins*, Connor Johnson, B., Ed., Academic Press, New York, 1983, 61.
22. **Nishizuka, Y.,** Protein kinases in signal transduction, *Trends Biochem. Sci.*, 9, 163, 1984.
23. **Roach, P. J. and Goldman, M.,** Modification of glycogen synthase activity in isolated rat hepatocytes by tumour-promoting phorbol esters: evidence for differential regulation of glycogen synthase and phosphorylase, *Proc. Natl. Acad. Sci. U.S.A.*, 80, 7170, 1983.
24. **Rylatt, D. B., Aitken, A., Bilham, T., Condon, G. D., Embi, N., and Cohen, P.,** Glycogen synthase from rabbit skeletal muscle. Amino acid sequence at the sites phosphorylated by glycogen synthase kinase-3, and extension of the N-terminal sequence phosphorylated by phosphorylase kinase, *Eur. J. Biochem.*, 107, 529, 1980.
25. **Weinstein, I. B.,** Protein kinase, phospholipid and control of growth, *Nature (London)*, 302, 750, 1983.
26. **Fujiki, H., Tanaka, Y., Mikaye, R., Kikkawa, U., Nishizuka, Y., and Sugimura, T.,** Activation of calcium activated phospholipid dependent protein kinase, protein kinase C, by new classes of tumour promoters teleocidin and debromoaplysia toxin, *Biochem. Biophys. Res. Commun.*, 120, 339, 1984.

27. **Mikaye, R., Tanaka, Y., Tsuda, T., Kaibuchi, K., Kikkawa, U., and Nishizuka, Y.,** Activation of protein kinase C by the non-phorbol tumour promoter, mezerein, *Biochem. Biophys. Res. Commun.*, 121, 649, 1984.
28. **Wightman, P. D. and Raetz, C. R. H.,** The activation of protein kinase C by biologically active lipid moieties of lipopolysaccharide, *J. Biol. Chem.*, 259, 10048, 1984.
29. **Lapetina, E. G.,** Regulation of arachidonic acid production: role of phospholipases C and A_2, *Trends Pharmacol. Sci.*, 3, 115, 1982.
30. **Mahadevappa, V. G. and Holub, B. J.,** Degradation of different molecular species of phosphatidylinositol in thrombin-stimulated human platets. Evidence for preferential degradation of 1-acyl-2-arachidonyl species, *J. Biol. Chem.*, 258, 5337, 1983.
31. **Prescott, S. M. and Majerus, P. W.,** Characterization of 1,2-diacylglycerol hydrolysis in human platelets. Demonstration of an arachidonyl-monoacylglycerol intermediate, *J. Biol. Chem.*, 258, 764, 1983.
32. **Fisher, S. K., Van Rooijen, L. A. A., and Agranoff, B. W.,** Renewed interest in the polyphosphoinositides, *Trends Biochem. Sci.*, 9, 53, 1984.
33. **Nishizuka, Y.,** Phospholipid degradation and signal translation for protein phosphorylation, *Trends Biochem. Sci.*, 8, 13, 1983.
34. **Edwards, M. C., Evans, F. J., Barrett, M. L., and Gordon, D.,** Structural correlations of phorbol ester induced stimulation of PGE_2 production by human rheumatoid synovial cells, *Inflammation*, 9, 33, 1985.
35. **Billah, M. M., Lapetina, E. G., and Cuatrecasas, P.,** Phospholipase A_2 and phospholipase C activities of platelets. Differential substrate specificity, Ca^{2+} requirement, pH dependence and cellular location, *J. Biol. Chem.*, 255, 10227, 1980.
36. **Joseph, S. K.,** Inositol trisphosphate: an intracellular messenger produced by Ca^{2+} mobilizing hormone, *Trends Biochem. Sci.*, 9, 420, 1984.
37. **Streb, H., Irvine, R. F., Berridge, M. J., and Schulz, I.,** Release of Ca^{2+} from nonmitochondrial intracellular store in pancreatic acinar cells by inositol-1,4,5-trisphosphate, *Nature (London)*, 306, 67, 1983.
38. **Naka, M., Nishikawa, M., Adelstein, R. S., and Hidaka, H.,** Phorbol ester-induced activation of human platelets is associated with protein kinase C phosphorylation of myosin light chains, *Nature (London)*, 306, 490, 1983.
39. **Jacobs, S., Sahyoun, N. E., Salteil, A. R., and Cuatrecasas, P.,** Phorbol esters stimulated the phosphorylation of receptors for insulin and somatomedin C, *Proc. Natl. Acad. Sci. U.S.A.*, 80, 6211, 1984.
40. **Machicao, E. and Wieland, O. H.,** Evidence that the insulin receptor-associated protein kinase acts as a phosphatidylinositol kinase, *FEBS Lett.*, 175, 113, 1984.
41. **Sawyer, S. T. and Cohen, S.,** Enhancement of calcium uptake and phosphatidylinositol turnover by epidermal growth factor in A-431 cells, *Biochemistry*, 20, 6280, 1981.
42. **Habernicht, A. J. R., Glomset, J. A., King, W. C., Nist, C., Mitchell, C. D., and Ross, R.,** Early changes in phosphatidylinositol and arachidonic acid metabolism in quiescent Swiss 3T3 cells stimulated to divide by platelet derived growth factor, *J. Biol. Chem.*, 256, 12329, 1981.
43. **Hunter, T.,** The epidermal growth factor receptor gene and its product, *Nature (London)*, 311, 414, 1984.
44. **Michell, R. H.,** A link between lithium, lipids and receptors?, *Trends Biochem. Sci.*, 7, 387, 1982.
45. **Berridge, M. J., Downes, C. P., and Hanley, M. R.,** Lithium amplifies agonist-dependent phosphatidylinositol responses in brain and salivary glands, *Biochem. J.*, 206, 587, 1982.
46. **Kaibuchi, K. Sano, K., Hoshijima, M., Takai, Y., and Nishizuka, Y.,** Phosphatidylinositol turnover in platelet activation: calcium metabolism and protein phosphorylation, *Cell Calcium*, 3, 323, 1982.
47. **Ashendel, C. L., Staller, J. M., and Boutwell, R. K.,** Protein kinase activity associated with a phorbol ester receptor purified from mouse brain, *Cancer Res.*, 43, 4333, 1983.
48. **Niedel, J. E., Kuhn, L. J., and Vandenbark, G. R.,** Phorbol diester receptor copurifies with protein kinase C, *Proc. Natl. Acad. Sci. U.S.A.*, 80, 36, 1983.
49. **Leach, K. L., James, M. L., and Blumberg, P. M.,** Characterisation of a specific phorbol ester aporeceptor in mouse brain cytosol, *Proc. Natl. Acad. Sci.* U.S.A., 80, 4208, 1983.
50. **Parker, P. J., Stabel, S., and Waterfield, M. D.,** Purification to homogeneity of protein kinase C from bovine brain-identity with the phorbol ester receptor, *EMBO J.*, 3, 953, 1984.
51. **Kikkawa, U., Takai, Y., Minakuchi, R., Inohara, S., and Nishizuka, Y.,** Calcium-activated, phospholipid-dependent protein kinase from rat brain. Subcellular distribution, purification and properties, *J. Biol. Chem.*, 257, 13341, 1982.
52. **Wise, B. C., Raynor, R. L., and Kuo, J. F.,** Phospholipid-sensitive Ca^{2+}-dependent protein kinase from heart. I. Purification and general properties. II. Substrate specificity and inhibition by various agents, *J. Biol. Chem.*, 257, 8481 and 8489, 1982.
53. **Iwasa, Y. and Hosey, M. M.,** Phosphorylation of cardiac sarcolemma proteins by the calcium-activated phospholipid-dependent protein kinase, *J. Biol. Chem.*, 259, 534, 1984.
54. **Kaibuchi, K., Takai, Y., and Nishizuka, Y.,** Cooperative role of various membrane phospholipids in the activation of calcium-activated phospholipid dependent protein kinase, *J. Biol. Chem.*, 256, 7146, 1981.

55. **Feinstein, M. B. and Hadjian, R. A.,** Effects of the calmodulin antagonist trifluoperazine on stimulus induced calcium mobilisation, aggregation, secretion and protein phosphorylation in platelets, *Mol. Pharmacol.*, 21, 422, 1982.
56. **Mori, T., Takai, Y., Minakuchi, R., Yu, B., and Nishizuka, Y.,** Inhibitory action of chlorpromazine, dibucaine and other phospholipid interacting drugs on calcium activated phospholipid dependent protein kinase, *J. Biol. Chem.*, 255, 8378, 1980.
57. **Schatzman, R. C., Raynor, R., Fritz, R. B., and Kuo, J. F.,** Purification to homogeneity, characterisation and monoclonal antibodies of phospholipid-sensitive Ca^{2+}-dependent protein kinase from spleen, *Biochem. J.*, 209, 435, 1983.
58. **Takai, Y., Kishimoto, A., Iwasa, M., Kawahara, Y., Mori, T., and Nishizuka, Y.,** Calcium-dependent activation of a multifunctional protein kinase by membrane phospholipids, *J. Biol. Chem.*, 254, 3692, 1979.
59. **Kikkawa, U., Takai, Y., Tanaka, Y., Miyake, R., and Nishizuka, Y.,** Protein kinase C as a possible receptor protein of tumor promoting phorbol esters, *J. Biol. Chem.*, 258, 11442, 1983.
60. **Sharkey, N. A., Leach, K. L., and Blumberg, P. M.,** Competitive inhibition by diacylglycerol of specific phorbol ester binding, *Proc. Natl. Acad. Sci. U.S.A.*, 81, 607, 1984.
61. **Kraft, A. S., Anderson, W. B., Cooper, H. L., and Sando, J. J.,** Decrease in cytosolic calcium/phospholipid-dependent protein kinase activity following phorbol ester treatment of EL4 thymoma cells, *J. Biol. Chem.*, 257, 13193, 1982.
62. **Qi, D. F., Schatzman, R. C., Mazzei, G. J., Turner, R. S., Raynor, R. L., Liao, S., and Kuo, J. F.,** Polyamines inhibit phospholipid-sensitive and calmodulin sensitive Ca^{2+}-dependent protein kinases, *Biochem. J.*, 213, 281, 1983.
63. **Kishimoto, A., Kajikawa, N., Shiota, M., and Nishizuka, Y.,** Proteolytic activation of calcium activated phospholipid dependent protein kinase by a calcium dependent neutral protease, *J. Biol. Chem.*, 258, 1156, 1983.
64. **Kraft, A. S. and Anderson, W. B.,** Phorbol esters increase the amount of Ca^{2+}, phospholipid dependent protein kinase associated with plasma membrane, *Nature (London)*, 301, 621, 1983.
65. **Sando, J. J. and Young, M. C.,** Identification of high affinity phorbol ester receptor in cytosol of EL4 thymoma cells: requirement for calcium, magnesium and phospholipids, *Proc. Natl. Acad. Sci. U.S.A.*, 80, 2642, 1983.
66. **Edwards, M. C., Nouri, A. M. E., Gordon, D., and Evans, F. J.,** Tumour-promoting and nonpromoting proinflammatory esters act as human lymphocyte mitogens with different sensitivities to inhibition by cyclosporin A, *Mol. Pharmacol.*, 23, 703, 1983.
67. **Hirata, F., Schiffman, E., Venkatasubramanian, K., Solomon, D., and Axelrod, J.,** A phospholipase A_2 inhibitory protein in rabbit neutrophils induced by glucocorticoids, *Proc. Natl. Acad. Sci. U.S.A.*, 77, 2533, 1980.
68. **Farrar, J. J., Mizel, S. B., Fuller-Farrar, J., Farrar, W. L., and Hiljiker, M. L.,** Macrophage-independent activation of helper-T-cells. I. Production of interleukin-2, *J. Immunol.*, 125, 793, 1980.
69. **Shackelford, D. A. and Trowbridge, I. S.,** Induction of expression and phosphorylation of the human interleukin 2 receptor by a phorbol diester, *J. Biol. Chem.*, 259, 11706, 1984.
70. **Evans, F. J. and Schmidt, R. J.,** The succulent euphorbias of Nigeria. III. Structure and potency of the aromatic ester diterpenes of *Euphorbia poissonii* Pax, *Acta Pharmacol. Toxicol*, 45, 181, 1979.
71. **Westwick J., Williamson, E. M., and Evans, F. J.,** Structure activity relationships of 12-deoxyphorbol esters on human platelets, *Thromb. Res.*, 20, 683, 1980.
72. **Driedger, P. E. and Blumberg, P. M.,** Structure-activity relationships in chick embryo fibroblasts for phorbol esters show anomalous activities *in vivo*, *Cancer Res.*, 40, 339, 1980.
73. **Williamson, E. M., Westwick, J., Kakkar, V. V., and Evans, F. J.,** Studies on the mechanism of action of 12-deoxyphorbolphenylacetate, a potent platelet aggregating tigliane ester, *Biochem. Pharmacol.*, 30, 2691, 1981.
74. **Edwards, M. C., Taylor, S. E., Williamson, E. M., and Evans, F. J.,** New phorbol and deoxyphorbol esters: Isolation and relative potencies in inducing platelet aggregation and erythema of skin, *Acta Pharmacol. Toxicol.*, 53, 177, 1983.
75. **Flockhart, D. A. and Corbin, J. D.,** Regulatory mechanisms in the control of protein kinases, *CRC Crit. Rev. Biochem.*, 133, 1982.
76. **Glass, D. B. and Krebs, E. G.,** Protein phosphorylation catalysed by cyclic AMP and cyclic GMP-dependent protein kinases, *Annu. Rev. Pharmacol. Toxicol.*, 20, 363, 1980.
77. **Aitken, A., Hemmings, B. A., and Hofmann, F.,** Identification of the residues on cyclic-GMP-dependent protein kinase that are autophosphorylated in the presence of cyclic AMP and cyclic GMP, *Biochim. Biophys. Acta*, 790, 219, 1984.

78. **Sano, K., Kaibuchi, K., Hoshijima, M., Yamanishi, J., Kikkawa, U., Takai, Y., and Nishizuka, Y.**, Phospholipid turnover as transmembrane signalling for protein phosphorylation and platelet activation, in *Posttranslational Covalent Modifications of Proteins*, Connor Johnson, B., Ed., Academic Press, New York, 1983, 105.
79. **James, R. and Bradshaw, R. A.**, Polypeptide growth factors, *Annu. Rev. Biochem.*, 53, 259, 1984.
80. **Carpenter, G. and Cohen, S.**, Peptide growth factors, *Trends Biochem. Sci.*, 9, 169, 1984.
81. **Weinberg, R. A.**, Cellular oncogenes, *Trends Biochem. Sci.*, 9, 131, 1984.
82. **Hunter, T., Ling, N., and Cooper, J. A.**, Protein kinase C phosphorylation of the EGF receptor at a threonine residue close to the cytoplasmic face of the plasma membrane, *Nature (London)*, 311, 480, 1984.
83. **Davis, R. J. and Czech, M. P.**, Tumour-promoting phorbol diesters mediate phosphorylation of the epidermal growth factor receptor, *J. Biol. Chem.*, 259, 8545, 1984.
84. **Iwashita, S. and Fox, C. F.**, Epidermal growth factor and potent phorbol tumour promoters induce epidermal growth factor receptor phosphorylation in a similar but distinctively different manner in human epidermoid carcinoma A431 cells, *J. Biol. Chem.*, 259, 2559, 1984.
85. **Cochet, C., Gill, G. N., Meisenhelder, J., Cooper, J. A., and Hunter, T.**, C-kinase phosphorylates the epidermal growth factor and reduces its epidermal growth factor-stimulated tyosine protein kinase activity, *J. Biol. Chem.*, 259, 2553, 1982.
86. **Gilmore, T. and Martin, G. S.**, Phorbol ester and diacylglycerol induce protein phosphorylation at tyrosine, *Nature (London)*, 306, 487, 1983.
87. **Akers, R. F., and Routtenberg, A.**, Brain protein phosphorylation *in vitro:* selective substrate action of insulin, *Life Sci.*, 35, 809, 1984.
88. **Aloyo, V. J., Zwiers, H., and Gispen, W. H.**, Phosphorylation of B-50 protein by calcium-activated, phospholipid-dependent protein kinase and B-50 protein kinase, *J. Neurochem.*, 41, 649, 1983
89. **Trevillyan, J. M., Kulkarni, R. K., and Byus, C. V.**, Tumour-promoting phorbol esters stimulate the phosphorylation of ribosomal protein S6 in quiescent Reuber H35 hepatoma cells, *J. Biol. Chem.*, 259, 897, 1984.
90. **Wu, V. S. and Byus, C. V.**, The induction of ornithine decarboxylase by tumour promoting phorbol esters ester analogues in Reuber H35 hepatoma cells, *Life Sci.*, 29, 1855, 1981.
91. **Werth, D. K., Neidel, J. E., and Pastan, I.**, Vinculin, a cytoskeletal substrate of protein kinase C, *J. Biol. Chem.*, 258, 11423, 1983.
92. **Werth, D. K. and Pastan, I.**, Vinculin phorphorylation in response to calcium and phorbol esters in intact cells, *J. Biol. Chem.*, 259, 5264, 1984.
93. **Feuerstein, N. and Cooper, H. L.**, Rapid phosphorylation induced by phorbol ester in HL-60 cells: unique alkali-stable phosphorylation of a 17,000 dalton protein detected by two-dimensional gel electrophoresis, *J. Biol. Chem.*, 258, 10768, 1983.
94. **Feuerstein, N. and Cooper, H. L.**, Rapid phosphorylation-dephosphorylation of specific proteins induced by phorbol ester in HL-60 cells, further characterisation of the phosphorylation of 17-kilodalton and 27-kilodalton protein in myeloid leukemic cells and human monocytes, *J. Biol. Chem.*, 259, 2782, 1984.
95. **Moolendar, W. H., Tsien, R. Y., van der Saag, P. T., and de Laat, S. W.**, Na^+/H^+ exchange and cytoplasmic pH in the action of growth factors in human fibroblasts, *Nature (London)*, 304, 645, 1983.
96. **Rosoff, P. M., Stein, L. F., and Cantley, L. C.**, Phorbol esters induce differentiation in a pre-B-lymphocyte cell line by enhancing Na^+/H^+ exchange, *J. Biol. Chem.*, 259, 7056, 1984.
97. **Burns, C. P. and Rozengurt, E.**, Serum, platelet-derived growth factor, vasopressin, and phorbol esters increase intracellular pH in Swiss 3T3 cells, *Biochem. Biophys. Res. Commun.*, 116, 931, 1983.
98. **Lyons, R. M. and Atherton, R. M.**, Characterisation of a platelet protein phosphorylated during the thrombin-induced release reaction, *Biochemistry*, 18, 544, 1979.
99. **Mastro, A. M. and Smith, M. C.**, Calcium-dependent activation of lymphocytes by ionophore, A23187 and a phorbol ester tumour promoter, *J. Cell. Physiol.*, 116, 51, 1983.
100. **Malaisse, W. J., Lebrun, P., Herchuelz, A., Sener, A., and Malaisse-Lagaie, F.**, Synergistic effect of a tumour-promoting phorbol ester and a hypoglycemic sulphonylurea upon insulin release, *Endocrinology*, 113, 1870, 1983.
101. **Boyton, A. L., Whitfield, J. F., and Isaacs, R. J.**, Calcium-dependent stimulation of BALB/c 3T3 mouse cell DNA synthesis by a tumour-promoting phorbol ester (PMA), *J. Cell. Physiol.*, 87, 25, 1976.
102. **Kojima, I., Lippes, H., Kojima, K., and Rasmussen, H.**, Aldosterone secretion: effect of phorbol ester and A23187, *Biochem. Biophys. Res. Commun.*, 116, 555, 1983.
103. **White. J. R., and Huang, C.-K., Hill, J. M., Naccache, P. H., Becker, E. L., and Sha'afi, R. I.**, Effect of phorbol 12-myristate 13-acetate and its analogue 4α-phorbol 12,13-didecanoate on protein phosphorylation and lysosomal enzyme release in rabbit neutrophils, *J. Biol. Chem.*, 259, 8605, 1984.
104. **Tanaka, C., Taniyama, K., and Kushunoki, M.**, A phorbol ester and A23187 act synergistically to release acetylcholine from the guinea pig ileum, *FEBS Lett.* 175, 165, 1984.

105. **Cuidad, C., Camici, M., Ahmad, Z., Wang, Y., DePaoli-Roach, A., and Roach, P. J.,** Control of glycogen synthase phosphorylation in isolated rat hepatocytes by epinephrine, vasopressin and glucagon, *Eur. J. Biochem.,* 142, 511, 1984.
106. **Ahmad, Z., Lee, F.-T., DePaoli-Roach, A., and Roach, P. J.,** Phosphorylation of glycogen synthase by the Ca^{2+}- and phospholipid-activated protein kinase (protein kinase C), *J. Biol. Chem.,* 259, 8743, 1984.
107. **Cohen, P., Yellowlees, D., Aitken, A., Donella-Deana, A., Hemmings, B. A., and Parker, P. J.,** Separation and characterisation of glycogen synthase kinase 3, glycogen synthase kinase 4 and glycogen synthase kinase 5 from rabbit skeletal muscle, *Eur. J. Biochem.,* 124, 21, 1982.
108. **Picton, C., Aitken, A., Bilham, T., and Cohen, P.,** Multisite phosphorylation of glycogen synthase from rabbit skeletal muscle. Organisation of the seven sites in the polypeptide chain, *Eur. J. Biochem.,* 124, 37, 1982.
109. **Imazu, M., Strickland, W. G., Chrisman, T. D., and Exton, J. H.,** Phosphorylation and inactivation of liver glycogen synthase by liver protein kinases, *J. Biol. Chem.,* 259, 1813, 1984.
110. **Garrison, J. C., Johnson, D. E., and Campanile, C. P.,** Evidence for the role of phosphorylase kinase, protein kinase C and other Ca^{2+} sensitive protein kinases in the response of hepatocytes to angiotensin II and vasopressin, *J. Biol. Chem.,* 259, 3283, 1984.
111. **Kelleher, D. J., Pessin, J. E., Ruoho, A. E., and Johnson, G. L.,** Phorbol ester induces desensitisation of adenylate cyclase and phosphorylation of the β-adrenergic receptor in turkey erythrocytes, *Proc. Natl. Acad. Sci. U.S.A.,* 81, 4316, 1984.
112. **Gomperts, B. D.,** Involvement of guanine nucleotide-binding protein in the gating of Ca^{2+} by receptors, *Nature (London),* 306, 64, 1983.
113. **Rinaldi, M. L., Le Peuch, C. J., and Demaille, J. G.,** The epinephrine-induced activation of the cardiac slow Ca^{2+} channel is mediated by the cAMP-dependent phosphorylation of calciductin, A 23,000 M_r sarcolemmal protein, *FEBS Lett.,* 129, 277, 1981.
114. **Fein, A., Payne, R., Corson, D. W., Berridge, M. J., and Irvine, R. F.,** Photoreceptor excitation and adaptation by inositol 1,4,5-trisphosphate, *Nature (London),* 311, 157, 1984.
115. **Brown, J. E., Rubin, L. J., Ghalayini, A. J., Tarver, A. P., Irvine, R. F., Berridge, M. J., and Anderson, R. E.,** *Myo*-inositol polyphosphate may be a messenger for visual excitation in *Limulus* photoreceptors, *Nature (London),* 311, 160, 1984.
116. **Turner, R. S., Chou, C.-H., J., Mazzei, G. J., Dembure, P., and Kuo, J. F.,** Phospholipid sensitive Ca^{2+}-dependent protein kinase preferentially phosphorylates serine-115 of bovine myelin basic protein, *J. Neurochem.,* 43, 1257, 1984.
117. **Orellana, S. A., Solski, P. A., and Brown, J. H.,** Phorbol ester inhibits phosphoinositide hydrolysis and calcium mobilisation in cultured *Astrocytoma* cells, *J. Biol. Chem.,* 260, 5236, 1985.
118. **Naccache, P. H., Molski, T. F. P., Borgeat, P., White, J. R., and Sha'afi, R. I.,** Phorbol esters inhibit the fMet-Leu-Phe- and Leukotriene B_4-stimulated calcium mobilisation and enzyme secretion in rabbit neutrophils, *J. Biol. Chem.,* 260, 2125, 1985.
119. **Gabrielli, B., Wettenhall, R. F. H., Kemp, B. E., Quinn, M., and Bizonova, L.,** Phosphorylation of ribosomal protein S6 and a peptide analogue of S6 by a protease-activated kinase isolated from rat liver, *FEBS Letts.,* 175, 219, 1984.
120. **Hardie, D. G., Carling, D., Ferrari, S., Guy, P. S., and Aitken, A.,** Characterisation of the phosphorylation of rat mammary ATP-citrate lyase and acetyl-CoA carboxylase-dependent protein kinase, *Eur. J. Biochem.,* submitted.
121. **Hardie, D. G.,** personal communication.

INDEX

A

B

C

D

E

F

G

I

J

K

L

M

N

O

P

Q

R

S

T